LabVIEW™

Student Edition

Thousands of successful engineers, scientists, and technicians use LabVIEW to create solutions for their demanding application needs. LabVIEW is a revolutionary graphical programming development environment based on the G programming language for data acquisition and control, data analysis, and data presentation. LabVIEW gives you the flexibility of a powerful programming language without the associated difficulty and complexity because its graphical programming methodology is inherently intuitive to scientists and engineers.

With LabVIEW, you build graphical programs called virtual instruments (VIs) instead of writing text-based programs. You quickly create front panel user interfaces that give you interactive control of your system. To add functionality to the user interface, you intuitively assemble block diagrams – a natural design notation for engineers and scientists.

"Providing our students with the tools to be competitive is the goal of all education, but it is a constant battle for schools to keep up with technology – money and training time are the problems. The LabVIEW Student Edition addresses these difficulties, placing the price of the newest technology at a level where every school can afford it."

- Clark Gedney, Ph.D., Instructional Computing Department of Biological Sciences, Purdue University

"Over 75% of our students use LabVIEW skills in their first job."

- Paul Dixon, Assistant Professor of Physics, CSUSB

"The use of LabVIEW has contributed towards developing additional engineering course laboratory packages, with topics such as fluids, machine design, and vibrations."

- Al Wicks, Ph.D., Associate Professor of Mechanical Engineering, Virginia Polytechnic Institute

"Our experience has been that the learning curve with LabVIEW is amazingly short: two lab sessions were more than enough for an interested student to get his code working."

- Prof. Roger Bengston, Department of Physics, University of Texas at Austin

Using Graphical Programming for...

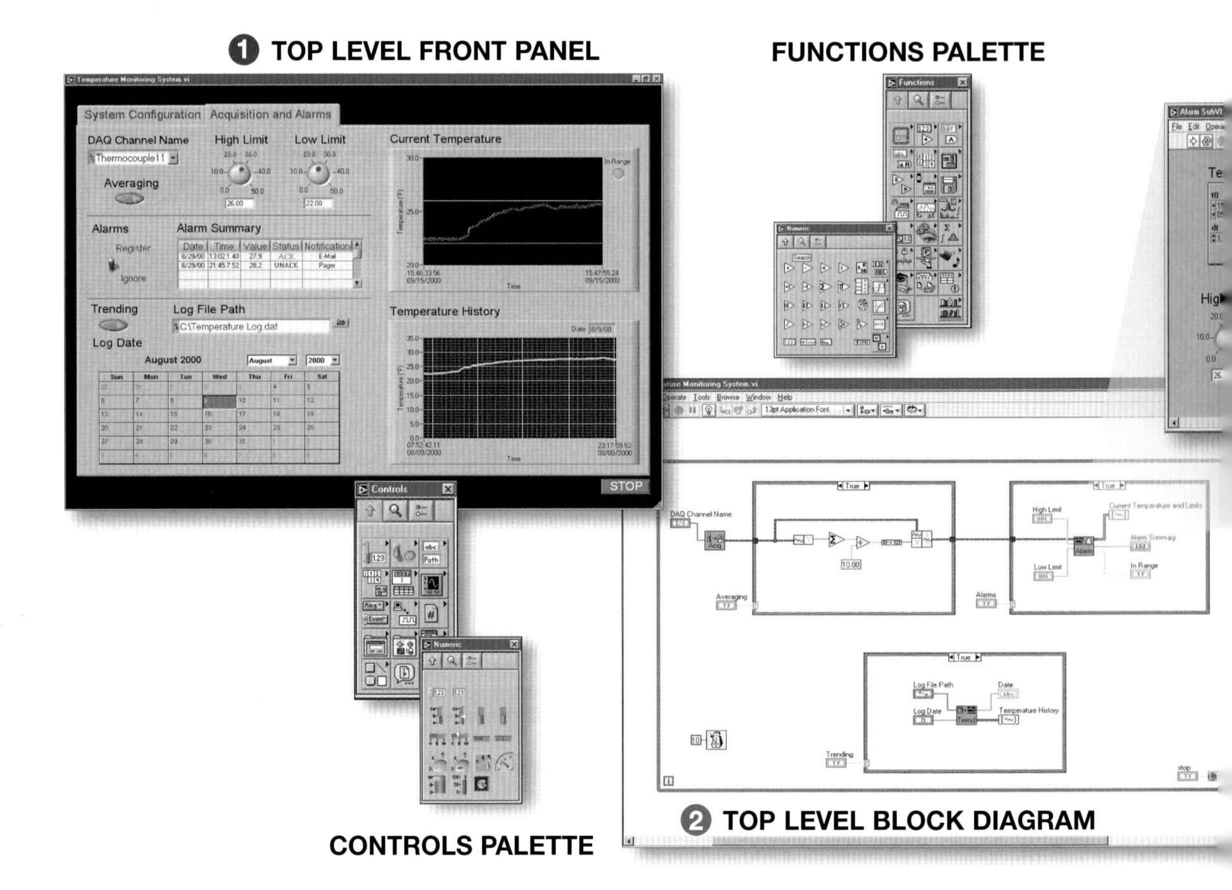

2 Construct the Graphical Block Diagram

- Construct the block diagram without worrying about the syntactical details of text-based programming languages
- Selecting objects (icons) from the Functions palette and connect them together with wires to transfer data among block diagram objects
- Choose objects such as simple arithmetic functions, advanced acquisition and analysis routines, network and file I/O operations, and more

1 Create the Front Panel

- Place the controls and data displays by selecting objects from the Controls palette
- Choose from numeric displays, meters, gauges, thermometers, tanks, LEDs, charts, and graphs
- Use the front panel to control your system by moving a slide, zooming in on a graph, or entering a value with the keyboard

3 Modularity and Hierarchy

- Use a VI by itself or as part of another VI
- Create VI icons to design a hierarchy of VIs and subVIs that serve as application building blocks
- Modify, interchange, and combine VIs to meet changing application needs
- Reuse parts of your applications by simple cut and paste

Measurement and Automation

❸ SUBVI PANEL

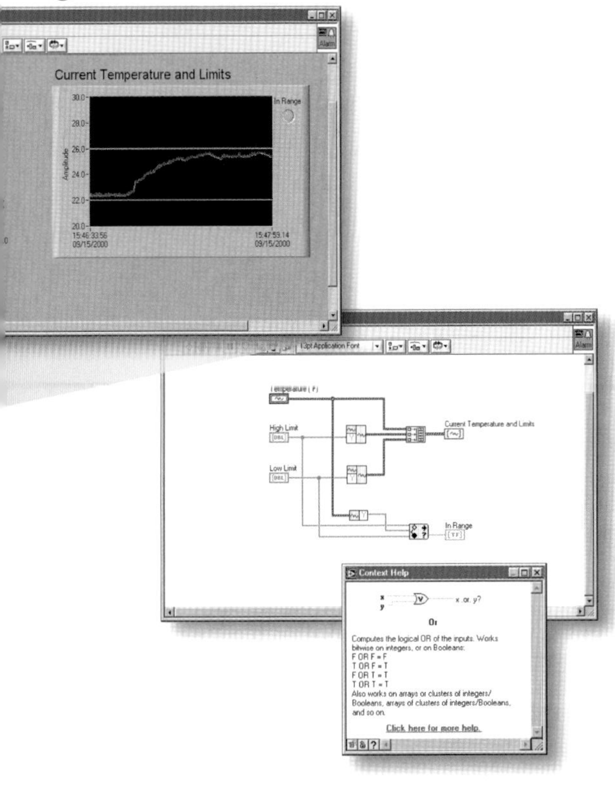

Dataflow Programming

- Utilize a patented dataflow programming model
- Create block diagrams that execute multiple operations in parallel. Consequently, LabVIEW is a multitasking system capable of running multiple execution threads and multiple VIs in parallel

Graphical Compiler

- Make use of execution speeds comparable to compiled C programs
- Analyze and optimize time-critical operations
- Increase your productivity with graphical programming without sacrificing execution speed

The Many Faces of LabVIEW

The use of LabVIEW in university laboratories spans a broad range of disciplines and applications. Virtually any laboratory that makes measurements and analyzes acquired data can benefit from LabVIEW. Typical uses of LabVIEW include:

Electrical and Computer Engineering: basic electrical measurements, digital communications, control theory, signal processing, computer fundamentals, and programming languages

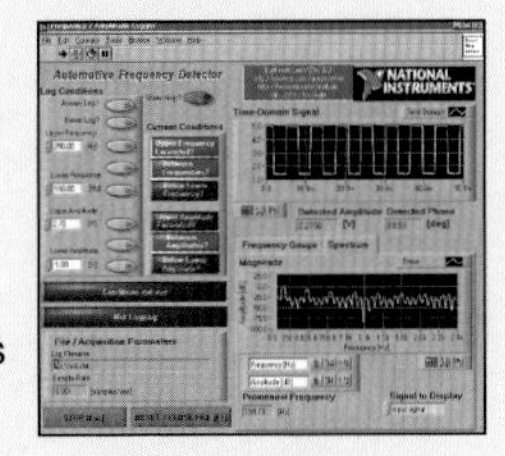

Mechanical Engineering: measurements and automation, fluid mechanics, heat and mass transfer, automatic control systems

Physics: introductory physics laboratory, electrical and timing measurements, measurements of force, pressure, weight, and temperature, and examination of electromagnetic fields and properties

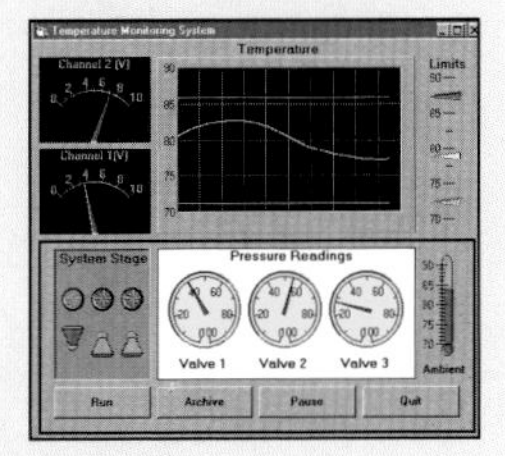

Biology/Physiology: genetic population modeling, physiological data acquisition, oxygen amplifier, pH meters, EKG, heart sounds, respiration, action potential and muscle contraction

Chemistry and Chemical Engineering: themistor circuits for calorimetry, digital titration, half-life of isotopes, spectrophotometer, topography, connective heat transfer, and Laser Doppler velocimetry

Internet Connectivity: enable VIs for the internet, view VIs across the internet from a Web browser, send e-mail from applications, send files to an FTP server

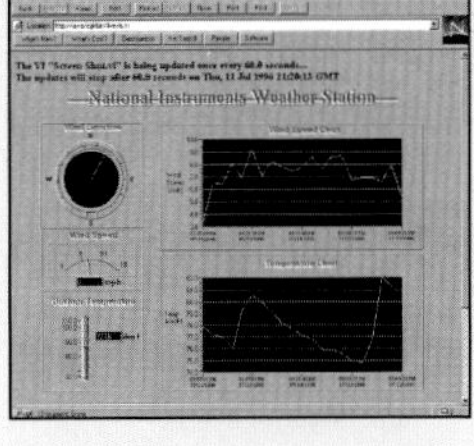

Math Functions: write equations on the front panel, plus advanced data visualization features such as surface mesh plots, contour plots and animation, more than 100 Math VIs for ordinary differential equations, optimization, root solving, integration, differentiation, transforms and more

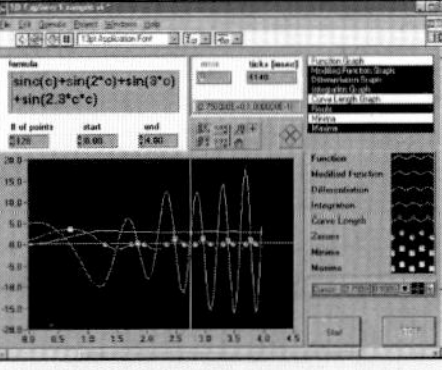

Keyboard Shortcuts

File

Ctrl-N	New VI (skips **New** dialog box)
Ctrl-O	Opens file
Ctrl-W	Closes file
Ctrl-S	Saves VI
Ctrl-P	Prints
Ctrl-I	Displays VI properties
Ctrl-Q	Quits LabVIEW

Edit

Ctrl-V	Pastes object
Ctrl-Shift-F	Displays search results
Ctrl-B	Removes broken wires
Ctrl-C	Copies an object
Ctrl-D	Allows you to redraw (**VI Hierarchy** window only)
Ctrl-F	Finds a terminal, local variable, reference, invoke node or property node
Ctrl-X	Cuts object
Ctrl-Z	Allows you to undo last action
Ctrl-Shift-Z	Allows you to redo last action

Operate

Ctrl-R	Runs VI
Ctrl-M	Changes to run/edit mode
Ctrl-.	Aborts VI

Tools

Ctrl-Y	Adds to VI Revision History

Window

Ctrl-E	Displays block diagram/front panel
Ctrl-L	Displays error list
Ctrl-T	Tiles the block diagram and front panel windows
Ctrl-/	Adjusts window to full size

Help

Ctrl-H	Displays context help
Ctrl-?	Displays help contents and index
Ctrl-Shift-L	Locks context help

Font

Ctrl-0	Displays **Font** dialog box
Ctrl-1	Changes Application font
Ctrl-2	Changes System font
Ctrl-3	Changes Dialog font
Ctrl-4	Changes Current font

Other Shortcuts

Ctrl-A Adds a comment (**VI Revision History** window only)
Shows all VIs (**VI Hierarchy** window only)
Performs last alignment

LabVIEW Web Resources

Support

www.ni.com/support

- KnowledgeBase – searchable database of tips, common questions, and more
- Troubleshooting Wizards
- Application notes and white papers
- Wishlist (online suggestions)

Training

www.ni.com/custed

- Course schedules, descriptions, and registration information
- Self-paced training information

Consulting

(Alliance Program Members)
www.ni.com/alliance

Instrument Drivers

www.zone.ni.com/idnet

Additional LabVIEW-Related Sites

www.vimarket.com
www.ltrpub.com

Developer Resources

zone.ni.com

- Resource Library – example programs, technical presentations, and tutorials
- Developer Exchange
- Product Advisor
- Measurement Glossary

Mac Os (PowerPC only)
Mac OS 7.6.1 or later
32 MB RAM, (64 MB recommeded)
100 MB disk space for minimum LabVIEW installation, 225 for full LabVIEW installation*
PowerPC processor

Windows 2000/NT/Me/9x
For Windows NT, use Windows NT 4.0 Service Pack 3 or later
32 MB RAM, (64 MB recommended)
65 MB disk space for mininum LabVIEW installation, 200 MB for full LabVIEW installation*
Pentium processor or equivalent recommended

*Driver software requires addtitional hard disk space for installation

FONTANA

LabVIEW Student Edition 6i

Robert H. Bishop
The University of Texas at Austin

Prentice Hall
Upper Saddle River, NJ 07458

Library of Congress Cataloging-in-Publication Data

CIP data on file.

Vice President and Editorial Director, ECS: *Marcia J. Horton*
Acquisitions Editor: *Laura Curless*
Editorial Assistant: *Erin Katchmar*
Vice President of Production and Manufacturing, ESM: *David W. Riccardi*
Executive Managing Editor: *Vince O'Brien*
Managing Editor: *David A. George*
Production Editor: *Lakshmi Balasubramanian*
Director of Creative Services: *Paul Belfanti*
Creative Director: *Carole Anson*
Cover Designer: *John Christiana*
Manufacturing Manager: *Trudy Pisciotti*
Manufacturing Buyer: *Pat Brown*
Marketing Manager: *Holly Stark*
Marketing Assistant: *Karen Moon*

Prentice-Hall, Inc.
Upper Saddle River, New Jersey 07458

Printed in the United States of America
10 9 8 7 6 5 4 3 2 1

ISBN 0-13-032550-3

Instructional Material Disclaimer
The programs presented in this book have been included for their instructional value. They have been tested with care but are not guaranteed for any particular purpose. Neither the publisher nor the author offer any warranties or representations, nor do they accept any liabilities with respect to the programs.

Prentice-Hall International (UK) Limited, *London*
Prentice-Hall of Australia Pty. Limited, *Sydney*
Prentice-Hall Canada Inc., *Toronto*
Prentice-Hall Hispanoamericana, S.A., *Mexico*
Prentice-Hall of India Private Limited, *New Delhi*
Prentice-Hall of Japan, Inc., *Tokyo*
Pearson Education Asia Pte. Ltd., *Singapore*
Editora Prentice-Hall do Brasil, Ltda., *Rio de Janeiro*

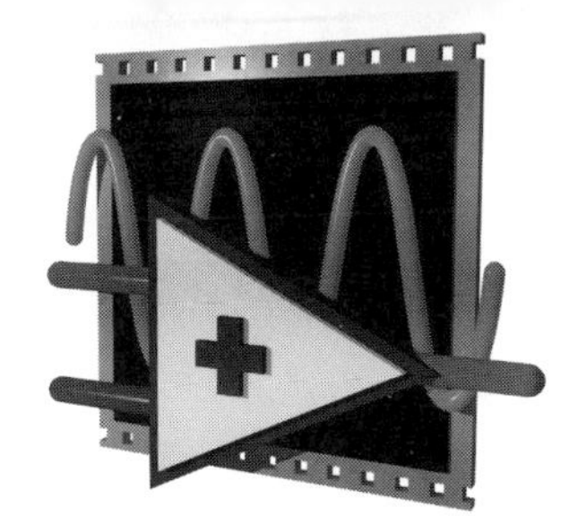

CONTENTS

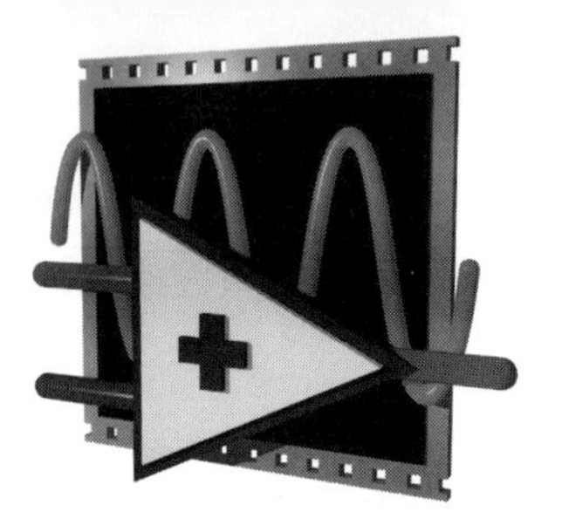

PREFACE

Learning with LabVIEW, 2nd Edition is the textbook that accompanies *LabVIEW Student Edition 6.0* from National Instruments, Inc. This textbook has been updated from the 1st Edition to include the latest features of LabVIEW. As you read through the book and work through the examples and play the games, we hope you will agree that this book is more of a personal tour guide than a software manual.

LabVIEW is a graphical programming language that has been widely adopted throughout industry, academia, and government labs as the standard for data acquisition, instrument control software, and analysis software. It is ideal for science and engineering applications—and is fun to use! The *LabVIEW Student Edition 6.0* affords students the opportunity for self-paced learning and independent project development.

The goal of this book is to help students learn to use LabVIEW on their own. With that goal in mind, this book is very art-intensive with over three-hundred and fifty figures in all. That means that there are numerous screen captures in each section taken from a typical LabVIEW session. The figures contain additional labels and pointers added to the LabVIEW screen captures to help students understand what they are seeing on their computer screens as they follow along in the book.

The most effective way to use *Learning with LabVIEW* is to have a concurrent LabVIEW session in progress on your computer and to follow along with the steps in the book. A directory of virtual instruments has been developed by the author exclusively for use by students using *Learning with LabVIEW*. These virtual instruments complement the material in the book. In most situations, the students are asked to develop the virtual instrument themselves following instructions given in the book, and then compare their solutions with the solution provided by the author to obtain immediate feedback. In other cases, students are asked to run a specified virtual instrument as a way to demonstrate an important LabVIEW concept.

THE *LABVIEW STUDENT EDITION 6.0*

With the renewed emphasis in higher education on hands-on laboratory experience, many educational institutions are beginning to improve their laboratory facilities with the goal of increasing student exposure to practical problems. Educators continue to be encouraged by industry to produce college graduates with experience in acquiring and analyzing data and with experience in constructing computer-based simulations of physical systems. LabVIEW takes advantage of the fact that computers are everywhere and are much more flexible than standard laboratory instruments. Using LabVIEW you can create your own virtual instruments (VIs). An effective way to improve laboratory instrumentation is by modifying and improving a LabVIEW computer program (that is, a virtual instrument that emulates a standard instrument), rather than to retrofit a laboratory with new hardware equipment. In LabVIEW, the software is the instrument.

The *LabVIEW Student Edition 6.0* software package is a powerful and flexible instrumentation and analysis software system for PCs running Microsoft Windows or Apple MacOS. The student edition is designed to give students early exposure to the many uses of graphical programming. LabVIEW not only helps reinforce basic scientific, mathematical, and engineering principles, but it encourages students to explore advanced topics as well. Students can run LabVIEW programs designed to teach a specific topic, or they can use their skills to develop their own applications. LabVIEW provides a real-world, hands-on experience that complements the entire learning process.

WHAT'S NEW WITH THE *LABVIEW STUDENT EDITION 6.0?*

The demand for LabVIEW in colleges and universities has led to the development of *LabVIEW Student Edition 6.0* based on the industry version of LabVIEW 6i. This is a new and significant software revision of *LabVIEW Student Edition 5.0* that delivers all of the graphical programming capabilities of the professional edition in a package that is even easier to use and learn. With the student edition, students can design graphical programming solutions for their classroom problems and laboratory experiments on their personal computers. The *LabVIEW Student Edition 6.0* features include the following:

- Compatibility with all National Instruments data acquisition and instrument control hardware.
- Support for all data types used in the LabVIEW industry version.
- Mathematical analysis library that has hundreds of VIs for statistics, time- and frequency-domain analysis, regression, linear algebra, and more.
- Extensive library of numerical analysis and visualization software.

- Capability to run MATLAB m-files from within the LabVIEW environment.
- Ability to utilize *simulated* data acquisition to learn about and practice with data acquisition in situations where actual hardware is unavailable.

The limitations of the *LabVIEW Student Edition 6.0* include:

- No integration of ActiveX Controls.
- User cannot link to external code via DLLs and Code Interface Nodes.
- User cannot build stand-alone executables.

ORGANIZATION OF *LEARNING WITH LABVIEW*

This textbook serves as a LabVIEW resource for students. The pace of instruction is intended for both undergraduate and graduate students. The book is comprised of twelve chapters and should be read sequentially when first learning LabVIEW. For more experienced students, the book can be used as a reference book by using the index to find the desired topics. The twelve chapters are as follows:

CHAPTER 1: LabVIEW Basics—This chapter introduces the LabVIEW environment and helps orient students when they open a virtual instrument. Concepts such as windows, toolbars, menus, and palettes are discussed.

CHAPTER 2: Virtual Instruments—The components of a virtual instrument are introduced in this chapter: front panel, block diagram, and icon/connector pair. This chapter also introduces the concept of controls (inputs) and indicators (outputs) and how to wire objects together in the block diagram.

CHAPTER 3: Editing and Debugging Virtual Instruments—Resizing, coloring, and labeling objects are just some of the editing techniques introduced in this chapter. Students can find errors using execution highlighting, probes, single-stepping, and breakpoints, just to name a few of the available debugging tools.

CHAPTER 4: SubVIs—This chapter emphasizes the importance of reusing code and illustrates how to create a VI icon/connector. It also shows parallels between LabVIEW and text-based programming languages.

CHAPTER 5: Structures— This chapter presents loops, case structures, and sequence structures that govern the execution flow in a VI. The Formula Node is introduced as a way to implement complex mathematical equations. The MATLAB Node is also introduced as a way to run m-files from within the LabVIEW environment.

CHAPTER 6: Arrays and Clusters—This chapter shows how data can be grouped, either with elements of the same type (arrays) or elements of a different type (clusters). This chapter also illustrates how to create and manipulate arrays and clusters.

CHAPTER 7: Charts and Graphs—This chapter shows how to display and customize the appearance of single and multiple charts and graphs.

CHAPTER 8: Data Acquisition—The basic characteristics of analog and digital signals are discussed in this chapter, as well as the factors students need to consider when acquiring and generating these signals. This chapter introduces students to the Measurement and Automation Explorer (MAX). All examples use the Easy I/O DAQ VIs.

CHAPTER 9: Strings and File I/O—This chapter shows how to create and manipulate strings on the front panel and block diagram. This chapter also explains how to write data to and read data from ASCII, spreadsheet, and binary files.

CHAPTER 10: Instrument Control—The components of an instrument control system using a GPIB or serial interface are presented in this chapter. Students are introduced to the notion of instrument drivers and of using the Measurement and Automation Explorer (MAX) to detect and install instrument drivers.

CHAPTER 11: Analysis—LabVIEW can be used in a variety of ways to support analysis of signals and systems. Several important analysis topics are discussed in this chapter, including how to use LabVIEW for signal generation, signal processing, linear algebra, curve fitting, formula display on the front panel, differential equations, finding roots (zero finder), and integration and differentiation.

CHAPTER 12: Other LabVIEW Applications—The concluding chapter discusses briefly other features of LabVIEW, including how to share data and VIs across a network using DataSocket technology. Also, a brief introduction is presented for interfacing with HiQ.

The important pedagogical elements in each chapter include the following:

1. A brief table of contents and a short preview of what to expect in the chapter.
2. A list of chapter goals to help focus the chapter discussions.
3. Margin icons that focus attention on a helpful hint or on a cautionary note.

4. An end-of-chapter summary and list of key terms.

5. Sections entitled **Building Blocks** near the end of each chapter present the continuous development and modification of a virtual instrument for measuring volume. The student is expected to construct the VIs based on the instructions given in the sections. The same VI is used as the starting point and then improved in each subsequent chapter as a means for the student to practice with the newly introduced chapter concepts. The VI for measuring volume was not used in Building Block sections whenever it proved more effective to use a different instrument to illustrate the important chapter concepts.

6. Many worked examples are included in each chapter. In most cases, students construct the VIs discussed in the examples by following a series of instructions given in the text. In the early chapters, the instructions for building the VIs are quite specific, but in the later chapters, students are expected to construct the VIs without precise step-by-step instructions. Of course, in all chapters, working versions of the VIs are provided for all examples in the Learning directory included as part of the *LabVIEW Student Edition 6.0*. Here is a sample of the worked examples:
   - Temperature system demonstration.
   - Solving a set of linear differential equations.
   - Building your first virtual instrument.
   - Computing a baseball batting average.
   - Computing and graphing the time value of money.
   - Studying chaos using the logistic difference equation.
   - Acquiring data and using the Measurement and Automation Explorer (MAX) to configure the system.
   - Writing ASCII data to a file.

7. A section entitled Relaxing Reading that describes how LabVIEW is being utilized to solve interesting real-world problems. The material is intended to give students a break from the technical aspects of learning LabVIEW and to stimulate thinking about how LabVIEW can be used in various other situations. Addresses and websites are given to help students continue to explore the topic more fully.
8. End-of-chapter exercises and problems reinforce the main topics of the chapter and provide practice with LabVIEW.

ORIGINAL SOURCE MATERIALS

Learning with LabVIEW, 2nd Edition was developed with the aid of several important references. Each reference is a manual published by National Instruments and one manual is actually used in a hands-on classroom environment taught by National Instruments instructors in their own classroom. Specifically, the following materials were used as primary sources:

- *LabVIEW Basics*, Course Software Version 6.0, Copyright ©2000.
- *LabVIEW User Manual*, Copyright ©2000.
- *LabVIEW Measurements Manual*, Copyright ©2000.

By design, there is a strong correlation between the material contained in the NI LabVIEW manuals and the material presented in this book. The information contained in the manuals was reduced in scope and refined to make it more accessible to students learning LabVIEW on their own.

LABVIEW STUDENT EDITION 6.0 SOFTWARE

It is assumed that the reader has a working knowledge of either the Windows or the MacOS operating systems. If your computer experience is limited, you may first want to spend some time familiarizing yourself with your computer and understanding the operation of your Mac or PC compatible. You should know how to access pull-down menus, open and save files, install software from a CD, and use a mouse. You will find previous computer programming experience helpful—but not necessary.

A set of virtual instruments has been developed by the author for this book. If you purchased the *LabVIEW Student Edition 6.0* package, which includes both the software and the book *Learning with LabVIEW*, you will find everything you need to begin the learning process. The library of virtual instruments is included on the accompanying CD. When you install the student edition, a directory entitled Learning will automatically appear and will include folders for each chapter containing the VIs used in that chapter. A library of instructional VIs for chemistry, physics, electrical and mechanical engineering, mathematics, statistics, computer science, and games is also included in the folder Instructional VIs found in the Learning directory.

On the other hand, if you purchased *Learning with LabVIEW* as a stand-alone book without the software CD, you will need to obtain the Learning directory from the Prentice Hall website. The LabVIEW page on the PH website is as follows: **http://www.prenhall.com/bishop**. You may also want to visit the National Instruments website at **http://www.ni.com/labviewse** for more information.

All of the VI examples in this book were tested by the author on a PC-compatible Dell Dimensions XPS D300 running Windows 98. Obviously, it is not possible to verify each VI on all the available PC-compatible and Macintosh platforms that are compatible with LabVIEW—so if you encounter platform-specific difficulties, please let us know.

Prentice Hall will replace a defective CD free of charge to registered users only. Make sure to fill out and mail in the registration card that comes with your *LabVIEW Student Edition 6.0* package.

If you would like information on upgrading to the LabVIEW 6i Professional Version, please write to

National Instruments
att.: Academic Sales
11500 North Mopac Expressway
Austin, TX 78759

or visit the National Instruments website **http://www.ni.com**.

LIMITED WARRANTY

The software and the documentation are provided "as is" without warranty of any kind, and no other warranties, either expressed or implied, are made with respect to the software. National Instruments does not warrant, guarantee, or make any representations regarding the use or the results of the use of the software or the documentation in terms of correctness, accuracy, reliablility, or otherwise and does not warrant that the operation of the software will be uninterrupted or error-free. This software is not designed with components and testing for a level of reliability suitable for use in the diagnosis and treatment of humans or as critical components in any life-support systems whose failure to perform can reasonably be expected to cause significant injury to a human. National Instruments expressly disclaims any warranties not stated herein. Neither National Instruments nor Prentice Hall shall be liable for any direct or indirect damages. The entire liability of National Instruments and its dealers, distributors, agents, or employees are set forth above. To the maximum extent permitted by applicable law, in no event shall National Instruments or its suppliers be liable for any damages including any special, direct, indirect, incidental, exemplary, or consequential damages, expenses, lost profits, lost savings, business interruption, lost business information, or any other damages arising out of the use or inability to use the software or thc documentation even if National Instruments has been advised of the possibility of such damages.

ACKNOWLEDGEMENTS

A special thanks to Ravi Marawar and Rick Francis at National Instruments for their assistance and input during the development of *Learning with LabVIEW, 2nd Edition*. Ravi provided the guiding hand and worked behind the scenes to insure a smooth working environment for Rick and me. A very special thanks to Rick for providing valuable insight into the inner workings of LabVIEW, for sharing his technical expertise as I developed the Learning directory VIs, and especially for helping me organize the chapters and for providing careful and accurate reviews of the manuscript. Finally, I wish to express my appreciation to Lynda Bishop for assisting me with the manuscript preparation, for providing valuable comments on the text, and for handling the day-to-day activities associated with the entire production.

KEEP IN TOUCH!

The author and the staff at Prentice Hall and at National Instruments would like to establish an open line of communication with the users of the *LabVIEW Student Edition 6.0*. We encourage students to e-mail the author or Prentice Hall with comments and suggestions for this and future editions.

Keep in touch!

Robert H. Bishop
bishop@csr.utexas.edu

Ravi Marawar, National Instruments Educational Product Manager
ravi.marawar@ni.com

CHAPTER 1

LabVIEW Basics

Welcome to the *Student Edition of LabVIEW*! **LabVIEW** is a powerful and complex programming environment. Once you have mastered the various concepts introduced in this book you will have the ability to develop applications in a graphical programming language and to develop virtual instruments for data acquisition, signal analysis, and instrument control. This introductory chapter provides a basic overview of LabVIEW and its components.

GOALS

1. Installation of the *Student Edition of LabVIEW*.
2. Familiarization with the basic components of LabVIEW.
3. Introduction to front panels and block diagrams, short cut and pull-down menus, palettes, VI libraries, and on-line help.

1.1 SYSTEM CONFIGURATION REQUIREMENTS

The *LabVIEW Student Edition* is distributed on CD-ROM and contains versions for Windows 2000/9X/NT, and Macintosh OS.

Windows 2000/9X On a PC compatible running Windows 2000/9X you will need:

- At least 16 MB of RAM (32 MB of RAM is recommended).
- At least 220 MB free hard disk space for a complete installation.

Windows NT On a PC compatible running Windows NT, you will need:

- Windows NT 4.0 Service Pack 3 or later.
- A Windows NT 80*x*86 computer (will not run on other processors, such as DEC Alpha, MIPS or PowerPC).
- At a minimum you should have a 486/DX processor, but a Pentium processor is strongly recommended.
- At least 16 MB of RAM (32 MB of RAM is recommended).
- At least 220 MB free hard disk space for a complete installation.

MacOS On an Apple running MacOS, you will need:

- A PowerPC.
- System 7.6.1 or higher.
- At least 32 MB of RAM (32 MB of RAM is recommended).
- At least 275 MB free hard disk space for a complete installation.

LabVIEW is Year-2000 compliant. The change to 2000 did not affect any internal storage of dates because LabVIEW has never stored two-digit years.

1.2 INSTALLING THE *STUDENT EDITION OF LABVIEW*

Windows After confirming that your system satisfies the system requirements, you can install the Windows 2000/9X/NT version of the LabVIEW Student Edition 6.0.

If you are running any programs when you insert the CD, please close them before installing LabVIEW.

1. Once you have placed the CD in the CD-ROM drive, double click on the computer icon on your desktop (usually labeled **My Computer**). Then double click on the CD-ROM icon.

2. Double click on the LabVIEW folder and then double click on setup.exe to start the installation. If you do not have a file by this name, double click on the file setup with the computer icon.
3. A dialog box for LabVIEW 6i should appear. Click on the **Next** button to continue the installation procedure.
4. The software license will then be displayed. Upon reviewing the license agreement, click the **I accept the License Agreement** button and the **Next** button to continue.
5. The subsequent dialog box allows you to change the default install directory. If you do not have a preference, keep the default directory. Otherwise you can press the **Browse** button to select an alternate directory. When ready, click the **Next** button to continue.
6. The default installation option is the **Complete Install**. The complete installation is the recommended option since it will install all the components for the *LabVIEW Student Edition 6.0*. If you are very low on disk space, you can select the **Custom Option** to select which features will be installed. Press the **Next** button to continue.
7. If you are performing a complete installation, you will have a choice of data acquisition (DAQ) drivers. The choices are **Real**, **Simulated**, or **None**. If you have a data acquisition board in your computer select **Real**. If you do not have a board, but would like to be able to have a software simulation then select **Simulated**. If you are not going to use DAQ you can select **None**. After you make your selection, click on **Next** to continue.
8. The installation should start installing files to your hard drive. Click on the **Finish** button when the installation has completed.
9. It is necessary to reboot the system if a DAQ driver is installed. A dialog box will appear prompting you to click **Yes** to restart your computer. If you click **No** you will need to restart the system manually before the DAQ driver will function correctly.

*When you first open LabVIEW, a startup screen will appear that includes a button to run the **LabVIEW Tutorial.** You can choose to view the tutorial for a quick introduction to LabVIEW.*

After the LabVIEW install is complete, you can install HiQ, mathematical software which you can use to create 2D and 3D graphs as well as lab reports and homework. To Install HiQ:

1. Make sure the **LabVIEW Student Edition 6.0** CD is in your CD ROM drive.
2. Double click on the My Computer icon on your Windows desktop.
3. Double click on the icon representing the CD-ROM.

4. Double click on the HiQ folder.
5. Double click on the setup.exe icon. A window entitled **NI HiQ 4.5 Setup** will appear. Click on the **Next** button to begin the installation.
6. Read the license agreement, then click on the **I accept the License Agreement** button and click the **Next** button to continue.
7. Once you have agreed to the terms of the license, enter the following user registration:

Name:	Your Name
Company:	University or college name
Serial number:	S79E27423

   Click on the **Next** button to continue.
8. You will not need to register online, so click on the **Next** button.
9. You may change the default install directory by clicking on the **Browse** button, or click the **Next** button to accept the default installation directory.
10. Click the **Next** button to start installing the files. You should see several files copied to your system followed by a dialog informing you that NI HiQ 4.5 has been successfully installed. When the installation is done, click on the **Finish** button.

*You must run HiQ once manually before LabVIEW will be able to launch HiQ automatically. Select **Start≫Programs≫National Instruments≫HiQ≫HiQ 4.5**. HiQ will launch and automatically open **Notebook1**. Then you can select **File≫Exit** to close HiQ.*

Macintosh

After confirming that your system satisfies the system requirements, you can install the Power Macintosh version of the LabVIEW Student Edition.

1. Once you have placed the CD in the CD-ROM drive, double click on the CD icon labeled LabVIEW 6 Student Edition on your desktop.
2. Double click on the LabVIEW 6 Student Edition icon in the **LabVIEW SE** folder.
3. A **LabVIEW Student Edition** dialog box will appear. Press the **Continue** button to continue.
4. The software license will be displayed, once you are done reviewing the license click the **Accept** button to continue.
5. The readme file for Version 6.0 of LabVIEW will be displayed. Once you are done reading press the **Continue** button.

6. This dialog box allows you to specify the installation type and location. In the upper left corner there is a drop down menu with two options: **Easy Install** and **Custom**. The **Easy Install** is recommended since it will install all the components for the *LabVIEW Student Edition 6.0*. If you are very low on disk space you can select the **Custom Option** and check which features will be installed.
7. If you selected the **Easy Install** in Step 6, you will have a choice of data acquisition (DAQ) drivers to install. The choices are **NI-DAQ**, **Simulated DAQ**, or **No DAQ**. If you have a data acquisition board in your computer select **NI-DAQ**. If you do not have a board, but would like to be able to have a software simulation then select **Simulated DAQ**. If you are not going to use DAQ you can select **No DAQ**.
8. At this point the software is being installed. Do not press the **Stop** button or the software installation will be halted before it is complete.
9. You should get a dialog box claiming that the installation was successful. At this point you can press the **Quit** button and run LabVIEW.

***HiQ** is not supported on Macintosh.*

1.3 THE LABVIEW ENVIRONMENT

LabVIEW is short for **Lab**oratory **V**irtual **I**nstrument **E**ngineering **W**orkbench. It is a powerful and flexible instrumentation and analysis software development application created by the folks at National Instruments—a company that creates hardware and software products that leverage computer technology to help engineers and scientists take measurements, control processes, and analyze and store data. National Instruments was founded over twenty-five years ago in Austin, Texas by James Truchard (known as Dr. T), Jeffrey Kodosky, and William Nowlin. At the time, all three men were working on sonar applications for the U.S. Navy at the Applied Research Laboratories at The University of Texas at Austin. Searching for a way to connect test equipment to DEC PDP-11 computers, Dr. T decided to develop an interface bus. He recruited Jeff and Bill to join him in his endeavor, and together they successfully developed LabVIEW and the notion of a "virtual instrument." In the process they managed to infuse their new company—National Instruments—with an entrepreneurial spirit that still pervades the company today.

Engineers and scientists in research, development, production, test, and service industries as diverse as automotive, semiconductor, aerospace, electronics, chemical, telecommunications, and pharmaceutical have used and continue to use LabVIEW to support their work. LabVIEW is a major player in the area of testing and measurements, industrial automation, and data analysis. For example, scientists at NASA's Jet Propulsion Laboratory used LabVIEW to analyze and display Mars Pathfinder Sojourner rover engineering data, including the position and temperature of the rover, how much power remained in the rover's battery, and generally to monitor Sojourner's overall health. This book is intended to help you learn to use LabVIEW as a programming tool and to serve as an introduction to data acquisition, instrument control, and data analysis.

LabVIEW programs are called **Virtual Instruments**, or VIs for short. LabVIEW is different from text-based programming languages (such as Fortran and C) in that LabVIEW uses a graphical programming language, known as the G programming language, to create programs relying on graphic symbols to describe programming actions. LabVIEW uses a terminology familiar to scientists and engineers, and the graphical icons used to construct the G programs are easily identified by visual inspection. You can learn LabVIEW even if you have little programming experience, but you will find knowledge of programming fundamentals helpful. If you have never programmed before (or maybe you have programming experience but have forgotten a few things) you may want to review the basic concepts of programming before diving into the G programming language.

LabVIEW provides an extensive library of virtual instruments and functions to help you in your programming. LabVIEW also contains application-specific libraries for data acquisition (discussed in Chapter 8), file input/output (discussed in Chapter 9), GPIB and serial instrument control (discussed in Chapter 10), and data analysis (discussed in Chapter 11). It includes conventional program debugging tools with which you can set breakpoints, single-step through the program, and animate the execution so you can observe the flow of data. Editing and debugging VIs is the topic of Chapter 3.

LabVIEW has a good set of VIs for data presentation on various types of charts and graphs. Chapter 7 discusses the process of presenting data on charts and graphs. Also, the *Student Edition of LabVIEW* is packaged with HiQ, a data visualization, report generation, and analysis program which can be used for creating lab reports. A brief introduction to HiQ is given in Chapter 12.

The LabVIEW system consists of the LabVIEW application executable files and many associated files and folders. LabVIEW uses files and directories to store information necessary to create your VIs. Some of the more important files and directories are:

1. The LabVIEW executable. Use this to launch LabVIEW.

2. The vi.lib directory. This directory contains libraries of VIs such as data acquisition, instrument control, and analysis VIs; it must be in the same

directory as the LabVIEW executable. Do not change the name of the vi.lib directory, because LabVIEW looks for this directory when it launches. If you change the name, you cannot use many of the controls and library functions.

3. The examples directory. This directory contains many sample VIs that demonstrate the functionality of LabVIEW.
4. The user.lib directory. This directory is where you can save VIs you have created, and they will appear in the LabVIEW **Functions palette**.
5. The instr.lib directory. This directory is where your instrument driver libraries are placed if you want them to appear in the **Functions palette**.
6. The Learning directory. This file contains a library of VIs that you will use with the *Learning with LabVIEW* book.

After the installation is complete, open the Learning *directory and create a folder called* Users Stuff*. This is where you will save your work as you progress through the book.*

1.4 THE STARTUP SCREEN

When you launch LabVIEW by double-clicking on its icon, the startup screen appears as in Figure 1.1.

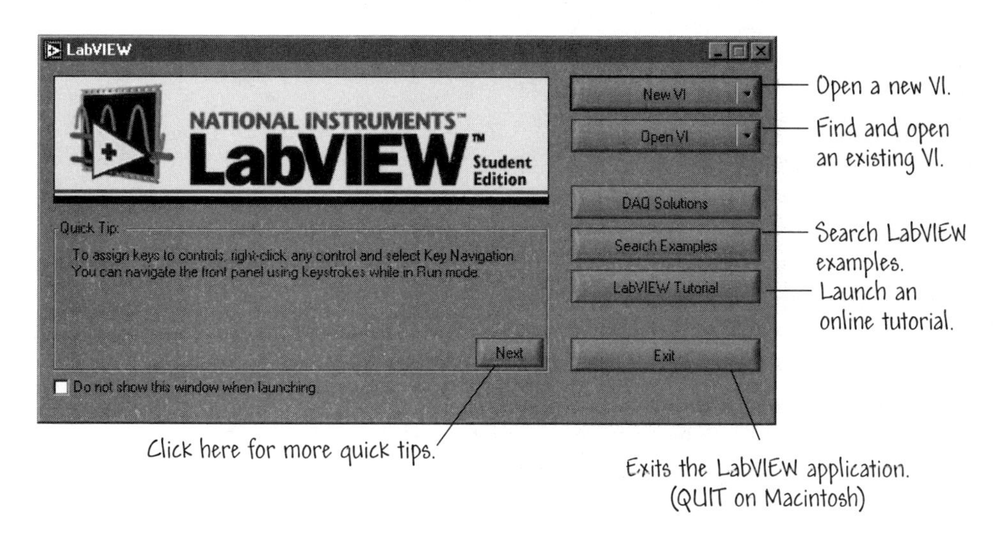

FIGURE 1.1
The startup screen.

Throughout this book, use the left mouse button (if you have one) unless we specifically tell you to use the right one.

At the beginning of each LabVIEW session the startup screen will contain a small box at the lower left hand corner that you can check if you want to start up without showing the start screen. If you choose this option, LabVIEW will always open up with a new VI named Untitled 1.vi rather than with the startup screen.

The startup screen includes a new quick tip each session. You can view more tips by clicking the **Next** button. After the LabVIEW session starts (for example, after you have opened a VI) a switch will appear at bottom of the startup screen window that lets you choose whether to see this large version of the startup screen or an abbreviated window with just the options to exit LabVIEW, open an existing VI, or create a new VI.

Searching the LabVIEW Examples

In this exercise you will search through the list of example VIs and demonstrations that are included with the *Student Edition of LabVIEW*. Open the LabVIEW application and get to the startup screen. The search begins at the LabVIEW startup screen by clicking on the **Search Examples** button, as shown in Figure 1.1.

After initiating the search, the LabVIEW examples shown are listed by category:

- **Fundamentals**
- **I/O Interfaces**
- **Measurements**
- **Communication**
- **Advanced**
- **Demonstrations**

The six categories, shown in Figure 1.2 include many examples. In this exercise, you will pursue a search in the **Demonstrations** group. In particular, look inside the **Analysis** group. Clicking on **Analysis** opens up the **Analysis Demonstration** page from where you can select the demonstration of your choice. Clicking on the **Demonstrations** selection opens the first window shown in Figure 1.3.

There are currently five demonstrations available:

1. **Temperature System Demo**
2. **Signal Generation and Processing**
3. **Vibration Analysis**
4. **Limit Testing for Unevenly Sampled Data**
5. **Basic Level Triggering Example**

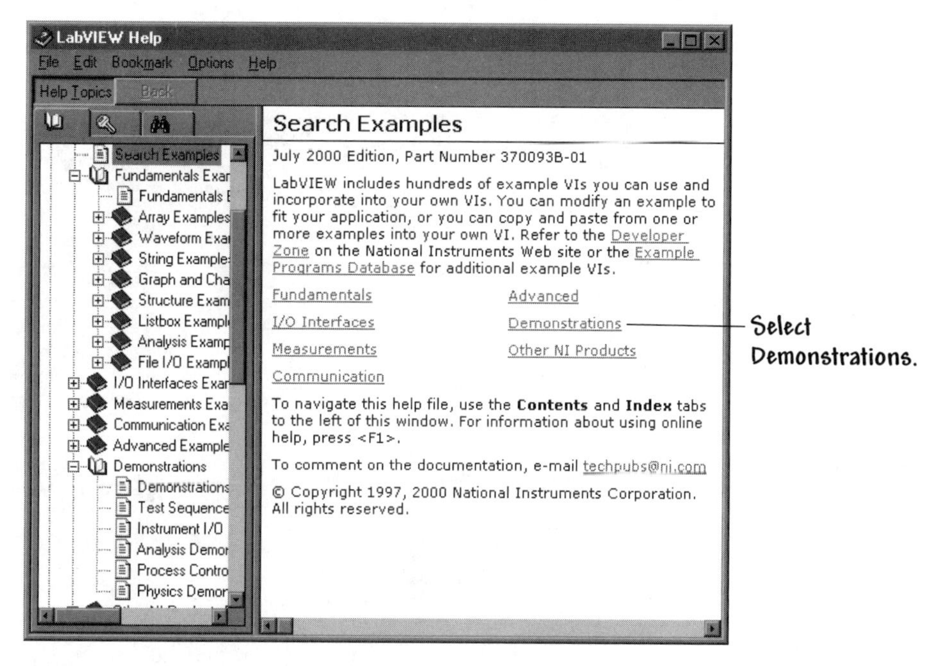

FIGURE 1.2
LabVIEW examples.

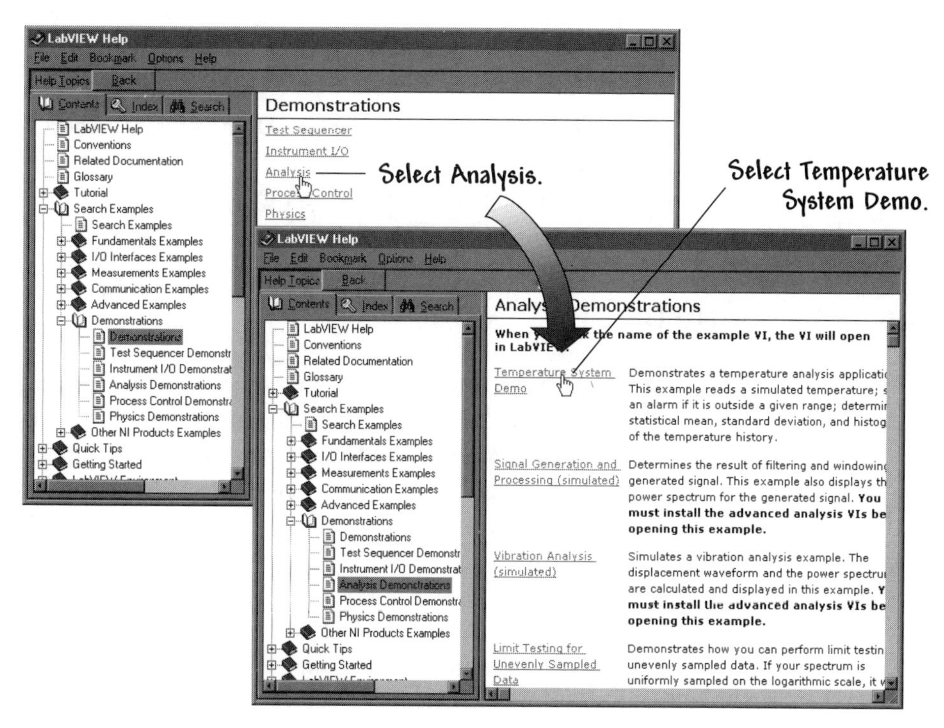

FIGURE 1.3
The **Analysis Demonstrations** screen.

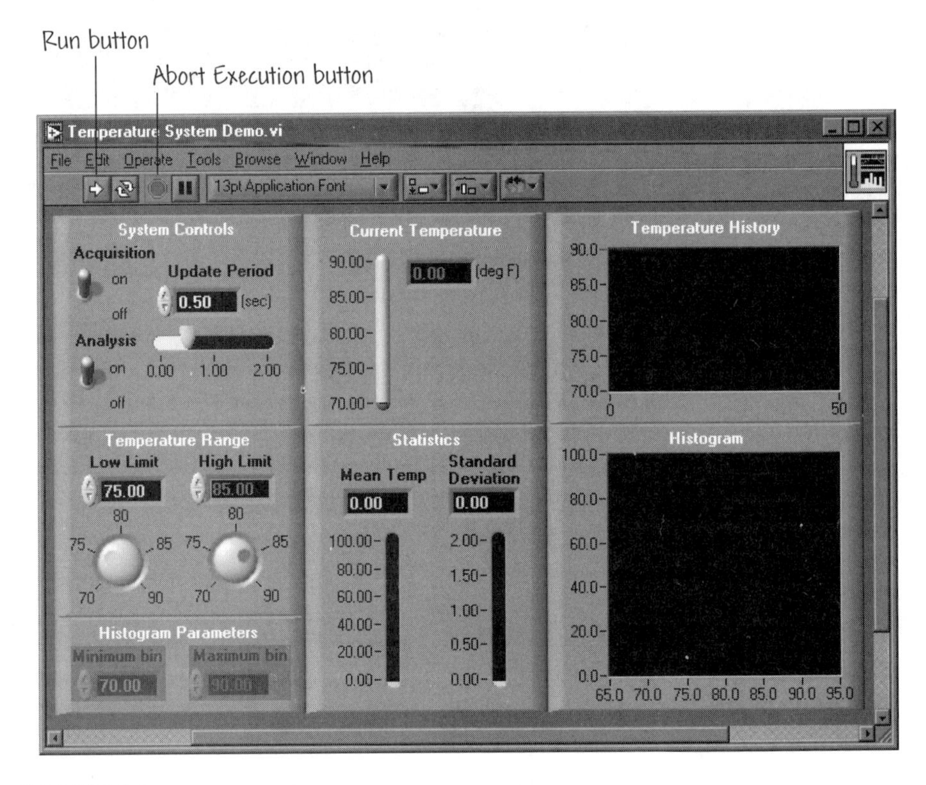

FIGURE 1.4
The temperature system demonstration front panel.

Selecting the **Temperature System Demo** opens up the associated virtual instrument (VI) (more on VIs in Chapter 2). Just for fun you can start the VI running and see what happens. Start the VI by clicking on the **Run** button, as shown in Figure 1.4. Stop the VI by clicking on the **Abort Execution** button. Try it! ◆

1.5 PANEL AND DIAGRAM WINDOWS

An untitled front panel window appears when you select **New VI** from the startup screen. The front panel window is the interface to your VI code and is one of the two LabVIEW windows that comprise a virtual instrument. The other window—the block diagram window—contains program code that exists in a graphical form (such as icons, wires, etc.).

Front panels and block diagrams consist of graphical objects that are the G programming elements. Front panels contain various types of controls and indicators (that is, inputs and outputs, respectively). Block diagrams contain terminals corresponding to front panel controls and indicators, as well as constants, functions, subVIs, structures, and wires that carry data from one object to another. Structures are program control elements (such as For Loops and While Loops).

Figure 1.5 shows a front panel and its associated block diagram. You can find the virtual instrument FirstVI.vi shown in Figure 1.5 in Chapter1 folder

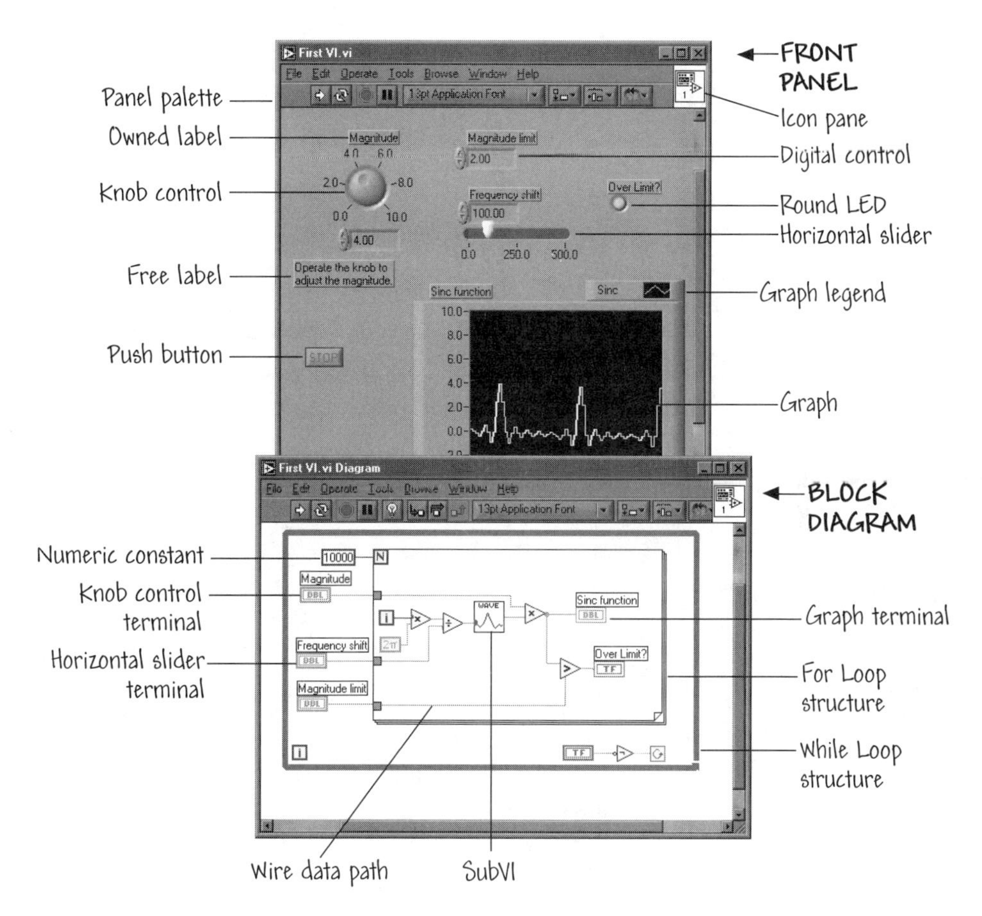

FIGURE 1.5
A front panel and the associated block diagram.

within the directory Learning. That VI can be located by choosing **Open VI** on the **Startup** screen and navigating to the Chapter1 folder in the Learning directory and then selecting FirstVI.vi. Once you have the VI front panel open, find the **Run** button on the panel toolbar and click on it. Your VI is now running. You can turn the knob and vary the different inputs and watch the output changes reflected in the graph. Give it a try! If you have difficulty getting things to work, then just press ahead with the material in the next sections and come back to this VI when you are ready.

1.5.1 Front Panel Toolbar

A toolbar of command buttons and status indicators that you use for controlling VIs is located on both the front panel and block diagram windows. The front panel toolbar and the block diagram toolbar are different, although they do each contain some of the same buttons and indicators. The toolbar that appears at the top of the front panel window is shown in Figure 1.6.

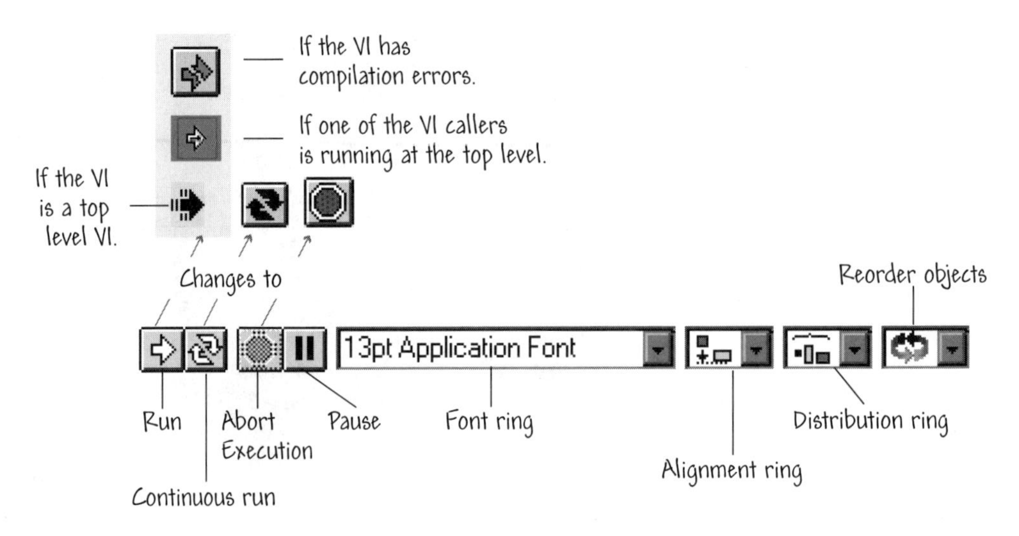

FIGURE 1.6
The front panel toolbar.

› While the VI is executing, the **Abort Execution** button appears. Although clicking on the abort button terminates the execution of the VI, as a general rule you should avoid terminating the program execution this way and either let the VI execute to completion or incorporate a programmatic execution control (that is, an on-off switch or button) to terminate the VI from the front panel.

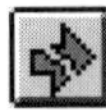

› The **Broken Run** button replaces the **Run** button when the VI cannot compile and run due to coding errors. If you encounter a problem running your VI, just click on the **Broken Run** button, and a window will automatically appear on the desktop that lists all the detected program errors. And then, if you double click on one of the specific errors in the list, you will be taken automatically to the location in the block diagram (that is, to the place in the code) where the error exists. This is a great debugging feature! More discussion on the issue of debugging VIs can be found in Chapter 3.

› Clicking on the **Continuous Run** button leads to a continuous execution of the VI. Clicking on this button again disables the continuous execution—the VI stops when it completes normally. The behavior of the VI and the state of the toolbar during continuous run is the same as during a single run started with the **Run** button.

› The **Pause/Continue** button pauses VI execution. To continue program execution after pausing, press the button again, and the VI resumes execution.

The **Font ring**, shown in Figure 1.7, sets font options—font type, size, style, and color.

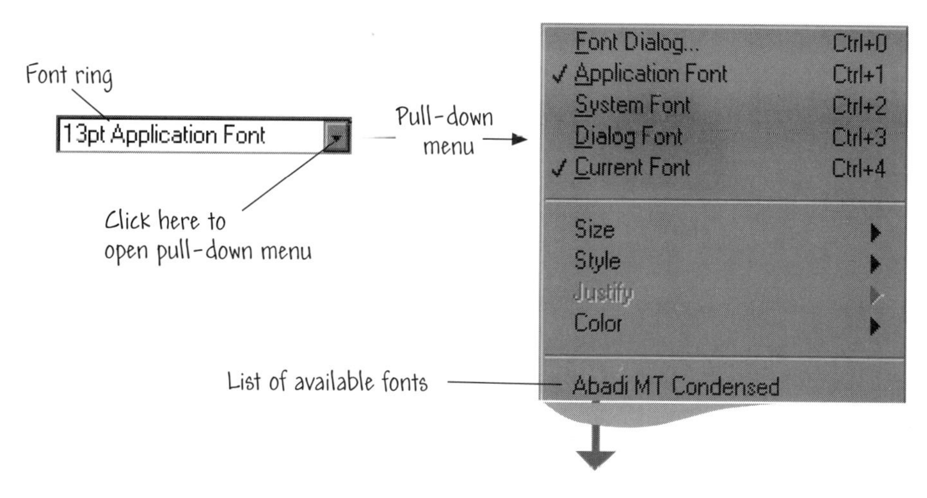

FIGURE 1.7
The **Font ring** pull-down menu.

The **Alignment ring** sets the preferred alignment of the various objects on either the front panel or on the block diagram. After selecting the desired objects for alignment, you can set the preferred alignment for two or more objects. For example, you can align objects by their left edges or by their top edges. The various alignment options are illustrated in the **Alignment ring** pull-down menu shown in Figure 1.8. Aligning objects is very useful in organizing the VI front panel (and the block diagram, for that matter). On the surface, it may appear that aligning the front panel objects, while making things "neat and pretty," does not contribute to the goal of a functioning VI. As you gain experience with constructing VIs, you will find that they are easier to debug and verify if the interface (front panel) and the code (block diagram) are organized to allow for easy visual inspection.

The **Distribution ring**, shown in Figure 1.9, sets the preferred distribution options for two or more objects. For example, you can evenly space selected objects, or you can remove all the space between the objects.

1.5.2 Block Diagram Toolbar

The block diagram toolbar contains many of the same buttons as the front panel toolbar. Four additional program debugging features are available on the block diagram toolbar and are enabled via the buttons, as shown in Figure 1.10.

› Clicking on the **Highlight Execution** button enables execution highlighting. As the program executes, you will be able to see the data flow through the code on the block diagram. This is extremely helpful for debugging and verifying proper

› execution. In the execution highlighting mode, the button changes to a brightly lit light bulb.

LabVIEW debugging capabilities allow you to single-step through the VI node to node. A **node** is an execution element, such as a For Loop or subVI. You

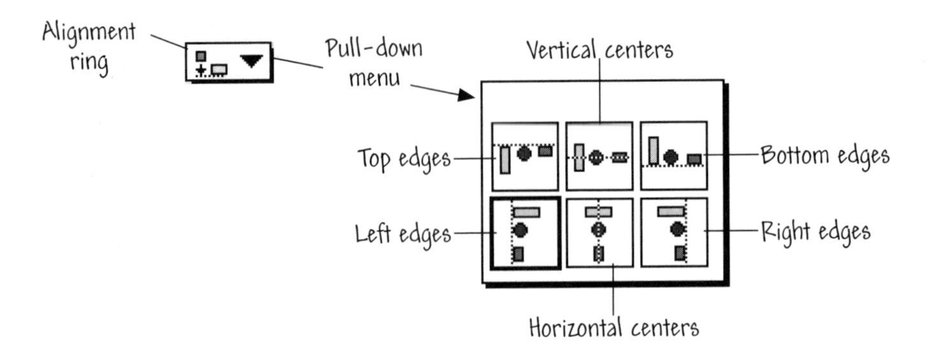

FIGURE 1.8
The **Alignment ring** pull-down menu.

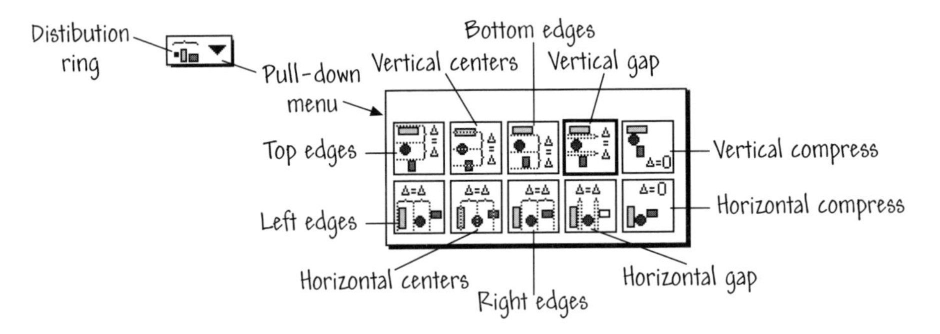

FIGURE 1.9
The **Distribution ring** pull-down menu.

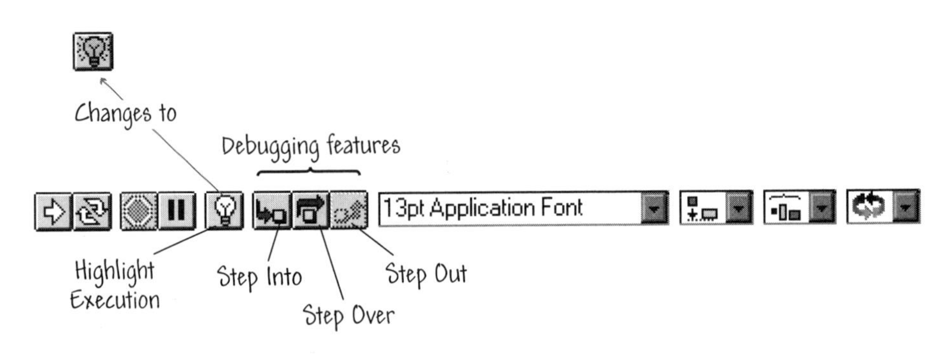

FIGURE 1.10
The block diagram toolbar.

will learn more about the different types of nodes as you proceed through the book, but for the time being you can think of a node as a section of the computer code that you want to observe executing. Each node blinks to show it is ready to execute.

› The **Step Over** button steps over a node. You are in effect executing the node without single stepping through the node.

› The **Step Into** button allows you to step into a node. Once you have stepped into the node, you can single step through the node.

› The **Step Out** button allows you to step out of a node. By stepping out of a node, you can complete the single stepping through the node and go to the next node.

› The **Warning** indicator only appears when there is a potential problem with your block diagram. The appearance of the warning does not prevent you from executing the VI. You can enable the Warning indicator on the Options...≫ Debugging menu in the **Tools** pull-down menu.

1.6 SHORT CUT MENUS

LabVIEW has two types of menus—**pull-down** menus and **short cut** menus. We will focus on short cut menus in this section and on pull-down menus in the next section. Our discussions here in Chapter 1 are top-level; we reserve the detailed discussions on each menu item for later chapters as they are used.

To access a short cut menu, position the cursor on the desired object on the front panel or block diagram and click the right mouse button on a PC-compatible or hold down the <Command> key and then click the mouse button on the Mac. In most cases, a short cut menu will appear since most LabVIEW objects have short cut menus with options and commands. This process is called "popping up," and short cut menus are also know as pop-up menus. You will also find that you can pop-up on the empty front panel and block diagram space, giving you access to the **Controls** and **Functions** palettes and other important palettes. The options presented to you on short cut menus depend on the selected object—popping up on a numeric control will open a different short cut menu than popping up on a For Loop. When you construct a program in G, you will use short cut menus extensively!

Many short cut and pull-down menus contain submenus, as shown in Figures 1.11 and 1.12.

On a PC-compatible, right mouse click on the object to open a short cut menu. On a Mac, press <Command> and simultaneously click on the object.

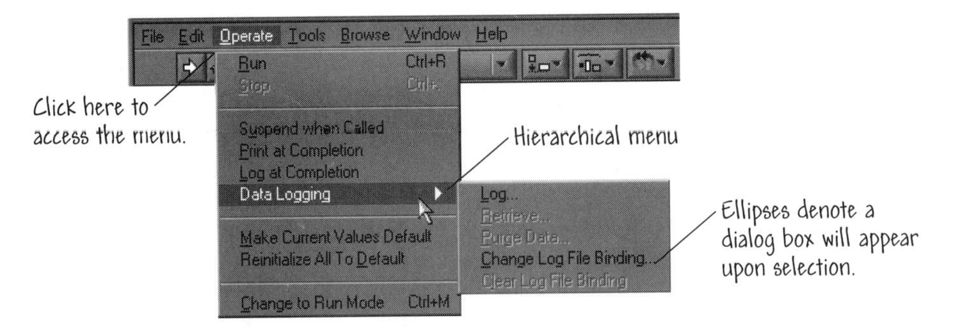

FIGURE 1.11
An example of a pull-down menu expanding into a submenu.

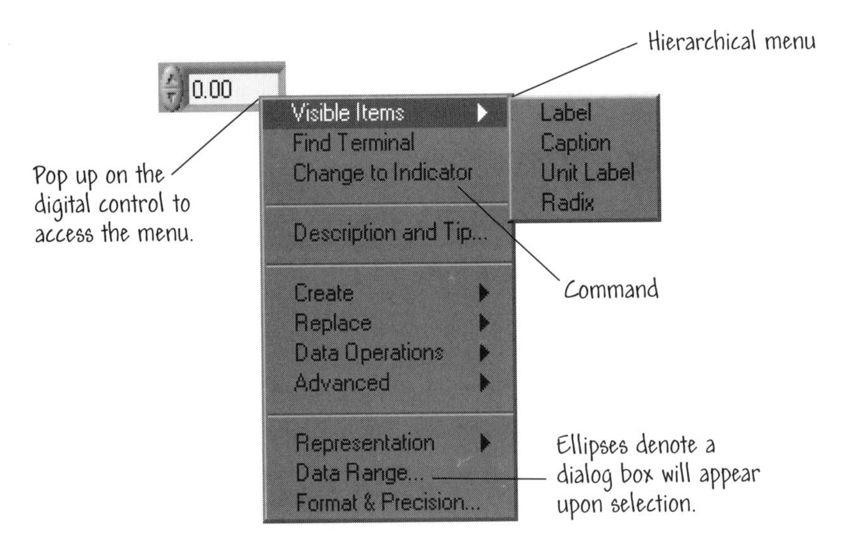

FIGURE 1.12
An example of a short cut menu.

Menu items that expand into submenus are called **hierarchical** menus and are denoted by a right arrowhead on the menu. Hierarchical menus will present you with different types of options and commands. One typical option is the so-called mutually exclusive option. This means that if you select the option (by clicking on it), a check mark will appear next, indicating the option is selected; otherwise the option is not selected.

Popping up on different areas of an object may lead to different short cut menus. If you pop up and do not see the anticipated menu selection, then pop up somewhere else on the object.

Another type of menu item opens dialog boxes containing options for you to use to modify and configure your program elements. Menu items leading to dialog boxes are denoted by ellipses (...). Menu items without right arrowheads or ellipses are generally commands that execute immediately upon selection. **Create**

Constant is an example of a command that appears in many short cut menus. In some instances, commands are replaced in the menu by their inverse commands when selected. For example, after you choose **Change to Indicator**, the menu selection is replaced by **Change to Control**.

1.7 PULL-DOWN MENUS

The menu bar at the top of the LabVIEW screen, shown in Figure 1.13, contains the important pull-down menus. In this section we will introduce the pull-down menus: **File**, **Edit**, **Operate**, **Tools**, **Browse**, **Window**, and **Help**. The level of discussion on pull-down windows given here is consistent with our previous discussions. The main goal is to introduce the menus.

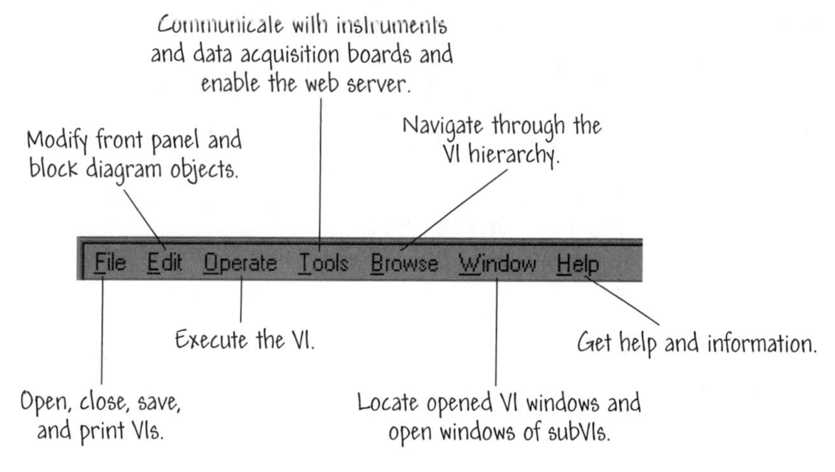

FIGURE 1.13
The menu bar.

1.7.1 File Menu

The **File** pull-down menu, shown in Figure 1.14, contains commands associated with file manipulations. For example, you can create new VIs or open existing ones from the **File** menu. You use options in the **File** menu primarily to open, close, save, and print VIs. As you observe each pull-down window, notice that selected commands and options have shortcuts listed beside them. These shortcuts are keystroke sequences that can be used to choose the desired option without pulling down the menu. For example, you can open a new VI by typing and entering Ctrl+O at the keyboard, or you can access the **File** pull-down menu and select **Open**.

1.7.2 Edit Menu

The **Edit** menu, shown in Figure 1.15, is used to modify front panel and block diagram objects of a VI. The **Undo** and **Redo** options are very useful when you

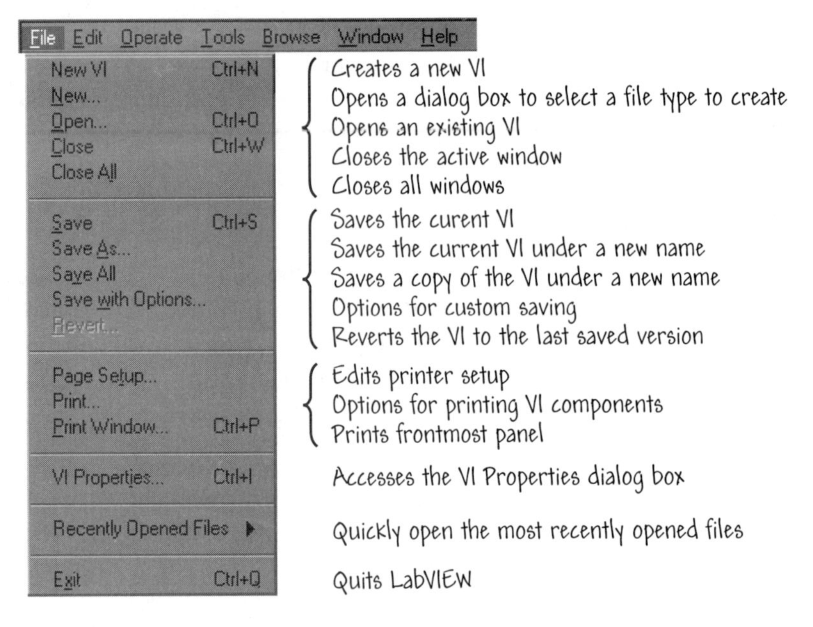

FIGURE 1.14
Pull-down menus—**File**.

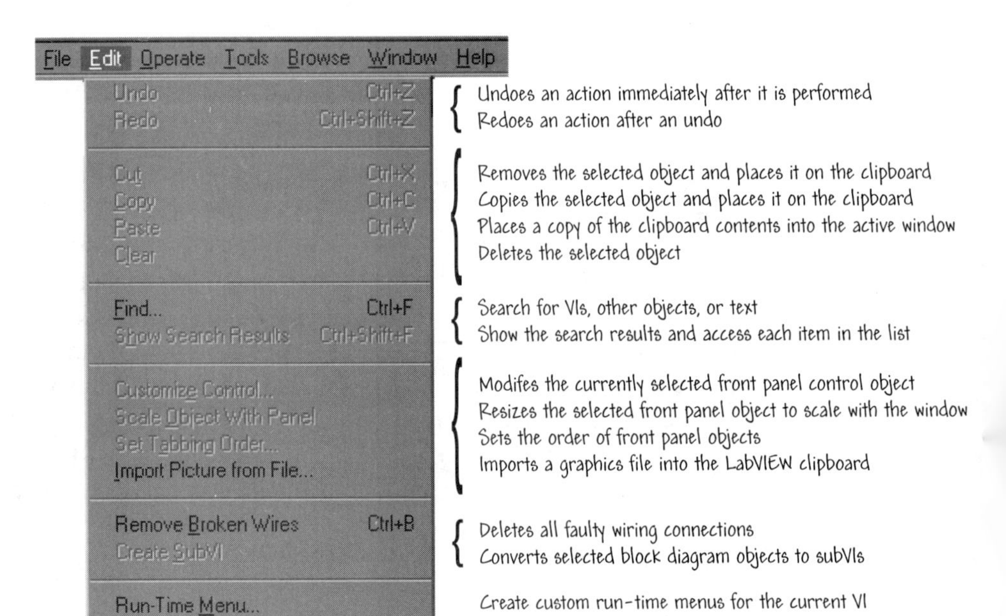

FIGURE 1.15
Pull-down menus—**Edit**.

are editing because it allows you to undo an action after it is performed, and once you undo an action you can redo it. By default, the maximum number of undo steps per VI is 8—you can increase or decrease this number if desired.

1.7.3 Operate Menu

The **Operate** menu, shown in Figure 1.16, can be used to run or stop your VI execution, change the default values of your VI, and switch between the run mode and edit mode.

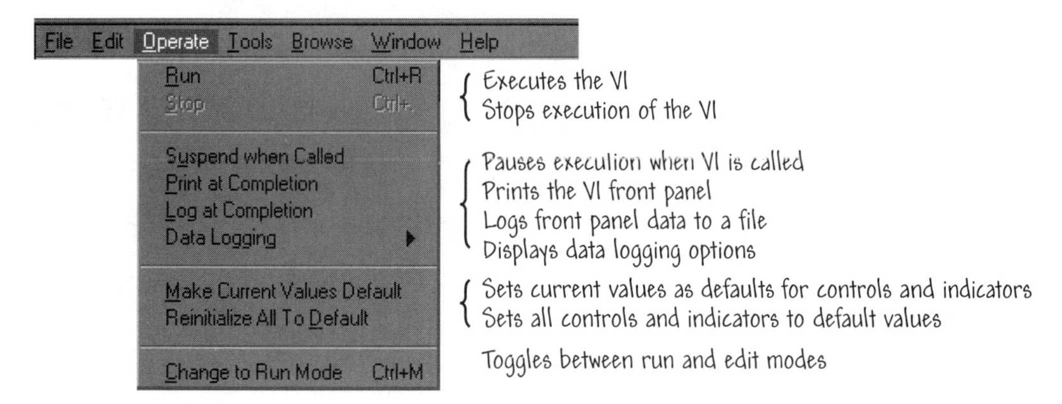

FIGURE 1.16
Pull-down menus—**Operate**.

1.7.4 Tools Menu

The **Tools** menu, shown in Figure 1.17, is used to communicate with instruments and data acquisition boards, compare VIs, build applications, enable the Web server, and access other options of LabVIEW.

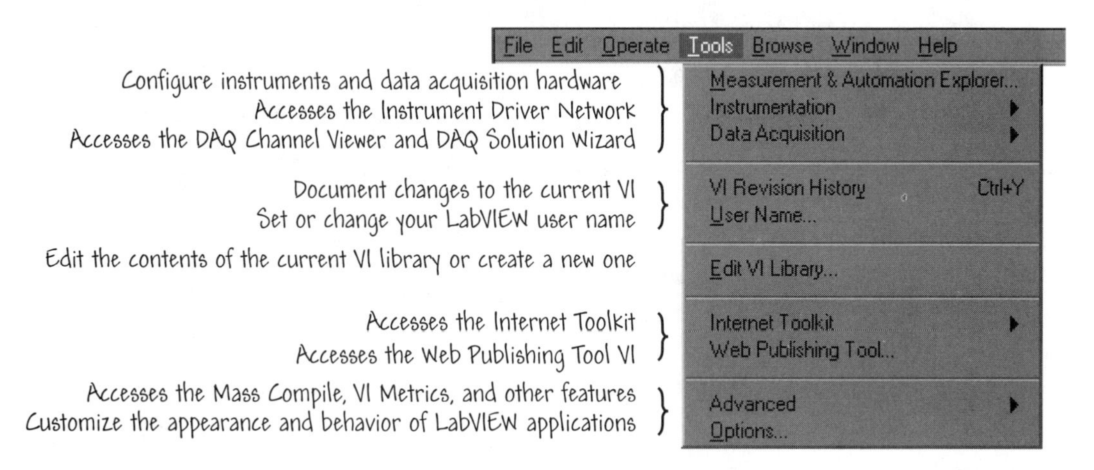

FIGURE 1.17
Pull-down menus—**Tools**.

An important link is to the main National Instruments website, where you can obtain general information about the company and its products. If you use LabVIEW to control external instruments, you will be interested in the **Instrument Driver Network...** link, which connects you to over 600 LabVIEW-ready instrument drivers. This can be found in the Tools≫Instrumentation hierarchichal menu. Refer to Chapter 10 for more information on instrument drivers.

1.7.5 Browse Menu

You use the **Browse** menu to navigate through the VI hierarchy. The **Browse** menu is depicted in Figure 1.18.

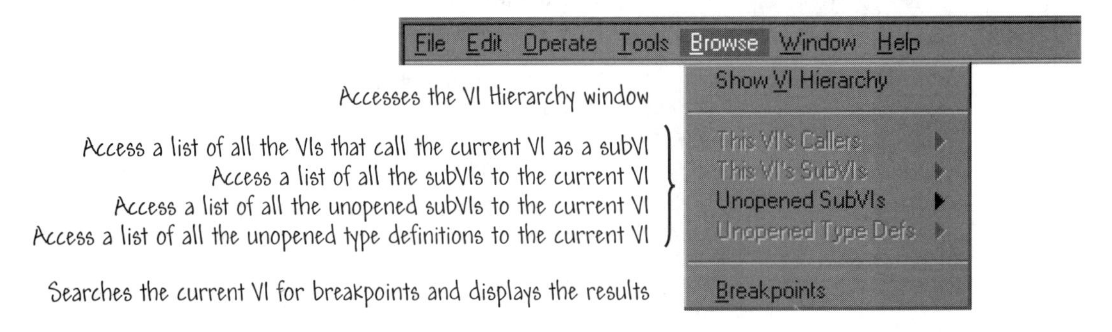

FIGURE 1.18
Pull-down menus—**Browse**.

1.7.6 Windows Menu

The **Windows** menu, shown in Figure 1.19, is used for a variety of activities. You can toggle between the panel and diagram windows, and you can "tile" both

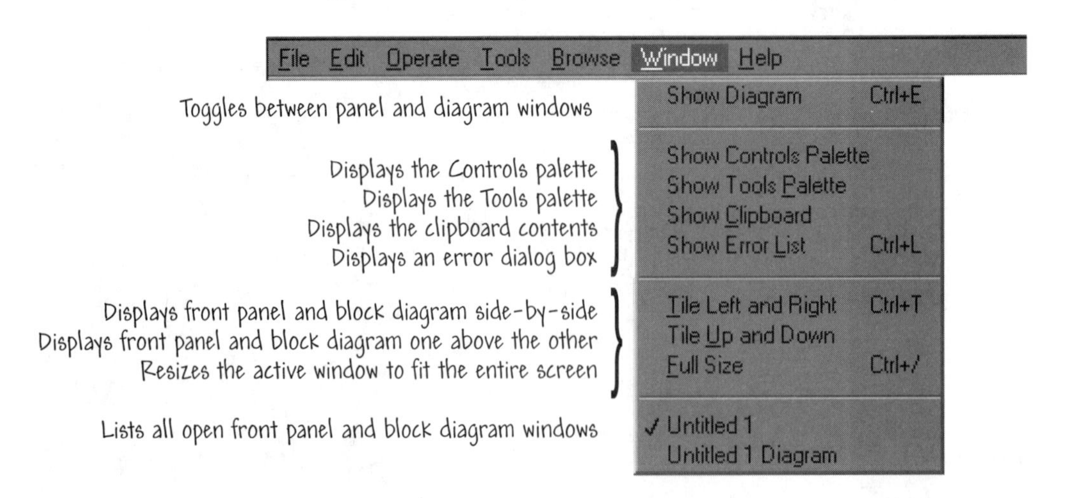

FIGURE 1.19
Pull-down menus—**Windows**.

windows so you can see them at the same time (one above the other or side-by-side). All the open VIs are listed in the menu (at the bottom), and you can switch between the open VIs. Also, if you want to show the palettes on the desktop, you can use the **Windows** menu to select either (or both) palettes (more on palettes in the next section).

1.7.7 Help Menu

The **Help** menu, shown in Figure 1.20, provides access to the extensive LabVIEW online help facilities. You can view information about panel or diagram objects, activate the online reference utilities, and view information about your LabVIEW version number and computer memory. The direct pathways to the Internet are provided by **Web Resources** and **Student Edition Web Resources**, which connect you directly to the main source of information concerning the LabVIEW Student Edition available on the Web. You should make a point to surf this website—this is where the latest and greatest information, news, and updates on the both the software *Student Edition of LabVIEW* and on the book *Learning with LabVIEW* will be posted.

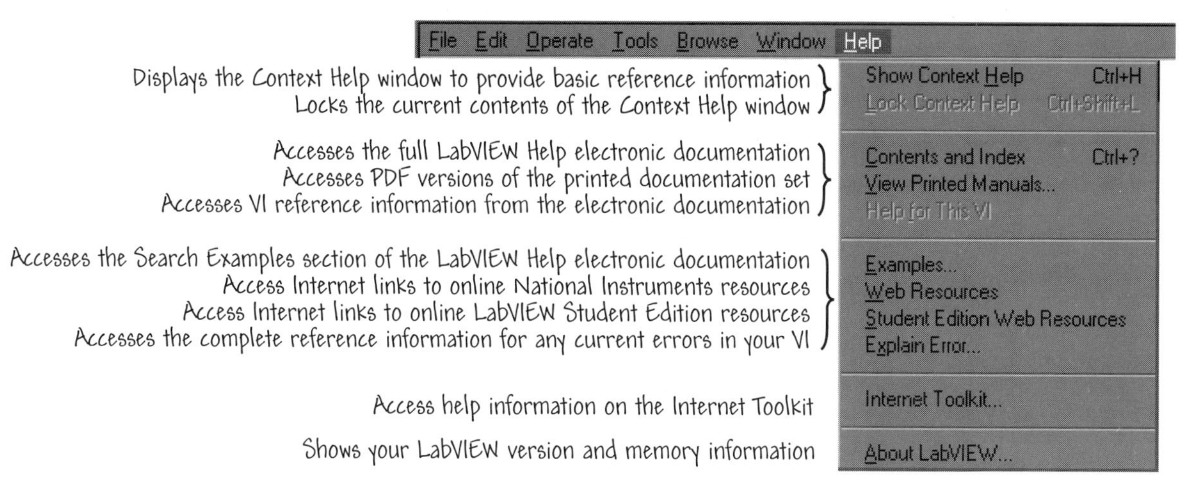

FIGURE 1.20
Pull-down menus—**Help**.

1.8 PALETTES

Palettes are graphical panels that contain various tools and objects used to create and operate VIs. You can move the palettes anywhere on the desktop that you want—preferably off to one side so that they do not block objects on either the front panel or block diagram. It is sometimes said that the palettes float. The three main palettes are the **Tools**, **Controls**, and **Functions** palettes.

1.8.1 Tools Palette

A **tool** is a special operating mode of the mouse cursor. You use tools to perform specific editing functions, similar to how you would use them in a standard paint program. You can create, modify, and debug VIs using the tools located in the floating **Tools** palette, shown in Figure 1.21. If the **Tools** palette is not visible, select **Show Tool Palette** from the **Windows** pull-down menu to display the palette. After you select a tool from this menu, the mouse cursor changes to the appropriate shape.

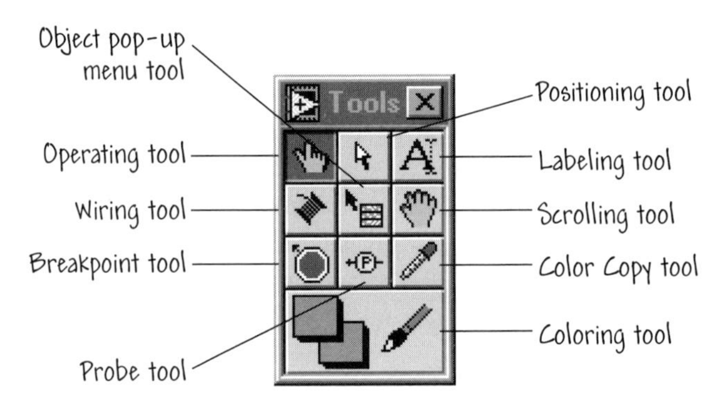

FIGURE 1.21
The **Tools** palette.

One way to access the online help is to place any tool found in the **Tools** palette over the object of interest in the block diagram window. If online help exists for that object, it will appear in a separate Help window. This process requires that you first select **Show Context Help** from the **Help** pull-down menu.

*On a **Windows** platform, a shortcut to accessing the **Tools** palette is to press the <shift> and the right mouse button on the panel or the diagram window. On a **Macintosh** platform, you can access the **Tools** palette by pressing <command-shift> and the mouse button on the panel or diagram window.*

1.8.2 Controls Palette

The **Controls** palette, shown in Figure 1.22, consists of top-level icons representing subpalettes, which contain a full range of available objects that you can use in creating a front panel.

You can access the subpalettes by clicking on the desired top-level icon. An example top-level icon is **Numeric** (see Figure 1.22), which when opened, will reveal the various numeric controls (by control we mean "input") and indicators (by indicator we mean "output") that you will utilize on your front panel as a way to move data into and out of the program code (on the block diagram). The topic of numeric controls and indicators is covered in Chapter 2. Each item on the **Controls** palette is discussed in more detail in the chapter in which it is first utilized.

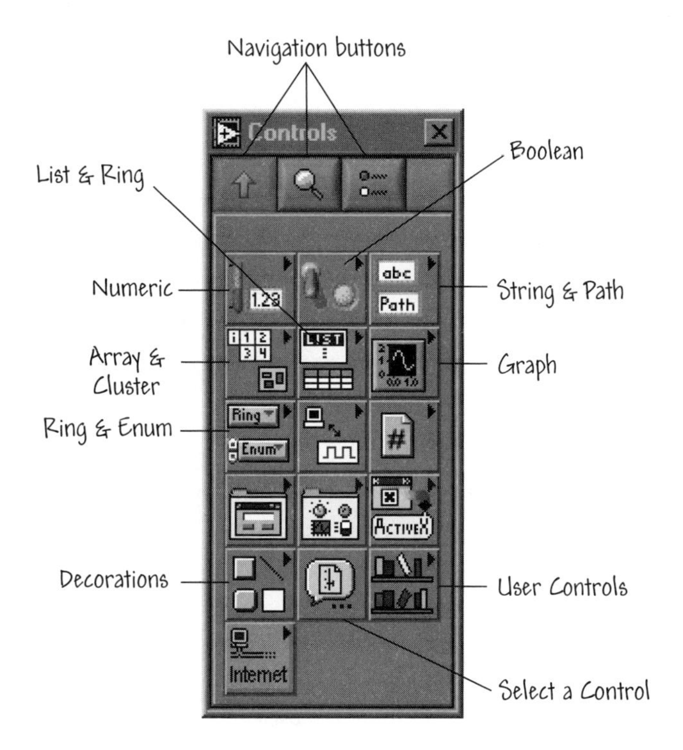

FIGURE 1.22
The **Controls** palette.

If the **Controls** palette is not visible, you can open the palette by selecting **Show Controls Palette** from the **Windows** pull-down menu. You also can access the **Controls** palette by popping up on an open area in the front panel window.

*The **Controls** palette is available only when the front panel window is active.*

1.8.3 Functions Palette

The **Functions** palettes, shown in Figure 1.23, works in the same general way as the **Controls** palette. It consists of top-level icons representing subpalettes, which contain a full range of available objects that you can use in creating the block diagram. You access the subpalettes by clicking on the desired top-level icon. Many of the G program elements are accessed through the **Functions** palette. For example, the subpalette **Structures** (see Figure 1.23) contains For Loops, While Loops, and Formula Nodes—all of which are common elements of VIs.

If the **Functions** palette is not visible, you can open the palette by selecting **Show Functions Palette** from the **Windows** pull-down menu. You can also access the **Functions** palette by popping up on an open area in the block diagram window.

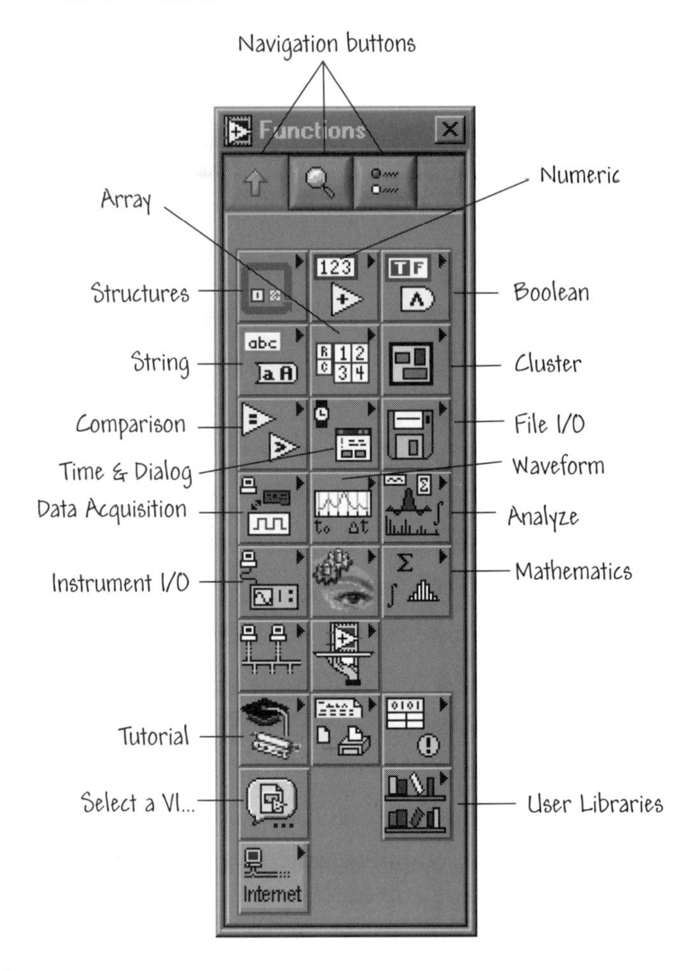

FIGURE 1.23
The **Functions** palette.

*The **Functions** palette is available only when the block diagram window is active.*

1.9 LOADING AND SAVING VIS

You load a VI into memory by choosing the **Open** option from the **File** menu. When you choose that option, a dialog box similar to the one in Figure 1.24 appears. VI directories and VI libraries appear in the dialog box next to a representative symbol. VI libraries contain multiple VIs in a compressed format.

You can open a VI library or directory by clicking on its icon and then on **Open**, or by double clicking on the icon. The dialog box opens VI libraries as if they were directories. Once the directory or library is opened, you can locate your VI and load it into memory by clicking on it and then on **OK**, or by double-clicking on the VI icon.

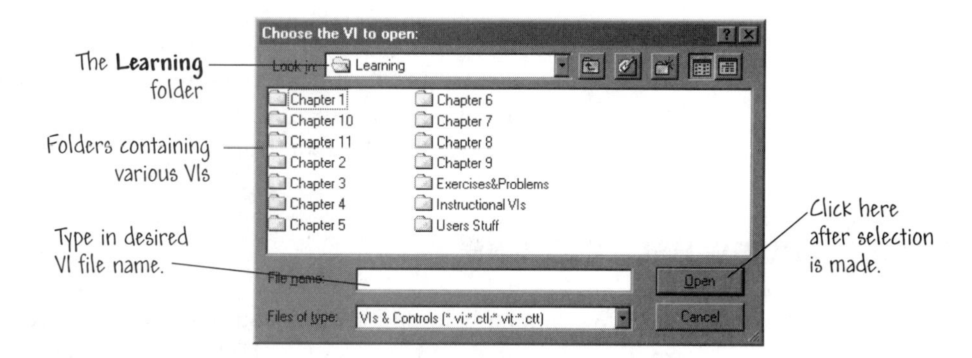

FIGURE 1.24
Locating the desired VI.

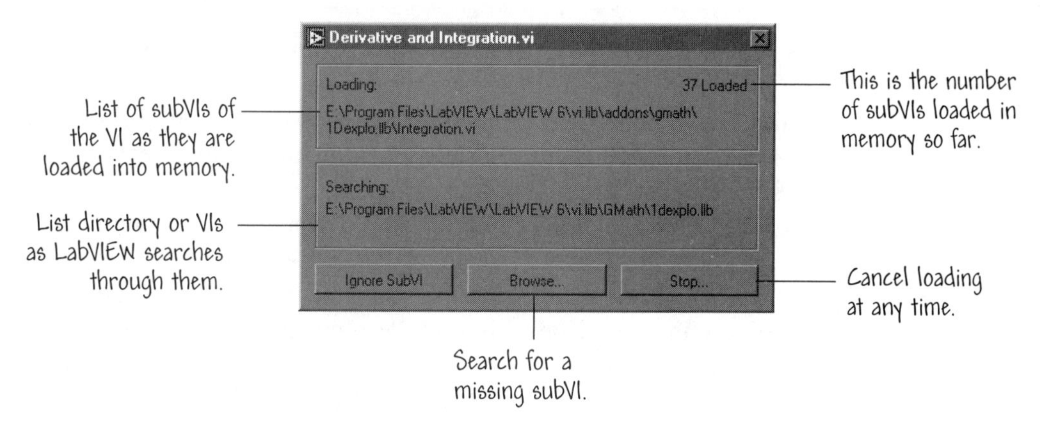

FIGURE 1.25
Status dialog box for VI loading.

The status dialog box shown in Figure 1.25 appears as the VI loads. The *Loading* field lists the subVIs of your VI as they are loaded into memory. *Number Loaded* is the number of subVIs loaded into memory so far. You can cancel the load at any time by clicking on **Stop...**

If LabVIEW cannot immediately locate a subVI (think of this as a subroutine) called by the VI, it begins searching through all directories specified by the VI Search Path (Tools≫Options≫Paths). The searching field lists directories or VIs as LabVIEW searches through them. You can have LabVIEW ignore the subVI by clicking on **Ignore SubVI**, or you can click on **Browse** to search for the missing subVI using the status dialog box shown in Figure 1.25.

You can save your VI to a regular directory or VI library by selecting **Save**, **Save As...**, **Save All...**, or **Save with Options...** from the **File** menu. You also can transfer VIs from one platform to another (for example, from LabVIEW for **Macintosh** to LabVIEW for **Windows**). LabVIEW automatically translates and recompiles the VIs on the new platform. Because VIs are files, you can use any file transfer method or utility to move your VIs between platforms. Porting VIs over networks using FTP protocol, Z- or X-Modem protocol, and other similar utilities eliminates the need for additional file translation software. If you port your VIs via magnetic media (such as floppy disks), you will need to use a file transfer utility (such as MacDisk or Transfer Pro).

1.10 LABVIEW HELP OPTIONS

The two common help options that you will use as you learn about LabVIEW programming are the **Show Context Help** and the **Contents and Index**. Both help options can be accessed in the **Help** pull-down menu.

1.10.1 Context Help Window

To display the help window, choose **Show Context Help** from the **Help** pull-down menu. If you have already placed objects on the front panel or block diagram, you can find out more about those objects by simply placing one of the tools from the **Tools** palette on block diagram and panel objects. This process causes the **Context Help Window** to appear showing the icon associated with the selected object and displaying the wires attached to each terminal. As you will discover in the next chapter, some icon terminals must be wired and others are optional. To help you locate terminals that require wiring, in the help window required terminals are labeled in bold, recommended connections in plain text, and optional connections are gray. The example in Figure 1.26 displays a help window in the so-called **Simple Context Help** mode.

On the lower left-hand side of the help window is a button to switch between the simple and detailed context help modes. The simple context emphasizes the

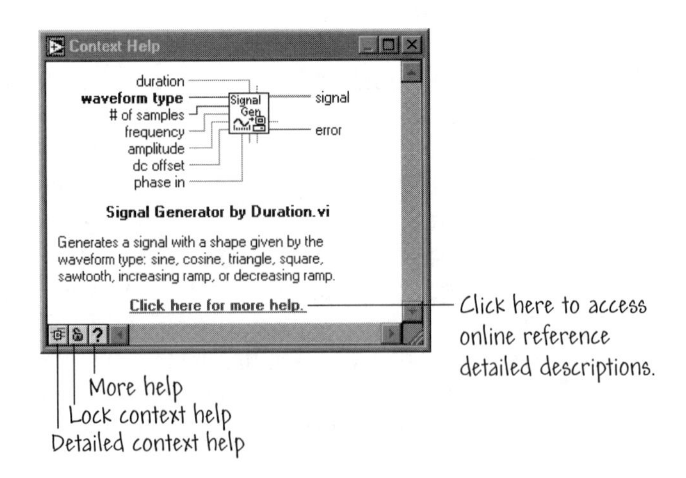

FIGURE 1.26
A simple context help window.

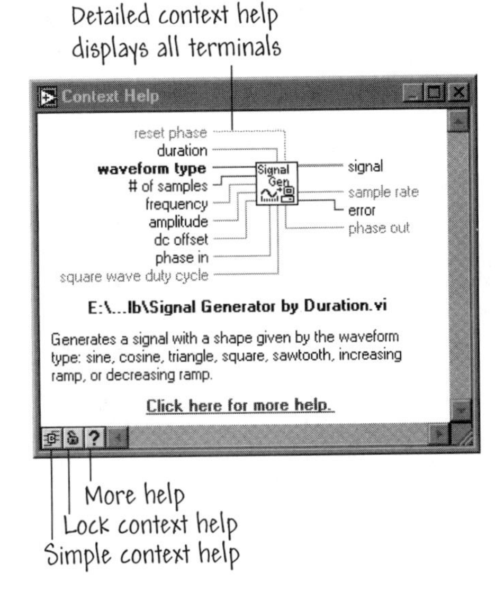

FIGURE 1.27
A detailed context help window.

important terminal connections—de-emphasized terminals are shown by wire stubs. The detailed help displays all terminals, as illustrated in Figure 1.27.

On the lower left-hand side of the help window is a lock icon that locks the current contents of the help window, so that moving the tool over another function or icon does not change the display. To unlock the window, click again on the lock icon at the bottom of the help window.

The **More Help** icon is the question mark located in the lower left-hand portion of the context help window. This provides a link to the description of the object in the online reference documentation, which features detailed descriptions of most block diagram objects.

1.10.2 Contents and Index

The LabVIEW online reference contains detailed descriptions of most block diagram objects. This information is accessible either by clicking on the More Help icon in the Context Help window, choosing **Contents and Index** from the **Help** menu, or clicking on the sentence **Click here for more help** in the Context Help window.

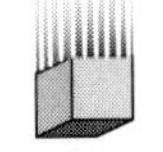

BUILDING BLOCK

1.11 BUILDING BLOCKS: TRAJECTORY ANALYSIS

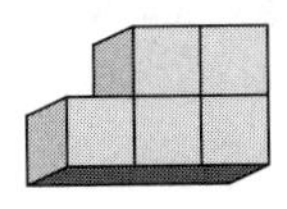

In each chapter of this book you will find a "Building Blocks" section. The purpose of this section is to give you the chance to apply the main principles of the chapter. In some cases the building block exercise of one chapter will continue in the next chapter; in other cases, the exercise will be new. The exercises will be short, and you will be asked to do all the work!

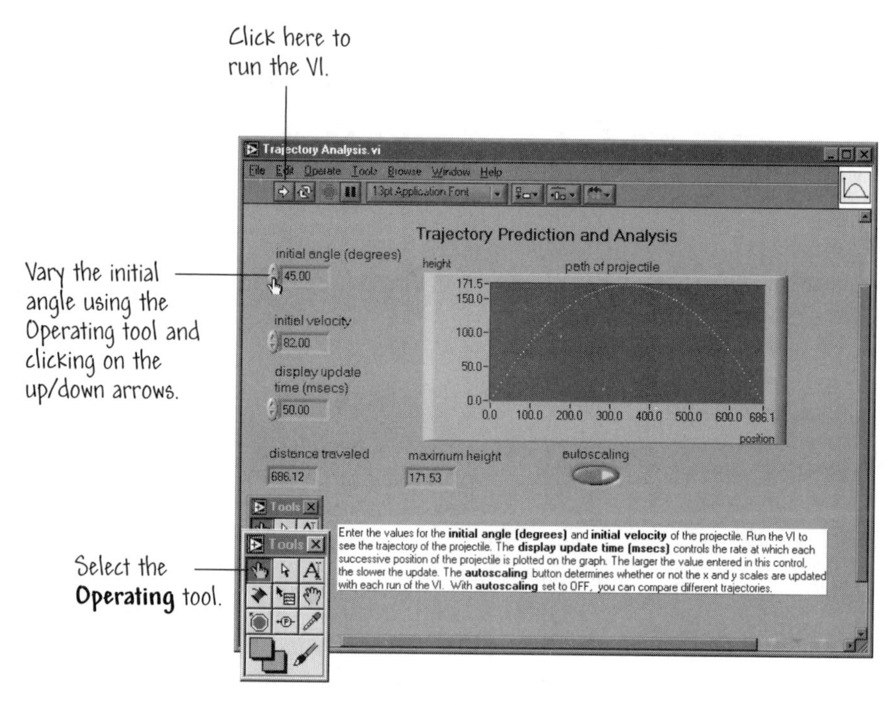

FIGURE 1.28
The Trajectory Analysis.vi front panel.

This first building block is an exercise in opening and running a VI. The VI that you should open is called Trajectory Analysis.vi and is included in the Building Blocks folder in the Learning directory. Find the VI and open it. The front panel is shown in Figure 1.28. Make sure the **Tools** palette is on the desktop and select the **Operating** tool. Run the VI and observe the path of the projectile in the graph. With the **Operating** tool, change the initial angle to 20 degrees and run the VI again. What is the maximum height achieved by the projectile? What is the distance traveled? Vary the initial velocity and run a few more numerical experiments.

1.12 RELAXING READING: MEASURING MUSIC WITH LABVIEW

At the end of each chapter you will find a section with a title that begins with the two words **Relaxing Reading**. What is the purpose of these sections? As the title implies, these sections are for fun! A time to read and think about how LabVIEW is being used in other academic and laboratory situations. Each story told comes from a source of real-world application stories called the **User's Solutions** which are catalogued by National Instruments and made available on their website. Other stories that you will find at the end of the chapters in this book include solar car racing, exploring the sense of smell, learning analytical chemistry, and more. If a particular story strikes your fancy, you can read the full-length story on the NI website, or you can contact the original author of the story at the address given at the end of each section.

The violin is an old instrument, dating back at least 250 years ago, when it gained the shape and sound that it essentially retains today. One of the finest violins is the so-called Stradivarius, named after Antonio Stradivarius (c.1644–1737). The characteristics that separate the Stradivarius from the mass-produced instruments of modern production plants have been the subject of study for many years. Faculty members and students at Virginia Tech and East Carolina University are using LabVIEW to archive measurements of the sound of the finest violins to digitally preserve the dynamic vibration characteristics. The experimental setup is shown in Figure 1.29. The violins are stimulated in a controlled fashion while measuring the response of the instrument. A LabVIEW virtual instrument (VI) was developed to control the input forcing functions, index the X-Y mirrors for scanning and acquisition, and process and store the response data (see

FIGURE 1.29
Faculty members at Virginia Tech and East Carolina University use LabVIEW for research on preserving the characteristics of a Stradivarius violin.

Figure 1.30). For more information, read the *User Solutions* entitled "Measuring Music and More with LabVIEW," which can be found on the National Instruments website, or contact:

Dr. Al Wicks
Virginia Polytechnic Institute
Department of Mechanical Engineering
Blacksburg, VA 24061-0238

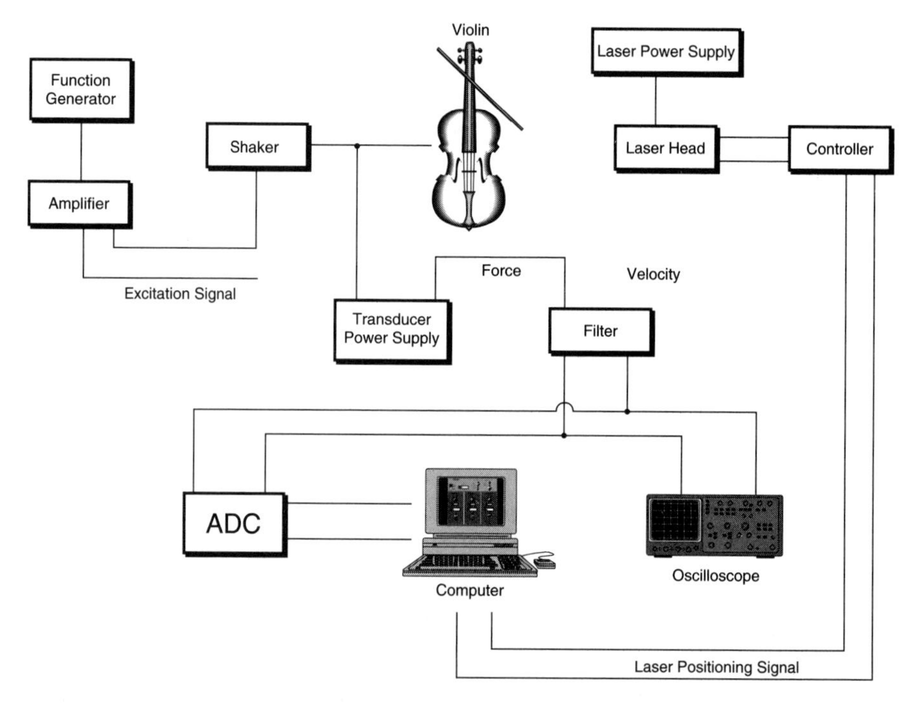

FIGURE 1.30
Configuration of the LabVIEW system used to measure the dynamic response of the violin.

1.13 SUMMARY

LabVIEW is a powerful and flexible instrumentation and analysis software system. It uses the graphical programming language G to create programs called virtual instruments (VIs) in a flowchartlike form called a block diagram. The user interacts with the program through the front panel. LabVIEW has many built-in functions to facilitate the programming process. The next chapters will teach you how to make the most of LabVIEW's many features.

KEY TERMS

Block diagram: Pictorial representation of a program or algorithm. In G, the block diagram, which consists of executable icons, called nodes, and wires that carry data between the nodes, is the source code for the VI.

Context Help window: Special window that displays the names and locations of the terminals for a function or subVI, the description of controls and indicators, the values of universal constants, and descriptions and data types of control attributes. The window also accesses the **Online Reference**.

Controls palette: Palette containing front panel controls and indicators.

Front panel: The interactive interface of a VI. Modeled from the front panel of physical instruments, it is composed of switches, slides, meters, graphs, charts, gauges, LEDs, and other controls and indicators.

Functions palette: Palette containing block diagram structures, constants, and VIs.

Hierarchical menus: Menu items that expand into submenus.

LabVIEW: **Lab**oratory **V**irtual **I**nstrument **E**ngineering **W**orkbench. It is a powerful and flexible instrumentation and analysis software development application.

Nodes: Execution elements of a block diagram consisting of functions, structures, and subVIs.

Palette: Menu of pictures that represent possible options.

Pop up: To call up a special menu by clicking an object with the right mouse button (on **Windows** platforms) or with the command key and the mouse button (on **Macintosh** platforms).

Pull-down menu: Menus accessed from a menu bar. Pull-down menu options are usually general in nature.

Short cut menu: Menus accessed by popping up, usually on an object. Menu options pertain to that object specifically.

Tool: A special operating mode of the mouse cursor.

Tools palette: Palette containing tools you can use to edit and debug front panel and block diagram objects.

Toolbar: Bar containing command buttons to run and debug VIs.

Virtual instrument (VI): Program in LabVIEW; so-called because it models the appearance and function of a physical instrument.

EXERCISES

E1.1 In this exercise we want to open and run an existing VI. In LabVIEW, open VibrationAnalysis.vi. This VI is located in Examples\Apps\demos.llb. The front panel should look like the one shown in Figure E1.1.

(a) Run the VI by clicking on the **Run** button.

(b) Vary the Acquisition Rate on the vertical pointer slide control.

(c) Vary the desired velocity on the Set Velocity [km/hr] dial and verify that the actual velocity, as indicated on the Actual Velocity [km/hr] gauge, matches the desired velocity.

E1.2 Referring to VibrationAnalysis.vi from E1.1, we can inspect the block diagram and watch it execute using **Highlight Execution**. Under the **Window** pull-down menu, select **Show Diagram**. The panel should switch to the block diagram shown in Figure E1.2.

(a) Click on the **Highlight Execution** button.

(b) Run the VI by clicking on the **Run** button.

(c) Watch as the data flows through the code.

E1.3 On the LabVIEW Startup screen, select **Search Example** to view the examples available in the various categories (see Figure E1.3). Look through and run several other VIs.

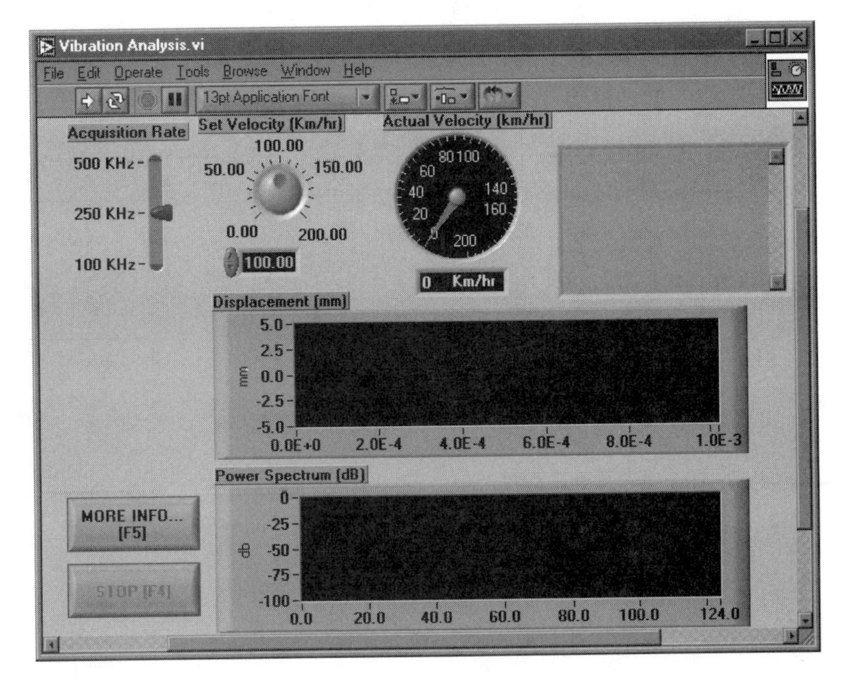

FIGURE E1.1
The Vibration Analysis.vi front panel.

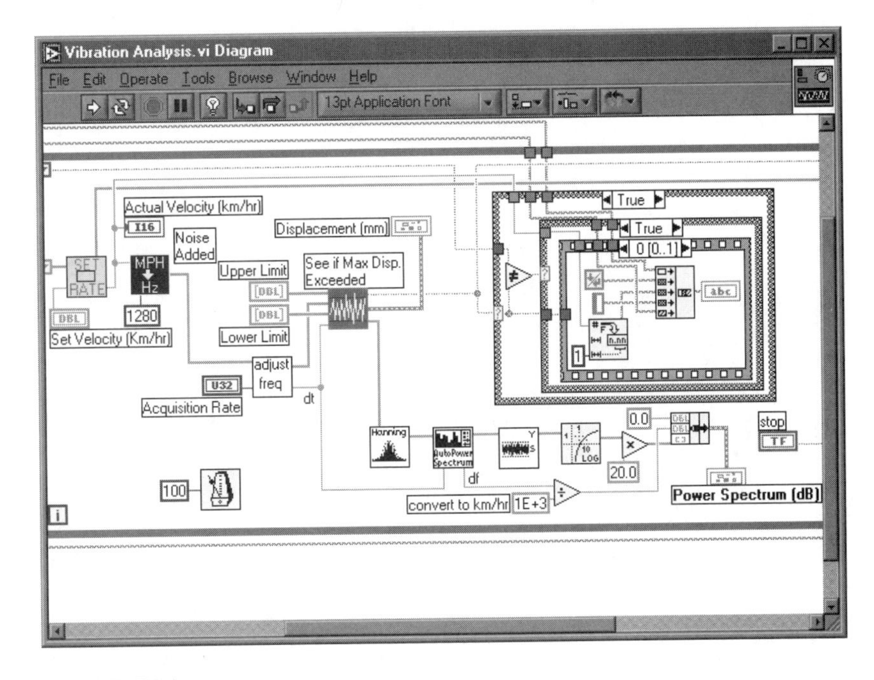

FIGURE E1.2
The Vibration Analysis.vi block diagram.

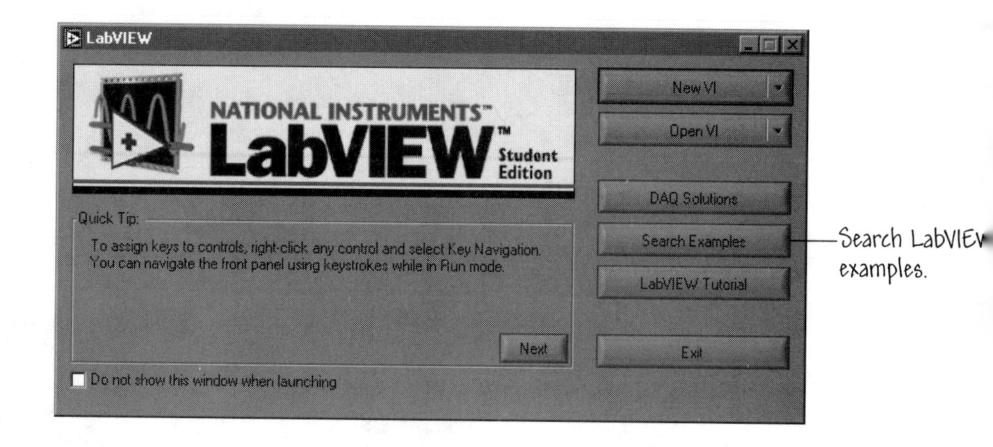

FIGURE E1.3
Select **Search Example** on the LabVIEW startup screen.

PROBLEMS

P1.1 Complete the crossword puzzle.

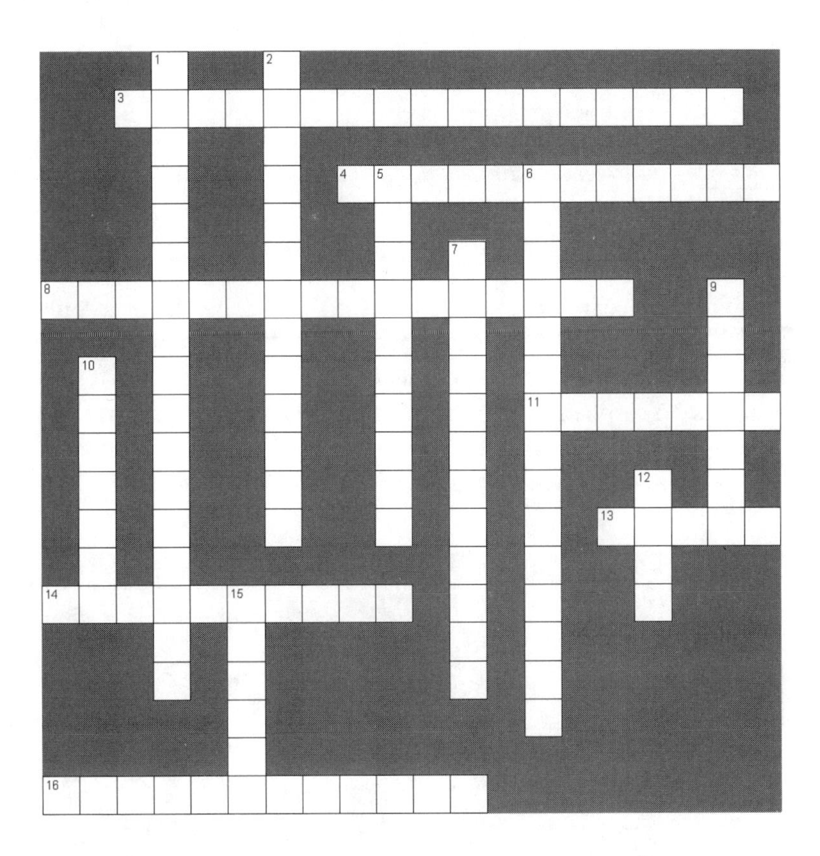

Across

3. Program in LabVIEW.
4. Menus accessed by popping up on an object.
8. Palette containing block diagram structures, constants, communication features, and VIs.
11. A powerful and flexible instrumentation and analysis software development application.
13. Execution elements of a block diagram consisting of functions, structures, and subVIs.
14. The interactive interface of a VI.
16. Bar containing command buttons to run and debug VIs.

Down

1. Menu items that expand into submenus.
2. Menus accessed from a menu bar.
5. Special window that displays helpful infomation.
6. Palette containing front panel controls and indicators.
7. Pictorial representation of a program or algorithm.
9. Menu of pictures that represent possible options.
10. Palette containing tools for editing and debugging.
12. A special operating mode of the mouse cursor.
15. To call up a special menu by clicking on an object.

P1.2 In the problem we want to open an existing VI from the Learning directory. You can open the VI by either selecting **Open VI** from the **Startup** screen, or if you are already in LabVIEW, you can use the **File** pull-down menu (see Figure 1.14) and select **Open...**. In both cases, you must navigate through your local file structure to find the desired VI. Find, open, and run Running Dog.vi located in Learning\Instructional VIs\CompSci.llb.

*This VI is only available on the **Windows** platform. If you are on a **Macintosh** platform, locate, open, and run Control Mixer Process.vi located in the library Examples\Apps\demos.llb.*

P1.3 You can construct games using LabVIEW! In this problem you will open a game VI. The Game.vi is located in Chapter1 in the Learning directory. Upon opening the VI you should see a front panel similiar to the one shown in Figure P1.1. You place your "bet" then "pull" the handle down by clicking on the slot machine arm with the **Operating** tool. The cards stop randomly when the handle is released. How long does it take to win?

FIGURE P1.1
A LabVIEW game.

CHAPTER 2

Virtual Instruments

Virtual instruments (VI) are the building blocks of LabVIEW programming. We will see in this chapter that VIs have three main components: the front panel, the block diagram, and the icon and connector pair. We will revisit the front panel and block diagram concepts first introduced in Chapter 1. An introduction to wiring the elements together on the block diagram is presented, although many of the debugging issues associated with wires are left to the next chapter. The important notion of data flow programming is also discussed in this chapter. Finally—you will have the opportunity to build your first VI!

GOALS

1. Gain experience by running more worked examples.
2. Understand the three basic components of a virtual instrument.
3. Begin the study of programming in G.
4. Understand the notion of data flow programming.
5. Build your first virtual instrument.

2.1 WHAT ARE VIRTUAL INSTRUMENTS?

LabVIEW programs are called virtual instruments (VIs) because they have the look and feel of physical systems or instruments. The illustration in Figure 2.1 shows an example of a front panel. A VI and its components are analogous to main programs and subroutines from text programming languages like C and Fortran. VIs have both an interactive user interface—known as the front panel—and the source code—represented in graphical form on the block diagram. LabVIEW provides mechanisms that allow data to pass easily between the front panel and the block diagram.

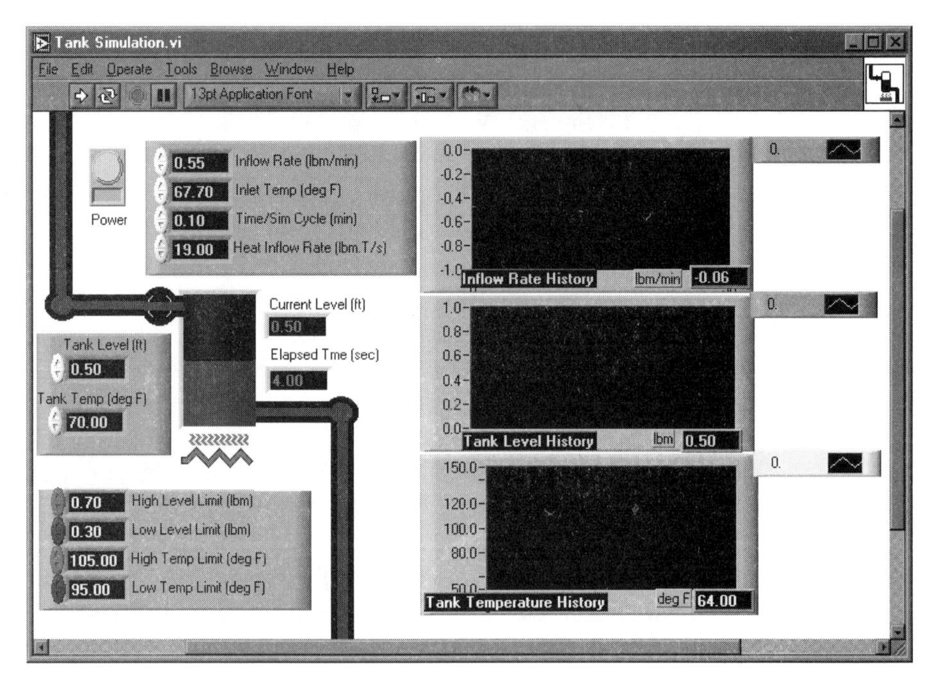

FIGURE 2.1
A virtual instrument front panel.

The block diagram is a pictorial representation of the program code. The block diagram associated with the front panel in Figure 2.1 is shown in Figure 2.2. The block diagram consists of executable icons (called nodes) connected (or **wired**) together. We will discuss wiring later in this chapter. The important concept to remember is that, in the G programming language, the block diagram is the source code.

The art of successful programming in G is an exercise in **modular programming**. After dividing a given task into a series of simpler subtasks (in G these subtasks are called subVIs and are analogous to subroutines), you then construct

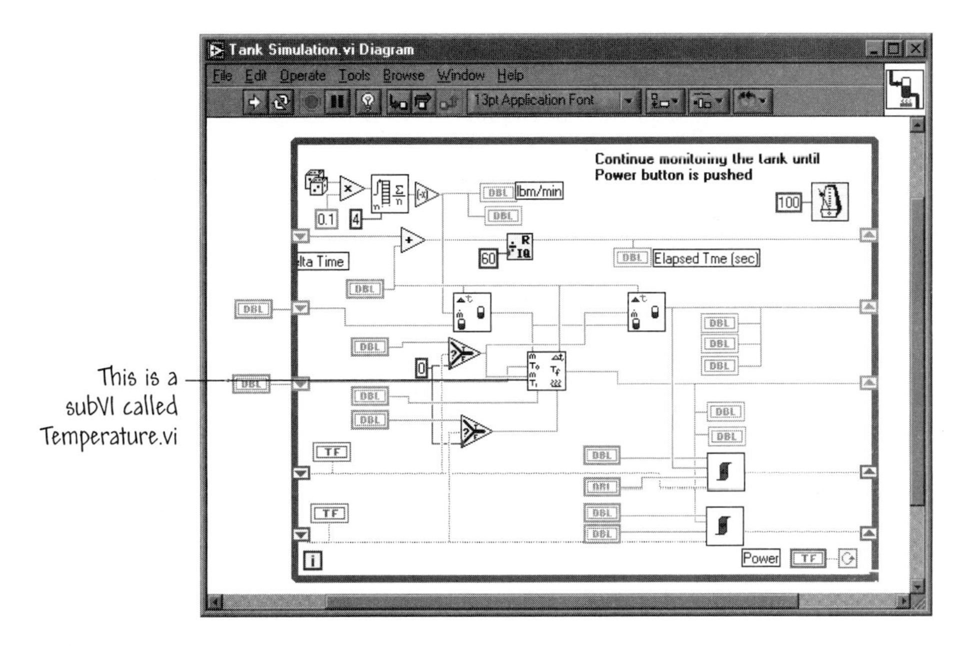

FIGURE 2.2
The virtual instrument block diagram associated with the front panel in Figure 2.1.

a virtual instrument to accomplish each subtask. Chapter 4 focuses on building subVIs. The resulting subtasks (remember, these are called subVIs) are then assembled on a top-level block diagram to form the complete program. Modularity means that you can execute each subVI independently, thus making debugging and verification easier. Furthermore, if your subVIs are general purpose programs, you can use them in other programs.

VIs (and subVIs) have three main parts: the front panel, the block diagram, and the icon/connector. The front panel is the interactive user interface of a VI—a window through which the user interacts with the code. When you run a VI, you must have the front panel open so you can pass inputs to the executing program and receive outputs (such as data for graphical display). The front panel is indispensable for viewing the program outputs. It is possible, as we will discuss in Chapter 9, to write data out to a file for subsequent analysis, but generally you will use the front panel to view the program outputs. The front panel contains knobs, push buttons, graphs, and many other controls (the term *controls* is interchangeable with *inputs*) and indicators (the term *indicators* is interchangeable with *outputs*).

The block diagram is the source code for the VI. The source code is "written" in the G programming language. We use the term *written* loosely, since in fact the code is made up of graphical icons, wires, and such, rather than the more common "lines of code." The block diagram is actually the executable code. The **icons** of a block diagram represent lower-level VIs, built-in functions, and program control structures. These icons are wired together to allow the data flow. As you will learn later in this chapter, the execution of a G program is governed by the data flow and not by a linear execution of lines of code. This concept is known as **data flow programming**.

The **icons** and **connectors** specify the pathways for data to flow into and out of VIs. The icon is the graphical representation of the VI in the block diagram and the connector defines the inputs and outputs. All VIs have an icon and a connector. As previously mentioned, VIs are hierarchical and modular. You can use them as top-level (or calling) programs or as subprograms (or subVIs) within other programs. The icon and connector are shown in Figure 2.3 for the subVI Temperature.vi. This subVI can be found in the center of the Tank Simulation.vi diagram in Figure 2.2.

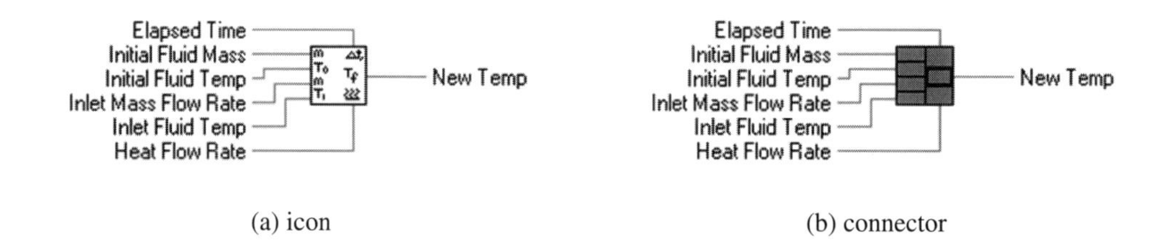

FIGURE 2.3
The icon and connector of the Temperature.vi subVI shown in Figure 2.2.

The subVI Temperature.vi has six inputs and one output. For proper operation of the subVI, the data flow must pass the (1) mass and (2) temperature of the initial fluid, the (3) flow rate and (4) temperature of the inlet fluid, the (5) heat flow rate, and the (6) elapsed time to the subVI. Once all the necessary input data to the subVI is available, the new temperature of the fluid is computed within the subVI and the result is output—the data "flows" out.

2.2 SEVERAL WORKED EXAMPLES

Before you construct your own VI, we will open several existing LabVIEW programs and run them to see how LabVIEW works. The first VI example—Temperature System Demo.vi—can be found in the suite of examples provided as part of LabVIEW. The second VI example—ODE Example.vi—illustrates how LabVIEW can be used to simulate linear systems. In this example, the motion of a mass-spring-cart system is simulated, and you can observe the effects of changing any of the system parameters of the resulting motion of the cart.

Temperature System Demo

In this example, you will open and run the virtual instrument called Temperature System Demo.vi. At the LabVIEW startup window, select **Open VI**. If you previously selected the option to bypass the startup window upon opening LabVIEW, you will see an empty front panel entitled **Untitled 1** appear instead of the startup panel. In the latter case, select **Open** from the **File** menu and navigate to the examples folder.

Double click on examples folder to open it and locate the apps folder within. Then double click on apps and after it opens up, double click on the library tempsys.llb. A file dialog box will appear listing the available VIs and subVIs:

- Temperature System Demo.vi
- histogram+.vi
- Temperature Status.vi
- Update Statistics.vi

Double click on Temperature System Demo.vi to open it. The front panel window appears and should resemble the one shown in Figure 2.4. The front panel contains numeric controls, Boolean switches, slide controls, knob controls, charts, graphs, and a thermometer indicator.

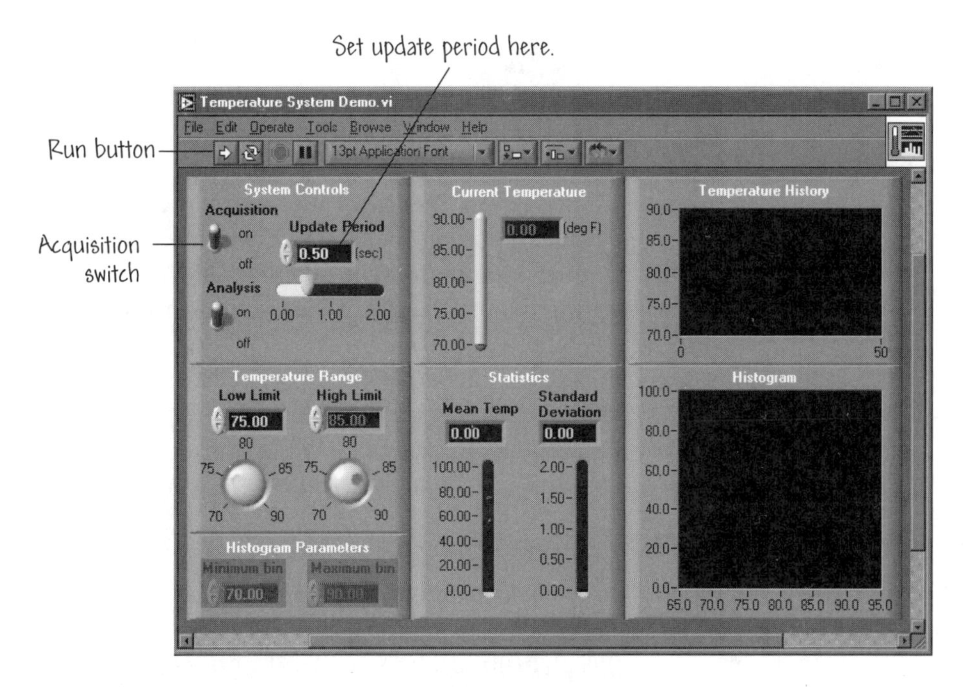

FIGURE 2.4
Temperature system demonstration front panel.

› Run the VI by clicking on the **Run** button. The front panel toolbar changes as the VI switches from edit mode to run mode. For example, once the VI begins executing, the stop button will change appearance on the front panel toolbar (it changes from a shaded symbol to a red stop sign). Also, the run button changes appearance to indicate that the VI is running.

This temperature system VI simulates a temperature monitoring application. The VI takes (simulated) temperature readings and displays them in both the thermometer indicator and on the chart. The simulated temperature readings are actually obtained from a pre-stored array of measurements, but with LabVIEW it is would be easy to modify the VI to acquire and process real temperature data.

The Update Period slide controls the speed at which the VI acquires the new temperature readings. The VI plots high and low temperature limits on the chart (high limit is in red, and low limit is in blue). These limits can be varied utilizing the Temperature Range knobs (located middle, left-hand side on the front panel). If the current temperature reading is out of the set range, Over Temp or Under Temp indicators will appear and light up next to the thermometer. You can also turn the data analysis on and off using the switch in the System Controls area of the front panel (upper left-hand side). The analysis section shows you a running calculation of the mean and standard deviation of the temperature values and a histogram of the temperature values. Once the VI begins to run, it will continue until you set the Acquisition switch to off.

› While the VI is running, use the **Operating** tool to change the values of the high and low limits. If you are in edit mode, select the **Operating** tool by clicking on it in the **Tools** palette; otherwise the tool will appear automatically while the VI is running. Then highlight the current high or low value of the limits, either by clicking twice on the value you want to change, or by clicking and dragging

› across the value with the tool. Type in the new value and click on the **Enter** button, located next to the **Run** button on the front panel toolbar.

Another input that you can modify is the Update Period, which is controlled by the slide—locate the slide control on the front panel in Figure 2.4. Place the **Operating** tool on the slider, click and drag it to a new location, and then run the VI. What changes do you observe? You can also operate slide controls by clicking on a point on the slide to snap the slider to that location, by clicking on the slider and moving it to the desired location, or by clicking in the slide's digital display and entering a number. When you are finished experimenting with the VI, terminate the execution by setting the Acquisition switch to off.

*LabVIEW does not accept values in digital displays until you press the **Enter** button or click the mouse in an open area of the window.*

Switch to the block diagram of Temperature System Demo.vi by choosing **Show Diagram** from the **Window** pull-down menu. The block diagram shown in Figure 2.5 is the underlying code for the VI. At this point in the learning process, you may not understand all of the block diagram elements depicted in Figure 2.5—but you will eventually!

As previously discussed, most VIs are hierarchical and modular. After creating a VI, you can (with a little work configuring the icon and connector) use the VI as a subVI (similar to a subroutine) in the block diagram of another VI. By creating subVIs, you can construct modular block diagrams which make your VIs easier to debug. The **Temperature System Demo VI** uses several subVIs. Open the Temperature Status subVI by double clicking on the appropriate subVI icon (see Figure 2.5). The icon for the subVI is labeled History Status. The front panel shown in Figure 2.6 should appear.

The icon and connector provide the graphical representation and parameter definitions required to use a VI as a subVI in the block diagrams of other VIs. The icon and connector are located in the upper right corner of the VI front panel (see Figure 2.6). The icon is a graphical representation of the VI when used as a component in a G program, that is, when used in the block diagram of other VIs. An icon can be a pictorial representation or a textual description of the VI, or a combination of both. The icon for the subVI Temperature Status includes both text (History Status) and a graphical representation of a thermometer.

Every VI (and subVI) has a connector. The connector is a set of terminals that correspond to its controls and indicators. When you show the connector for

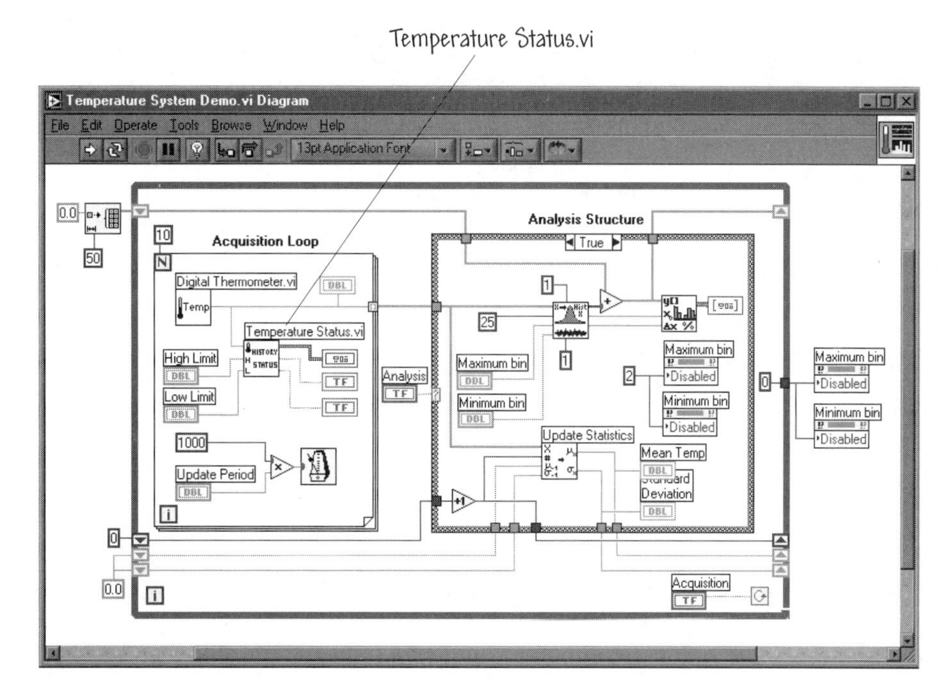

FIGURE 2.5
Temperature system demonstration block diagram—the code.

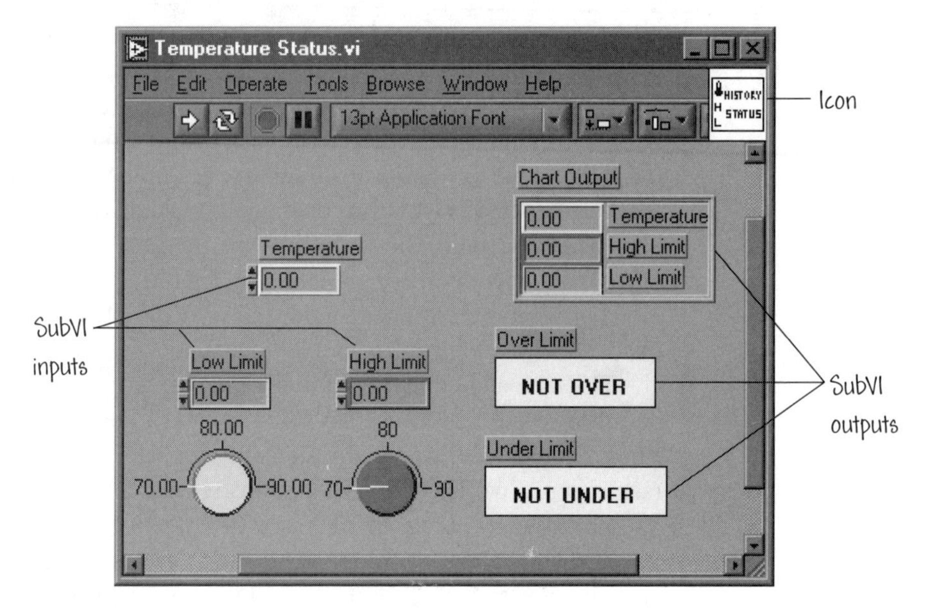

FIGURE 2.6
Temperature Status subVI.

the first time, LabVIEW will suggest a connector pattern that has one terminal for each control or indicator on the front panel—you can choose a different pattern. In Chapter 3, you will learn how to associate front panel controls and indicators with connector terminals. The connector terminals determine where you must wire the inputs and outputs on the icon. These terminals are analogous to parameters of a subroutine. You might wonder where the icon is located relative to the connector. It is at the same location—the icon sits on top of the connector pattern. The icon and connector of the Temperature Status subVI are shown in Figure 2.7.

Every VI has a default icon, which is displayed in the icon panel in the upper right corner of the front panel and block diagram windows in Figure 2.8. You will learn how to edit the VI icon in the next chapter. You may want to personalize VI icons so that they transmit information by visual inspection about the contents of the underlying VI.

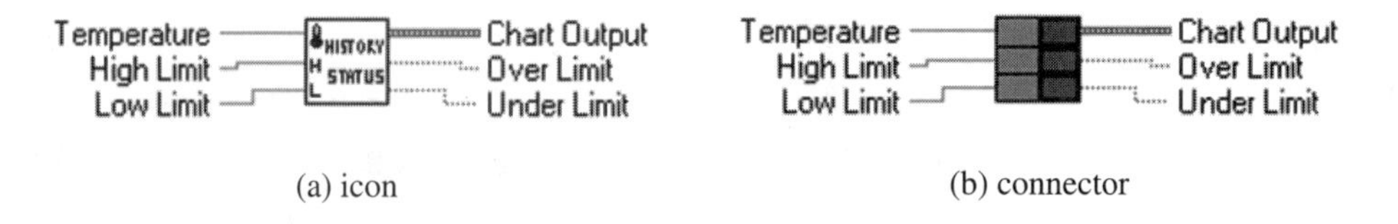

FIGURE 2.7
The icon and connector of Temperature Status subVI.

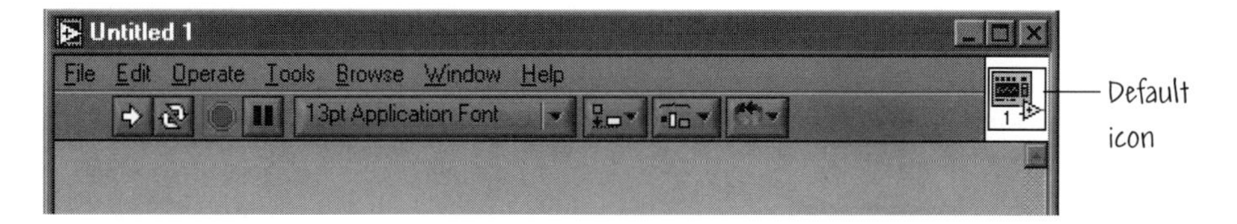

FIGURE 2.8
The default icon.

When you are finished experimenting with the Temperature System Demo.vi, close the VI and subVI by selecting **Close** from the **File** pull-down menu on each open front panel. Remember—do not save any changes!

*Selecting **Close** from the **File** pull-down menu of a block diagram closes the block diagram window only. Selecting **Close** on a front panel window closes both the front panel and the block diagram.* ◆

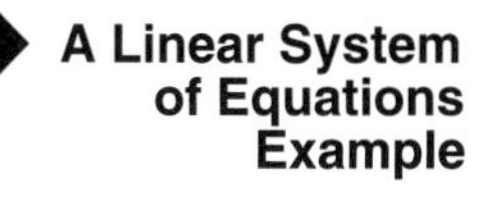

A Linear System of Equations Example

In this example we use LabVIEW to solve a set of linear, constant coefficient, ordinary differential equations. Many physical systems can be modeled mathematically as a set of linear, constant coefficient differential equations of the form

$$\dot{\mathbf{x}} = \mathbf{A}\mathbf{x}$$

with initial conditions $\mathbf{x}(0) = \mathbf{x}_o$, where the matrix $\mathbf{A}$ is a constant matrix.

Suppose we want to model the motion of a mass-spring-damper system, as shown in Figure 2.9. Let m represent the mass of the cart, k represent the spring constant, and b the damping coefficient. The position of the cart is denoted by y, and the velocity is the time derivative of the position, that is, the velocity

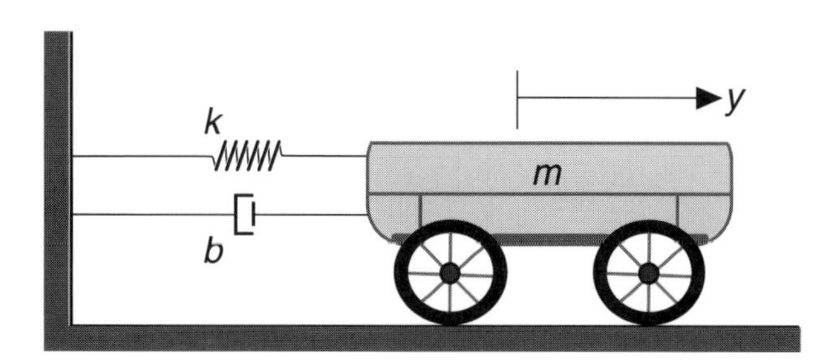

FIGURE 2.9
A simple mass-spring-damper system.

is $\dot{y}$. Equating the sum of forces to the mass times acceleration (using Newton's Second Law) we obtain the equation of motion:

$$m\ddot{y} + b\dot{y} + ky = 0 \,,$$

with the initial conditions $y(0) = y_o$ and $\dot{y}(0) = \dot{y}_o$. The motion of the cart is described by the solution of the second-order linear differential equation above. For this simple system we can obtain the solution analytically. For more complex systems it is usually necessary to obtain the solution numerically using the computer. In this exercise we seek to obtain the solution numerically using LabVIEW.

It is sometimes convenient for obtaining the numerical solution to rewrite the second-order linear differential equation as two first-order differential equations. We first define the state vector of the system as

$$\mathbf{x} = \begin{pmatrix} y \\ \dot{y} \end{pmatrix} ,$$

where the components of the state vector are given by

$$x_1 = y \quad \text{and} \quad x_2 = \dot{y} \,.$$

Using the definitions of the state vector and the equation of motion, we obtain

$$\begin{aligned} \dot{x}_1 &= x_2 \\ \dot{x}_2 &= -\frac{k}{m}x_1 - \frac{b}{m}x_2 \,. \end{aligned}$$

Writing in matrix notation yields

$$\dot{\mathbf{x}} = \mathbf{A}\mathbf{x} \,,$$

where

$$\mathbf{A} = \begin{bmatrix} 0 & 1 \\ -k/m & -b/m \end{bmatrix} .$$

In this example, let

$$\frac{k}{m} = 2 \quad \text{and} \quad \frac{b}{m} = 4 \,.$$

Choose the initial conditions as

$$x_1 = 10 \quad \text{and} \quad x_2 = 0 \,.$$

Now are ready to compute the solution of the system of ordinary differential equations numerically with LabVIEW.

- Select **Open VI** from the startup screen or by using the **File** menu to open the VI.
- Open the ODE Example.vi located in the folder Chapter 2 in the directory Learning.

The front panel depicted in Figure 2.10 will appear. Verify that the initial conditions $\mathbf{x}_o$ are set correctly to $x_1 = 10$ and $x_2 = 0$. Run the VI by clicking on the **Run** button. About how long does it take for the cart position to come to rest? Change the initial position to $x_1 = 20$. Now how long does it take for the cart to come to rest? What is the maximum value of the cart velocity?

Open and examine the block diagram by choosing **Show Diagram** from the **Window** menu. The code is shown in Figure 2.11. This is a relatively complex VI—most of the ones that you develop in this book will be simpler! But if you need to solve a set of linear ordinary differential equations, you can start with this VI and modify it as necessary. The idea of starting with a VI that solves a related problem is a good approach in the early stages of learning LabVIEW. Access the **Help** pull-down menu and select **Show Context Help**. Move the cursor over various objects on the block diagram and read what the online help has to say.

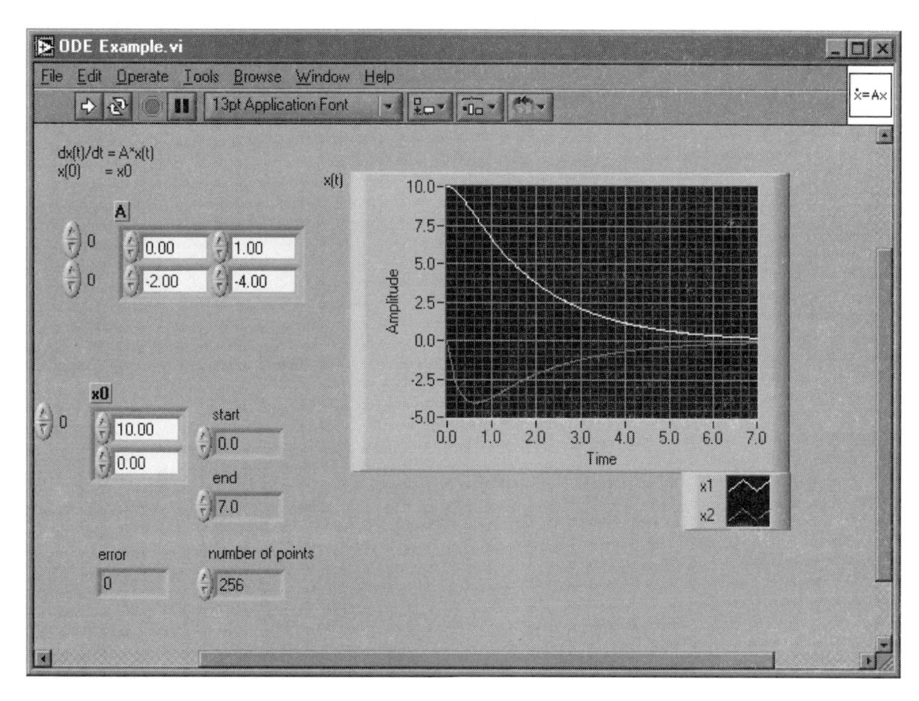

FIGURE 2.10
Linear system of equations front panel.

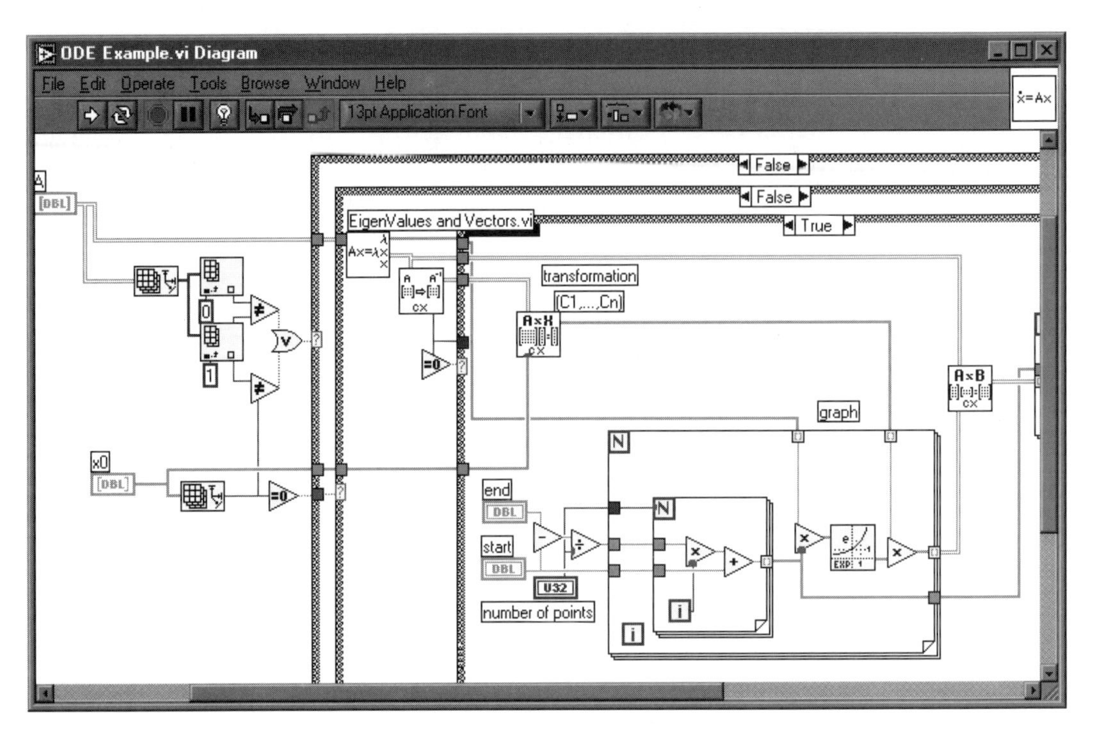

FIGURE 2.11
Linear system of equations block diagram.

For example, if you are familiar with eigenvalues and eigenvectors, you should recognize the online help discussion related to EigenValues and Vectors.vi.

When you are finished experimenting, close the VI by selecting **Close** from the **File** menu.

◆

2.3 THE FRONT PANEL

The front panel of a VI is a combination of controls and indicators. Controls simulate the types of input devices you might find on a conventional instrument, such as knobs and switches, and provide a mechanism to move input from the front panel to the underlying block diagram. On the other hand, indicators provide a mechanism to display data originating in the block diagram back on the front panel. Indicators include various kinds of graphs and charts (more on this topic in Chapter 7), as well as numeric, Boolean, and string indicators. Thus, when we use the term *controls* we mean "inputs," and when we say *indicators* we mean "outputs."

You place controls and indicators on the front panel by selecting and "dropping" them from the **Controls** palette. Once you select a control (or indicator) from the palette and release the mouse button, the cursor will change to a "hand" icon, which you then use to carry the object to the desired location on the front panel and "drop" it by clicking on the mouse button again. Once an object is on the front panel, you can easily adjust its size, shape, and position (see Chapter 3). If the **Controls** palette is not visible, you can either pop up on an open area of the front panel window, or select **Show Controls Palette** from the **Window** pull-down menu.

*If you pop up on an open area of the front panel window, you easily access the **Controls** palette. Similarly, you access the **Functions** palette by popping up on an open area of the block diagram.*

2.3.1 Numeric Controls and Indicators

You access the numeric controls and indicators from the **Numeric** subpalette located on the **Controls** palette, as shown in Figure 2.12. As the figure shows, there are quite a large number of available numeric controls and indicators. The two most commonly used numeric objects are the digital control and the digital indicator. When you construct your first VI later in this chapter, you will get the chance to practice dropping digital controls and indicators on the front panel. Once a digital control is on the front panel, you click on the increment buttons (that is, the up and down arrows on the left hand-side of the control) with the **Operating** tool to enter or change the displayed numerical values. Alternatively, you can double click on the current value of the digital control with either the

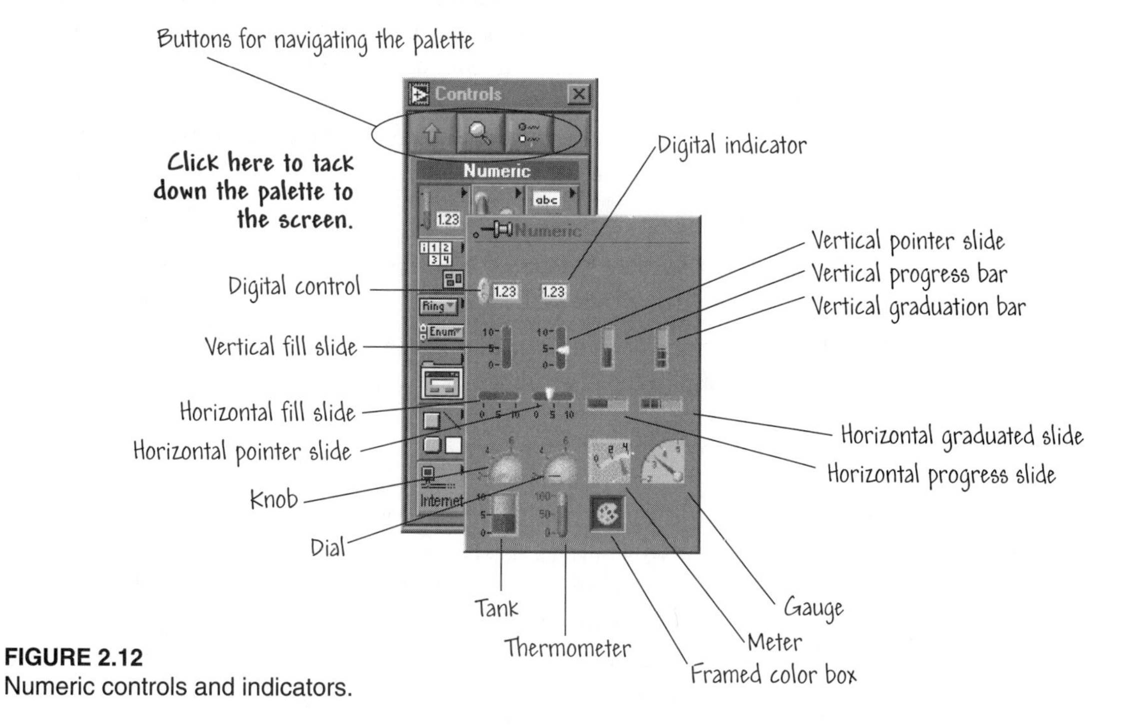

FIGURE 2.12
Numeric controls and indicators.

Labeling tool or the **Operating** tool, which will highlight the value, and you can then enter a different value.

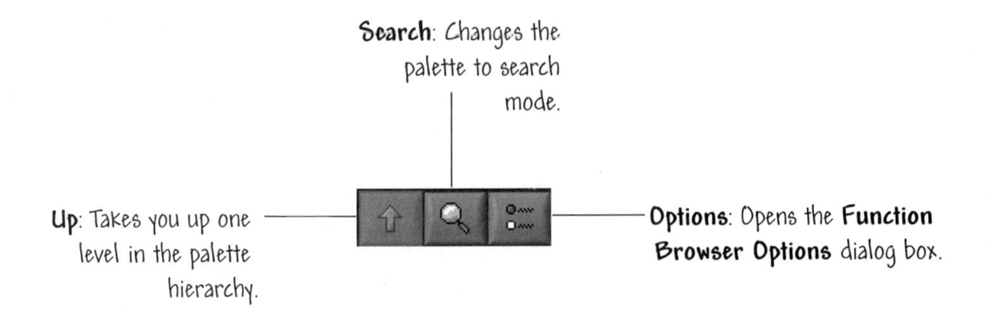

FIGURE 2.13
Navigating the **Functions** and **Controls** palettes.

*You can tack down the **Numeric** palette (and most other palettes) to the screen so they are visible at all times by clicking on the thumbtack on the top left corner of the palette.*

You can use the navigation buttons (see Figure 2.12 on the **Controls** and **Functions**) palettes to navigate and search for controls, VIs, and functions. When you left click a subpalette icon, the entire palette changes to the subpalette you selected. If you right click a subpalette icon, the subpalette appears, but the **Controls** (or **Functions**) palette remains in view. An example of a subpalette icon on the **Controls** palette is the **Numerics** subpalette. The **Controls** and **Functions** palettes contain three navigation buttons (as illustrated in Figure 2.13):

- **Up**—Takes you up one level in the palette hierarchy.
- **Search**—Changes the palette to search mode. In search mode, you can perform text-based searches to locate controls, VIs, or functions in the palettes.
- **Options**—Opens the **Function Browser Options** dialog box, from which you can configure the appearance of the palettes.

2.3.2 Boolean Controls and Indicators

You access the Boolean controls and indicators from the **Boolean** subpalette located on the **Controls** palette, as shown in Figure 2.14. As with the numeric controls and indicators, there are quite a large number of available Boolean controls and indicators. Boolean controls and indicators simulate switches, buttons, and LEDs and are used for entering and displaying Boolean (True-False) values. For example, you might use a Boolean LED in a temperature monitoring system

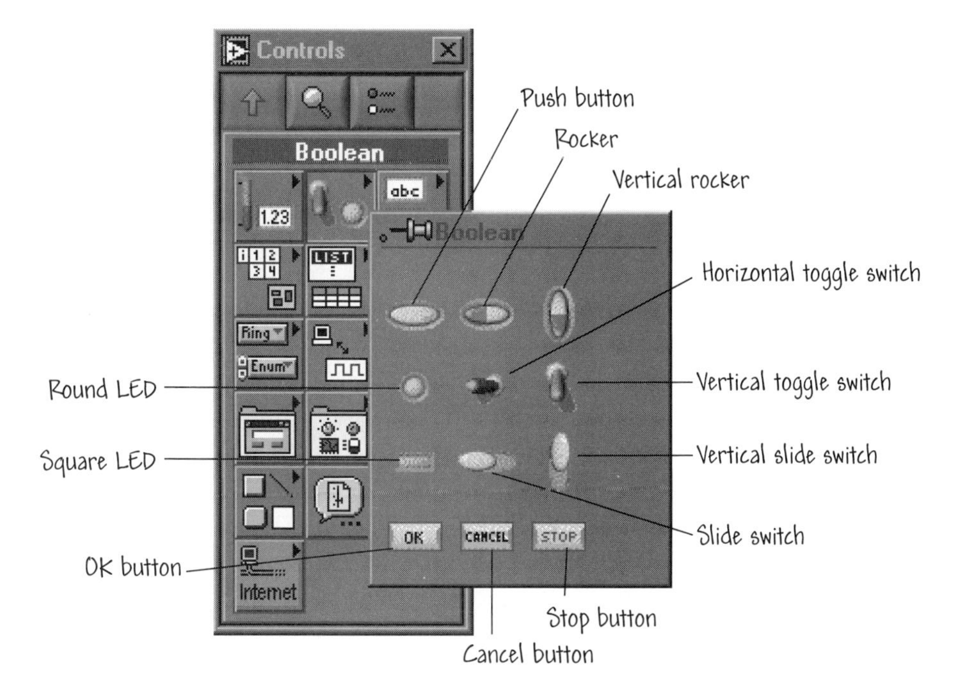

FIGURE 2.14
Boolean controls and indicators.

as a warning signal that a measured temperature has exceeded some predetermined safety limit. The measured temperature is too high, so an LED indicator on the front panel turns from green to red! Before continuing, take a few moments to familiarize yourself with the types of available digital and Boolean controls and indicators shown in Figures 2.12 and 2.14, so that when you begin to construct your own VIs, you'll have a feel for what's available on the palettes.

2.3.3 Configuring Controls and Indicators

Popping up on a digital control displays the short cut menu, as shown in Figure 2.15. You can change the defaults for controls and indicators using options from the short cut menus. For example, under the submenu **Representation** you will find that you can choose from 12 representations of the digital control or indicator, including 32-bit single precision, 64-bit double precision, signed integer 8-bit, and more. The representation indicates how much memory is used to represent a number. This choice is important if you are displaying large amounts of data and you want to conserve computer memory. Another useful item on the short cut menu is the capability to switch from control and indicator using **Change to Indicator** and vice-versa. We will discuss each item in the short cut menu on an as-needed basis.

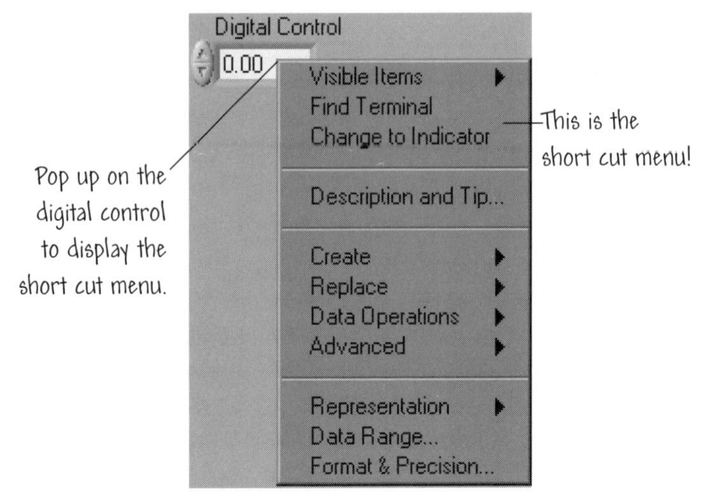

FIGURE 2.15
Configuring a digital control using the short cut menu.

2.4 THE BLOCK DIAGRAM

The graphical objects comprising the block diagram together make up what is usually call the source code. The block diagram (visually resembling a computer program flowchart) corresponds to the lines of text found in text-based programming languages. In fact, the block diagram is the actual executable code. The block diagram is built by wiring together objects that perform specific functions.

The components of a block diagram (a VI is depicted in Figure 2.16) belong to one of three classes of objects:

- **Nodes**: Program execution elements.
- **Terminals**: Ports through which data passes between the block diagram and the front panel and between nodes of the block diagram.
- **Wires**: Data paths between terminals.

2.4.1 Nodes

Nodes are analogous to statements, functions, and subroutines in text-based programming languages. There are three node types—**functions**, **subVI nodes**, and **structures**. Functions are the built-in nodes for performing elementary operations such as adding numbers, file I/O, or string formatting. The Add and Multiply functions in Figure 2.16 represent one type of node. SubVI nodes are VIs that you design and later call from the diagram of another VI. Structures—such as For Loops and While Loops—control the program flow.

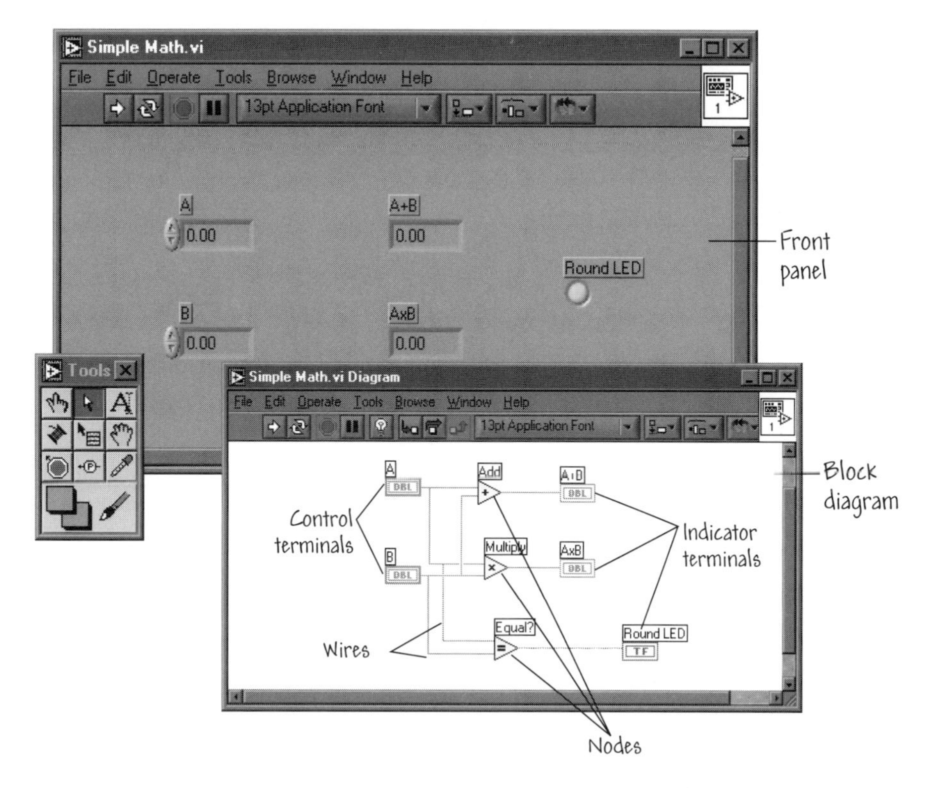

FIGURE 2.16
A typical VI illustrating nodes, terminals, and wires.

2.4.2 Terminals

Terminals are analogous to parameters and constants in text-based programming languages. There are different types of terminals—control and indicator terminals, node terminals, constants, and specialized terminals that you will find on various structures. In plain words, a terminal is any point to which you can attach a wire to pass data.

For example, in the case of control and indicator terminals, numeric data entered into a numeric control passes to the block diagram via the control terminals when the VI executes. When the VI finishes executing, the numeric output data passes from the block diagram to the front panel through the indicator terminals. Data flows in only one direction—from a "source" terminal to one or more "destination" terminals. In particular, controls are source terminals and indicators are destination terminals. The data flow direction is from the control terminal to the indicator terminal, and not vice-versa. Clearly, controls and indicators are not interchangeable.

Control and indicator terminals belong to front panel controls and indicators and are automatically created or deleted when you create or delete the corresponding front panel control or indicator. You cannot delete a terminal on the block diagram that belongs to a control or indicator—the terminal disappears only when you delete the associated control or indicator on the front panel. The block diagram of the VI in Figure 2.16 shows terminals belonging to four front panel controls and indicators. The Add and Multiply functions shown in the figure also have node terminals through which the input and output data flow into and out of the functions.

Control terminals have thick *borders and indicator terminal borders have* thin *borders. It is important to distinguish between thick and thin borders since they are not functionally equivalent.*

2.4.3 Wiring

Wires are data paths between terminals and are analogous to variables in conventional languages. How then can we represent different data types on the block diagram? Since the block diagram consists of graphical objects, it seems appropriate to utilize different wire patterns (shape, line style, color, etc.) to represent different data types. In fact, each wire possesses a unique pattern depending on the type of data (numeric, Boolean, string, etc) that flows through the wire. On a color monitor, each data type appears in a different color for emphasis. To determine the data types on a given wire, match up the colors and styles with the wire types are shown in Table 2.1.

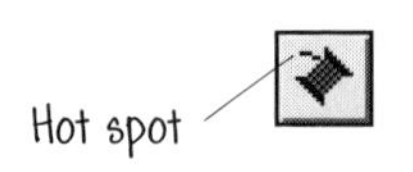

FIGURE 2.17
The hot spot on the **Wiring** tool.

The hot spot of the **Wiring** tool is the tip of the unwound wiring segment. To wire from one terminal to another, click the hot spot of the **Wiring** tool on the first terminal (you can start wiring at either terminal!), move the tool to the second terminal, and click on the second terminal. No need to hold down the mouse button while moving the **Wiring** tool from one terminal to the other. The wiring process is illustrated in Figure 2.17. When you wire two terminals, notice that moving the **Wiring** tool over one of the terminals causes that terminal to blink. This is an indication that clicking the mouse button will make the wire connection.

The VI shown in Figure 2.17 is easy to construct. It consists of one digital control and one digital indicator wired together. Open a new VI and try to build the VI! A working version can be found in the Chapter 2 folder in the Learning directory—it is called Wiring Demo. The function of the VI is to set a value for Input control on the front panel and to display the same input at the digital indi-

TABLE 2.1 Common wire types.

	Scalar	1D array	2D array	Color
Number				Orange (floating point) & Blue (integer)
Boolean				Green
String				Pink

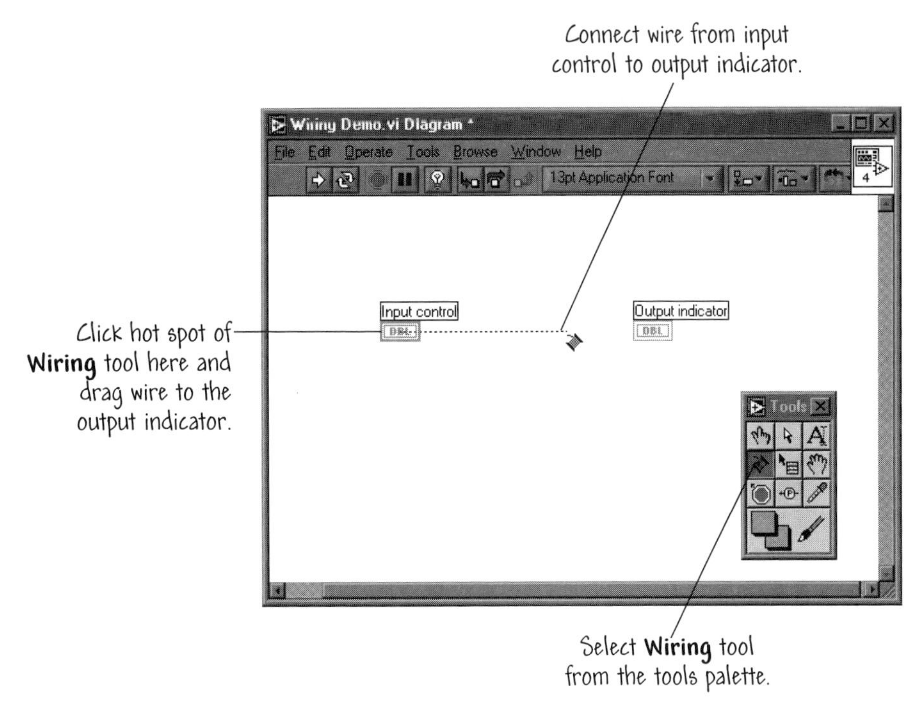

FIGURE 2.17
Wiring terminals.

cator Output indicator.

To delete a wire as you are wiring:

- **Windows**—Click the right mouse button or click on the origination terminal.
- **Macintosh**—Hold down <option> and click or click on the origination terminal.

When wiring two terminals together, you may want to bend the wire to avoid running the wire under other objects. This is accomplished during the wiring process by clicking the mouse button to tack the wire down at the desired location of the bend, and moving the mouse in a perpendicular direction to continue the wiring to the terminal. Another way to change the direction of a wire while wiring is to press the space bar while moving the **Wiring** tool.

***Windows**: All wiring is performed using the* left *mouse button.*

Tip strips make it easier to identify function and node terminals for wiring. When you move the **Wiring** tool over a terminal, a tip strip pops up, as illustrated in Figure 2.18. Tip strips are small text banners that display the name of each terminal. When you place the **Wiring** tool over a node, each input and output will show as a wire stub—a dot at the end of the wire stub indicates an input. Tip strips should help you wire the terminals correctly.

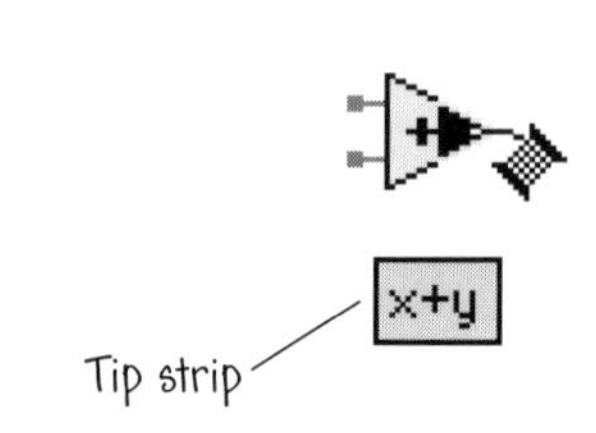

FIGURE 2.18
Tip strip.

It is possible to have objects wired automatically. A new feature of LabVIEW is the capability to automatically wire objects when you first drop them on the diagram. After you select a node from the **Functions** palette, move that node close to another node to which you want to wire the first node. Terminals containing similar datatypes with similar names will automatically connect. You can disable the automatic wiring feature by pressing the space bar. You can adjust the auto wiring settings from the Tools≫Options≫Block Diagram window.

Since it is important to correctly wire the terminals on functions, LabVIEW provides an easy way to show the icon connector to make the wiring task easier. This is accomplished by popping up on the function and choosing Visible Items≫Terminals from the short cut menu, as illustrated in Figure 2.19. To return to the icon, pop up on the function and deselect Visible Items≫Terminals.

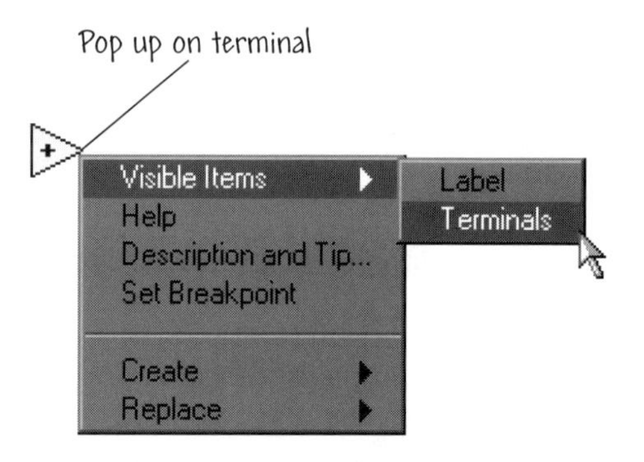

FIGURE 2.19
Showing terminals.

To use dots to denote wire junctions on your block diagram, enable the feature by selecting Options≫Block Diagram from the **Tools** menu and checking the first box entitled **Show dots at wire junctions**.

2.5 BUILDING YOUR FIRST VI

In this section you will create your first virtual instrument to perform the following functions:

- Add two input numbers and display the result.
- Multiply the same two input numbers and display the result.
- Compare the two input numbers and turn on an LED if the numbers are equal.

Begin by considering the front panel shown in Figure 2.20. It has two digital control inputs for the numbers A and B, two digital indicator outputs to display the results $A + B$ and $A \times B$, respectively, and a round LED that will turn on when the input numbers A and B are identical.

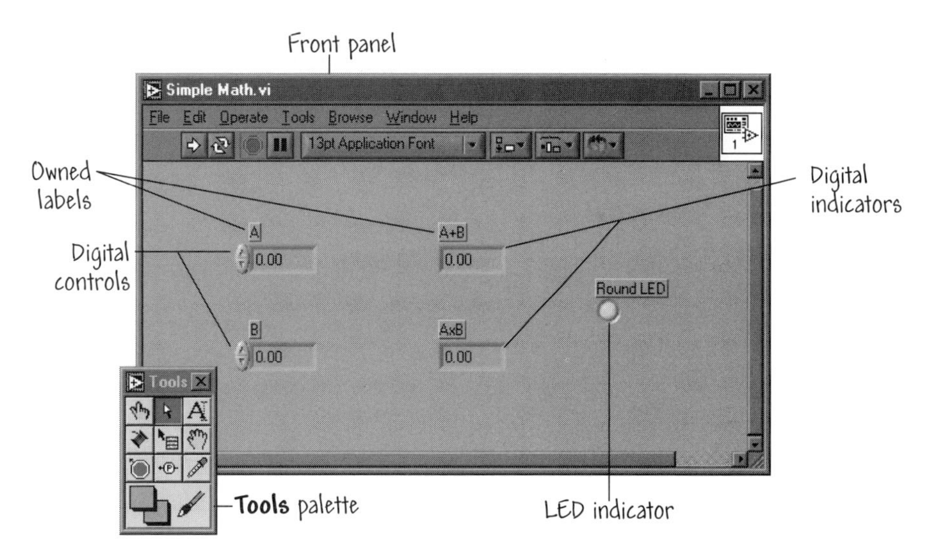

FIGURE 2.20
The front panel for your first VI.

As with the development of most sophisticated computer programs, constructing VIs is an art, and you will develop your own style as you gain experience with programming in G. With that in mind, you should consider the following steps as only one possible path to building a working VI that carries out the desired calculations and displays the results.

1. Open a new front panel by choosing **New** from the **File** menu. In **Windows**, if you previously closed all open VIs, you must select the **New VI** button from the startup screen.
2. **Create the numeric digital controls and indicators**. The two front panel controls are used to enter the numbers, and the two indicators are used to display the results of the addition and multiplication of the input numbers.
   (a) Select **Digital Control** from the **Numeric** subpalette of the **Controls** palette. If the **Controls** palette is not visible, pop up in an open area of the front panel to gain access to the palette.
   (b) Drop the control on the front panel, as illustrated in Figure 2.21. Drag the control to the desired location and then click the mouse button to complete the drop.

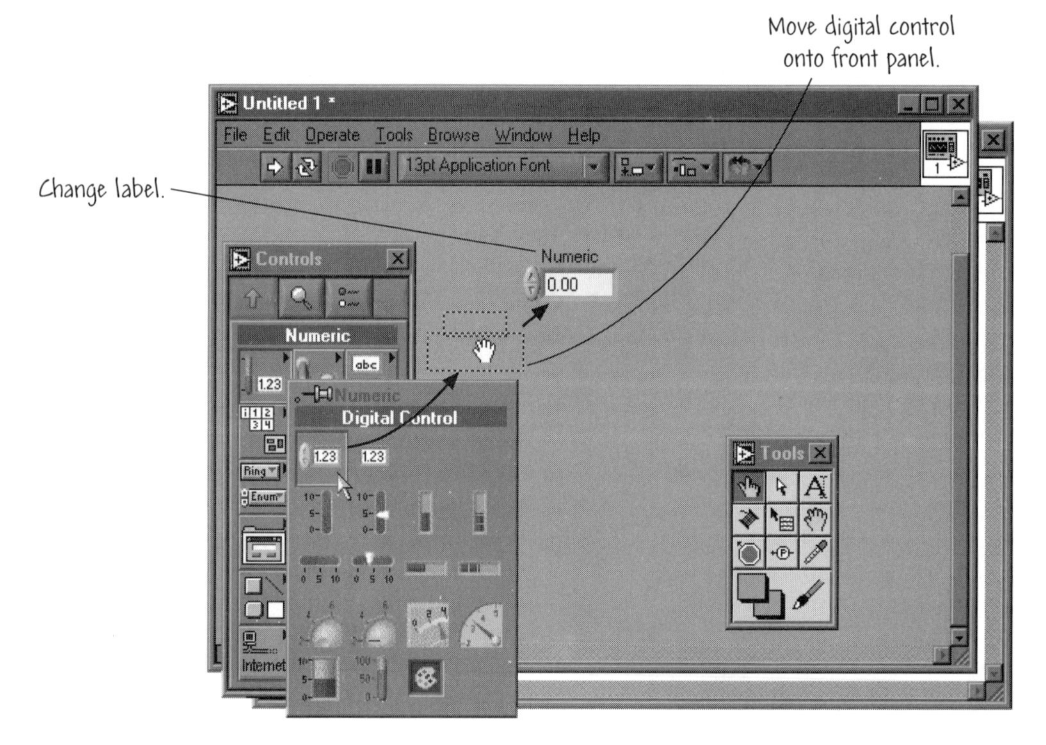

FIGURE 2.21
Placing the controls and indicators on the front panel.

   (c) Type the letter *A* inside the label box (which appears above the control) and press the **Enter** button on the front panel toolbar. If you do not type the control label before starting other programming actions (such as dropping the other control on the front panel), the label box will remain labeled with the default label Numeric. If the control or indicator

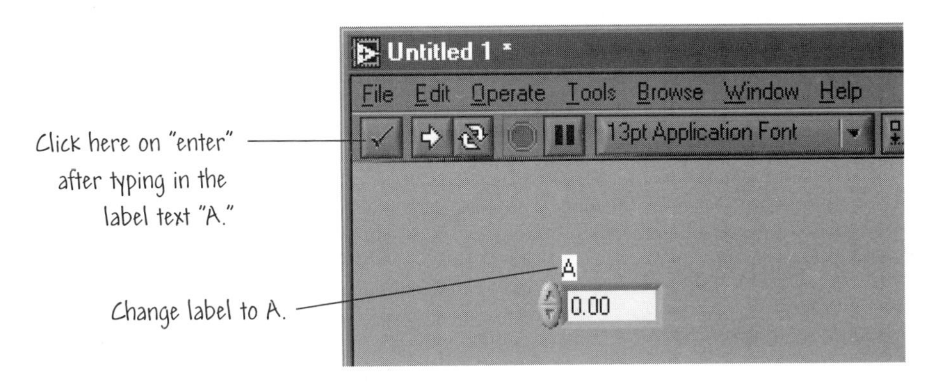

FIGURE 2.22
Labeling the digital control and indicators on the front panel.

does not have a label, you can pop up on the control and select **Label** from the **Show** menu. The label box appears, and you can then edit the text using the Labeling tool (see Figure 2.22).

(d) Repeat the above process to create the second numeric digital control and the two digital indicators. You can arrange the digital controls and indicators in any manner that you choose—although a neat and orderly arrangement is preferable. Add the labels to each control and indicator using Figure 2.20 as a guide.

3. **Create the Boolean LED**. This indicator will turn on if the two input numbers are identical, or remain off if they do not match.

   (a) Pop up in an open area of the front panel and select **Round LED** from the pop-up **Boolean** subpalette (see Figure 2.14 for the location of the LEDs). Place the indicator on the front panel, drag it to the desired location, and then click the mouse button to complete the process.

   (b) Type **Round LED** inside the label box and click anywhere outside the label when finished, or click on the **Enter** button.

   Each time you create a new control or indicator, LabVIEW automatically creates the corresponding terminal in the block diagram. The terminal symbols suggest the data type of the control and indicator. For example, a terminal with DBL appearing on the icon represents a double-precision floating-point number, and the TF represents a Boolean.

The Block Diagram

1. Switch your center of activity to the block diagram by selecting **Show Diagram** from the **Window** pull-down menu. The completed block diagram is shown in Figure 2.23. It may be helpful to display the front panel and block diagram simultaneously using either the **Tile Left and Right** or the **Tile Up**

and Down options found in the **Window** pull-down menu. For this example, the up and down option works better in the sense that the all the block diagram and front panel objects can be displayed on the screen without having to use the scrollbars.

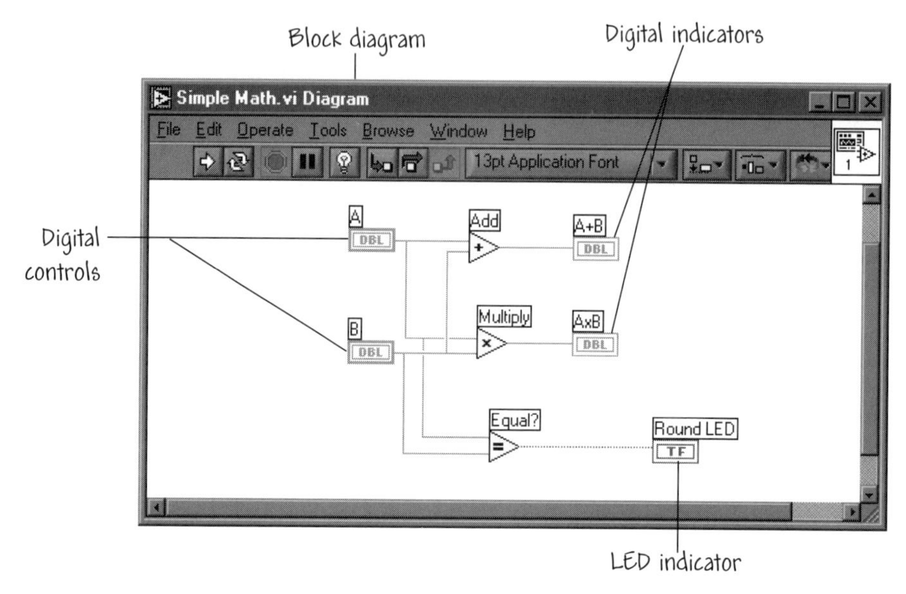

FIGURE 2.23
The block diagram window for your first VI.

2. Now we want to place the addition and multiplication functions on the block diagram. Select the Add function from the **Numeric** subpalette of the **Functions** palette. If the **Functions** palette is not visible, pop up on an open area of the block diagram to gain access to the palette. Drop the **Add** function on the block diagram in approximately the same position as shown in Figure 2.23. The label for the Add function can be displayed using the short cut menu and selecting Visible Items≫Label. This is illustrated in Figure 2.24. Following the same procedure, place the Multiply function on the block diagram and display the label.
3. Select the Equal? function from the **Comparison** subpalette of the **Functions** palette and place it on the block diagram, as shown in Figure 2.25. The Equal? function compares two numbers and returns TRUE if they are equal or FALSE if they are not. To get more information on this function, you can activate the online help by choosing **Show Context Help** from the **Help** menu. Then placing any of the tools on the Equal? function (or any of

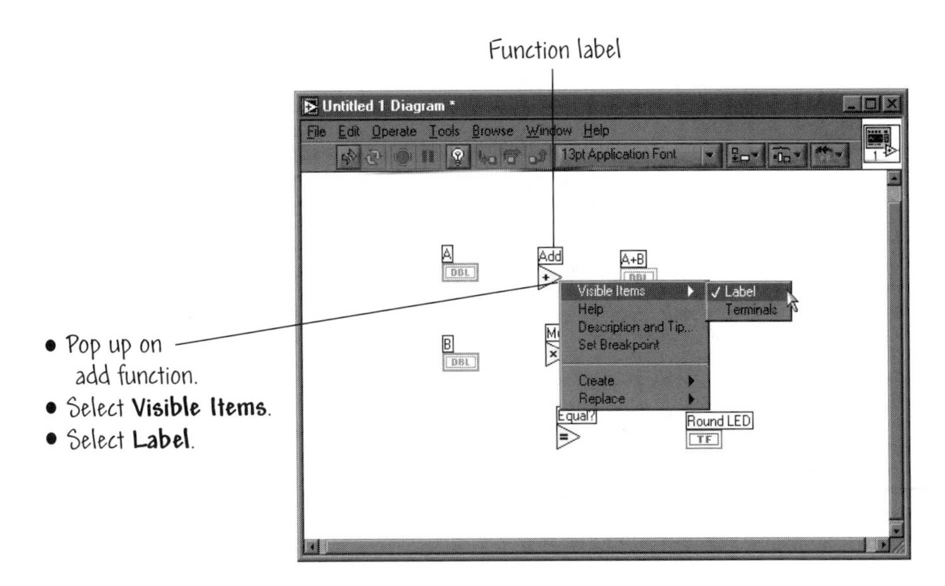

FIGURE 2.24
Pop up to select the Visible Items≫Label option.

the other functions on the block diagram) leads to the display of the online help information.

Also notice in the Figure 2.25 that the control terminals have thick borders and the indicator terminals have thin borders. A quick look at the terminal border tells you whether the terminal is an indicator or control!

4. Using the **Wiring** tool found on the **Tools** palette, wire the terminals as shown in Figure 2.23. As seen in Figure 2.26, to wire from one terminal to another, click the **Wiring** tool on the first terminal, move the tool to the second terminal, and click on the second terminal. Remember that it does not matter on which terminal you initiate the wiring. To aid in wiring, pop up on the three functions and choose Visible Items≫Terminals. Having the terminals shown explicity helps to wire more quickly and accurately. Once the wiring is finished for a given function, it is best to return to the icon by popping up on the function and choosing **Visible Items** and deselecting **Terminals**.
5. Switch back to the front panel window by clicking anywhere on it or by choosing **Show Panel** from the **Window** menu.
6. **Save the VI** as Simple Math.vi. Select **Save** from the **File** menu and make sure to save the VI in the Users Stuff folder within the Learning directory.

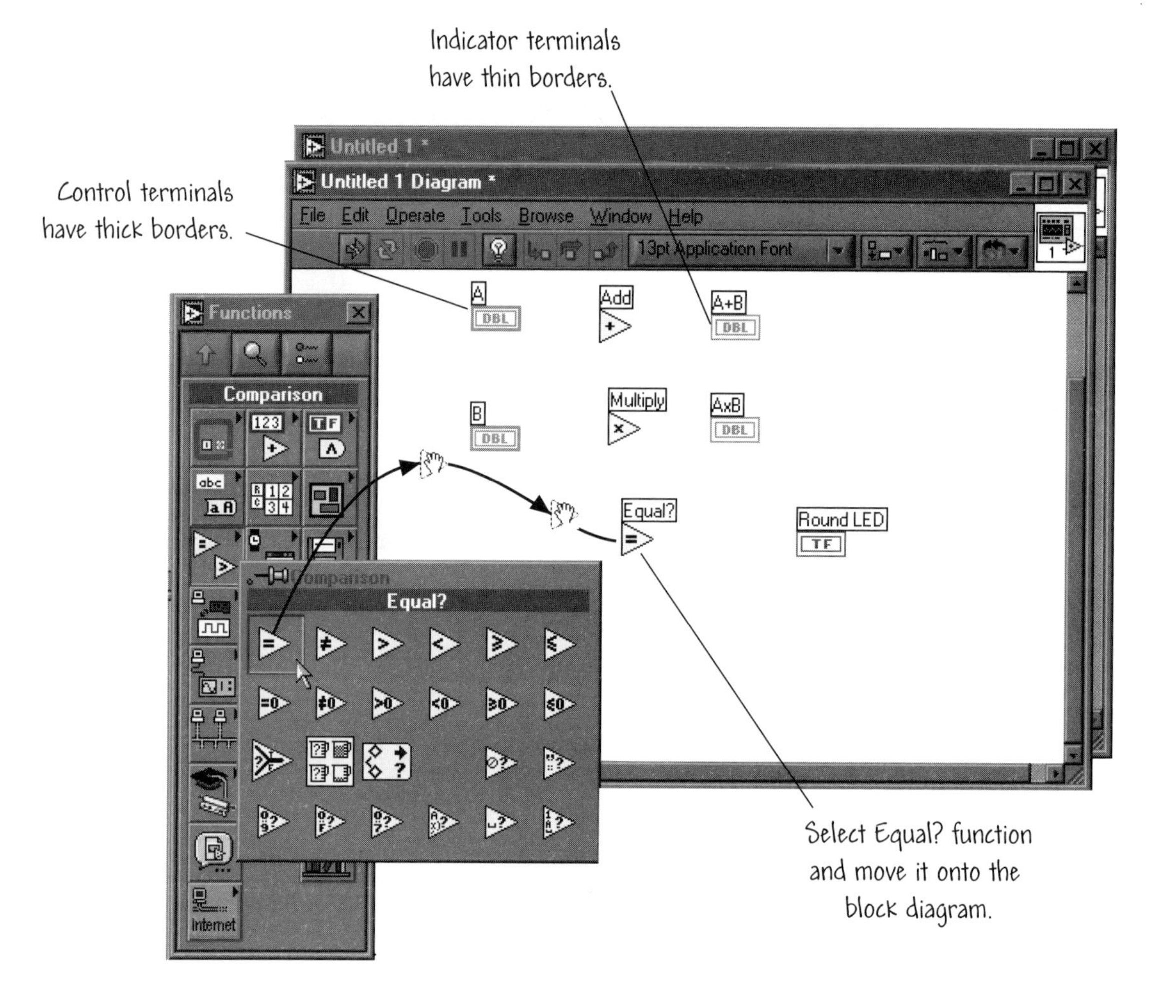

FIGURE 2.25
Adding the Equal? function to the VI.

In case you cannot get your VI to run properly, a working version of the VI (called Simple Math.vi*) is located in the* Chapter 2 *folder within the* Learning *directory.*

7. **Enter input data**. Enter numbers in the digital controls utilizing the **Operating** tool by double-clicking in the digital control box and typing in a number. The default values for A and B are 0 and 0, respectively. You can run the VI using these default values as a first try! When you use the default values, the LED should light up since $A = B$.
8. Run the VI by clicking on the **Run** button.
9. Experiment with different input numbers—make A and B identical, and verify that the LED does indeed turn on.

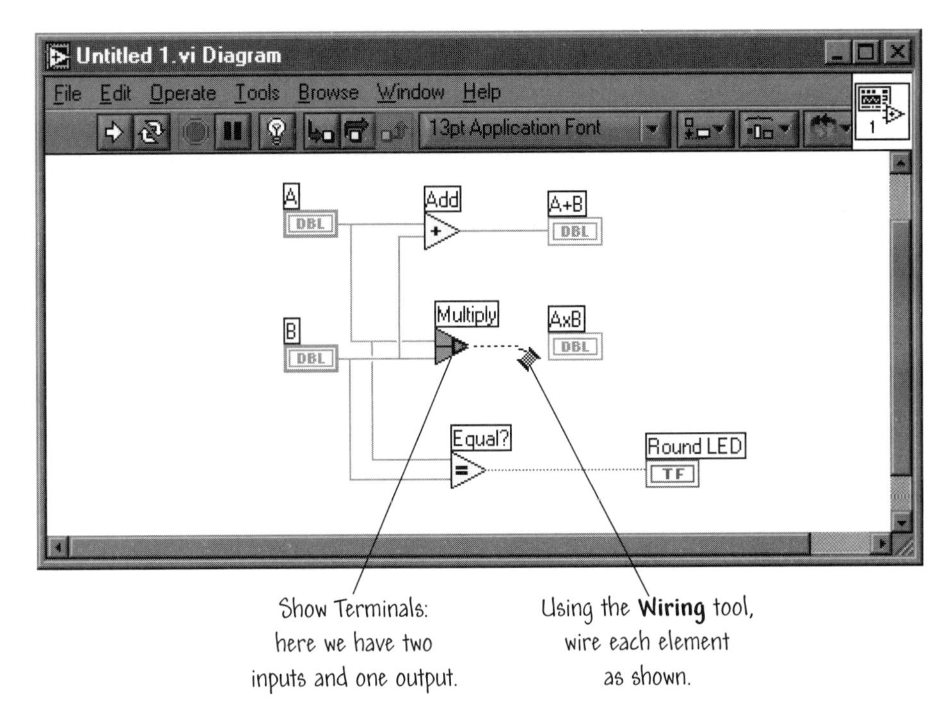

FIGURE 2.26
Wiring from one terminal to another.

10. When you are finished experimenting, close the VI by selecting **Close** from the **File** menu.

To increase the speed of incrementing (and decrementing) the input to a digital control, press <shift> while clicking on the increment and decrement buttons on the digital control, as seen in Figure 2.27.

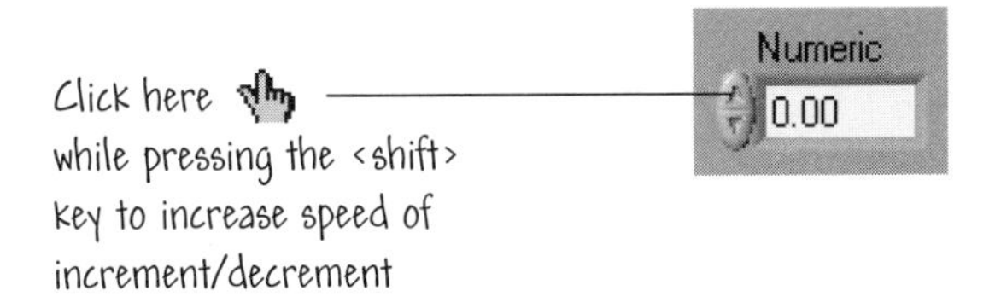

FIGURE 2.27
Press the shift key while incrementing and decrementing to increase the speed.

2.6 DATA FLOW PROGRAMMING

The principle that governs VI execution is known as data flow. Unlike most sequential programming languages, the executable elements of a VI execute only when they have received all required input data—in other words, data flows out of the executable element only after the code is finished executing. The notion of data flow contrasts with the control flow method of executing a conventional program, in which instructions execute sequentially in the order specified by the programmer. Another way to say the same thing is that the flow of traditional sequential code is instruction driven, while the data flow of a VI is data driven.

Consider the VI block diagram, shown in Figure 2.28, that adds two numbers and then computes the sine of the result. In this case, the block diagram executes from left to right, not because the objects are placed in that order, but because one of the inputs of the Sine & Cosine function is not valid until the Add function has added the numbers together and passed the data to the Sine & Cosine function. Remember that a node (in this case, the Sine function) executes only when data is available at all of its input terminals, and it supplies data to its output terminals only when it finishes execution. Open DataFlowA.vi located in the Chapter 2 folder in the Learning directory, press on execution highlighting and then run.

Consider the example in Figure 2.29. Which code segment would execute first—the one on the left or the one on the right? You cannot determine the answer just by looking at the codes. The one on the left does not necessarily execute first. In a situation where one code segment must execute before another, and there is no type of dependency between the functions, you must use a Sequence structure to force the order of execution. To observe the data flow on the code in Figure 2.29, open the DataFlowB.vi located in the folder Chapter 2 in the Learning directory. Before running the VI, click on the **Execution Highlighting** button and watch the flow of the execution.

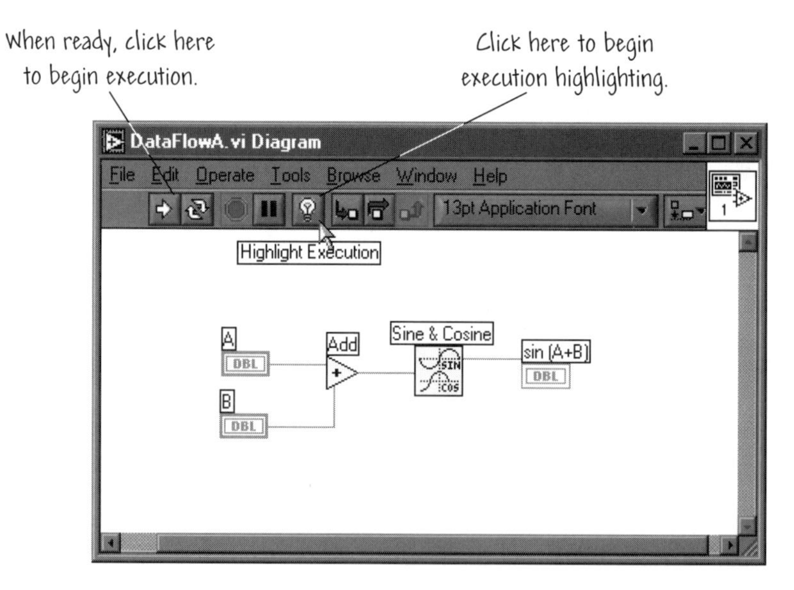

FIGURE 2.28
Block diagram that adds two numbers and then computes the sine of the result.

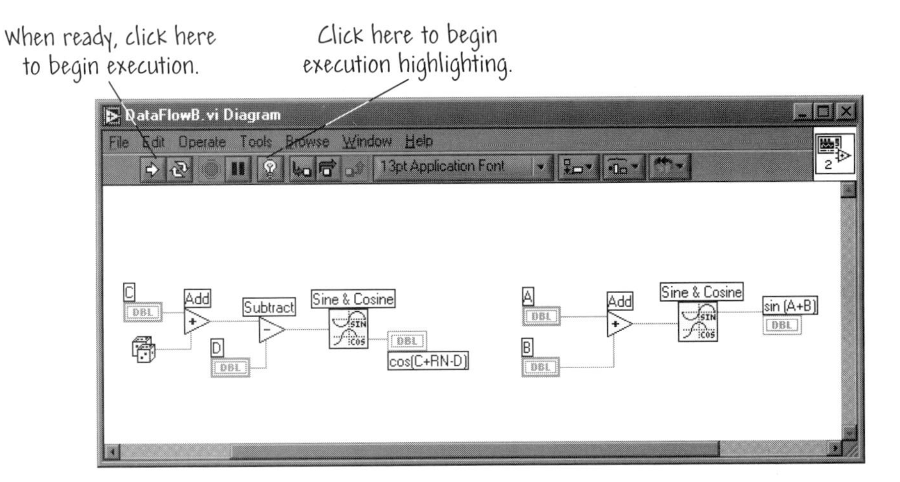

FIGURE 2.29
Which code executes first?

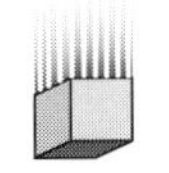

BUILDING BLOCK

2.7 BUILDING BLOCKS: DISPLACEMENT, VELOCITY, AND ACCELERATION

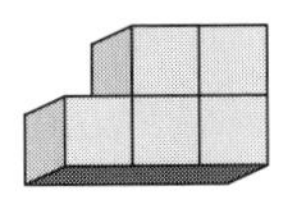

Like the first building block in Chapter 1, this is an exercise in opening and running an existing VI. The VI that you should open is called Displacement Velocity & Acc.vi and is included in the Building Blocks folder in the Learning directory. Find the VI and open it. The front panel is shown in Figure 2.30. This VI takes four input values, one of which is a string input: equation for displacement. We will discuss strings in Chapter 9 in detail. The default input string is $\sin(t)\exp(t/20)$. Run the VI and observe the output graphs. You should see an increasing-magnitude sinusoidal output for the displacement, velocity, and acceleration. Using the **Operating** tool, run several individual experiments by changing the input equation to the following:

- $\sin(t)\exp(-t/20)$
- $\sin(2t)/t$

To change the input equation, highlight the existing equation with the **Operating** tool, type in the desired equation, and then click on the **Enter** button on the front panel toolbar.

Switch to the block diagram window by selecting **Show Diagram** from the **Window** pull-down menu. The block diagram is depicted in Figure 2.30.

Locate the terminals associated with the four controls: number of points, time start, time end, and equation for displacement. Notice that the terminals all have thick borders and that the blue number of points terminal contains

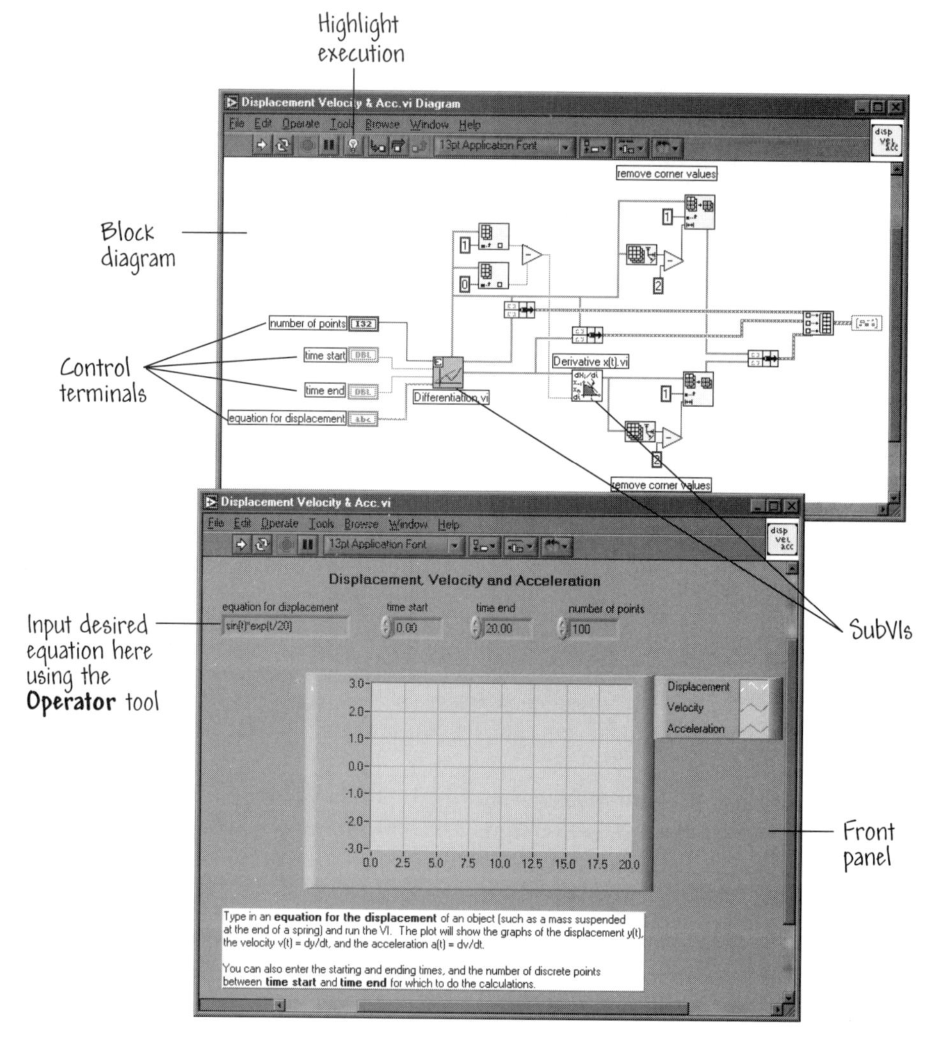

FIGURE 2.30
The Displacement Velocity & Acc.VI.

I32 to indicate it is a long integer. The two terminals time start and time end are orange and contain the label DBL indicating double-precision floating-point numbers. What color is the border of the string control equation for displacement? It should be pink and contain the label abc indicating a string.

The block diagram contains two subVIs: Differentiation.vi and Derivative x(t).vi. Find them on the block diagram. If you want to see what the subVI looks like internally, simply double-click on it's icon using the **Operating** tool. In case you accidentally change something in the subVI after opening it up, make sure to not save the changes when prompted upon closing the subVI!

Other interesting objects to note are the wires. Notice that some wires are solid, thick orange, some are thick pink and others are thin blue. Refer to Table 2.1 to determine the data type carried by each wire shown in Figure 2.30.

As a final experiment with this VI, on the block diagram select **Highlight Execution** and run the VI. Observe the data flow through the VI. Does the data flow strictly from left to right? You should see that the data flows at various points in the block diagram—and not left to right.

2.8 RELAXING READING: A SOLAR CAR DATA TELEMETRY SYSTEM

At Drexel University, the students and faculty are active in designing, building, and running solar-powered electric cars. The SunDragon Solar Racing Team is composed of electrical and computer, mechanical, and chemical engineering students at Drexel. Recent versions of their solar-powered car use a data telemetry system based on LabVIEW to help the students monitor and operate the car during races. The Drexel University team has won many races, including the Tour de Sol—a race and exhibition of electric and solar vehicle technology.

One recent incarnation of the solar car is the SunDragon IV. This car is a three-wheeled, one-passenger vehicle weighing about 675 pounds (including the weight of the driver and the batteries). The vehicle is constructed almost entirely of composite materials. An 8 m^2 solar panel provides energy to the motor and recharges the batteries. The car was once clocked at 98.3 km/h (67 mph) at the Indianapolis Motor Speedway.

The solar car data telemetry system transmits motor rpm, voltages, currents, and temperatures via radio to a chase vehicle. The foundation of the telemetry system is a Fluke Corporation Hydra 2625A data acquisition (DAQ) unit in the car and LabVIEW. The Hydra 2625A is capable of measuring 21 channels of voltage, frequency, resistance, and temperature. In high-resolution mode, the Hydra measures four channels per second. Nine channels of critical data are sent by spread-spectrum RF modem to the chase vehicle following the car. Users can operate both the modem and the Hydra using 12 VDC. The data is received in the chase vehicle via modem connection.

Using LabVIEW, the students at Drexel University created a "virtual dashboard" to display the data, as shown in Figure 2.31. Analog meter displays, digital displays, and colored status lights assist the students in identifying potentially dangerous operating conditions. Because the solar car has more gauges and displays than a normal vehicle, having a duplicate dashboard offloads some the effort needed to successfully operate the solar car from the driver to the chase team, allowing the driver to focus on driving. LabVIEW performs many important calculations in real time. For example, LabVIEW computes the generating speed using inputs of motor rpm, wheel diameter, and gear ratio. The program also saves all collected data to a spreadsheet-compatible file format.

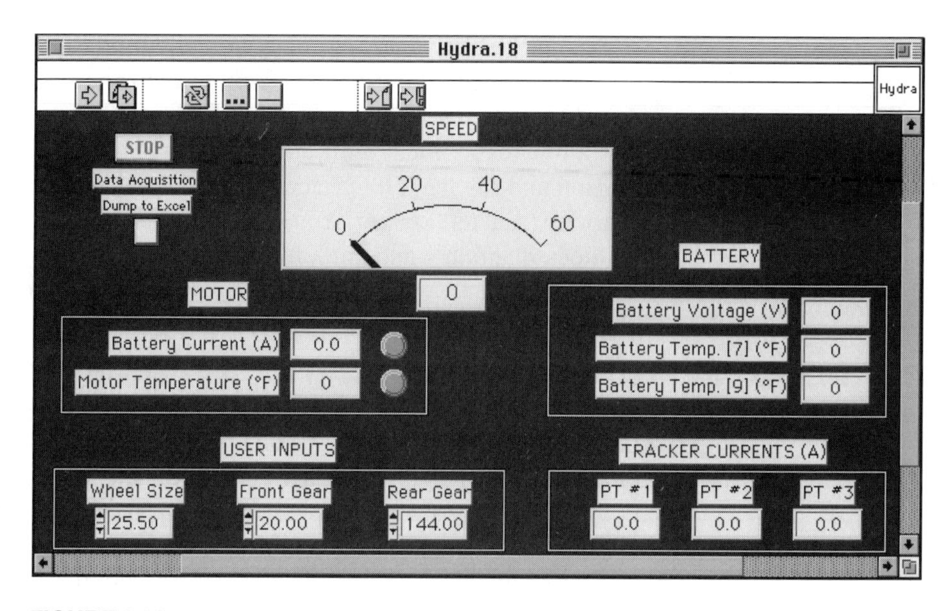

FIGURE 2.31
Students at Drexel University use a "virtual dashboard" to display telemetered data to a chase vehicle.

For more information on this exciting application of LabVIEW, read the *User Solutions* entitled "LabVIEW Helps Lead Solar Car to Victory," which can be found on the National Instruments website, or contact:

Kevin Scoles, Ph.D.
Department of Electrical and Computer Engineering
Drexel University, 32nd and Chestnut Street
Philadelphia, PA 19104
e-mail: scoles@ece.drexel.edu

2.9 SUMMARY

Virtual instruments (VIs) are the building blocks of LabVIEW. The graphical programming language is known as the G programming language. VIs have three main components: the front panel, the block diagram, and the icon and connector pair. VIs follow a data flow programming convention in which each executable node of the program executes only after all the necessary inputs have been received. Correspondingly, output from each executable node is produced only when the node has finished executing.

KEY TERMS

Boolean controls and indicators: Front panel objects used to manipulate and display or input and output Boolean (True or False) data.

Connector: Part of the VI or function node that contains its input and output terminals, through which data passes to and from the node.

Connector pane: Region in the upper right corner of the front panel that displays the VI terminal pattern. It underlies the icon pane.

Data flow programming: Programming system consisting of executable nodes in which nodes execute only when they have received all required input data and produce output automatically when they have executed.

G programming language: Graphical programming language used in LabVIEW.

Icon: Graphical representation of a node on a block diagram.

Icon pane: Region in the upper right corner of the front panel and block diagram that displays the VI icon.

Input terminals: Terminals that emit data. Sometimes called source terminals.

Modular programming: Programming that uses interchangeable computer routines.

Numeric controls and indicators: Front panel objects used to manipulate and display numeric data.

Output terminals: Terminals that absorb data. Sometimes called destination terminals.

String controls and indicators: Front panel objects used to manipulate and display or input and output text.

Terminals: Objects or regions on a node through which data passes.

Tip strips: Small yellow text banners that identify the terminal name and make it easier to identify function and node terminals for wiring.

Wire: Data path between nodes.

Wiring tool: Tool used to define data paths between source and sink terminals.

EXERCISES

E2.1 In this first exercise you get to play a drawing game. Open Spirograph.vi located in Chapter 2 of the Learning directory. The front panel should look like the one shown in Figure E2.1. Did you ever use a spirograph drawing kit as a kid? This VI is the computer version of that children's game. To play the drawing game,

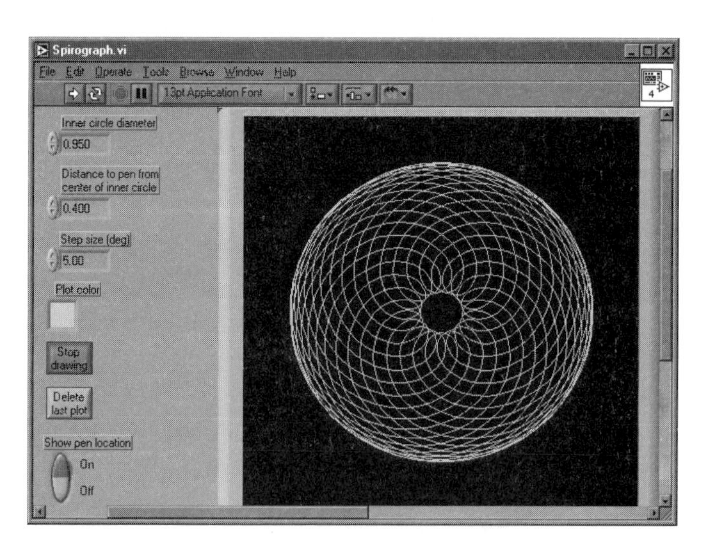

FIGURE E2.1
The Spirograph.vi front panel.

run the VI and experiment by varying the controls. After clicking on the **Run** button to start the VI execution, click on the Begin drawing push button on the front panel. Once the drawing starts, the Stop drawing label appears in the push button. The VI execution is halted by clicking on the Stop drawing button.

E2.2 Construct a VI that uses a round push button control to turn on a square light indicator whenever the push button is depressed. The front panel is very simple and should look something like the one shown in Figure E2.2.

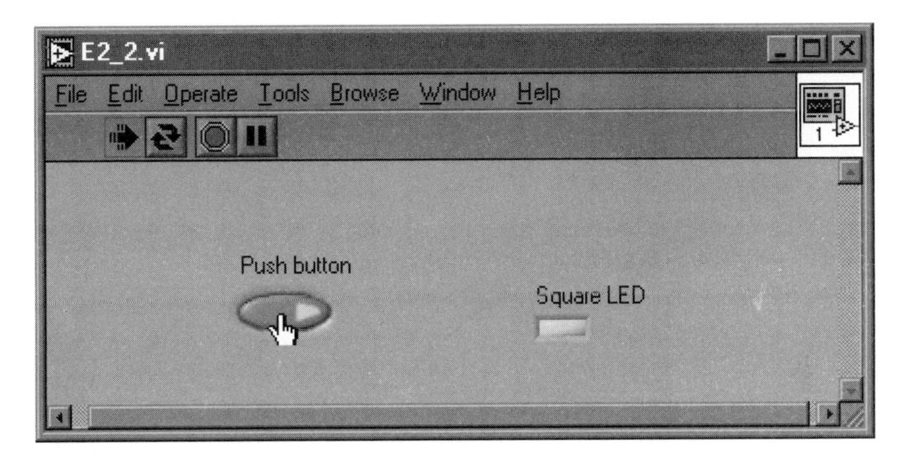

FIGURE E2.2
A front panel using a round push button control and a square light indicator.

E2.3 Open SimPhone.vi, located in Examples\Sound\sndExample adv.llb. The

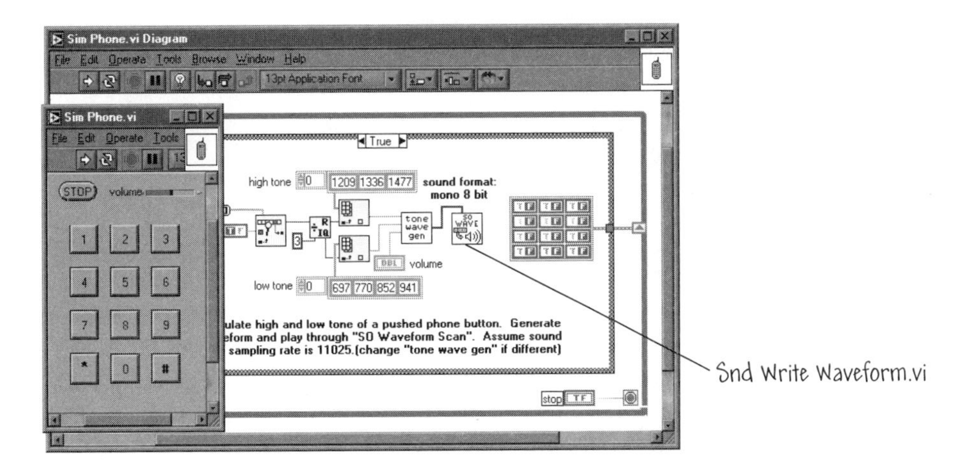

FIGURE E2.3
The SimPhone VI.

front panel and block diagram are shown in Figure E2.3. Using the online help, determine the inputs and outputs of the subVI Snd Write Waveform. Sketch the subVI icon and connector showing the inputs and outputs.

PROBLEMS

P2.1 Complete the crossword puzzle.

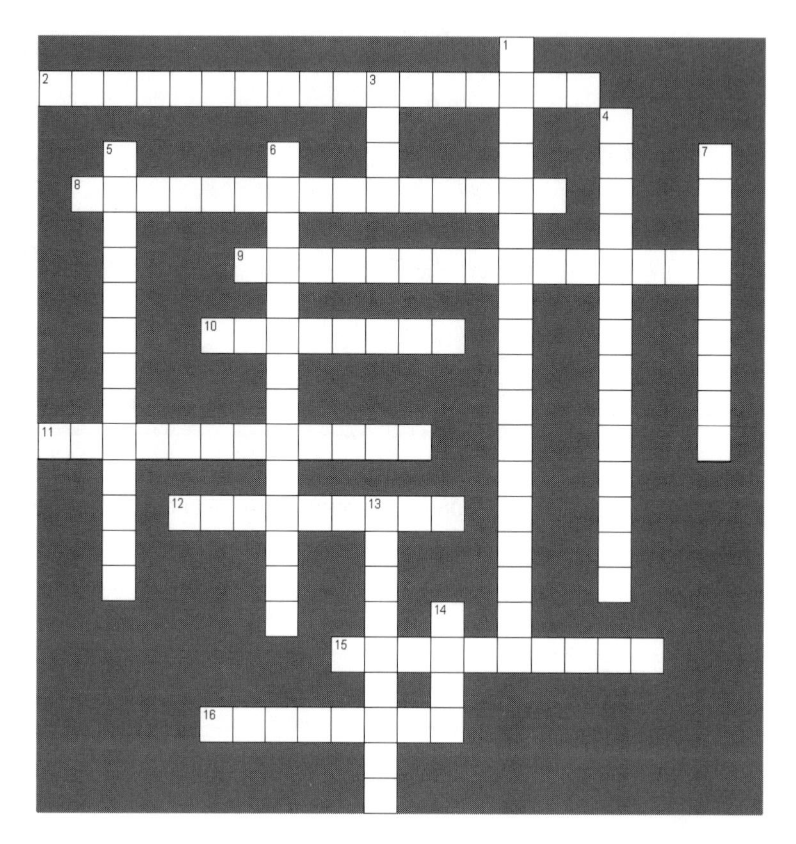

Across

2. Terminals that absorb data.
8. Front panel objects used to input Boolean data.
9. Terminals that emit data.
10. Programming system in which nodes execute only when they have received all required input data.
11. Graphical programming language used in LabVIEW.
12. Part of the VI or function node that contains terminals through which data passes to and from the node.
15. Tool used to define data paths between source and sink terminals.
16. Region in the upper right corner of the front panel and block diagram that displays the VI icon.

Down

1. Front panel objects used to manipulate and display numeric data.
3. Graphical representation of a node on a block diagram.
4. Front panel objects used to manipulate and display or input and output text.
5. Region in the upper right corner of the front panel that displays the VI terminal pattern.
6. Programming that uses interchangeable computer routines.
7. Small yellow text banners that identify the terminal name and make it easier for wiring.
13. Objects or regions on a node through which data passes.
14. Data path between nodes.

P2.2 Construct a VI that performs the following tasks:

- Takes two floating-point numbers as inputs on the front panel: X and Y.
- Subtracts Y from X and displays the result on the front panel.
- Divides X by Y and displays the result on the front panel.
- If the input $Y = 0$, a front panel LED lights up to indicate division by zero.

Name the VI Subtract and Divide.vi and save it in the Users Stuff folder in the Learning directory.

P2.3 Construct a VI that uses a vertical slide control for input and a meter indicator for output display. A front panel and block diagram that can be used as a guide are shown in Figure P2.3. Referring to the block diagram, you see a pair of dice,

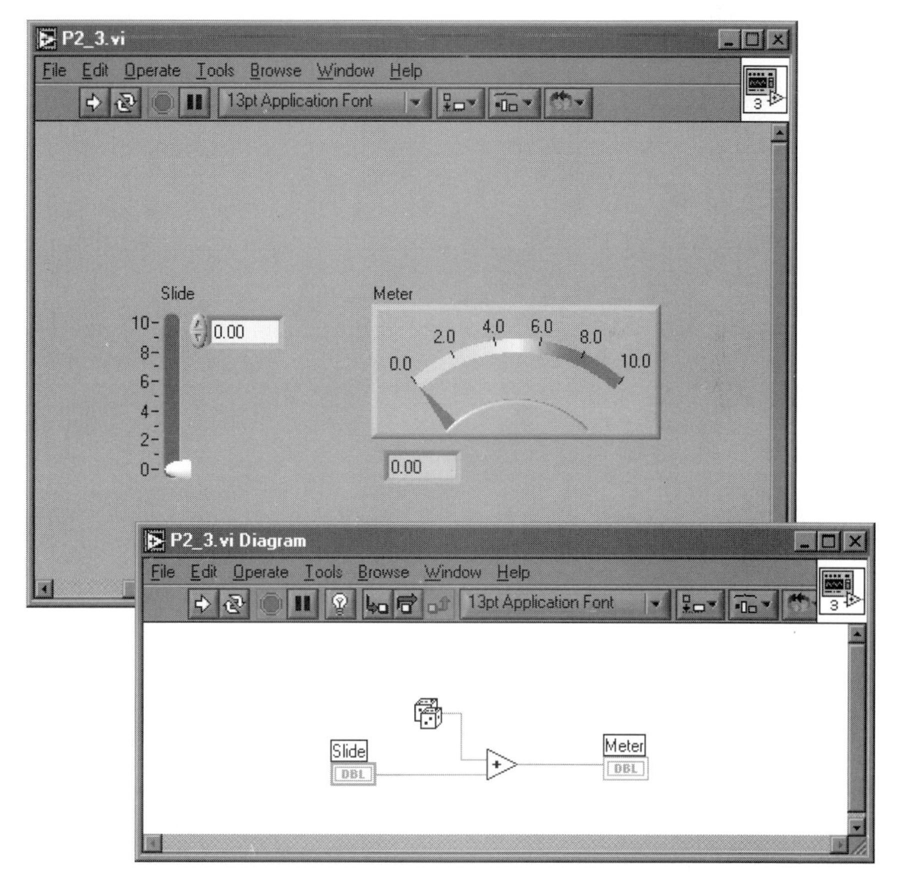

FIGURE P2.3
Using a vertical slide control and a meter indicator.

which is the icon for a random number function. You will find the random number function on the Functions≫Numeric palette. When running the VI, any input you provide via the vertical slide will be reflected on the meter indicator. The random number function adds “noise” to the input so that the meter output will not be exactly the same as the input. Run the VI in **Continuous Run** mode and vary the slide input.

P2.4 Create a VI that has a numeric control to input a number x, uses the add and multiply functions to calculate $3x^2 + 2x + 5.0$ and display the output using a numeric indicator.

CHAPTER 3

Editing and Debugging Virtual Instruments

Like text-based computer programs, virtual instruments are dynamic. VIs change as their applications evolve (usually increasing in complexity). For instance, a VI that initially only performs addition may at a later time be updated to add a multiplication capability. You need debugging and editing tools to verify and test VI coding changes. Since programming in G is graphical in nature, editing and debugging are also graphical, with options available in pull-down and short cut menus and various palettes. In this chapter, we discuss how to create, select, delete, move, and arrange objects on the front panel and block diagrams. The important topic of selecting and deleting wires and locating and eliminating bad wires is also presented. Debugging subjects covered include execution highlighting (you can watch the code run!), single-stepping through code, and inserting probes to view data as the VI executes.

GOALS

1. Learn to access and practice with VI editing tools
2. Learn to access and practice with VI debugging tools

3.1 EDITING TECHNIQUES

3.1.1 Creating Controls and Indicators on the Block Diagram

As discussed in previous chapters, when building a VI you can create controls and indicators on the front panel and know that their terminals will automatically appear on the block diagram. Switching to the block diagram, you can begin wiring the terminals to functions (such as addition or multiplication functions), subVIs, or other objects. In this section we present an alternative method to create and wire controls and indicators in *one* step on the block diagram. In the following discussions, you should open a new VI and follow along by repeating the steps as presented.

1. Open a new VI and switch to the block diagram.
2. Place the square root function located in the **Numeric** subpalette on the **Functions** palette on the block diagram, as shown in Figure 3.1.

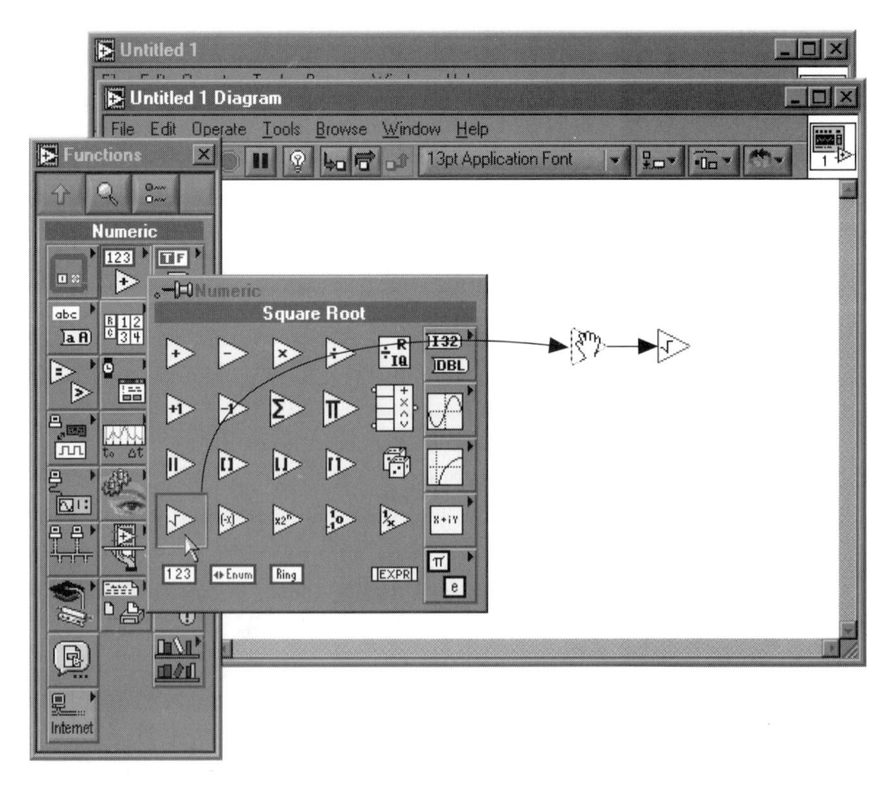

FIGURE 3.1
Add the square root function to the block diagram.

3. Now we want to add (and wire) a control terminal to the square root function. As shown in Figure 3.2, you pop up on the left side of the square root function and select **Control** from the **Create** menu. The result is that a control terminal is created and automatically wired to the Square Root function. Switch to the front panel and notice that a numeric control has appeared!

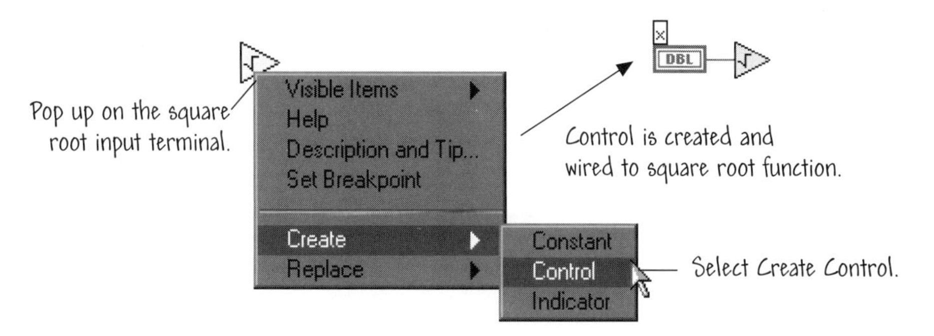

FIGURE 3.2
Front panel control created for the Square Root function.

*To change a control to an indicator (or vice versa), pop up on the terminal (on the block diagram) or on the object (on the front panel) and select **Change to Indicator** (or **Change to Control**).*

4. In fact, you can wire indicators and constants by popping up on the node terminal and choosing the desired selection. For example, popping up on the right side of the square root function and selecting **Indicator** from the **Create** menu creates and automatically wires a front panel indicator, as illustrated in Figure 3.3. In many situations, you may wish to create and automatically wire a constant to a terminal, and you accomplish this by popping up and choosing **Constant** from the **Create** menu.

*To change a control or indicator (for example, from a knob to a dial), pop up on the object on the front panel and select **Replace**. The **Controls** palette will appear, and you can navigate to the desired new object.*

5. The resulting control and indicator for the square root function are shown in Figure 3.4.

After you create the front panel control or indicator from the block diagram, you can delete the object only from the front panel.

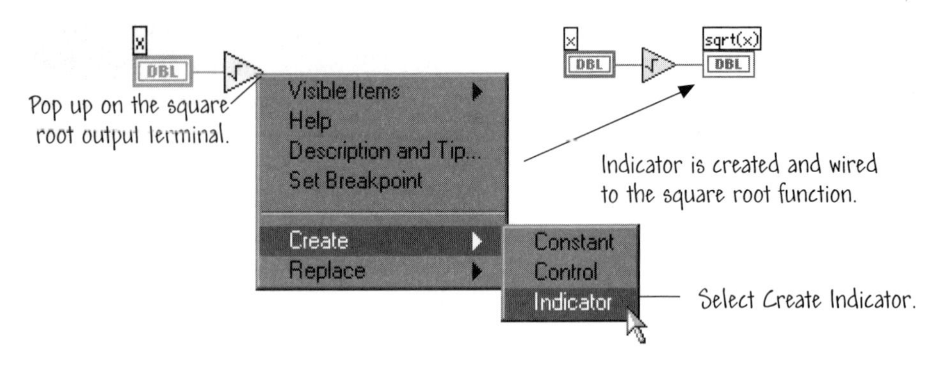

FIGURE 3.3
Front panel indicator created for the Square Root function.

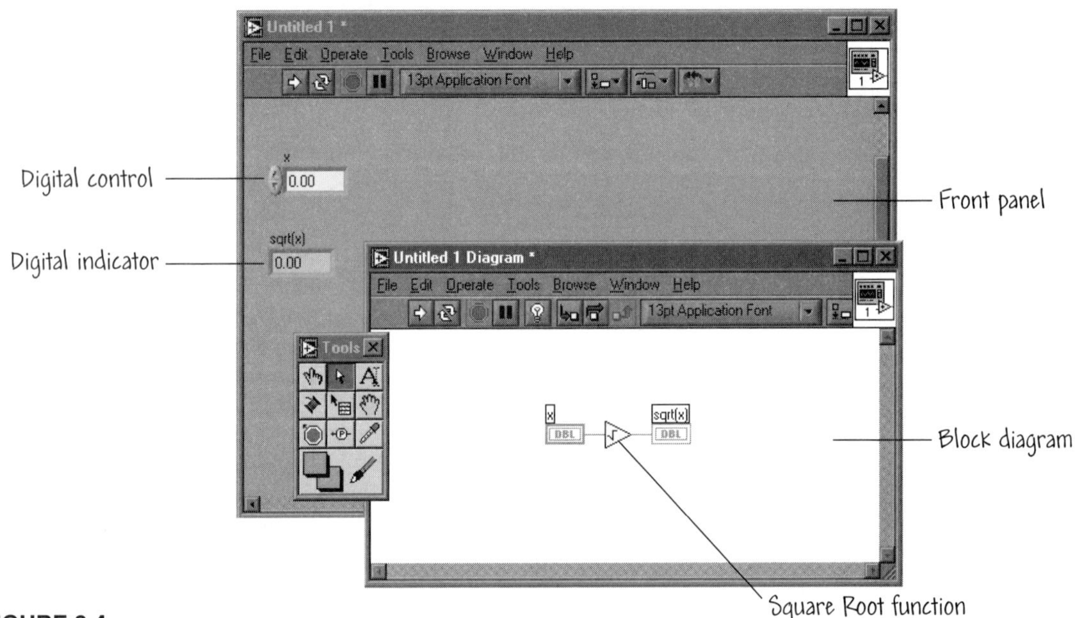

FIGURE 3.4
Front panel and block diagram for the Square Root function.

3.1.2 Selecting Objects

The **Positioning** tool selects objects in the front panel and block diagram windows. In addition to selecting objects, you use the **Positioning** tool to move and resize objects (more on these topics in the next several sections). To select an object, click the left mouse button while the **Positioning** tool is over the object. When the object is selected, a surrounding dashed outline appears, as shown in

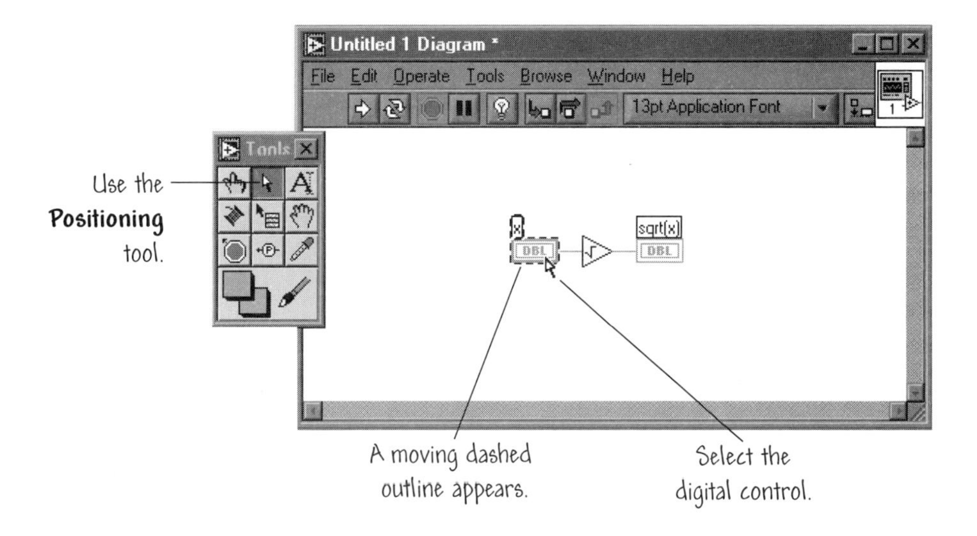

FIGURE 3.5
Selecting an object using the **Positioning** tool.

Figure 3.5. To select more than one object, shift-click (that is, hold down the <shift> key and simultaneously click) on each additional object you want to select. You also can select multiple objects by clicking in a nearby open area and dragging the cursor until all the desired objects lie within the selection rectangle that appears (see Figure 3.6). Sometimes after selecting several objects, you may want to deselect just one of the objects (while leaving the others selected). This is accomplished by shift-clicking on the object you want to deselect—the other objects will remain selected, as desired.

3.1.3 Moving Objects

You can move an object by clicking on it with the **Positioning** tool and dragging it to a desired location. The objects of Figure 3.6 are selected and moved as shown in Figure 3.7. Selected objects can also be moved using the up/down and right/left arrow keys. Pressing the arrow key once moves the object one pixel; holding down the arrow key repeats the action. In this manner, you can move and locate your objects very precisely.

The direction of movement of an object can be restricted to being either horizontal or vertical by holding down the <shift> key when you move the object. The direction you initially move decides whether the object is limited to horizontal or vertical motion.

If you change your mind about moving an object while you are in the midst of dragging it to another location, continue to drag until the cursor is outside all open

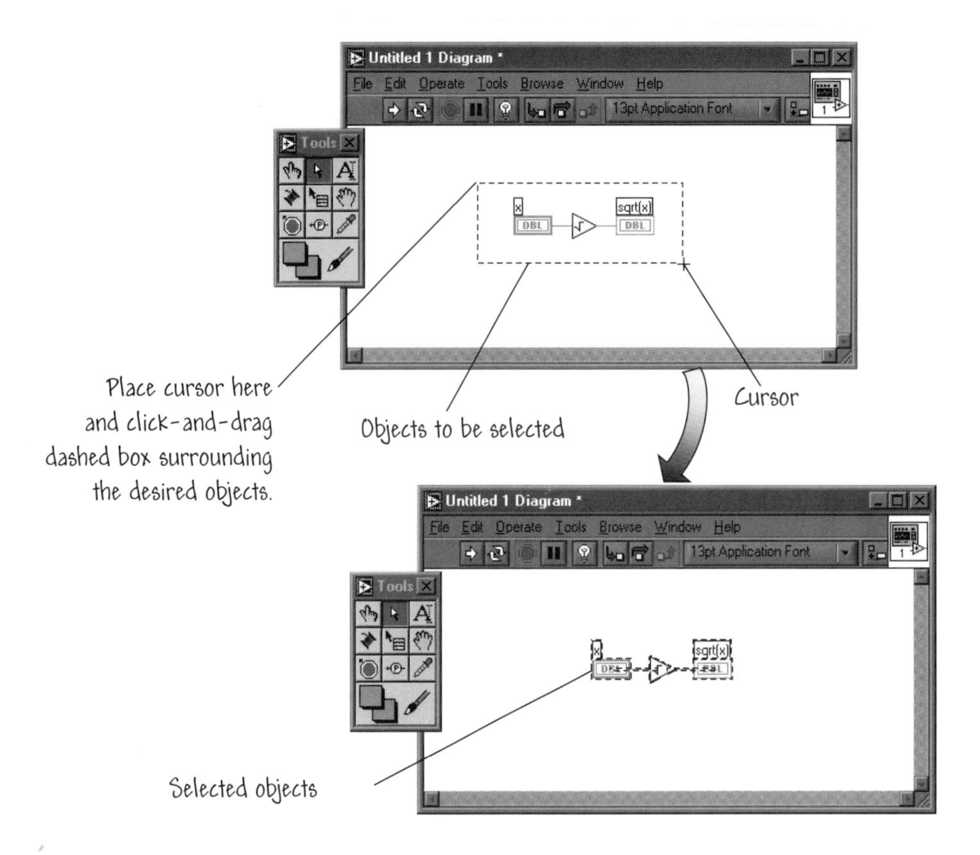

FIGURE 3.6
Selecting a group of objects.

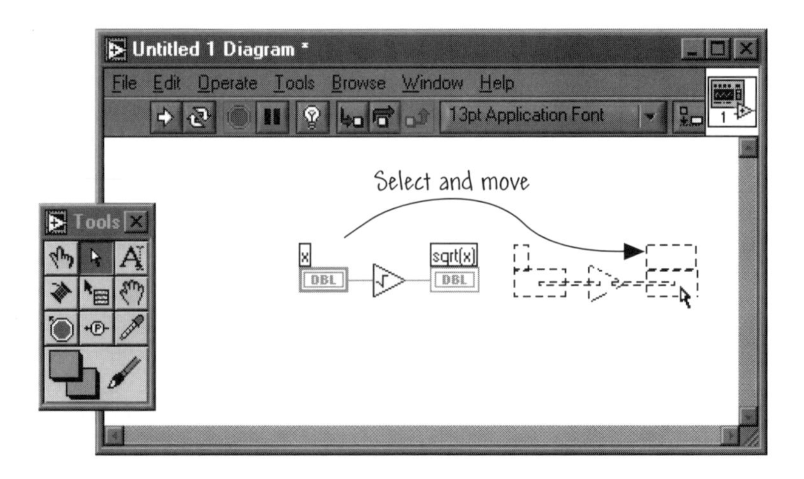

FIGURE 3.7
Selecting and moving a group of objects.

windows and the dashed line surrounding the selected object disappears—then release the mouse button. This will cancel the move operation and the object will not move. Alternatively, if the object is dragged and dropped to an undesirable location, you can select **Undo Move** from the **Edit** menu to undo the move operation.

3.1.4 Deleting and Duplicating Objects

You can delete objects by selecting the object(s) and choosing **Clear** from the **Edit** menu or pressing the <backspace> key in **Windows** or the <delete> key on the **Macintosh**. Most objects can be deleted; however you cannot delete certain components of a control or indicator, such as the label or digital display. You must hide these components by popping up and deselecting the Visible Items≫Label or Visible Items≫Digital Display from the short cut menu. Also remember that you can only delete controls and indicators on the front panel.

Most objects can be duplicated, and there are three ways to duplicate an object—by copying and pasting, by cloning, and by dragging and dropping. In all three cases, a complete new copy of the object is created, including, for example, the terminal belonging to a front panel control and the control itself.

You can copy text and pictures from other applications and paste them into LabVIEW.

To clone an object, click the **Positioning** tool over the object while pressing <ctrl> for **Windows** (or <option> on the **Mac**) and drag the object to its new location. After you drag the selection to a new location and release the mouse button, a copy of the object appears in the new location, and the original object remains in the original location. When you clone or copy objects, the copies are labeled by the same name with the word *copy* appended (or *copy 1*, *copy 2*, and so forth for copies of copies).

You also can duplicate objects using Edit≫Copy and then Edit≫Paste from the **Edit** menu. First, place the **Positioning** tool over the desired object and choose Edit≫Copy. Then click at the location where you want the duplicate object to appear and choose Edit≫Paste from the **Edit** menu. You can use this process to copy and paste objects within a VI or even copy and paste objects between VIs.

The third way to duplicate objects is to use the drag-and-drop capability. In this way, you can copy objects, pictures, or text between VIs and from other applications. To drag and drop an object, select the objects, pictures, or text file with the **Positioning** tool and drag it to the front panel or block diagram of the target VI. You can also drag VIs from the file system (in **Windows** and on the **Mac**) to the active block diagram to create subVIs (we will discuss subVIs in Chapter 4).

*On the **Macintosh**, you must have the Drag Manager in order to use the external drag-and-drop capability. The Drag Manager is built into System 7.5 and later; it is available for System 7.0 to 7.5 as an extension from Apple.*

3.1.5 Resizing Objects

You can easily resize most objects. **Resizing handles** appear when you move the **Positioning** tool over a resizable object, as illustrated in Figure 3.8. On rectangular objects the resizing handles appear at the corners of the object; resizing circles appear on circular objects. Passing the **Positioning** tool over a resizing handle transforms the tool into a resizing cursor. To enlarge or reduce the size of an object, place the **Positioning** tool over the resizing handle and click and drag the resizing cursor until the object is the desired size. When you release the mouse button, the object reappears at its new size. The resizing process is illustrated in Figure 3.8.

To cancel a resizing operation, continue dragging the frame corner outside the active window until the dotted frame disappears—then release the mouse button and the object will maintain its original size. Alternately, if the object has already been resized and you want to undo the resizing, you can use the **Undo Resize** command found in the **Edit** pull-down menu.

Some objects only change size horizontally or vertically, or keep the same proportions when you resize them (e.g., a knob). In these cases, the resizing

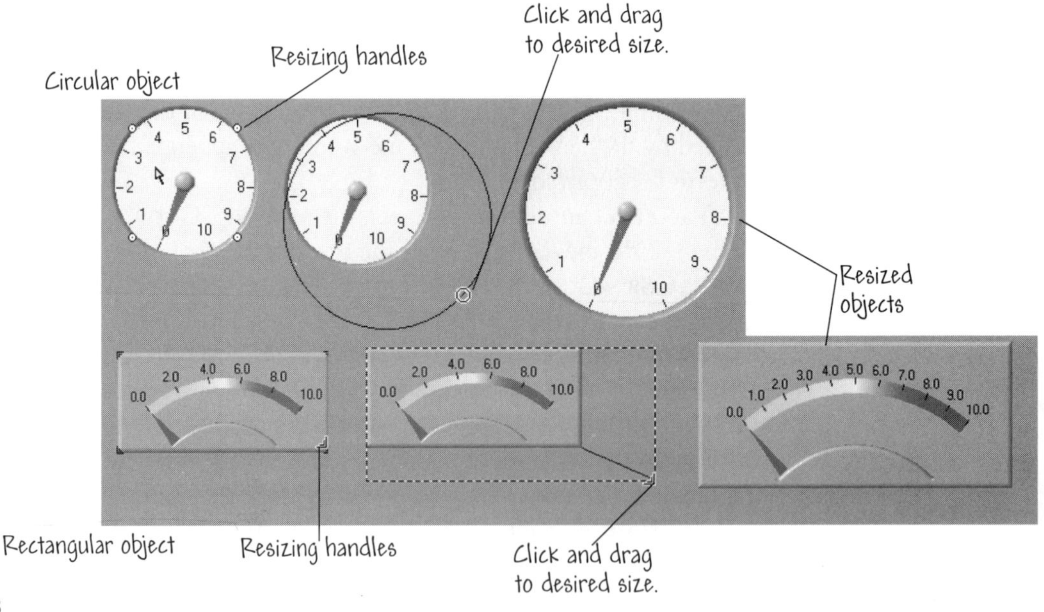

FIGURE 3.8
Resizing rectangular and circular objects.

cursor appears the same, but the dotted resize outline moves in only one direction. To restrict the resizing of any object in the vertical or horizontal direction (or to maintain the current proportions of the object) hold down the <shift> key as you click and drag the object.

3.1.6 Labeling Objects

Labels are blocks of text that annotate components of front panels and block diagrams. There are two kinds of labels—free labels and owned labels. Owned labels belong to and move with a particular object and describe that object only. You can hide these labels but you cannot copy or delete them independently of their owners. Free labels are not attached to any object, and you can create, move, or dispose of them independently. Use them to annotate your front panels and block diagrams. Free labels are one way to provide accessible documentation for your VIs.

You use the **Labeling** tool to create free labels or to edit either type of label. To create a free label, choose the **Labeling** tool from the **Tools** palette, and then click anywhere in an open area and type the desired text in the bordered box that appears. An example of creating a free label is shown in Figure 3.9.

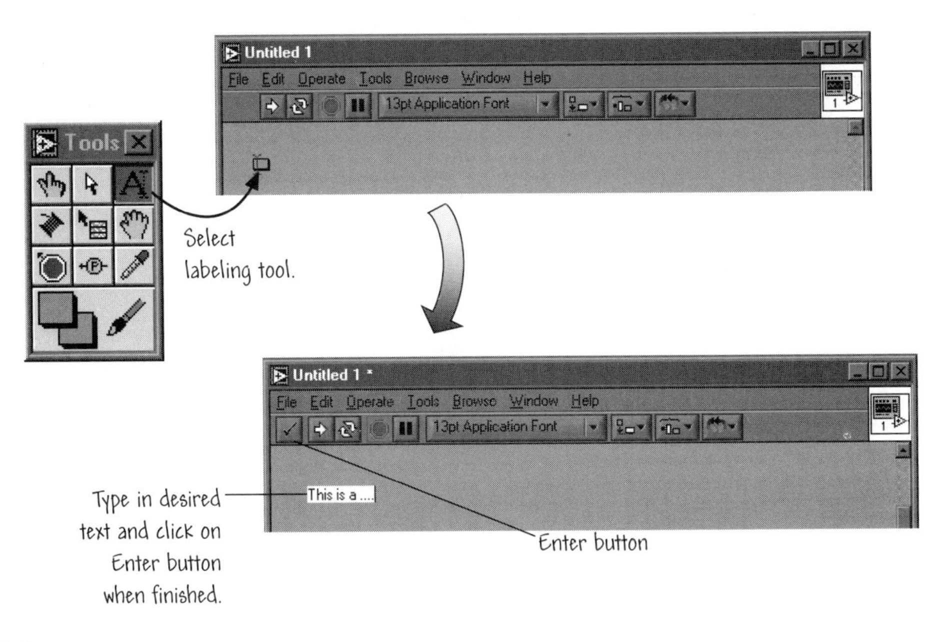

FIGURE 3.9
Creating a free label.

When finished entering the text, click on the **Enter** button on the toolbar, which appears on the toolbar to remind you to end your text entry. You can also end your text entry by pressing the <Enter> key on the numeric keypad (if you have a numeric keypad). Remember that if you do not type any text in the label, the label disappears as soon as you click somewhere else.

When you add a control or an indicator to the front panel, an owned label automatically appears. The label box is ready for you to enter the desired text. If you do not enter text immediately, the default label remains. To create an owned label for an existing object, pop up on the object and select Visible Items≫Label from the short cut menu (see Figure 1.12). not enter the text immediately, the label disappears.

If you place a label or any other object over (or partially covering) a control or indicator, it slows down screen updates and could make the control or indicator flicker. To avoid this problem, do not overlap a front panel object with a label or other objects.

You can copy the text of a label by double-clicking on the text with the **Labeling** tool or dragging the **Labeling** tool across the text to highlight the desired text. When the desired text is selected, choose Edit≫Copy to copy the text onto the clipboard. You can then highlight the text of a second label and use Edit≫Paste to replace the highlighted text in the second label with the text from the clipboard. To create a new label with the text from the clipboard, click on the screen with the **Labeling** tool where you want the new label positioned and then select Edit≫Paste.

The resizing technique described in the previous section also works for labels. You can resize labels as you do other objects by using the resizing cursor. Labels normally autosize; that is, the label box automatically resizes to contain the text you enter. If for some reason you don't want the text in labels to automatically resize to the entered text, pop up on the label and select **Size to Text** to toggle autosizing off.

The text in a label remains on one line unless you enter a carriage return to resize the label box. By default, the <enter> (or <return>) key is set to add a new line. This can be changed in Tools≫ Options≫ Front Panel so that the text input is terminated with the <enter> (or <return>) key.

3.1.7 Changing Font, Style, and Size of Text

Using the **Font ring** in the toolbar, you can change the font, style, size, and alignment of any text displayed in a label or on the display of controls and indicators. Certain controls and indicators display text in multiple locations—for example, on graphs (a type of indicator), the graph axes scale markers are made up of many numbers, one for each axes tick. You have the flexibility to modify each text display independently.

The font ring was discussed in Chapter 1 and is shown in Figure 1.7. Notice the word **Application** showing in the **Font ring** pull-down menu, which also contains the **System**, **Dialog**, and **Current** options. The last option in the ring—the **Current Font**—refers to the last font style selected. The predefined fonts are used for specific portions of the various interfaces:

- The **Application** font is the default font. It is used for the **Controls** and **Functions** palettes.
- The **System** font is the font used for menus.
- The **Dialog** font is the font used for text in dialog boxes.

These fonts are predefined so that they map "best" when porting your VIs to other platforms.

The **Font ring** has size, style, justify, and color options. Selections made from any of these submenus (that is, size, style, etc.) apply to all selected objects. For example, if you select a new font while you have a knob selected, the labels, scales, and digital displays all change to the new font. Figure 3.10 illustrates the situation of changing the style of all the text associated with a knob from plain text to bold text.

If you select any objects or text and make a selection from the **Font ring** pull-down menu, the changes apply to everything selected. The process of selecting

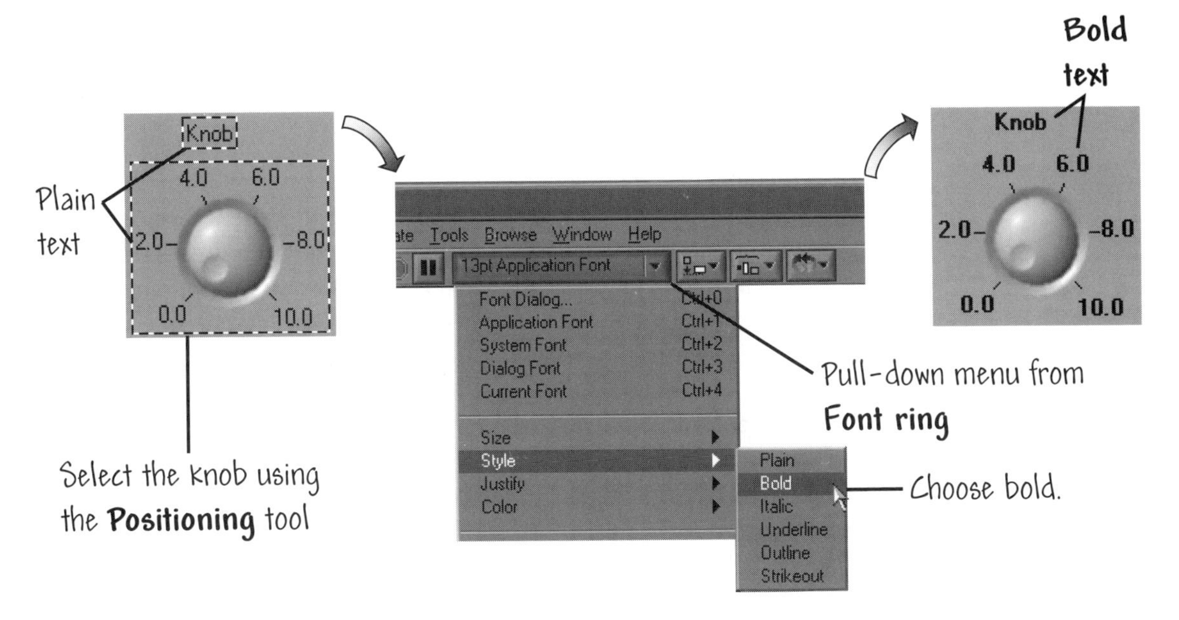

FIGURE 3.10
Changing the font style on a knob control.

just the text of an owned label of a knob is illustrated in Figure 3.11. Once the desired text is selected (remember to use the **Labeling** tool to select the desired text), you can make any changes you wish by selecting the proper pull-down submenu from the **Font ring**. If no text is selected, the font changes apply to the default font, so that labels created from that point on will reflect the new default font, while not affecting the font of existing labels.

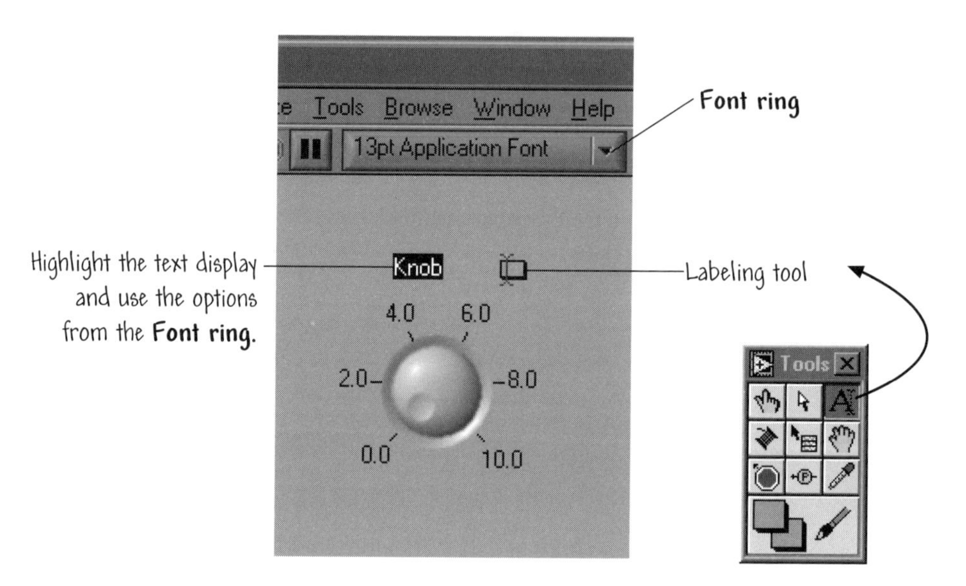

FIGURE 3.11
Selecting text for modification of style, font, size, and color.

When working with objects that have multiple pieces of text (e.g. slides and knobs), remember that text selections affect the objects or text currently selected. For example, if you select the entire knob while selecting bold text, the scale, digital display, and label all change to a bold font, as shown in Figure 3.10. As shown in Figure 3.12a, when you select the knob label, followed by selecting bold text from the **Style** submenu of the **Font ring** pull-down menu, only the knob label changes to bold. Similarly, when you select text from a scale marker while choosing bold text, all the markers change to bold, as shown in Figure 3.12b.

If you select **Font Dialog...** in the **Font ring** while a front panel is active, the dialog box shown in Figure 3.13 appears. If a block diagram is active instead, the **Panel Default** option at the bottom of the dialog box is checked. With either the **Panel Default** or **Diagram Default** checkbox selected, the other selections made in this dialog box will be used with new labels on the front panel or block diagram. In other words, if you click the **Panel Default** and **Diagram Default** checkboxes, the selected font becomes the current font for the front panel, the block diagram, or both. The current font is used on new labels. The checkboxes

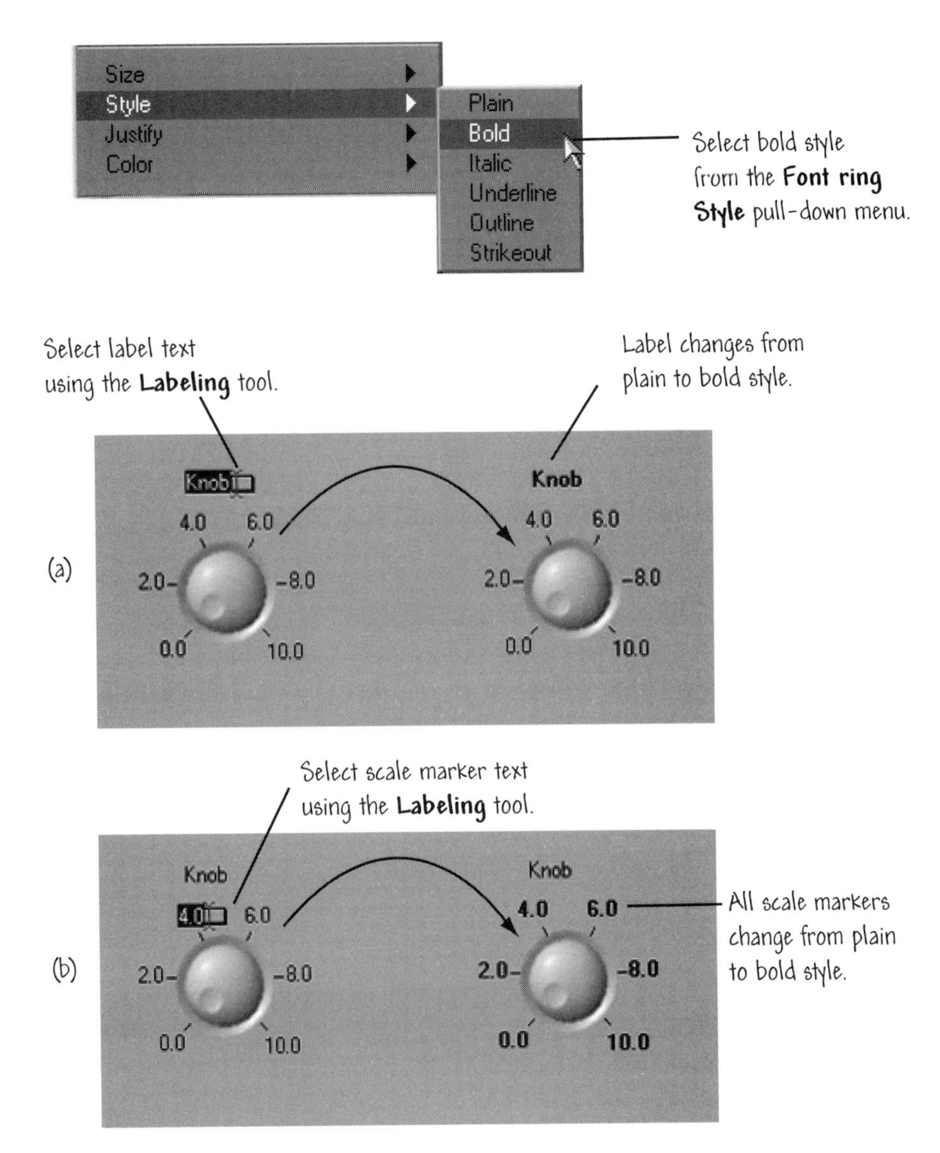

FIGURE 3.12
Changing text attributes on a knob.

allow you to set different fonts for the front panel and block diagram. For example, you could have a small font on the block diagram and a large one on the front panel.

3.1.8 Selecting and Deleting Wires

A single horizontal or vertical piece of wire is known as a **wire segment**. The point where three or four wire segments join is called a **junction**. A **wire branch**

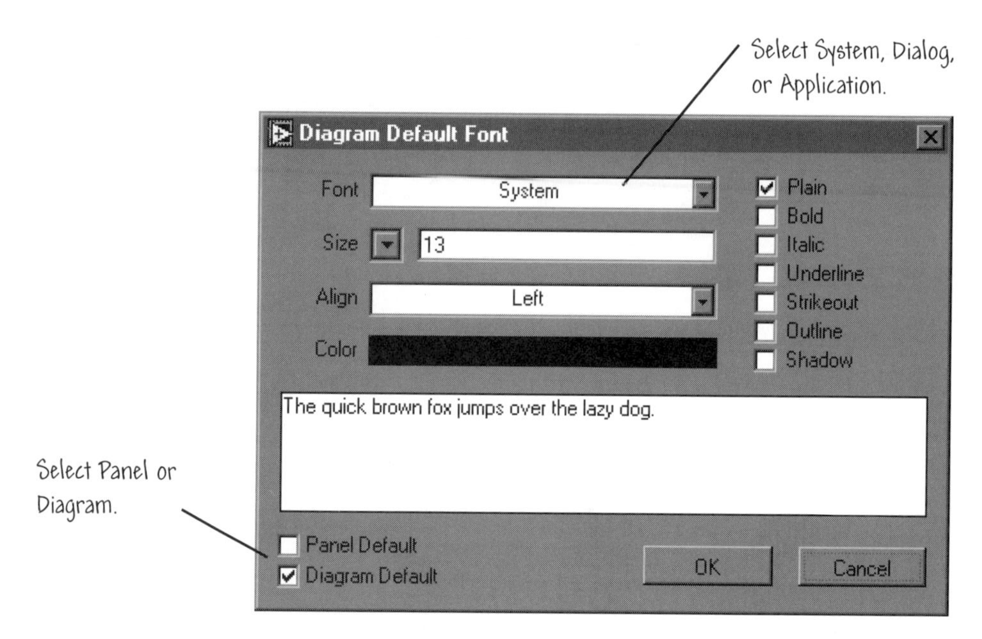

FIGURE 3.13
Using the **Font Dialog** box to change the font, size, alignment, color and style.

contains all the wire segments from one junction to another, from a terminal to the next junction, or from one terminal to another if there are no junctions in between. You select a wire segment by clicking on it with the **Positioning** tool. Clicking twice selects a branch, and clicking three times selects the entire wire. See Figure 3.14 for an example of selecting a branch, segment, or an entire wire.

3.1.9 Wire Stretching and Broken Wires

Wired objects can be moved individually or in groups by dragging the selected objects to a new location with the **Positioning** tool. The wires connecting the objects will stretch automatically. If you want to move objects from one diagram to another, the connecting wires will not move with the selected objects, unless you select the wires as well. After selecting the desired objects and the connecting wires, you can cut-and-paste the objects as a unit to another VI.

An example of stretching a wire is shown in Figure 3.15. In the illustration, the object (in this case, a numeric indicator) is selected with the **Positioning** tool and stretched to the desired new location.

Wire stretching occasionally leaves behind loose ends. Loose ends are wire branches that do not connect to a terminal. Your VI will not execute until you remove all loose ends. An example of a loose end is shown in Figure 3.16. Loose ends can be easily removed using the Edit≫Remove Broken Wires command.

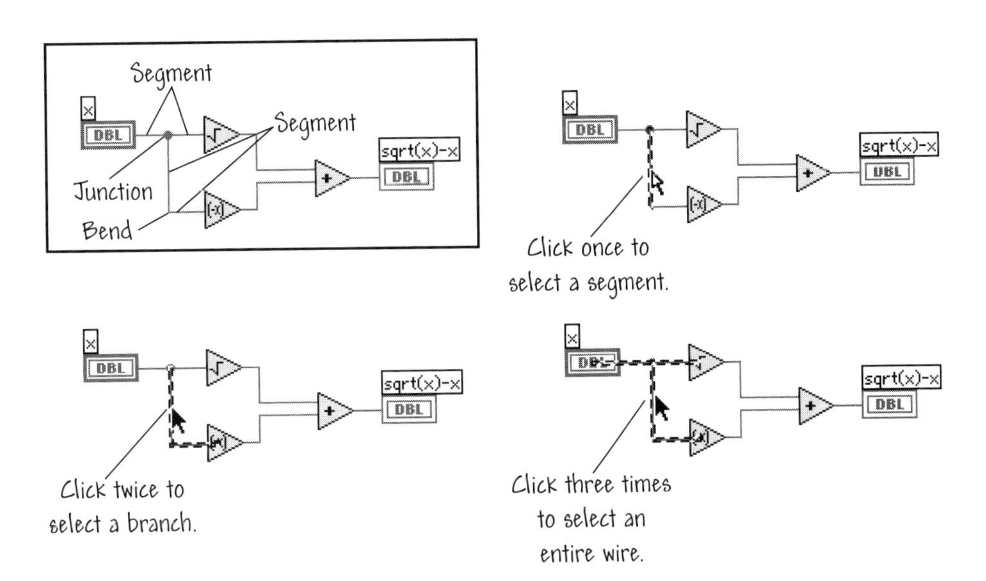

FIGURE 3.14
Selecting a segment, branch, or an entire wire.

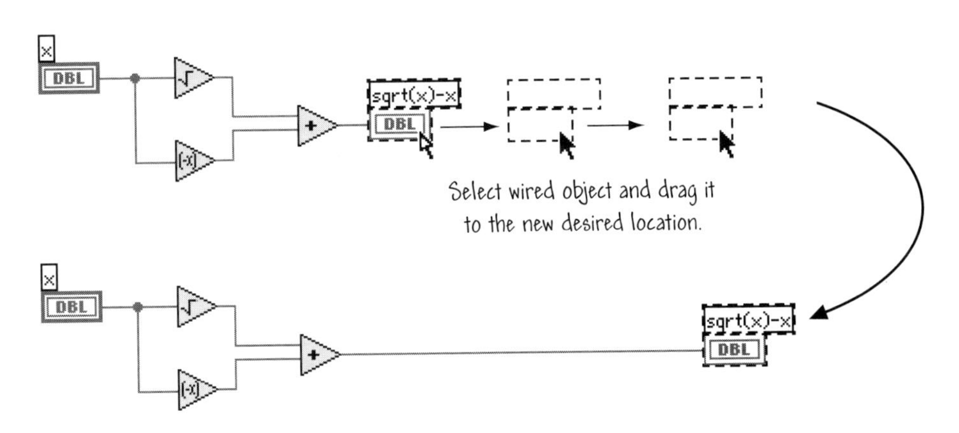

FIGURE 3.15
Moving wired objects.

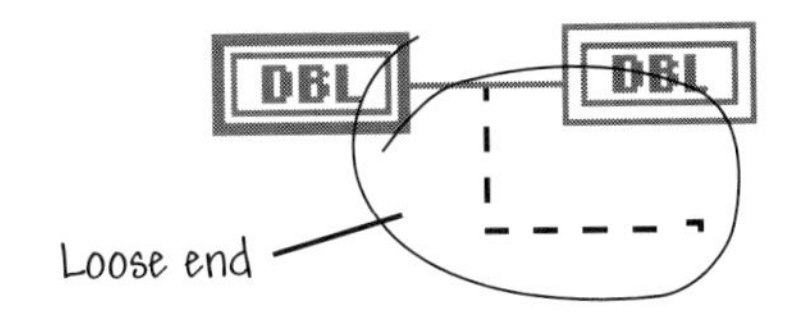

FIGURE 3.16
Loose ends.

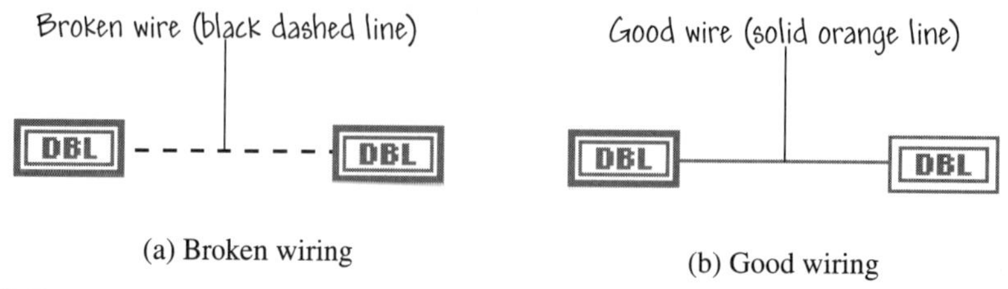

FIGURE 3.17
Locating broken wires.

When you make a wiring mistake, a broken wire (indicated by a dashed line) appears. Figure 3.17 shows a dashed line that represents a broken wire. It is inevitable that broken wires will occur in the course of programming in G. One common mistake is to attempt to connect two indicator terminals together or to connect a control terminal to an indicator terminal when the data types do not match (for example, connecting a Boolean to a numeric). If you have a broken wire, you can remove it by selecting it with the **Positioning** tool and eliminating it by pressing the <delete> key. If you want to remove all broken wires on a block diagram at one time, choose **Remove Broken Wires** from the **Edit** menu.

There are many different conditions leading to the occurrence of broken wires. Some examples are:

- **Wire type, dimension, unit, or element conflicts**—A wire type conflict occurs when you wire two objects of different data types together, such as a numeric and a Boolean, as shown in Figure 3.18a.

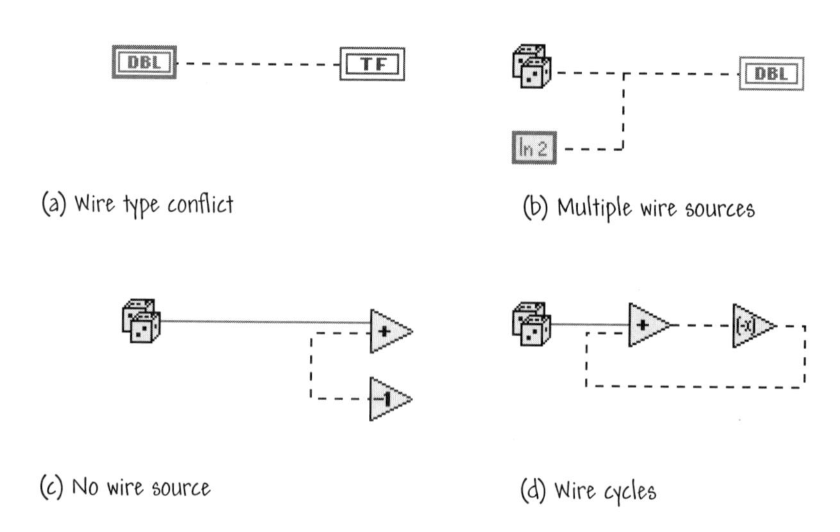

FIGURE 3.18
Typical wiring errors leading to broken wire indications.

- **Multiple wire sources**—You can wire a control to multiple output destinations (or indicators), but you cannot wire multiple data sources to a single destination. In the example shown in Figure 3.18b, we have attempted to wire two data sources (that is, the random number and the constant ln 2) to one indicator. This produces a broken wire and must be fixed by disconnecting the random number (represented by the dice) or disconnecting the ln 2 constant. Another example of a common multiple sources error occurs during front panel construction when you inadvertantly place a numeric control on the front panel when you meant to place a numeric indicator. If you do this and then try to wire an output value to the terminal of the front panel control, you will get a multiple sources error. To fix this error, just pop up on the terminal and select **Change To Indicator**.
- **No wire source**—An example of a wire with no source is shown in Figure 3.18c. This problem is addressed by providing a control. Another example of a situation leading to a no-source error is if you attempt to wire two front panel indicators together when one should have been a control. To fix this error, just pop up on a terminal and select **Change to Control**.
- **Wire cycles**—Wires must not form closed loops of icons or structures, as shown in Figure 3.18d. These closed loops are known as cycles. LabVIEW cannot execute cycles because each node waits on the other to supply it data before it executes (remember data flow!). Shift registers (discussed in Chapter 5) provide the proper mechanism to feed back data in a repetitive calculation.

Sometimes you have a faulty wiring connection that is not visible because the broken wire segment is very small or is hidden behind an object. If the **Broken Run** button appears in the toolbar, but you cannot see any problems in the block diagram, select Edit≫Remove Broken Wires—this will remove all broken wires in case there are hidden, broken wire segments. If the **Run** button returns, you have corrected the problem. If the wiring errors have not all been corrected after selecting Edit≫Remove Broken Wires, click on the **Broken Run** button to see a list of errors. Click on one of the errors listed in the **Error List** dialog box, and you will automatically be taken to the location of the erroneous wire in the block diagram. You can then inspect the wire, detect the error and fix the wiring problem.

3.1.10 Aligning and Distributing Objects

To align a group of objects, first select the desired objects. Then choose the axis along which you want to align them from the **Alignment** ring in the toolbar. You can align objects along the vertical axis using left, center, or right edge. You also can align objects along a horizontal axis using its top, center, or bottom edge. Open a new VI and place three objects (such as, three numeric controls) on the front panel and experiment with different aligning options. Figure 3.19 illustrates the process of aligning three objects by their left edges.

In a similar fashion, you can distribute a group of objects by selecting the objects and then choosing the axis along which you want to distribute the selected objects from the **Distribution** ring in the toolbar (see Figure 1.9). edges or centers equidistant, four menu items at the right side of the ring let you add or delete gaps between the objects, horizontally or vertically.

3.1.11 Coloring Objects

You can customize the color (or shades of gray on monochrome monitors) of many LabVIEW objects. However, the coloring of objects that convey information via their color is unalterable. For example, block diagram terminals of front panel objects and wires use color codes for the type and representation of data they carry, so you cannot change their color.

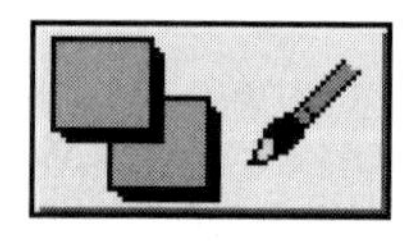

To change the color of an object (or the background of a window), pop up on the object of interest with the **Coloring** tool from the **Tools** palette, as seen in Figure 3.20. Choose the desired color from the selection palette that appears. If you keep the mouse button pressed as you move through the color selection palette, the object or background you are coloring redraws with the color the cursor currently is touching. This gives you a "real-time" preview of the object in the new color. If you release the mouse button on a color, the selected object retains the chosen color. To cancel the coloring operation, move the cursor out of the color selection palette before releasing the mouse button, or use the Edit≫Undo Color Change command after the undesired color change has been made.

If you select the box with a **T** in it, the object is rendered transparent. One use of the **T** (transparent) option is to hide the box around labels. Another interesting use is to create numeric controls without the standard three-dimensional border. Transparency affects the appearance but not the function of the object!

Some objects have both a foreground and a background color that you can set separately. The foreground color of a knob, for example, is the main dial area, and the background color is the base color of the raised edge. The display at the bottom of the color selection box indicates if you are coloring the foreground, the background, or both. Both the foreground and the background are selected by default. To change between foreground and background, you can press the <f> key for foreground and the <b> key for background. Pressing <a> for all selects both foreground and background. Pressing any other key also toggles the selection between foreground and background. The foreground and background

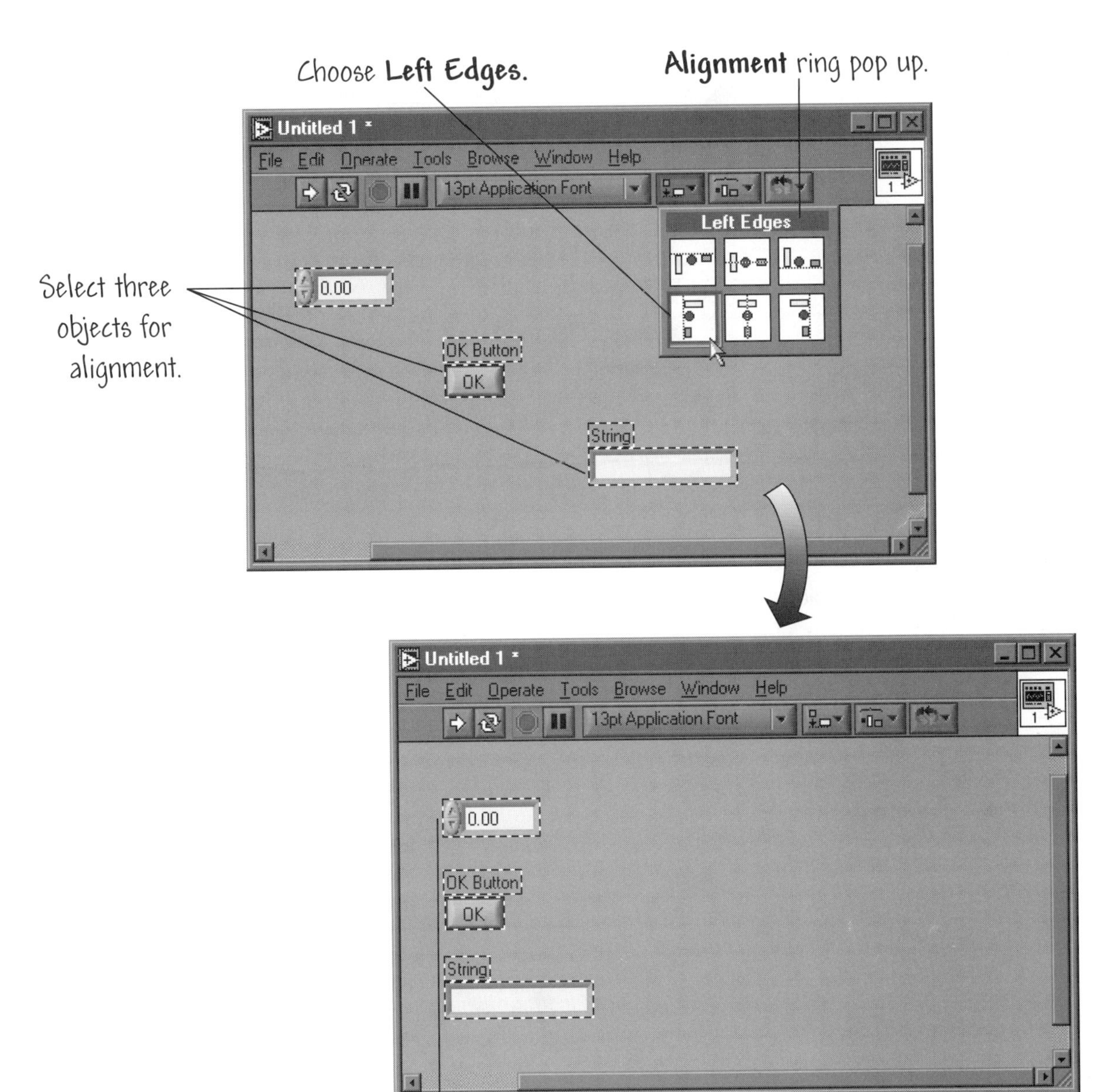

FIGURE 3.19
Aligning objects.

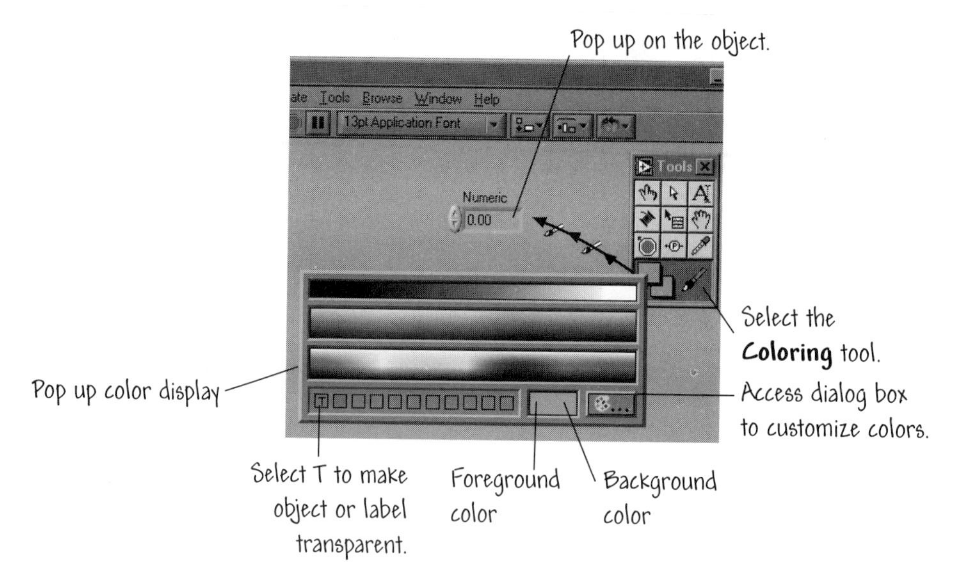

FIGURE 3.20
Customizing the color of objects.

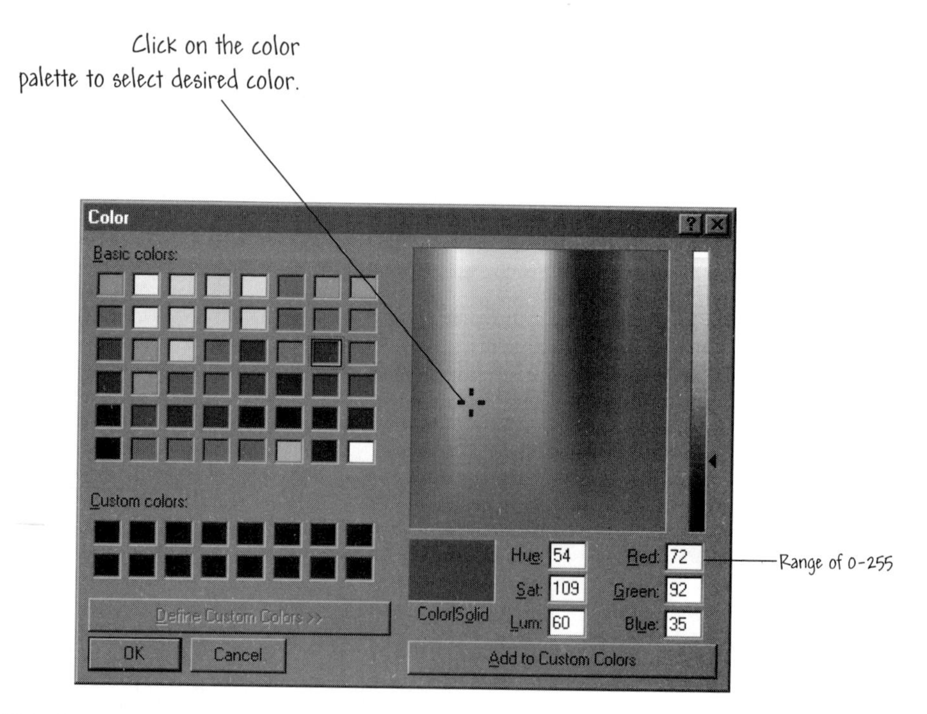

FIGURE 3.21
The **Color** dialog box.

indicators are shown in Figure 3.20. Selecting the button on the lower right-hand side in the palette (see Figure 3.21) accesses a dialog box with which you can customize the colors. Each of the three color components, red, green, and blue, describes eight bits of a 24-bit color (in the lower right-hand side of the **Color** dialog box). Therefore, each component has a range of 0 to 255. The last color you select from the palette becomes the current color. Clicking on an object with the **Color** tool sets that object to the current color.

Using the **Color Copy** tool, you can copy the color of one object and transfer it to a second object without using the **Color** palette. To accomplish this, click with the **Color Copy** tool on the object whose color you want to transfer to another object. Then, select the **Color** tool and click on another object to change it's color.

Practice with Editing

In this exercise you will edit and modify an existing VI to look like the panel shown in Figure 3.22. After editing the VI, you will wire the objects in the block diagram and run the program.

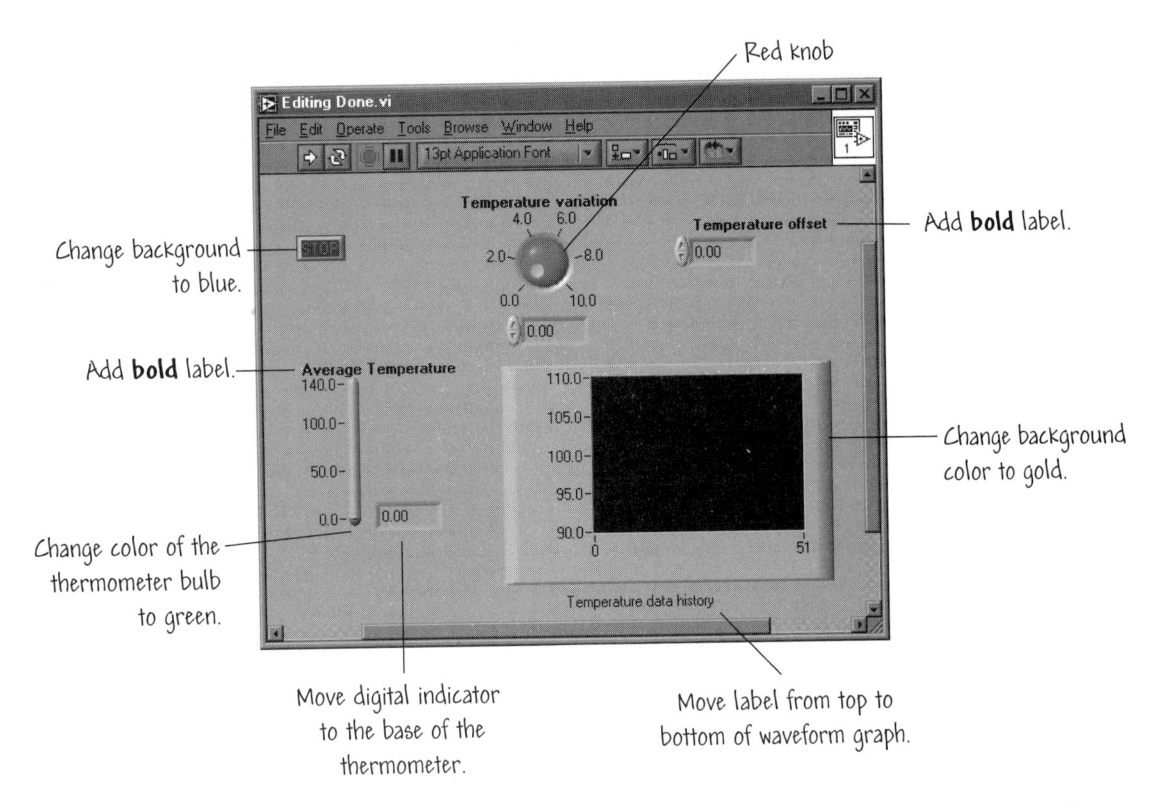

FIGURE 3.22
The front panel for the Editing VI.

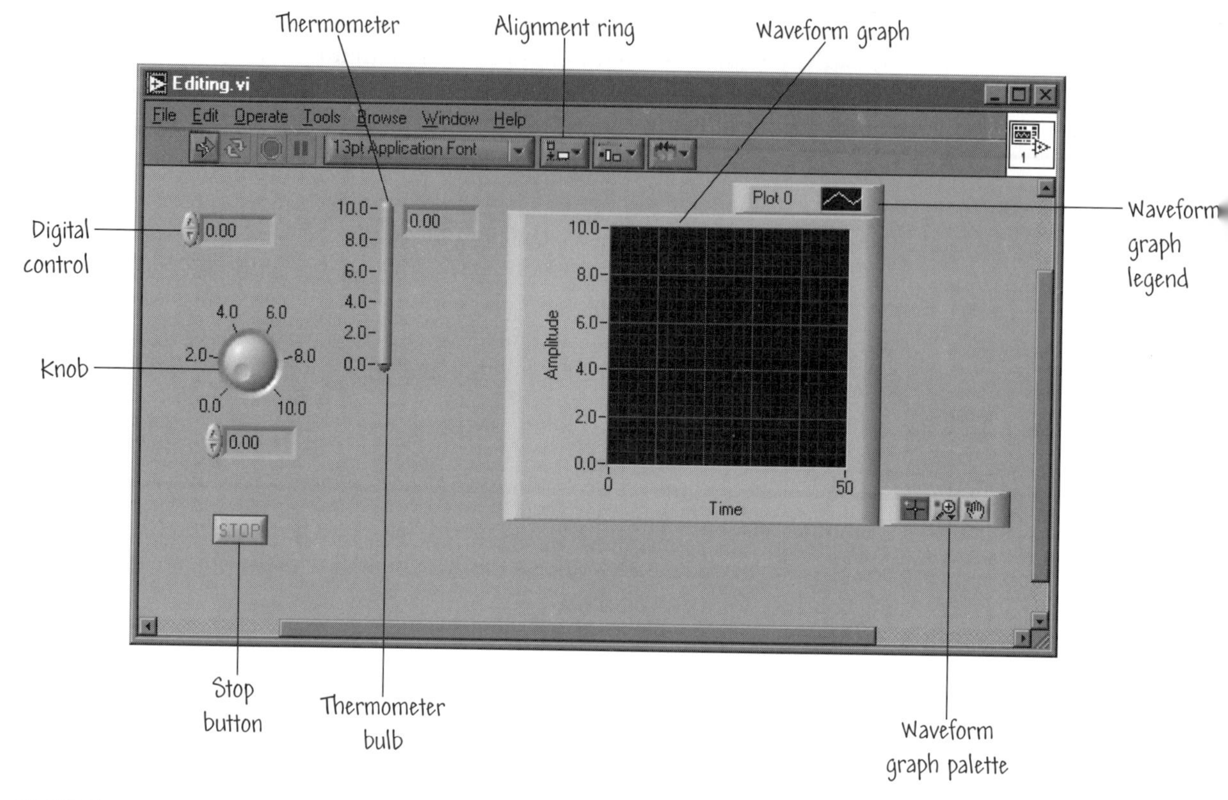

FIGURE 3.23
The Editing VI with various objects: knob, waveform chart, thermometer, and digital control.

1. Open the VI Editing by choosing **Open** from the **File** menu (or using the **Open VI** button on the startup window) and searching in the Chapter 3 folder in the Learning directory. The front panel of the Editing VI contains a number of objects depicted in Figure 3.23. The objective of this exercise is to make the front panel of the Editing VI look like the one shown in Figure 3.22. If you have a color monitor, you can see the final VI in color by opening the VI Editing Done located in the Chapter 3 folder in the Learning directory.
2. Add an owned label to the digital control using the **Positioning** tool by popping up on the digital control and selecting Visible Items≫Label from the pop up menu. Type in the text Temperature offset inside the bordered box and click the mouse outside the label or click the **Enter** button on the left-hand side of the toolbar.
3. Reposition the wavefrom graph and digital control.
   (a) Choose the **Positioning** tool from the **Tools** palette.
   (b) Click on the waveform graph and drag it to the lower center of the front panel (see Figure 3.22 for the approximate location). Then click on the

digital control and drag it to the upper right of the front panel (see Figure 3.22 for approximate location).

Notice that as you move the digital control, the owned label moves with the control. If the control is currently selected, click on a blank space on the front panel to deselect the control and then click on the label and drag it to another location. Notice that in this case the control does not follow the move. You can position an owned label anywhere relative to the control. If you move the owned label to an undesirable location, you can move it back by selecting Edit≫Undo Move to place the label back above the digital control.

4. Reposition the stop button, the knob, and the thermometer to the approximate locations shown in Figure 3.22.
5. Add labels to the knob and to the thermometer. To accomplish this task, pop up on the object and choose Visible Items≫Label. When the label box appears, type in the desired text. In this case, we want to label the knob Temperature variation and the thermometer Average Temperature.
6. Move the digital indicator associated with the thermometer to a location near the bottom of the thermometer. Then reposition the thermometer label so that it is better centered above the thermometer (see Figure 3.22).
7. In the next step, we will reformat the waveform graph.
   (a) To remove the waveform graph legend, pop up on the waveform graph and select Visible Items≫Plot Legend. On the pull-down menu, the check mark next to **Plot Legend** will disappear.
   (b) To remove the waveform graph palette, pop up on the waveform graph and select Visible Items≫Graph Palette. On the pull-down menu, the check mark next to **Graph Palette** will disappear.
   (c) Remove the waveform x-axis scale by popping up on the waveform graph and choosing Visible Items≫X Scale. On the pull-down menu, the check mark next to **X Scale** will disappear.
   (d) Add a label to the wavefrom graph—label the object Temperature data history. Select the owned label and move it to the bottom of the waveform graph, as illustrated in Figure 3.22.
8. Align the stop button and the thermometer.
   (a) Select both the stop button and the thermometer using the **Positioning** tool. Pick a point somewhere to the upper left of the stop button and drag the dashed box down until it encloses both the stop button and the thermometer. Upon release of the mouse button, both objects will be surrounded by moving dashed lines.
   (b) Click on the **Alignment** ring and choose **Left Edges**. The stop button and the thermometer will then align to the left edges. If the objects

appear not to move at all, this indicates that they were essentially aligned to the left edges already.

9. Align the stop button, the knob, and the digital control horizontally by choosing **Vertical Centers** axis from the **Alignment** ring in the toolbar. Remember to select all three objects beforehand!

10. Space the stop button, the knob, and the digital control evenly by choosing the **Horizontal Centers** axis from the **Distribution** ring in the toolbar. Again, remember to select all three objects beforehand!

11. Change the color of the stop button.

    (a) Using the **Coloring** tool, pop up on the stop button to display the color palette.

    (b) Choose a color from the palette. The object will assume the last color you selected. In the Editing Done VI you will see that a dark blue color was selected—you can choose a color that you like.

12. Change the color of the knob. Using the **Coloring** tool, pop up on the knob to display the color palette and then choose the desired color from the color palette. In the Editing Done VI you will find that a red/pink color was selected—you can choose a color that suits you.

13. Change the color of the waveform graph. In the Editing Done VI you will see that a gold color was selected.

14. Change the color of the thermometer bulb. Using the **Coloring** tool, pop up on the thermometer bulb (at the bottom of the thermometer) to display the color palette and then choose the desired color. In the Editing Done.vi you will find that a green color was selected.

15. Change the font style of the owned labels. Use the **Labeling** tool to highlight each of the labels and then select Style≫Bold from the **Font** ring. Do this for three of the owned labels: thermometer, knob, and digital control.

At this point, the front panel of the Editing VI should look very similar to the front panel shown in Figure 3.22.

The block diagram for the Editing.vi is shown in Figure 3.24 after the editing of the front panel is finished—but before wiring the block diagram. We can now wire together the various objects to obtain a working VI. Notice that several additional objects are on the block diagram: a uniform white noise subVI, a mean subVI that computes the mean (or average) of a signal, a While Loop, and an Add function. In later chapters, you will learn to use While Loops and learn how to use preexisting subVIs packaged with LabVIEW. For instance, the Uniform

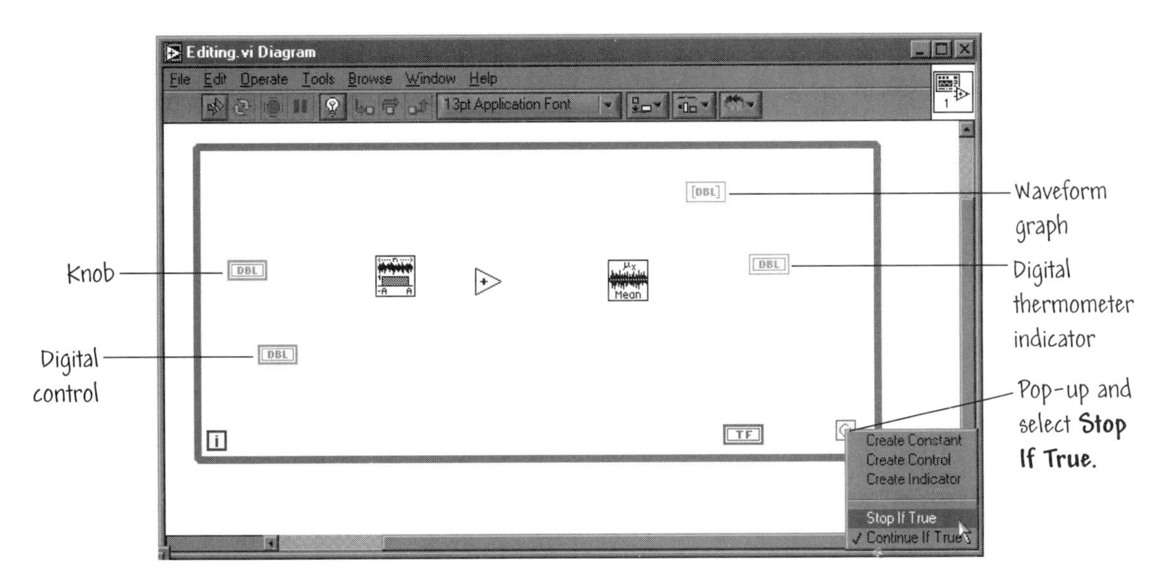

FIGURE 3.24
The block diagram for the **Editing.vi**—before wiring.

White Noise.vi is located on the **Functions** palette under the Analyze≫Signal Processing≫Signal Generation subpalette.

Go ahead and wire the Editing.vi block diagram so that it looks like the block diagram shown in Figure 3.25. If you run into difficulties, you can open and

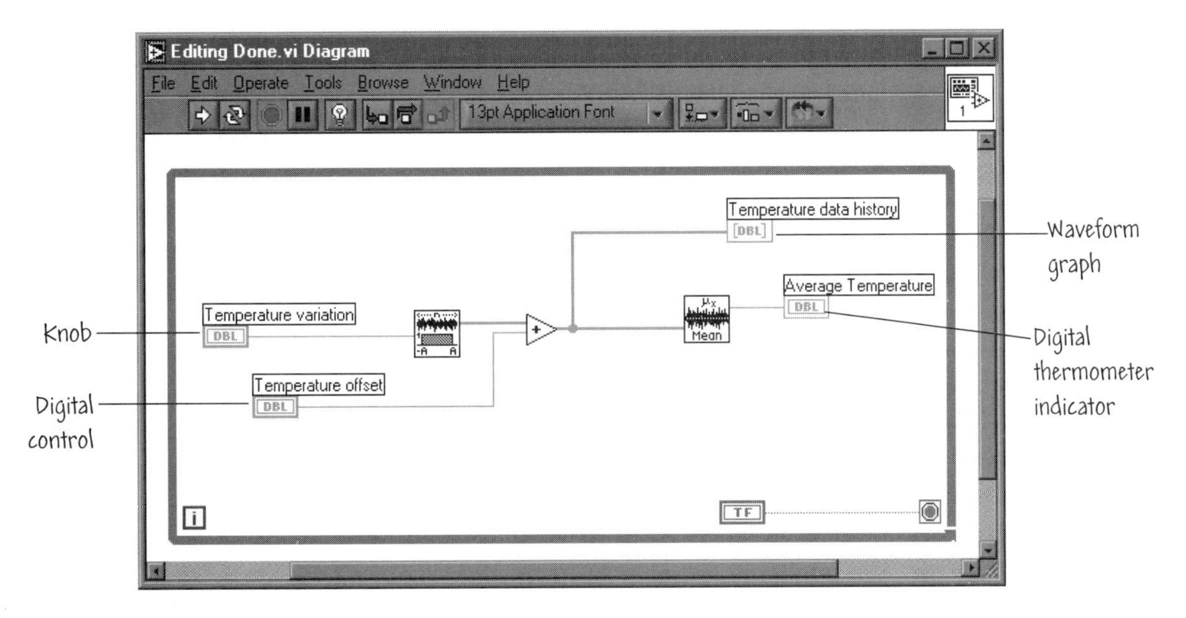

FIGURE 3.25
The block diagram for the **Editing Done.vi**—after wiring.

examine the block diagram for Editing Done.vi (which can be found in Chapter 3 of the Learning directory).

A few wiring tips:

1. To wire the objects together with the **Wiring** tool, click and release on the source terminal and drag the **Wiring** tool to the destination terminal. When the destination terminal is blinking, click and release the left mouse button.
2. To identify terminals on the Add function, pop up on the icon and select Visible Items≫Terminal to see the connector. When the wiring is finished, pop up again on the function and choose Visible Items≫Terminal to show the icon once again.
3. To bend the wires as you connect two objects, click with left mouse button on the bend location with the **Wiring** tool.

After you have finished wiring the objects together, switch to the front panel by selecting **Show Panel** from the **Windows** menu. Use the **Operating** tool to change the value of the front panel controls. Run the VI by clicking on the **Run** button on the toolbar.

The Average Temperature indicator should be approximately the value that you select as the Temperature offset. The amount of variation in the temperature history as shown on the Temperature data history waveform graph, should be about the same as the setting on the Temperature variation knob.

When you are finished editing and experimenting with your VI, save it by selecting **Save** from the **File** menu. Remember to save all your work in the Users Stuff folder in Learning directory. Close the VI by selecting **Close** from the **File** menu.

◆

3.2 DEBUGGING TECHNIQUES

In this section we discuss LabVIEW's basic debugging elements, which provide an effective programming debugging environment. Most features commonly associated with good interactive debugging environments are provided—and in keeping with the spirit of graphical programming, the debugging features are accessible graphically. Execution highlighting, single stepping, breakpoints, and probes helps debug your VIs easily by tracing the flow of data through the VI. You can actually watch your program code as it executes!

3.2.1 Finding Errors

When your VI cannot compile or run due to a programming error, a **Broken Run** button appears on the toolbar. Programming errors typically appear during VI development and editing stages and remain until you properly wire all the objects in the block diagram. You can list all your program errors by clicking on the **Broken Run** button. An information box called **Error List** appears listing all the errors. This box is shown in Figure 3.26 for the Editing Done VI with a broken wire.

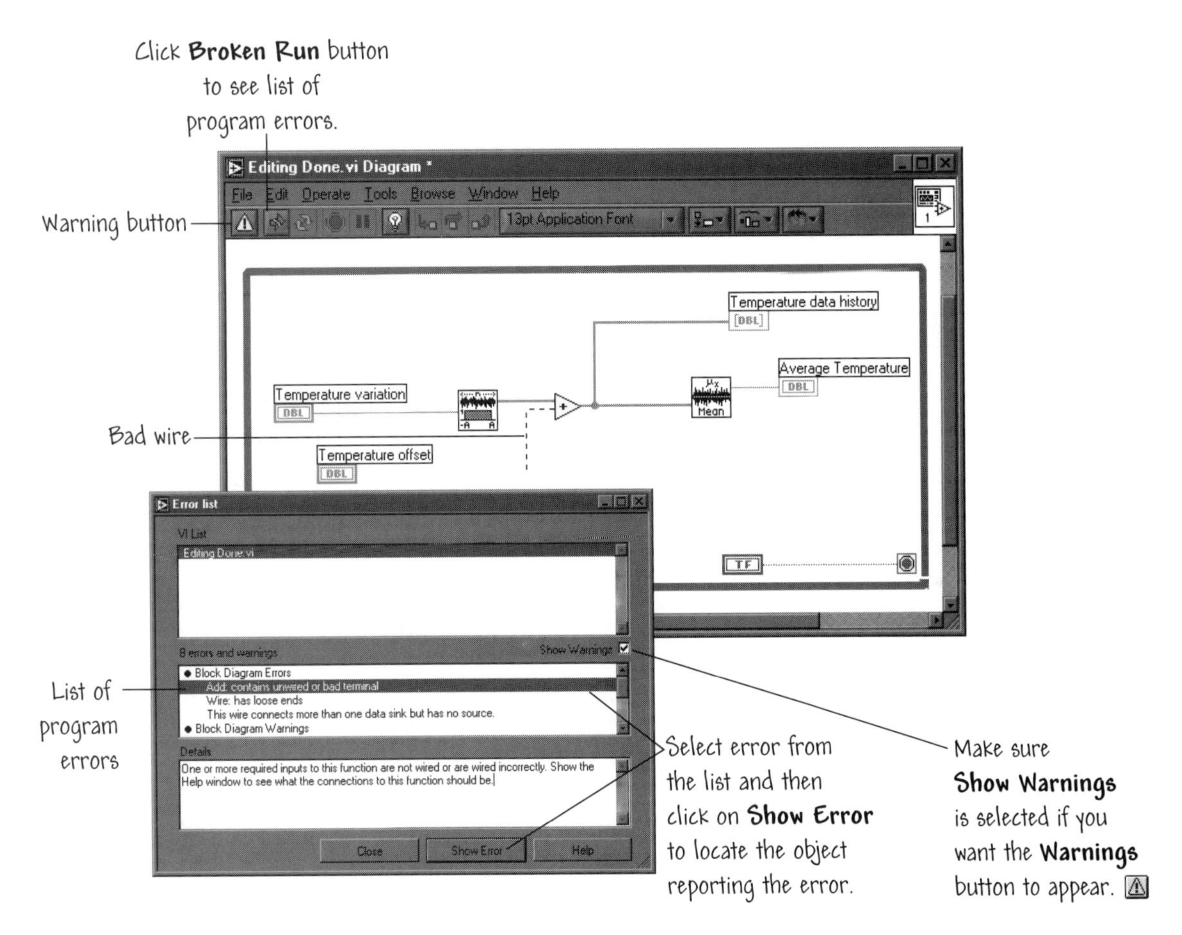

FIGURE 3.26
Locating program errors.

Warnings make you aware of potential problems when you run a VI, but they do not inhibit program execution. If you want to be notified of any warnings, click the **Show Warning** checkbox in the **Error List** dialog. A warning button then appears on the toolbar whenever a warning condition occurs.

If your program has any errors that prevent proper execution, you can search for the source of a specific error by selecting the error in the **Error List** (by clicking on it) and then clicking on **Show Error** (lower right-hand corner of the **Error**

List dialog box). This process will highlight the object on the block diagram that reported the error, as illustrated in Figure 3.26. Double clicking on an error in the error list will also highlight the object reporting the error.

Some of the most common reasons for a VI being broken during editing are:

1. A function terminal requiring an input is unwired. For example, an error will be reported if you do not wire all inputs to arithmetic functions.
2. The block diagram contains a broken wire because of a mismatch of data types or a loose, unconnected end.
3. A subVI is broken.

3.2.2 Highlight Execution

You can animate the VI block diagram execution by clicking on the **Highlight Execution** button located in the block diagram toolbar, shown in Figure 3.27.

For debugging purposes, it is helpful to see an animation of the VI execution in the block diagram, as illustrated in Figure 3.28. When you click on the **Highlight Execution** button, it changes to a bright light to indicate that the data flow will be animated for your visual observation when the program executes. Click on the **Highlight Execution** button at any time to return to normal running mode.

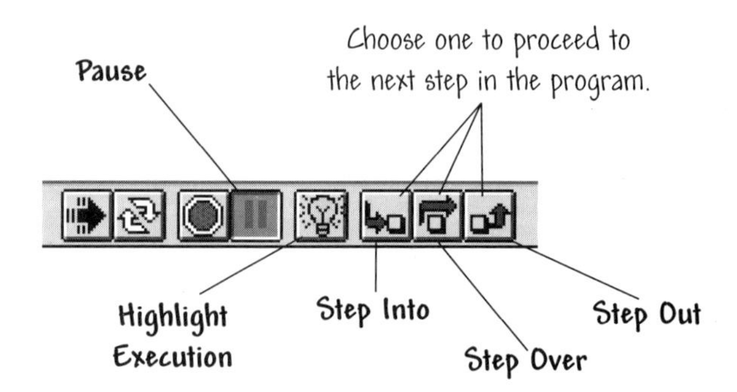

FIGURE 3.27
The **Highlight Execution** and step buttons located on the toolbar.

Execution highlighting is commonly used with single-step mode (more on single stepping in the next section) to trace the data flow in a block diagram in an effort to gain an understanding of how data flows through the block diagram. Keep in mind that when you utilize the highlight execution debugging feature, it greatly reduces the performance of your VI—the execution time increases significantly! The data flow animation shows the movement of data from one node to another using "bubbles" to indicate data motion along the wires. This process

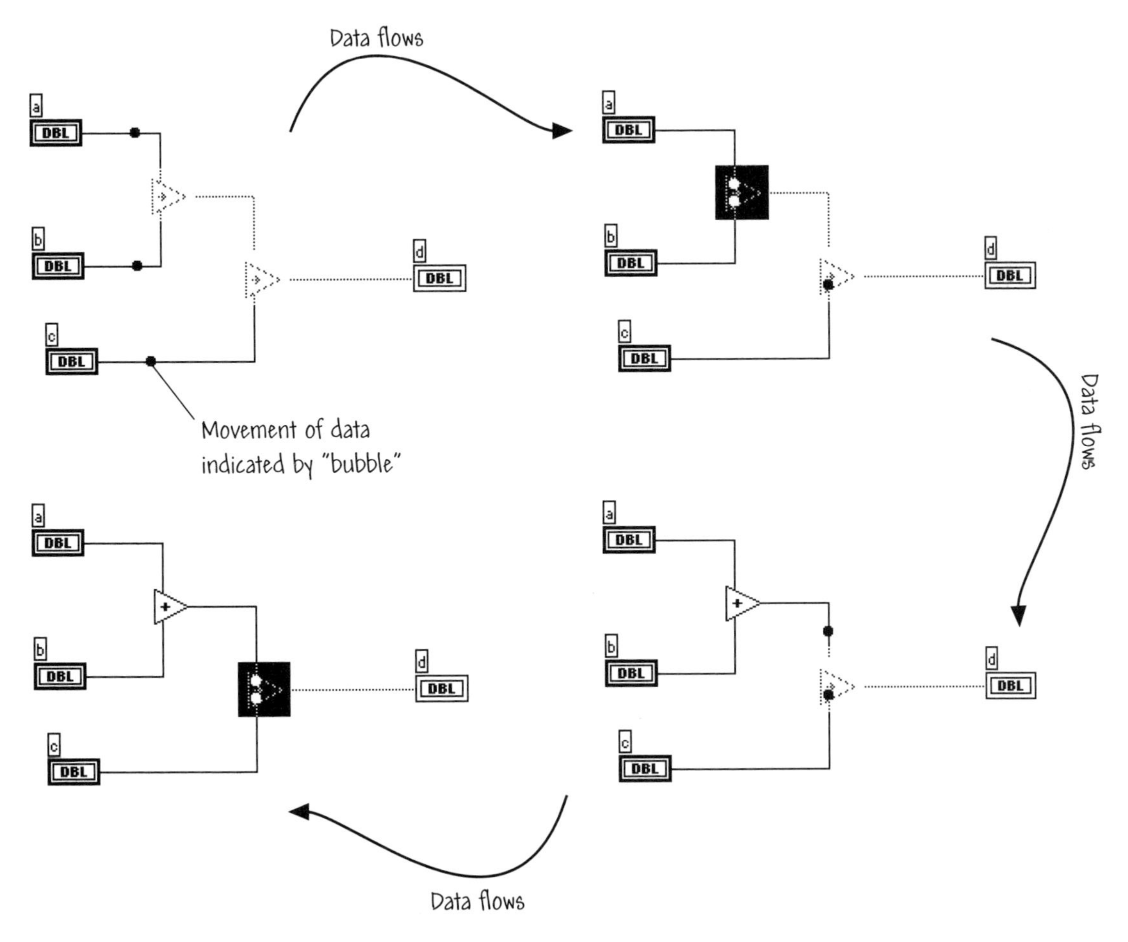

FIGURE 3.28
Using the highlight execution mode to watch the data flow through a VI.

is illustrated in Figure 3.28. Additionally, in single-step mode, the next node to be executed blinks until you click on the single-step button.

3.2.3 Single Stepping Through a VI and Its subVIs

For debugging purposes, you may want to execute a block diagram node by node. This is known as single stepping. To run a VI in single-step mode, press any of the debugging step buttons on the toolbar to proceed to the next step. The step buttons are shown on the toolbar in Figure 3.27. The step button you press determines where the next step executes. You click on either the **Step Into** or **Step Over** button to execute the current node and proceed to the next node. If the node is a structure (such as a While Loop) or a subVI, you can select the **Step Over** button to execute the node, but not single step through the node. For example, if the node is a subVI and you click on the **Step Over** button, you execute the subVI and proceed to the next node, but cannot see how the subVI node executed internally. To single step through the subVI you would select the **Step Into** button.

Click on the **Step Out** button to finish execution of the block diagram nodes or finish up the single-step debugging session. When you press any of the step buttons, the **Pause** button is pressed as well. You can return to normal execution at any time by releasing the **Pause** button.

If you place your cursor over any of the step buttons, a tip strip will appear with a description of what the next step will be if you press that button.

You might want to use highlight execution as you single-step through a VI, so that you can follow data as it flows through the nodes. In single-step mode and highlight execution mode, when a subVI executes, the subVI appears on the main VI diagram with either a green or red arrow in its icon, as illustrated in Figure 3.29. The diagram window of the subVI is displayed on top of the main VI diagram. You then single step through the subVI or let it complete executing.

You can save a VI without single stepping or highlight execution capabilities. This compiling method typically reduces memory requirements and increases performance by 1–2 %. To do this, pop up in the icon pane (upper-right corner

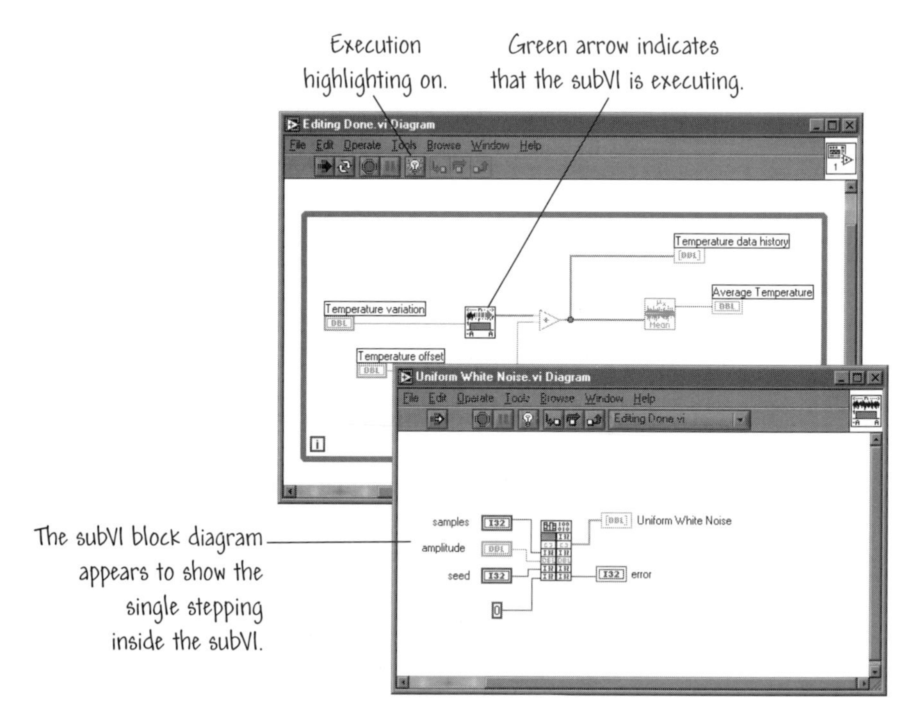

FIGURE 3.29
Single stepping into a subVI with execution highlighting selected.

of the front panel window) and select **VI Properties** from the **File** menu. As in Figure 3.30, from the **Execution** menu, deselect the **Allow Debugging** option to hide the **Highlight Execution** and **Single Step** buttons.

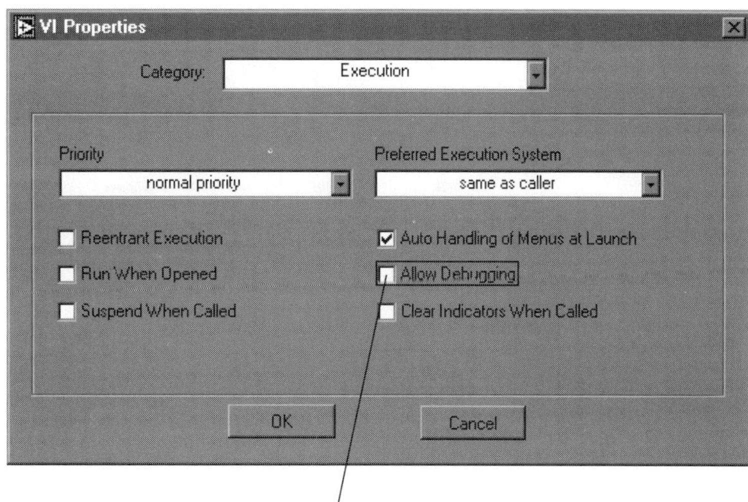

FIGURE 3.30
Turning off the debugging options using the **VI Setup** pop-up menu.

3.2.4 Breakpoints and Probes

You may want to halt execution (set **breakpoints**) at certain locations of your VI (for example, subVIs, nodes, or wires). Using the **Breakpoint** tool, click on any item in the block diagram where you want to set or clear a breakpoint. Breakpoints are depicted as red frames for nodes and red dots for wires.

You use the **Probe** tool to view data as it flows through a block diagram wire. To place a **probe** in the block diagram, click on any wire in the block diagram where you want to place the probe. Probes are depicted as numbered yellow boxes with associated pop-up windows in which the values passing through the wire at runtime are displayed. You can place probes all around the block diagram to assist in the debugging process.

Practicing with Debugging

A nonexecutable VI—named Debug.vi— has been developed for you to debug and to fix. You will get to practice using the single-step and execution highlighting modes to step through the VI, and to insert breakpoints and probes to regulate the program execution and to view the data values.

1. Open the Debug VI in Chapter 3 of the Learning directory by choosing **Open** from the **File** menu. (Remember—If you previously closed all open VIs, you must select the **Open VI** button from the startup window). Notice the **Broken Run** button in the toolbar indicating the VI is not executable.
2. Switch to the block diagram by choosing **Show Diagram** from the **Windows** pull-down menu. You should see the block diagram shown in Figure 3.31.

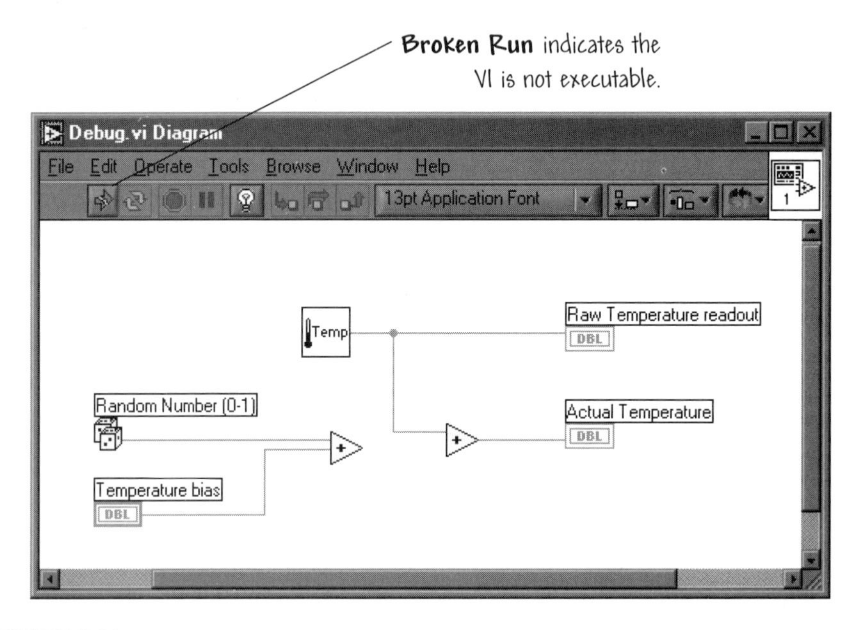

FIGURE 3.31
Block diagram showing the debugging exercise.

(a) The **Random Number (0-1)** function (represented by the two die) can be found in the **Numeric** subpalette and returns a random number between zero and one.

(b) The Add function (which can also be found in the **Numeric** subpalette) adds the random number to a bias number represented by the variable Temperature bias, which accepts input by the user on the front panel.

(c) The subVI Temp (which can be found in the **Tutorial** subpalette) simulates acquiring temperature data one value at a time from a data acquistion board. It uses the Demo Read Voltage subVI to acquire the simulated data. As you single step through the code, you will see the program execution inside both subVIs.

(d) The second Add function adds the simulated temperature data point to the random number plus the bias number to create a variable named Actual Temperature.

3. Investigate the source of the programming error by clicking on the **Broken Run** button to obtain a list of the programming errors. You should find one error—**Add: contains unwired or bad terminal**.
4. Highlight the error in the **Error List** and then click on **Show Error** to locate the source of the error within the block diagram. You should find that the second Add function (the one on the right side) is shown to be the source of the programming error.
5. Fix the error by properly wiring the two Add functions together. Once this step is successfully completed, the **Broken Run** button should reappear as the **Run** button.
6. With the **Operating** tool, select **Highlight Execution** and then run the VI by pressing the **Continuous Run** button. You can watch the data flow through the code! Returning to the front panel, you will see the waveform chart update as the new temperature is calculated and plotted. Notice that the simulation runs very slowly in highlight execution mode. Click the **Highlight Execution** button off to see everything move much faster!
7. Terminate the program execution by clicking on the **Abort Execution** button.
8. Enable single stepping by clicking on one of the step buttons. You can enable **Highlight Execution** if you want to see the data values as they are computed at each node.
9. Use the step buttons to single step through the program as it executes. Remember that you can let the cursor idle over the step buttons and a tip strip will appear with a description of what the next step will be if you press that button. Press the **Pause** button at any time to return to normal execution mode.
10. Enable the probe by popping up on the wire connecting the two Add functions and selecting **Probe**. Continue to single step through the VI and watch how the probe displays the data carried by the wire. Alternatively, you could have used the **Probe** tool to place the probe on the wire. Try placing a probe on the block diagram (say between the Temp subVI and the Add function) using the **Probe** tool.
11. Place a few more probes around the VI and repeat the single-stepping process to see how the probes display the data.
12. When you are finished experimenting with the probes, close all open probe windows.
13. Set a breakpoint by selecting the **Breakpoint** tool from the **Tools** palette and clicking on the wire between the two Add functions. You will notice that a red ball appears on the wire indicating that a breakpoint has been set at that location.
14. Run the VI by clicking on the **Continuous Run** button. The VI will pause each time at the breakpoint. To continue VI execution, click on the **Pause**

button. It helps to use highlight execution when experimenting with the breakpoints; otherwise the program executes too fast to easily observe the data flow.

15. Terminate the program execution by pressing the **Abort Execution** button. Remove the breakpoint by clicking on the breakpoint (that is, on the red ball) with the **Breakpoint** tool.
16. Save the working VI by selecting **Save** from the **File** menu. Remember to save the working VI in the folder Users Stuff within the Learning directory. Close the VI and all open windows by selecting **Close** from the **File** menu.

◆

3.3 A FEW SHORTCUTS

Frequently used menu options have equivalent command key shortcuts. For example, to save a VI you can choose **Save** from the **File** menu, or press the control key equivalent <ctrl-S> **(Windows)** or <command-S> **(Mac)**. Some of the main key equivalents are shown in Table 3.1. The shortcut access to the **Tools** palette on the front panel and block diagram is given by:

- **Windows**—Press <shift> and the right mouse button
- **Macintosh**—Press <Command-shift> and the mouse button

To rotate through common tools in the tool palette, use the <Tab> key. This is a quick way to change tools. For example, suppose you need the **Positioning** tool, but have the **Operating** tool selected. Press the <Tab> key once to rotate the selected tool to the **Positioning** tool. Also, in the front panel window, pressing

TABLE 3.1 Frequently used command key shortcuts.

Windows	Macintosh	Function
<Ctrl-S>	<Command-S>	Save a VI
<Ctrl-R>	<Command-R>	Run a VI
<Ctrl-E>	<Command-E>	Toggle between the front panel and the block diagram
<Ctrl-H>	<Command-H>	Toggle the **Context Help** window on and off
<Ctrl-B>	<Command-B>	Remove all bad wires
<Ctrl-W>	<Command-W>	Close the active window
<Ctrl-F>	<Command-F>	Find objects and VIs

the space bar toggles between the **Positioning** tool and the **Operating** tool. In the block diagram window, pressing the space bar toggles between the **Positioning** tool and the **Wiring** tool.

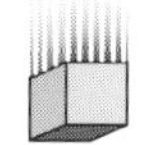

BUILDING BLOCK

3.4 BUILDING BLOCKS: MEASURING VOLUME

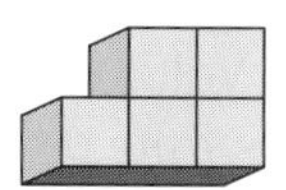

Two common measuring units for volume are liters and gallons. In this building block exercise you will construct and debug a VI that simulates reading a volume measurement in user-selectable units of liters or gallons.

The front panel and block diagram are shown in Figure 3.32. Using the figure as a guide, construct and debug a VI that simulates a volume measurement and

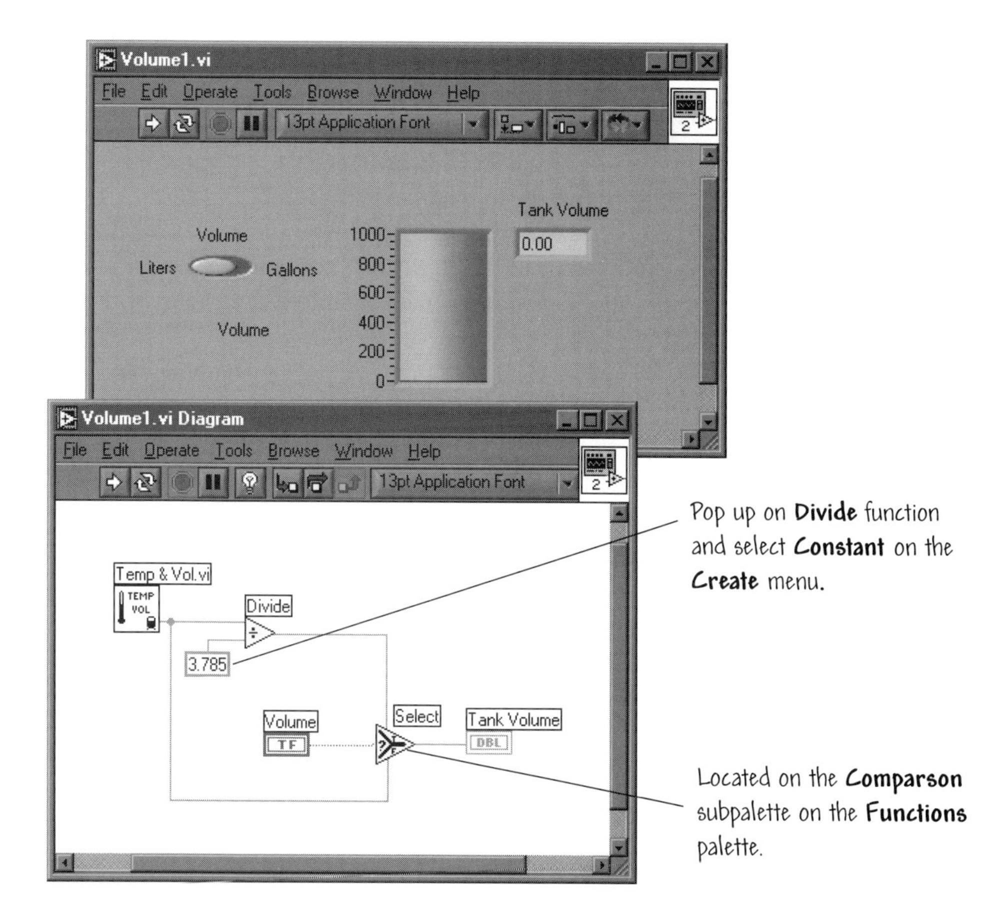

FIGURE 3.32
Front panel and block diagram for a volume measuring system.

allows the user to select the units of the measurement as either liters or gallons. The VI utilizes the subVI Temp & Vol to simulate the measurement of volume. The subVI is located in the Building Blocks folder found in the Learning directory.

Practice using the various debugging tools as you develop the VI— highlight execution, the step buttons, probes, and breakpoints. Also, notice that you will have to edit the front panel to add free labels to the horizontal switch indicating Liters on the left hand-side and Gallons on the right hand-side. Both the horizontal switch and the tank need to be labeled Volume and Tank Volume, respectively. The vertical scale of the tank should be changed to vary from 0 to 1000 (the default is 0 to 10).

On the block diagram, use the **Alignment** pull-down menu to organize the objects in a pleasing arrangement for ease in visualization. Remember that for more complex VIs it will become increasingly difficult to debug the code if the VI objects are arranged haphazardly with crossing wires and so forth.

When you are satisfied that your VI is wired properly and ready to run, begin the execution of the VI by choosing **Run**. With the **Operating** tool, select Liters and run the VI. Then, select Gallons and run the VI again. To view the process in a continuous run mode, select **Continuous Run** and while the VI is executing, switch the output from liters to gallons, and back.

When you are done experimenting with your new VI, save it as Volume1.vi in the Users Stuff folder in the Learning directory. You will use this VI as a building block in later chapters—so make sure to save your work!

A working version of Volume1.vi can be found in the Building Blocks folder of the Learning directory.

3.5 RELAXING READING: INSTRUMENTATION AND CONTROL

On the question of engineering education, industry feedback is unequivocal: new engineering graduates need communication, teamwork, and project skills to be prepared for todays team-based work environment. An important component of a student's education is "hands-on" experience gained in laboratory classes. Students need to obtain experience with modern instrumentation, including both plug-in data acquisition (DAQ) boards and computer-controlled, stand-alone instruments.

Lake Superior State University is addressing these needs by integrating modern, computer-based instrumentation and data acquisition instruction into their course work. Data acquisition is introduced to students in beginning courses, and its use expanded in other courses, including dynamics, vibrations, control systems, and digital signal processing, as well as appropriate senior design projects.

The senior design projects involve students from electrical engineering, mechanical engineering, manufacturing engineering technology, and environmental engineering technology working in cross-discipline teams. The senior student teams deliver an industrial-quality product design or automated manufacturing process that meets industrial specifications while managing a real dollar budget. Recent projects have included a computer-based test unit to verify proper operation of prototype automotive components in a development lab, a system to acquire the force and moment sensor data on an automotive wheel, a system to acquire and analyze test data on automotive safety restraint systems, and a system to acquire sound levels from a parking brake tester where the laptop data acquisition acquired sound measurement data while controlling the programmable logic controller that sequenced the test.

One of the earlier projects involved using LabVIEW in the process of making design modifications to an electric pump and updating the workstation where the operator completes the final pump assembly and tests the pump. The design team developed a prototype test stand with data acquisition and developed the operator test sequence using LabVIEW. The VI is shown in Figure 3.33. The operator enters the specific model number for the pump and the LabVIEW program accesses the test specifications for the pump. The VI then controls the test sequence and measures pump parameters while guiding the operator through the assembly and testing process.

For more information on this application of LabVIEW in the classroom, read the *User Solutions* entitled "Using LabVIEW in Student Projects for

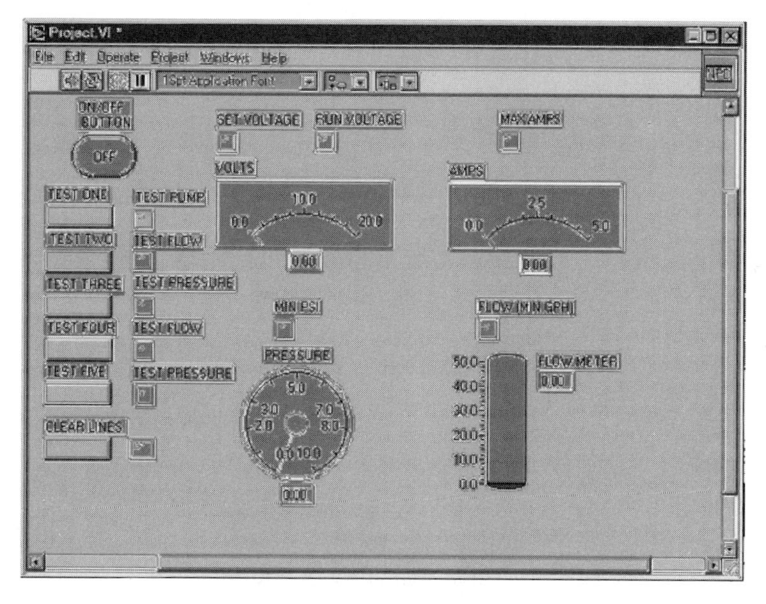

FIGURE 3.33
The VI for the pump stand.

Instrumentation and Control," which can be found on the National Instruments website, or contact:

Professor David McDonald
School of Engineering Technology and Mathematics
Lake Superior State University
Sault Ste. Marie, MI 49783
e-mail: dmcdonald@lakers.lssu.edu

3.6 SUMMARY

The subjects of editing and debugging VIs were the main subjects of this chapter. Just as you would edit a C program or debug a Fortran subroutine, you must know how to edit and debug VIs. We discussed how to create, select, delete, move, and arrange objects on the front panel and block diagrams. The important topic of selecting and deleting wires and locating and eliminating bad wires was also discussed. The program debugging topics covered included execution highlighting, and how to step into, through, and out of the code and how to use probes to view data as it flow through the code.

KEY TERMS

Breakpoint: A pause in execution used for debugging. You set a breakpoint by clicking a VI, node, or wire with the **Breakpoint** tool.

Breakpoint tool: Tool used to set a breakpoint on a node or wire.

Broken VI: A VI that cannot compile and run.

Coloring tool: Tool used to set foreground and background colors.

Execution highlighting: Debugging feature that animates the VI execution to illustrate the data flow within the VI.

Label: Text object used to name or describe other objects or regions on the front panel or block diagram.

Labeling tool: Tool used to create labels and enter text.

Operating tool: Tool used to enter data into controls and operate them—resembles a pointing finger.

Positioning tool: Tool used to move, select, and resize objects.

Probe: Debugging feature for checking intermediate values in a VI during execution.

Probe tool: Tool used to create probes on wires.

Resizing handles: Angled handles on the corners of objects that indicate resizing points.

EXERCISES

E3.1 Construct a VI to accept five numeric inputs, add them up and display the result on a gauge, and light up a round light if the sum of the input numbers is less than 8.0. The light should light up in green, and the gauge dial should be yellow. The VI in Figure E3.1 can be used as a guide.

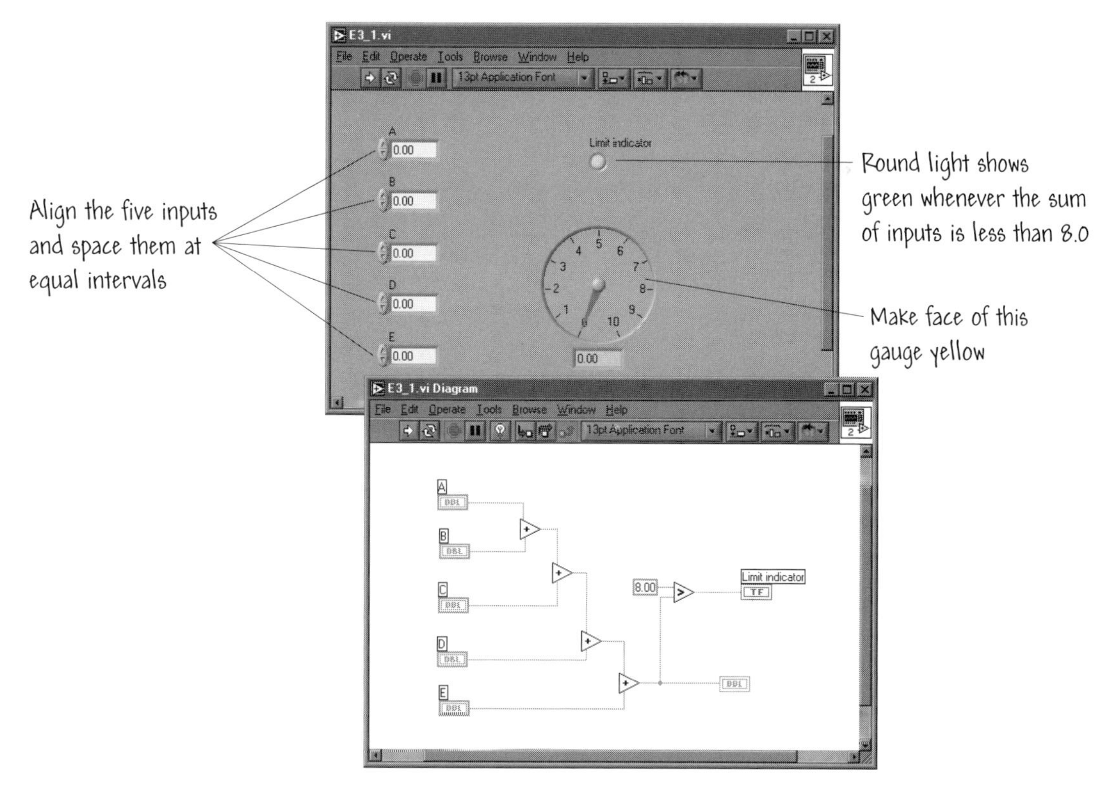

FIGURE E3.1
A VI to add five numeric inputs and light up a round LED if the sum is less than 8.0.

E3.2 Open Change to Indicator.vi in Learning\Exercises&Problems\Chapter 3. Click on the **Broken Run** button. Select the error, and click on the **Show Error** button. This wire connects more than one data source. The wire between the output of the Add function and the control Result is highlighted. Both the output and the control are "sources of data." You can only connect a data source to a data display. Pop up on the control Result and select **Change to Indicator**. The error disappears and you can run the VI. Input $x = 2$ and $y = 3$ and run the VI. Verify that the output result $= 5$.

E3.3 Open the virtual instrument Multiple Controls-1 Terminal.vi. You can find this VI in Learning\Exercises&Problems\Chapter 3. Click on the **Broken Run** button, select the error **This Wire Connects to More than One Data Source**, and click on the **Show Error** button. The wire from the two controls are highlighted. If you look closely, you'll see that both wires are going to the same terminal. That means two controls are defining the value of one input value. Delete the wire from the y control. Wire the y control to the other input terminal on the Add function. Both errors disappear and you can run the VI. Input $x = 1$ and $y = 4$ and run the VI. Verify that the output $x + y = 5$.

PROBLEMS

P3.1 Construct a VI that generates two random numbers (between 0 and 1) and displays both random numbers on meters. Label the meters Random number 1 and Random number 2, respectively. Make the face of one meter blue and the face of the other meter red. When the value of the random number on the red meter is greater than the random number on the meter with the blue face, have a square LED show green; otherwise have the LED show black. Run the VI several times and observe the results. On the block diagram select **Highlight Execution** and watch the data flow through the code.

P3.2 In this problem you will construct a stop light display. Create a dial control that goes from 0 to 2, with three LED displays: one green, one yellow, and one red. Have the VI turn the LED green when the dial is on 0, yellow when the dial is on 1, and red when the dial is on 2.

P3.3 Create a front panel that has 8 LED indicators and a vertical slider control that is an 8 bit unsigned integer. Display a digital indicator for the slider, and make sure that the LEDs are evenly spaced and aligned at the bottom. The problem is to turn the 8 LEDs into a binary (base 2) representation for the number in the slider. For example, if the slider is set to the number 10 (which in base 2 is $00001010 = 1 * (2^3) + 1 * (2^1)$), the LED's 1 and 3 should be on. To test your solution, check the number 131. LED's 0, 1 and 7 should be on since 131 is 10000011 in base 2.

P3.4 Complete the crossword puzzle.

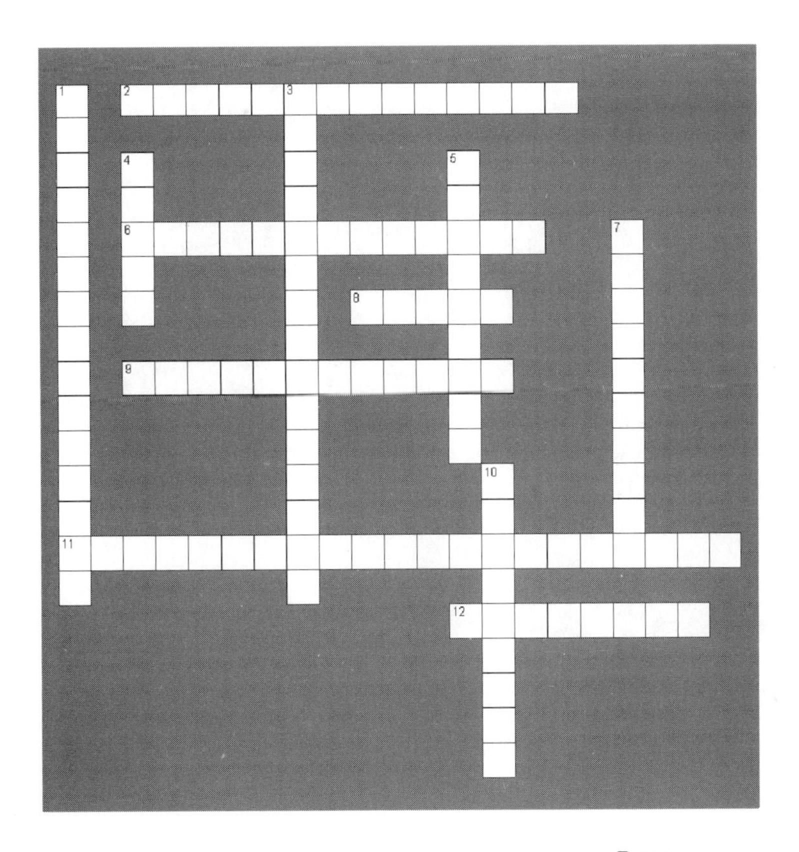

Across

2. Tool used to set a breakpoint on a node or wire.
6. A tool (resembling a pointing finger) used to enter data into controls and operate .
8. Text object used to name or describe other objects or regions on the front panel or block diagram.
9. Tool used to create labels and enter text.
11. Debugging feature that animates the VI execution to illustrate the data flow within the VI.
12. A VI that cannot compile and run.

Down

1. Angled handles on the corner of objects that indicate resizing points.
3. Tool used to move, select, and resize objects.
4. Debugging feature for checking intermediate values in a VI during execution.
5. Tool used to create probes on wires.
7. A pause in execution used for debugging. You set a breakpoint by clicking a VI, node, or wire with the Breakpoint tool.
10. Tool used to set foreground and background colors.

P3.5 Develop a VI that converts an input value in degrees to radians with four digits of precision. Refer to Figure P3.1 for help on changing the output display to four digits.

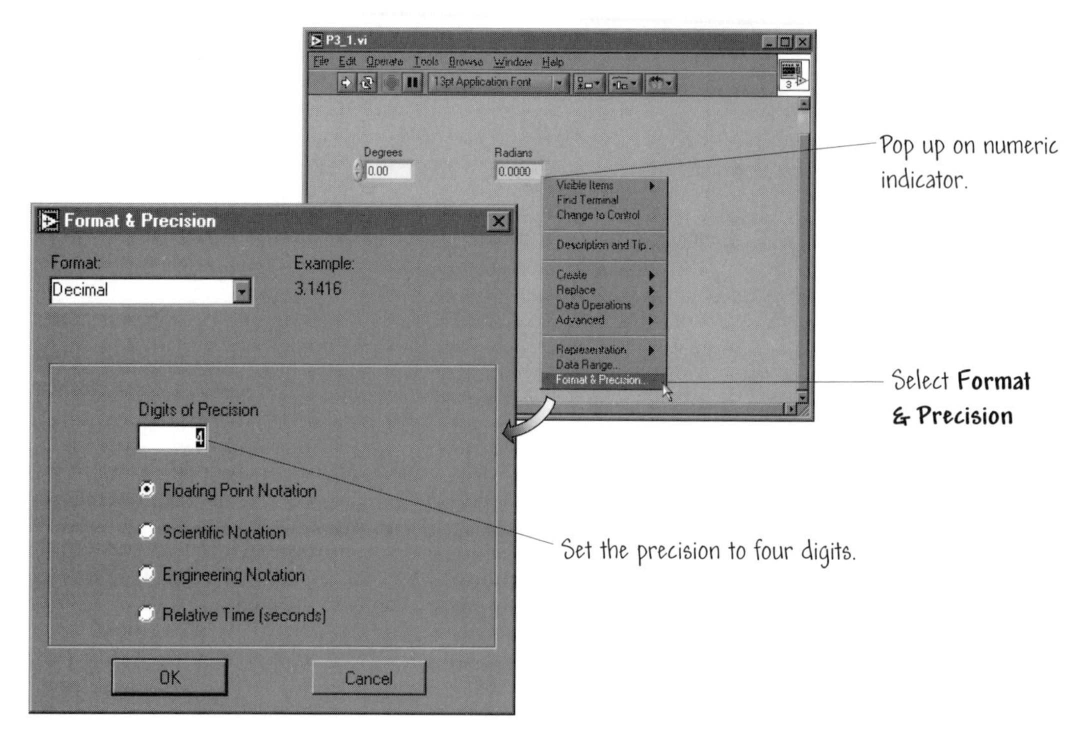

FIGURE P3.1
Converting degrees to radians with four digits of output precision.

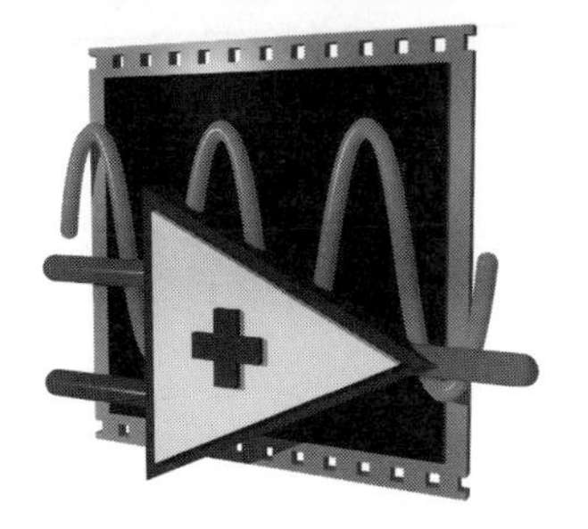

CHAPTER 4

SubVIs

In this chapter we learn to build subVIs. SubVIs are VIs used by other VIs—like subroutines. One of the keys to constructing successful VIs is understanding how to build and use subVIs. The hierarchical design of LabVIEW applications (that is, of virtual instruments) depends on the use of subVIs. We will discuss two basic ways to create and use a subVI: creating subVIs from VIs and creating subVIs from selections. The **Icon Editor** is presented as a way to personalize subVI icons so the information about the function of the subVI is apparent from visual inspection. The editor has a tool set similar to that of most common paint programs. The **Hierarchy Window** will be introduced as a helpful tool for managing the hierarchical nature of your programs.

GOALS

1. Learn to build and use subVIs.
2. Understand the hierarchical nature of VIs.
3. Practice with the **Icon Editor** and with assigning terminals.

4.1 WHAT IS A SUBVI?

It is important to understand and appreciate the hierarchical nature of virtual instruments. **SubVIs** are critical components of a hierarchical and modular VI that is easy to debug and maintain. A subVI is a stand-alone VI that is called by other VIs—that is, a subVI is used in the block diagram of a **top-level VI**. SubVIs are analogous to subroutines in text-based programming languages like C or Fortran, and the subVI node is analogous to a subroutine call statement. The **pseudocode** and block diagram shown in Figure 4.1 demonstrate the analogy between subVIs and subroutines. There is no limit to the number of subVIs you can use in

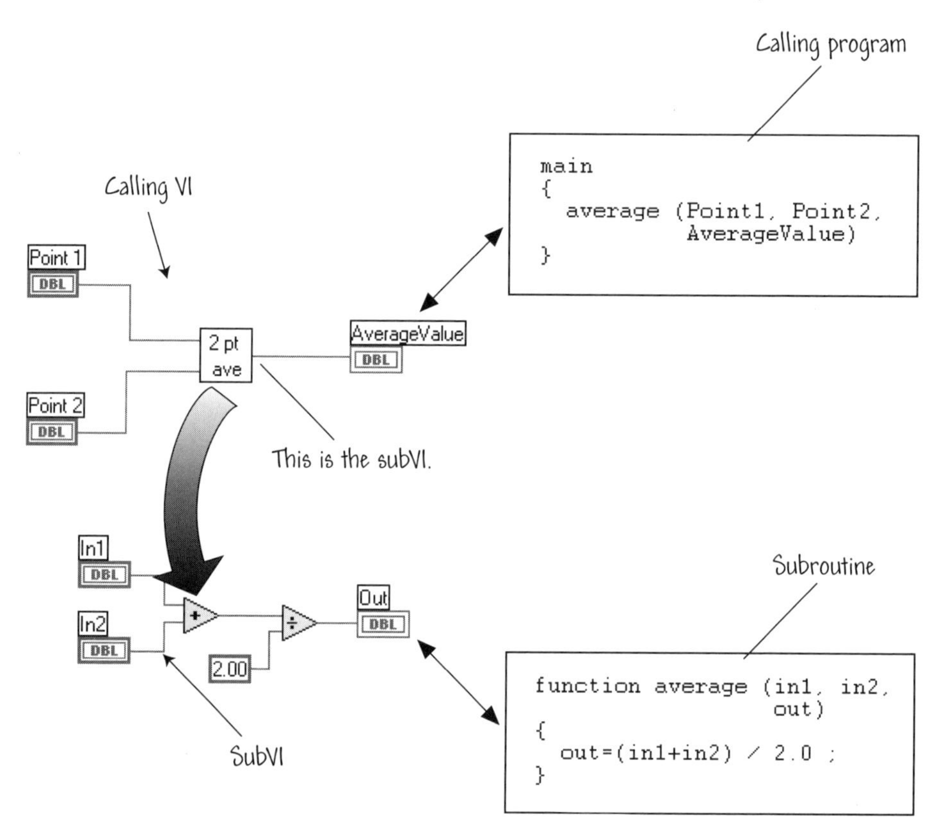

FIGURE 4.1
Analogy between subVIs and subroutines.

a calling VI. Using subVIs is an efficient programming technique in that it allows you to reuse the same code in different situations. The hierarchical nature of programming in G follows from the fact that you can call a subVI from within a subVI.

You can create subVIs from VIs, or create them from selections (selecting existing components of a VI and placing them in a subVI). When creating a new subVI from an existing VI, you begin by defining the inputs and outputs of the subVI, and then you "wire" the subVI connector properly. This allows calling VIs to send data to the subVI and to recieve data from the subVI. On the other hand, if an existing complex block diagram has a large number of icons, you may choose to group related **functions** and icons into a lower-level VI (that is, into a subVI) to maintain the overall simplicity of the block diagram. This is what is meant by using a modular approach in your program development.

4.2 REVIEW OF THE BASICS

Before moving on to the subject of subVIs, we present a brief review of some of the basics presented in the first three chapters as part of an exercise in constructing a VI. Your VI will ultimately be used as a subVI later in the chapter.

Everyone who follows baseball knows that keeping track of statistics is one of the main features of the game for players and fans alike. And one of the most important statistics is the batting average. In this section, we will build a VI that computes a batting average and later turn the VI into a subVI. As with most things in baseball, the calculation of a batting average, while conceptually simple, has many nuances that are not accounted for in our VI. The main aspects of calculating the batting average are retained, however, and the primary goal of learning about programming is enhanced.

If you are not familiar with the game of baseball, then view the following discussion as an exercise is constructing a VI that performs elementary mathematical calculations and is suitable for use as a subVI. It is not necessary to understand the game of baseball to understand how to construct a VI to compute batting averages!

During an appearance at the plate, we suppose that one of two things happens: the batter hits the ball, or the batter does not hit the ball. We consider first the situation where the batter does not hit the ball. In this case, one of three things occurs: (1) the batter strikes out, (2) the batter receives a base on balls (i.e., a walk), or (3) the batter is hit by the pitch. Even though the batter reaches first base safely, the walk and being hit by the pitch do not count as official appearances at the plate; hence they do not factor into the batting average. On the other hand, a strikeout is an official at-bat and reduces the batting average.

If the batter hits the ball, one of four situations can arise: (1) the batter gets a clean hit, (2) the batter reaches base due a fielding error by a defensive player, (3) the batter is put out (e.g., flies out or grounds out), or (4) the batter is out on a sacrifice (that is, a base runner advances on the play, but the batter is out nonetheless). A sacrifice does not count against the batting average, and only hitting safely increases the batting average. That means that hitting the ball does not necessarily improve the batting average—an error by the defensive player on a ball hit in play will actually reduce the batting average!

Let H denote the number of clean hits, K the number of strikeouts, E the number of errors committed by the defense, and O the number of fly-outs, ground outs, and other episodes wherein the batter is thrown out. The total number of times a batter appears at the plate is $N = H + K + E + O + S + BB + HP$; however, the number of walks (BB), hit by pitches (HP), and sacrifices (S) do not contribute to the number of official appearances at the plate; hence they do not contribute to the batting average computation. Therefore, according to our simplistic evaluation of how to compute the batting average, we utilize the formula

$$Batting\ Average = 1000\frac{H}{H + K + E + O}.$$

The factor of 1000 is incorporated into the batting average to account for the fact that, by tradition, a perfect batting average is 1000.

Your goal is to develop a VI to compute the batting average according to the formula above. The following list serves as a step-by-step guide. Feel free to deviate from the provided list and program according to your own style. As you proceed through this book you will find that the step-by-step lists will slowly disappear, and you will have to construct your VIs on your own (with help, of course!).

1. Open a new VI. Save the untitled VI as Batting Average.vi in the folder Users Stuff in the Learning directory. In this case you are saving the VI before any programming has actually occurred. This is a matter of personal choice, but saving a VI frequently during development may save you lots of rework if your system freezes up or crashes.
2. Add a digital **control** to the front panel and label it H. This will be where the number of hits is input.
3. Make three duplicates (or clones) of the digital control. Using the **Positioning** tool, select the digital control while pressing the <Ctrl> **(Windows)** or <Option> **(Macintosh)** and drag the digital control to a new location. When you release the mouse button, a copy labeled H 2 appears in the new location. Repeat this process until four digital controls are visible on the front panel.
4. Relabel the four digital controls: H, K, E, and O.
5. Place the digital controls in a column on the left side of the front panel. Using the **Alignment** tool, align the four digital controls by their left edges.
6. Add a digital **indicator** to the front panel and label it Batting Average. This will be where the batting average is displayed.
7. Switch to the block diagram. Arrange the control and indicator terminals so that the four control terminals appear aligned in a vertical column on the left side and the batting average indicator appears on the right side.

8. Program the batting average formula by wiring the block diagram. You can use the VI shown in Figure 4.2 as a guide.
9. The batting average should be rounded off and displayed as an integer value. To accomplish this in the VI, you need to change the format of the indicator. Pop up on the digital indicator and select **Representation**; then select **I32** as shown in Figure 4.3.
10. Once the block diagram has been wired, switch back to the front panel and input values into the various digital controls. For example, let $H = 66$, $K = 43$, $O = 98$, and $E = 10$.
11. Run the program with **Continuous Run** and select **Highlight Execution** to watch the data flow through the program. Change the various values of the input parameters to see how the batting average changes. When you are finished experimenting, you can stop the program execution.
12. Practice single stepping through the program using the step buttons on the block diagram.
13. Once the VI execution is complete, switch back to the front panel and pop up on the icon in the upper right corner. Select **Show Connector** from the menu. Observe the connector that appears—it should indicate four input terminals and one output terminal, as illustrated in Figure 4.4. We will be learning how to edit the icon and connector in the next section.

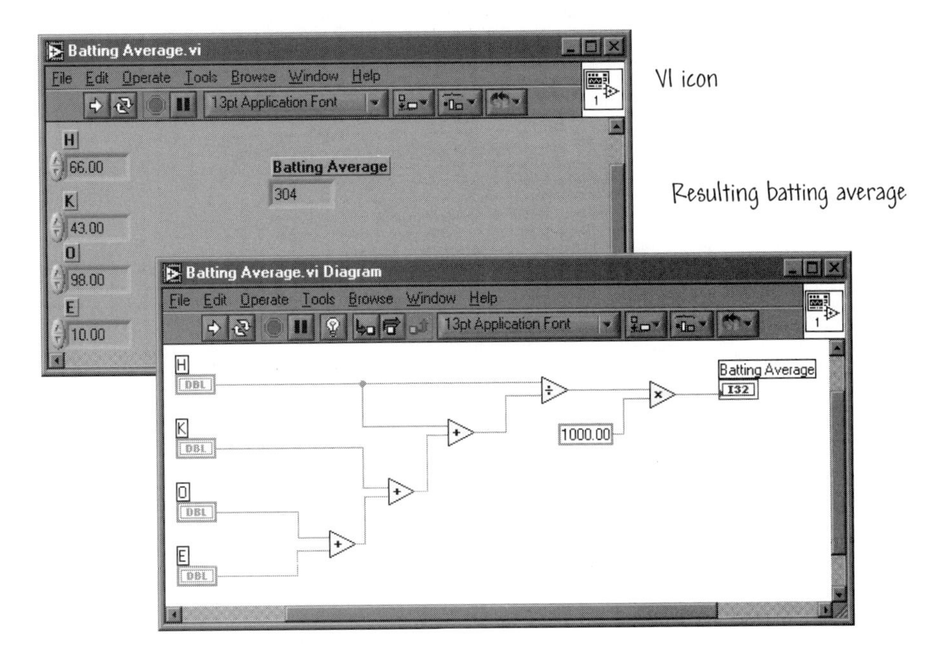

FIGURE 4.2
The Batting Average VI.

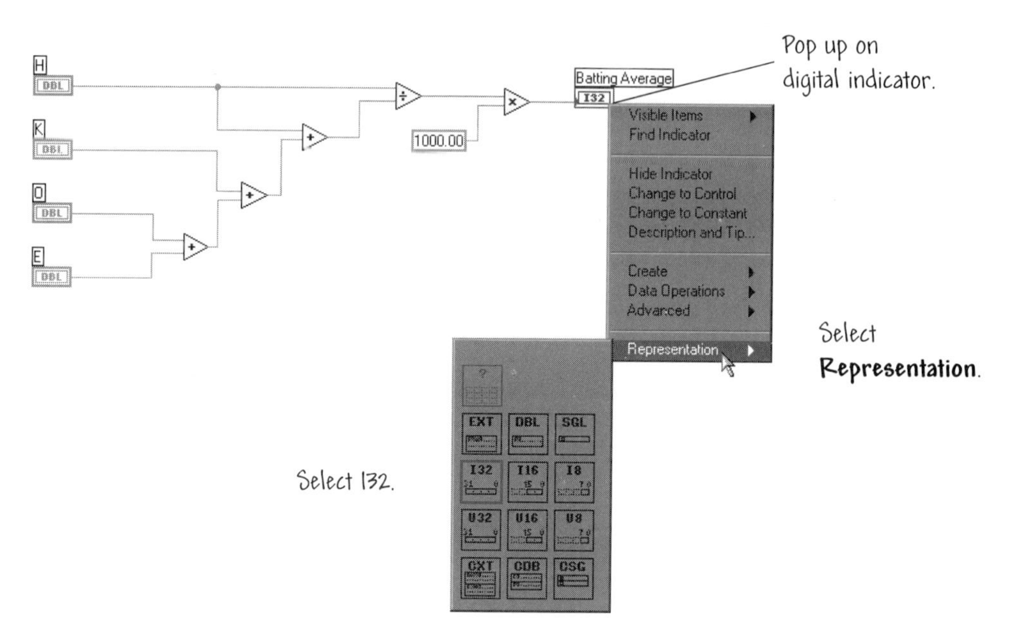

FIGURE 4.3
Changing the representation of the digital indicator to I32.

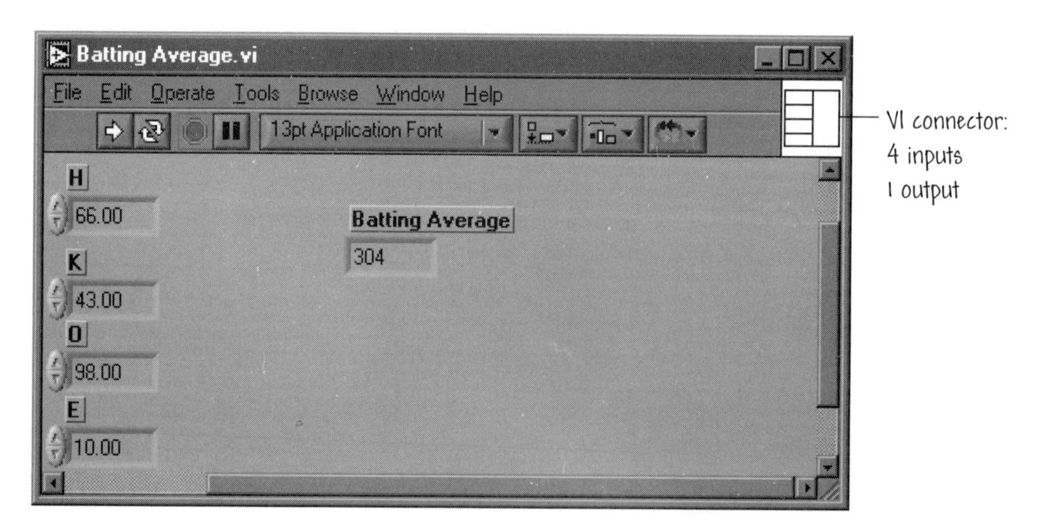

FIGURE 4.4
The Batting Average icon terminal.

14. Save the VI by selecting **Save** from the **File** menu.

This exercise provided the opportunity to practice placing objects on the front panel, wiring them on the block diagram, using editing techniques to arrange the objects for easy visual inspection of the code, and using debugging tools to verify proper operation of the VI. The Batting Average VI can be used as a subVI (requiring some effort to wire the VI connector properly), and that is the subject of the next sections.

A working version of this Batting Average VI can be found in the Chapter 4 *folder of the* Learning *directory. It is called* Batting Average.vi.

4.3 EDITING THE ICON AND CONNECTOR

A subVI is represented by an icon in the block diagram of a calling VI. The subVI also must have a connector with terminals properly wired to pass data to and from the top-level VI. In this section we discuss icons and connectors.

4.3.1 Icons

Every VI has a default icon displayed in the upper-right corner of the front panel and block diagram windows. The default icon is a picture of the LabVIEW logo and a number indicating how many new VIs you have opened since launching LabVIEW. You use the **Icon Editor** to customize the icon. To activate the editor, pop up on the default icon in the front panel and select **Edit Icon** as illustrated in Figure 4.5. You can also display the **Icon Editor** by double clicking on the default icon.

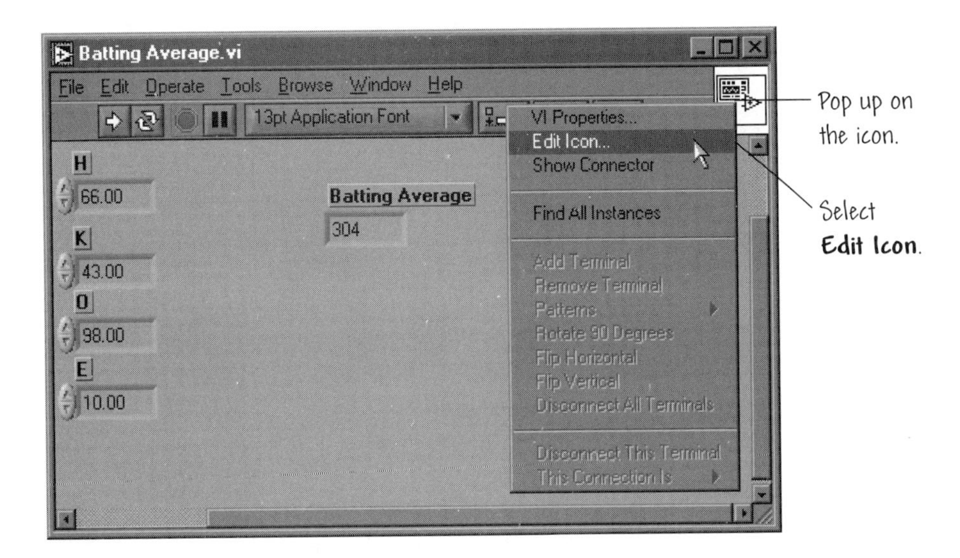

FIGURE 4.5
Activating the **Icon Editor**.

*You can only open the **Icon Editor** when the VI is not in the run mode. If necessary, you can change the mode of the VI by selecting **Change to Edit Mode** from the **Operate** menu.*

The **Icon Editor** dialog box is shown in Figure 4.6. You use the tools on the palette at the left of the dialog box to create the icon design in the pixel editing area. An image of the icon (in actual size) appears in a vertical column just right of center in the dialog box and to the right of the pixel editing area. Depending

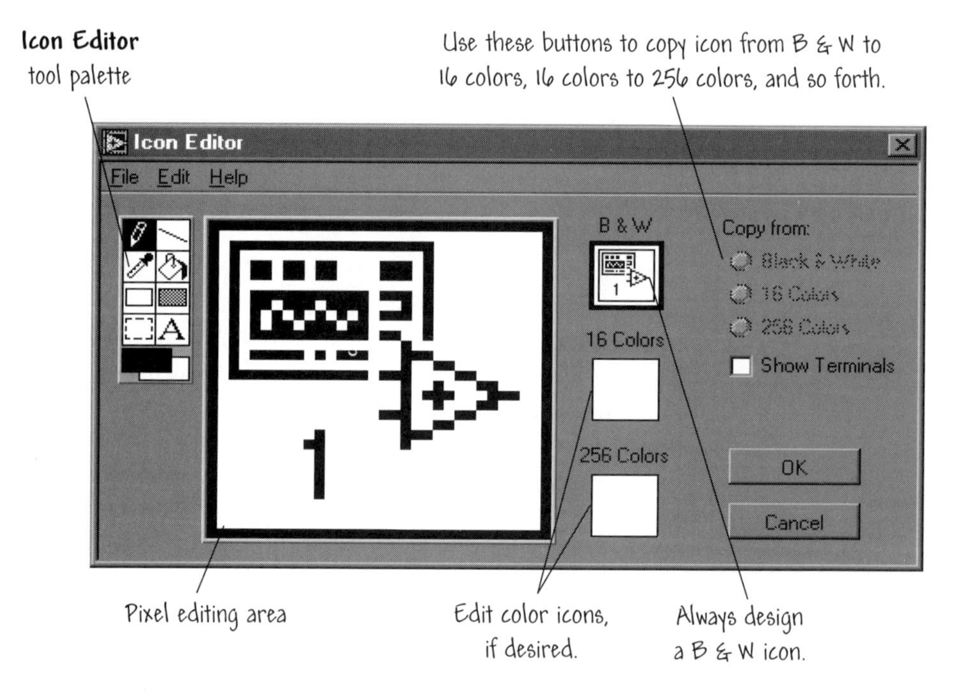

FIGURE 4.6
The **Icon Editor**.

on the type of monitor you are using, you can design and save a separate icon for B&W, 16-color, and 256-color mode. The editor defaults to **Black & White**, but you can click on one of the other color options to switch modes. To copy the icon from B&W to 16 colors, select the 16-color box and then click on the **Copy from: Black & White** button located at the upper right side of the **Icon Editor** dialog box. Similarly, you can use the appropriate **Copy from** menu items to copy icons from 16 colors to 256 colors, 16 colors to B&W, and so forth.

The tools in the **Icon Editor** palette to the left of the editing area perform many functions. If you have used any other paint program, you should be familiar with the tools available in the editor. Table 4.1 describes the various tools.

TABLE 4.1 Icon Editor tools.

	Draws and erases pixel by pixel.
	Draws straight lines. Use <shift> to restrict drawing to horizontal, vertical, and diagonal lines.
	Selects the foreground color from an element in the icon.
	Fills an outlined area with the foreground color.
	Draws a rectangle bordered with the foreground color. Double click on this tool to frame the icon in the foreground color.
	Draws a rectangle bordered with the foreground color and filled with the background color. Double click on this tool to frame the icon in the foreground color and fill it with the background color.
	Selects an area of the icon for moving, cloning, deleting, or performing other changes. Double click on this tool and press Delete on the keyboard to delete the entire icon at once.
A	Enters text into the icon. Double click on this tool to select a different font. Small fonts is commonly used as the fonts in icons.
	Displays the current foreground and background colors. Click on each to get a palette from which you can choose new colors.

The options at the right of the editing screen perform the following functions:

- **Show terminal**—Click on this option to display the terminal pattern of the connector overlayed on the icon.
- **OK**—Click on this button to save your drawing as the VI icon and return to the front panel window.
- **Cancel**—Click on this button to return to the front panel window without saving any changes.

You can use the **Icon Editor** menu to cut, copy, and paste images from and to the icon. When you paste an image and a portion of the icon is selected, the image is resized to fit into the selection.

As an exercise, open Batting Average.vi and edit the icon. Remember that you developed the Batting Average.vi at the beginning of this chapter—it should be saved in User Stuff. In case you did not save it before or cannot find it now, you can open the VI located in the Chapter 4 folder in the Learning directory. An example of an edited icon is shown in Figure 4.7. Can you replicate the new icon? Check out the Batting Average Edited.vi to see the edited icon shown in Figure 4.7.

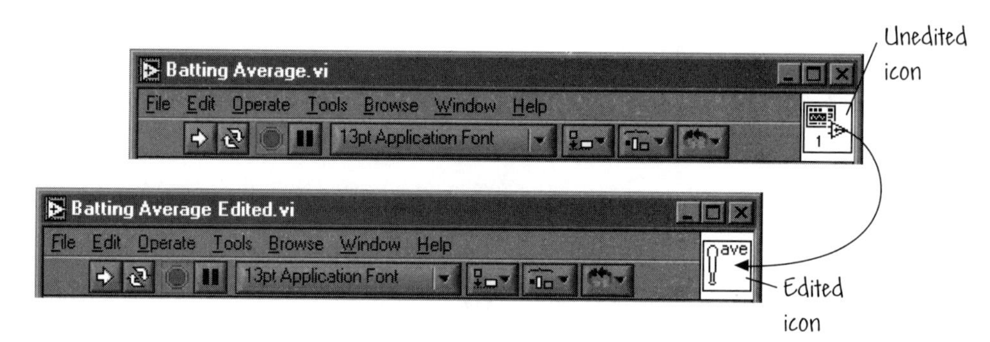

FIGURE 4.7
Editing the Batting Average.vi icon.

4.3.2 Connectors

The connector is a set of terminals that correspond to the VI controls and indicators. This is where the inputs and outputs of the VI are established so that it can be used as a subVI. A connector receives data at its input terminals and passes the data to its output terminals when the VI is finished executing. Each terminal corresponds to a particular control or indicator on the front panel. The connector terminals act like parameters in a subroutine parameter list of a function call.

If you use the front panel controls and indicators to pass data to and from subVIs, these controls or indicators need terminals on the connector pane. In this section, we will discuss how to define connections by choosing the number of terminals you want for the VI and assigning a front panel control or indicator to each of those terminals.

You can view and edit the connector pane from the front panel only.

To define a connector, you select **Show Connector** from the icon pane pop-up menu on the front panel window as shown in Figure 4.8. The connector replaces the icon in the upper-right corner of the front panel window. By default, LabVIEW displays a terminal pattern based on the number of controls and indicators on your front panel, as shown in Figure 4.8. Control terminals are on the left side of the connector pane, and indicator terminals are on the right. If desired, you can select a different terminal pattern for your VI.

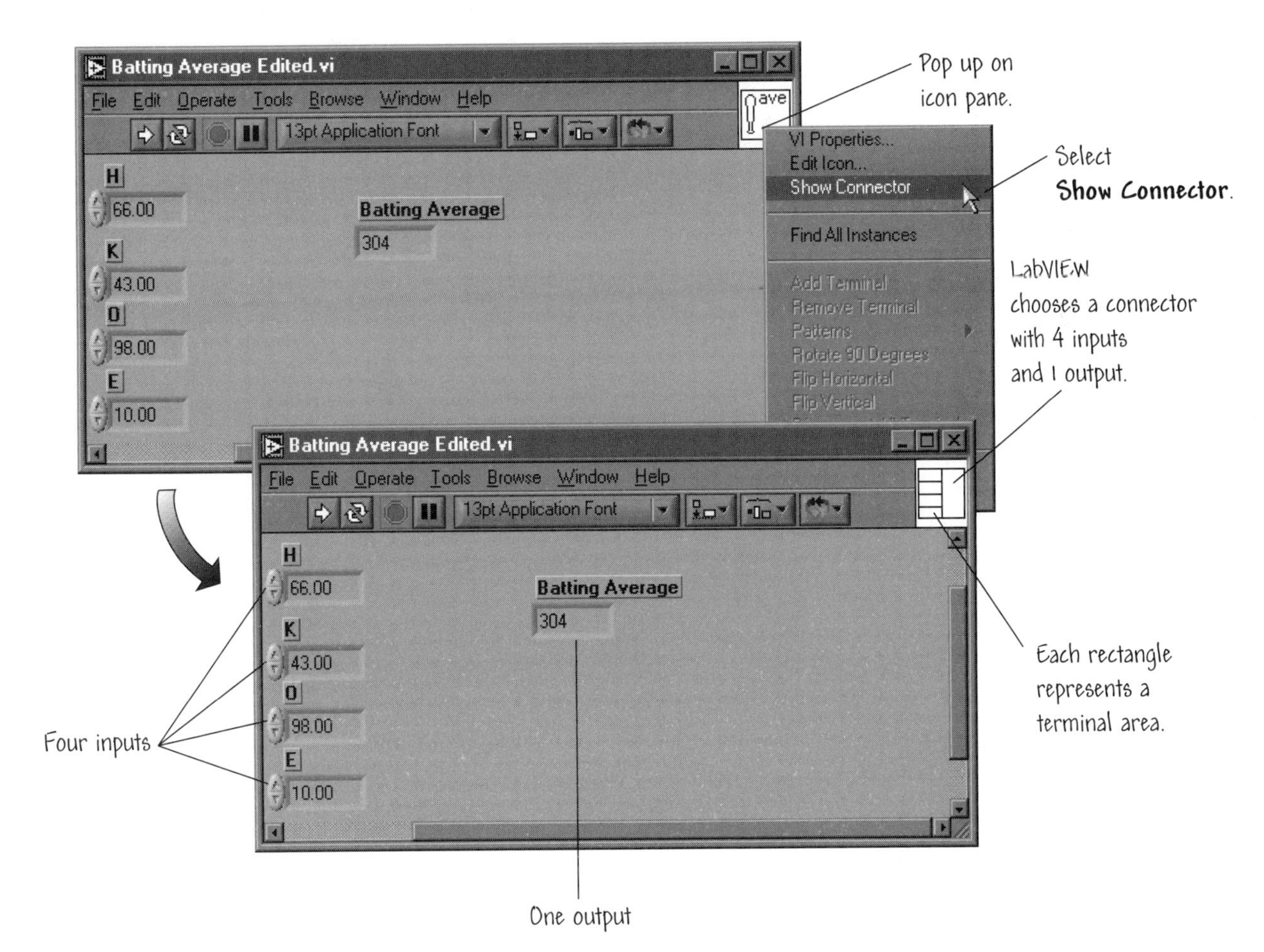

FIGURE 4.8
Defining a connector.

4.3.3 Selecting and Modifying Terminal Patterns

To select a different terminal pattern for your VI, pop up on the connector and choose **Patterns** from the pop-up menu. This process is illustrated in Figure 4.9. To change the pattern, click on the desired pattern on the palette. If you choose a new pattern, you will lose any assignment of controls and indicators to the terminals on the old connector pane.

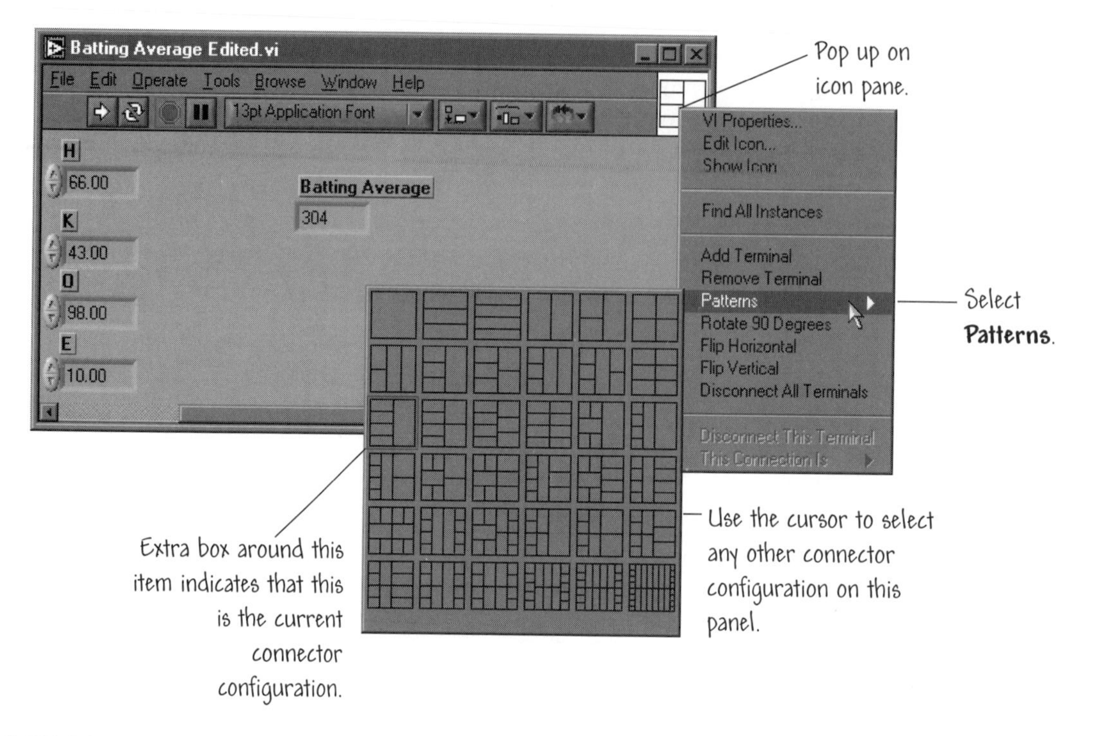

FIGURE 4.9
Changing your terminal pattern.

The maximum number of terminals available for a subVI is 28.

If you want to change the spatial arrangement of the connector terminal patterns, choose one of the following commands from the connector pane pop-up menu: **Flip Horizontal**, **Flip Vertical**, or **Rotate 90 Degrees**. If you want to add a terminal to the connector pattern, place the cursor where the terminal is to be added, pop-up on the connector pane window and select **Add Terminal**. If you want to remove an existing terminal from the pattern, pop up on the terminal and select **Remove Terminal**.

Think ahead and plan your connector patterns well. For example, select a connector pane pattern with extra terminals if you think that you will add additional inputs or outputs at a later time. With these extra terminals, you do not have to change the connector pane for your VI if you find you want to add another input or output. This flexibility enables you to make subVI changes with minimal effect on your hierarchical structure. Another useful hint is that if you create a group of subVIs that are often used together, consider assigning the subVIs a consistent connector pane with common inputs. You then can easily remember each input location without using the **Context Help** window—this will save you time. If you create a subVI that produces an output that is used as the input to another subVI, try to align the input and output connections. This technique simplifies your wiring patterns and makes debugging and program maintenance easier.

Place inputs on the left and outputs on the right of the connector when linking controls and indicators—this prevents complex wiring patterns.

4.3.4 Assigning Terminals to Controls and Indicators

Front panel controls and indicators are assigned to the connector terminals using the **Wiring** tool. The following steps are used to associate the connector pane with the front panel controls and indicators.

1. Click on the connector terminal with the **Wiring** tool. The terminal turns black, as illustrated in Figure 4.10a.
2. Click on the front panel control or indicator that you want to assign to the selected terminal. As shown in Figure 4.10b, a dotted-line **marquee** frames the selected control.
3. Position the cursor in an open area of the front panel and click. The marquee disappears, and the selected terminal takes on the data color of the connected object, indicating that the terminal is assigned. This process is illustrated in Figure 4.10c.

The connector terminal turns white to indicate that a connection was not made. If this occurs, you need to repeat steps 1 through 3 until the connector terminal takes on the proper data color.

4. Repeat steps 1–3 for each control and indicator you want to connect.

The connector terminal assignment process also works if you select the control or indicator first with the **Wiring** tool and then select the connector terminal. As already discussed, you can choose a pattern with more terminals than you need since unassigned terminals do not affect the operation of the VI. It is also true that you can have more front panel controls or indicators than terminals. Once the terminals have been connected to controls and indicators, you can disconnect them all at one time by selecting **Disconnect All Terminals** in the connector pane short cut menu. Note that although the **Wiring** tool is used to

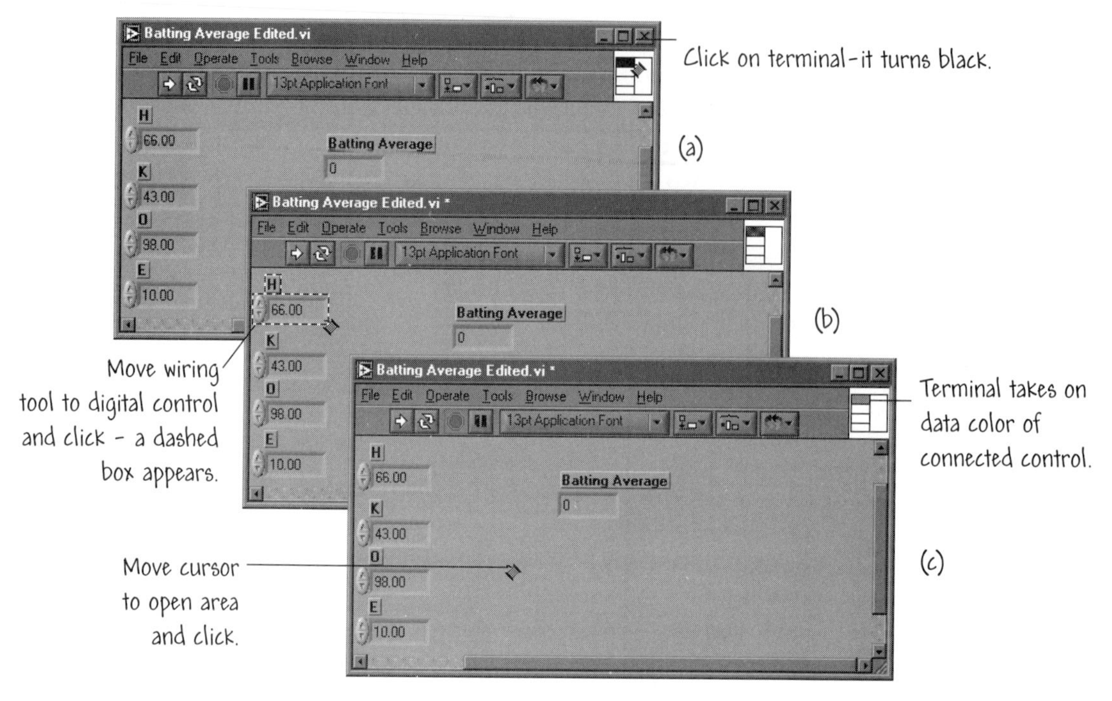

FIGURE 4.10
Assigning connector terminals to controls and indicators.

assign terminals on the connector to front panel controls and indicators, no wires are drawn.

4.4 THE HELP WINDOW

You enable the **Context Help** window by selecting Help≫Show Context Help. When you do this, you find that whenever you move an editing tool across a subVI node, the **Context Help** window displays the subVI icon with wires attached to each terminal. An example is shown in Figure 4.11 where the cursor was moved over the subVI Temp & Vol.vi, and the **Context Help** window appears and shows that the subVI has two outputs: Temp and Volume.

LabVIEW has a help feature that can keep you from forgetting to wire subVI connections—indications of required, recommended, and optional connections in the connector pane and the same indications in the **Context Help** window. For example, by classifying an input as **Required**, you can automatically detect whether you have wired the input correctly and prevent your VI from running if you have not wired correctly. To view or set connections as **Required**, **Recommended**, or **Optional**, click a terminal in the connector pane on the VI front panel and select **This Connection Is**. A checkmark indicates its status, as shown in Figure 4.12. By default, inputs and outputs of VIs you create are set to

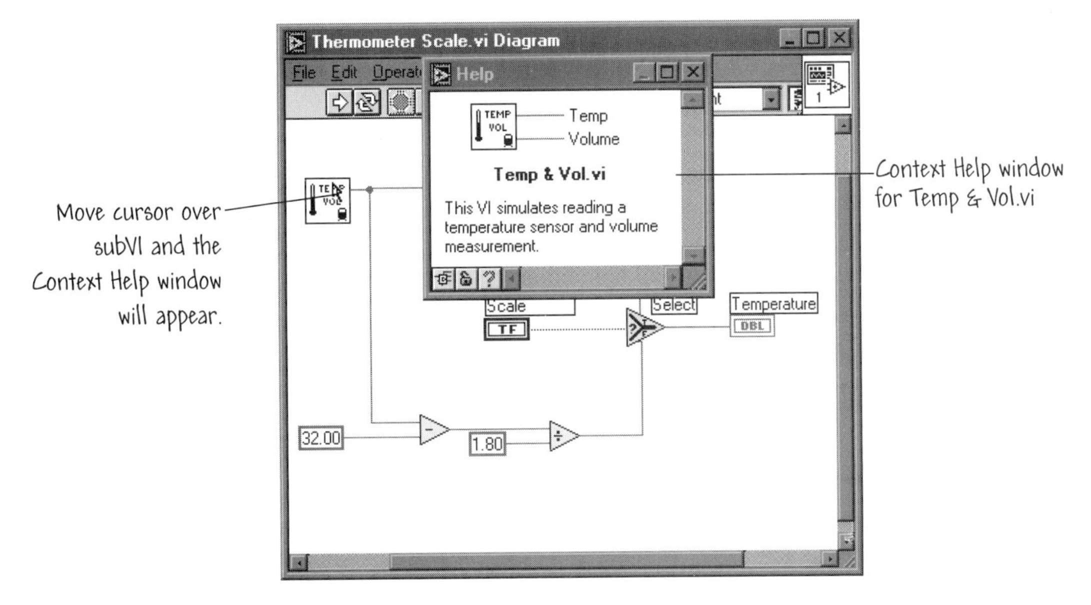

FIGURE 4.11
The **Context Help** window.

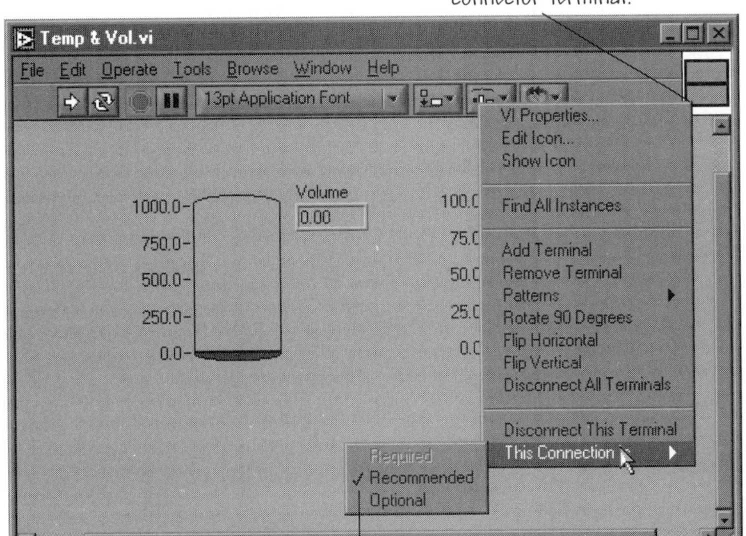

FIGURE 4.12
Showing the status of terminal connectors.

Recommended—if a change is desired, you must change the default to either **Required** or **Optional**.

When you make a connection **Required**, then you cannot run the VI as a subVI unless that connection is wired correctly. In the **Context Help** window, **Required** connections appear in bold text. When you make a connection **Recommended**, then the VI can run even if the connection is not wired, but the error list window will list a warning. In the **Context Help** window, **Recommended** connections appear in plain text. A VI with **Optional** connections can run without the optional terminals being wired. In the **Context Help** window, **Optional** connections are grayed text for detailed diagram help and hidden for simple diagram help.

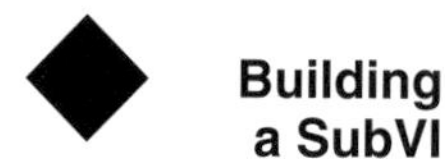

Building a SubVI

A VI designed to compute a baseball batting average was presented in Section 4.2. In this exercise you will assign the connector terminals of that VI to the digital controls and digital indicator so that the VI can be used as a subVI by other programs. Begin by opening Batting Average Edited.vi in the Chapter 4 folder of the Learning directory. Recall that the difference between the two VIs Batting Average Edited and Batting Average is that the edited version has a custom icon rather than the default icon. A subVI with an edited icon (rather than the default icon) enables easier visual determination of the purpose of the subVI because you can inspect the icon as it exists in the code.

When you have completed assigning the connector terminals, the front panel connector pane should resemble the one shown in Figure 4.13. To display the terminal connectors, pop up on the VI icon in the front panel and select **Show Connector**. You should notice that a terminal pattern appropriate for the VI has been automatically selected with four inputs on the left side of the connector pane and one output on the right. The number of terminals selected by default depends on the number of controls and indicators on the front panel.

Using the **Wiring** tool, assign the four input terminals to the four digital controls: H, K, O, and E. Similarly, assign the output terminal to the digital indicator Batting Average.

There are two levels of code documentation that you can pursue. It is important to document your code to make it accessible and understandable to other users. You can document a VI by choosing File≫VI Properties≫Documentation. A **Description** box will appear as illustrated in Figure 4.14. You type a description of the VI in the dialog box, and then whenever **Show Context Help** is selected and the cursor is placed over the VI icon, your description will appear in the **Context Help** window.

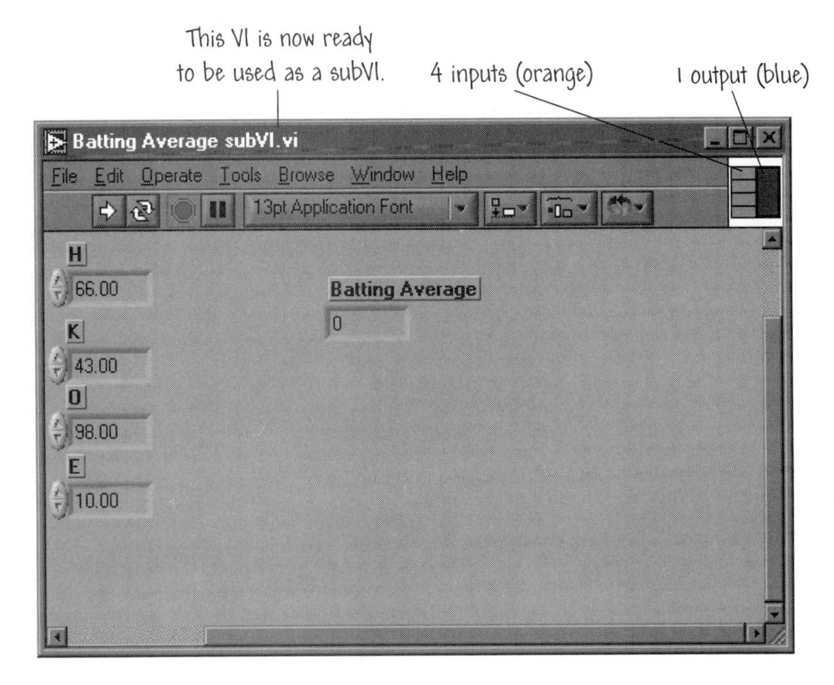

FIGURE 4.13
Assigning terminals for the Batting Average Edited.vi.

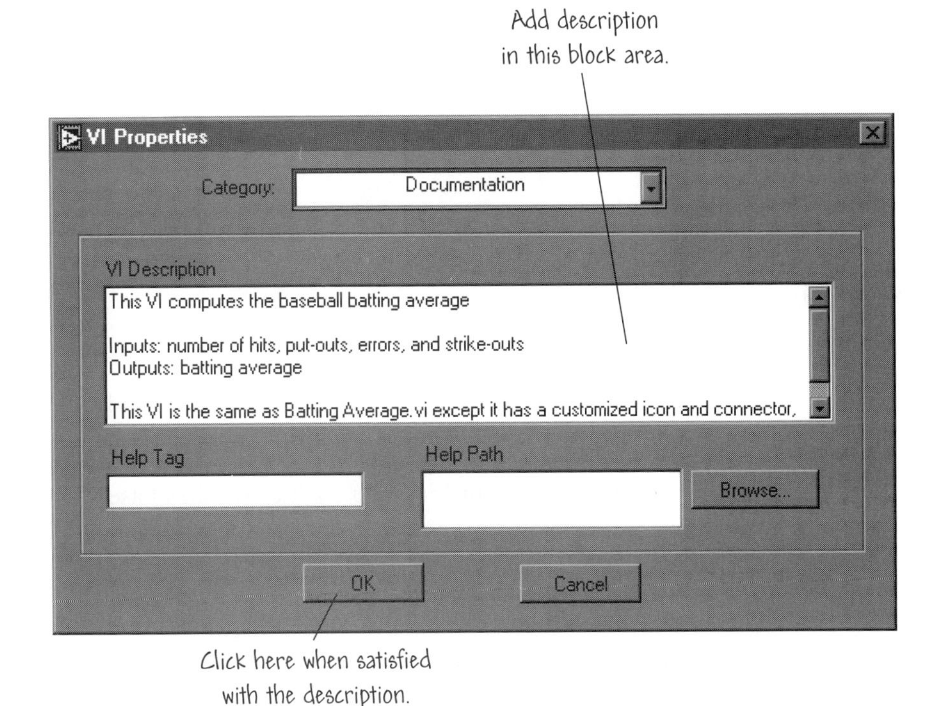

FIGURE 4.14
Typing the description for the subVI.

You can also document the objects on the front panel by popping up on an object and choosing **Description and Tip...** from the object short cut menu. Type the object description in the dialog box that appears as in Figure 4.15. You cannot edit the object description in the run mode—change to edit mode.

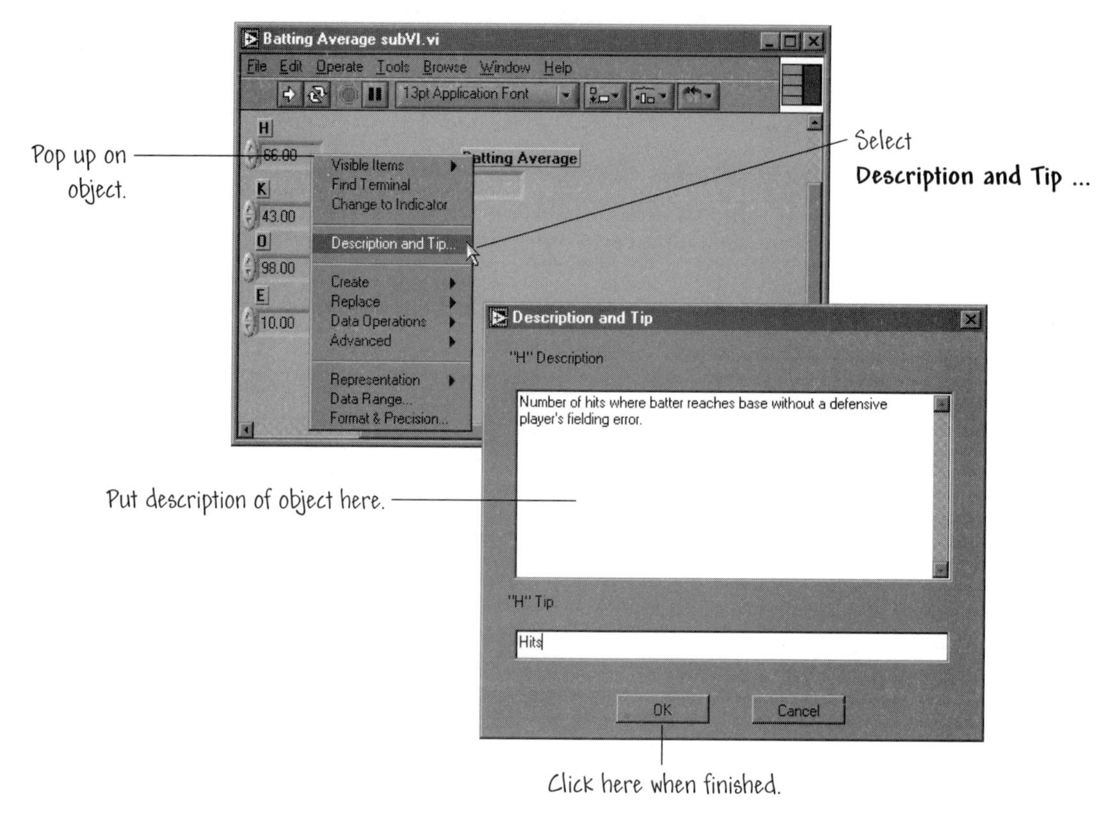

FIGURE 4.15
Documenting objects on a subVI.

Continuing with the exercise, you should add a description for each object on the front panel. Some suggested descriptions follow:

- *H*: Number of hits where batter reaches base without a defensive player's fielding error.
- *K*: Number of strikeouts.
- *O*: Number of put-outs, including ground outs, fly outs, pop ups, etc.
- *E*: Number of times the batter reaches base on a fielding error by a defensive player.
- *Batting Average*: Computed batting average.

When you have finished assigning the connector terminals to the controls and indicators, and you have documented the subVI and each of the controls and indicators, save the subVI as Batting Average subVI.vi. Make sure to save your work in the Users Stuff folder.

A working version of Batting Average subVI can be found in the Chapter 4 *folder in the* Learning *directory. You may want to open the VI, read the documentation, and take a look at the icon and the connector just to verify that your construction of the same subVI was done correctly.* ◆

4.5 USING A VI AS A SUBVI

There are two basic ways to create and use a subVI: creating subVIs from VIs and creating subVIs from selections. In this section we concentrate on the first method, that is, using a VI as a subVI. Any VI that has an icon and a connector can be used as a subVI. In the block diagram, you can select VIs to use as subVIs from Functions≫Select a VI... palette. Choosing this option produces a file **dialog box** from which you can select any available VI in the system, as shown in Figure 4.16.

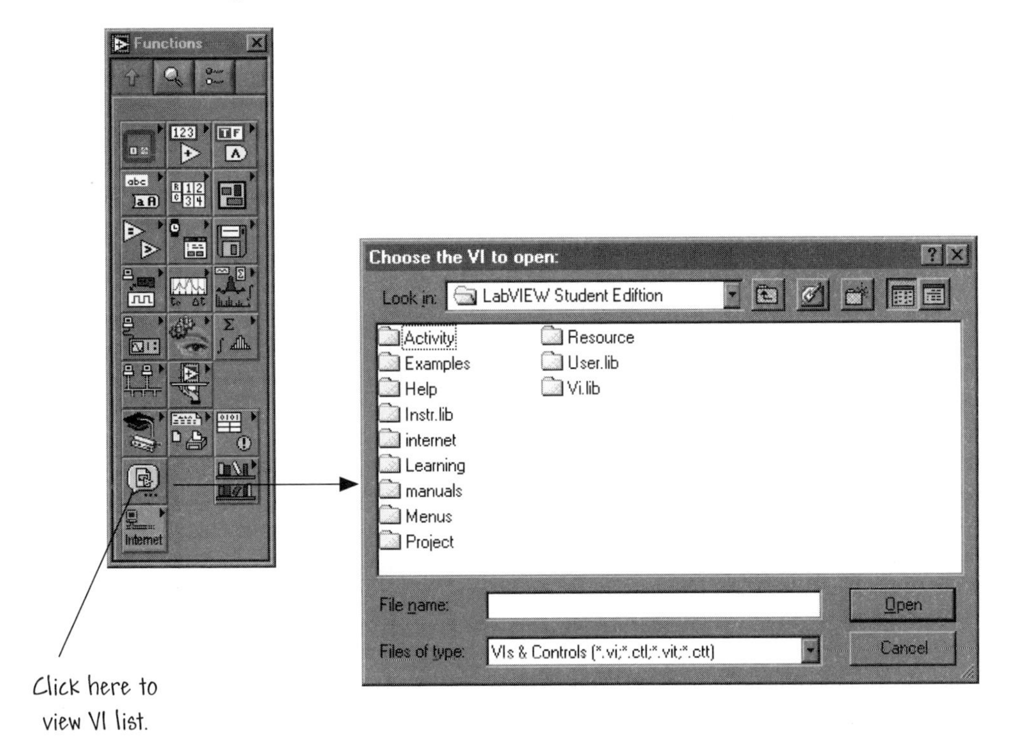

FIGURE 4.16
The **Select a VI...** palette.

The *LabVIEW Student Edition* comes with many ready-to-use VIs. In Chapter 1 you searched around the LabVIEW examples and found many of the example and demonstration VIs. In the following example you will use one of the preexisting VIs called Generate Waveform.vi as a subVI.

Using a VI as a SubVI

1. Open a new front panel.
2. Select a vertical switch control from the Controls≫Boolean palette and label it Temperature Scale. Place free labels on the vertical switch to indicate Fahrenheit and Celsius using the **Labeling** tool, as shown in Figure 4.17, and arrange the labels as shown in the figure.
3. Select a thermometer from Controls≫Numeric and place it on the front panel. Label the thermometer Temperature.
4. Change the range of the thermometer to accommodate values ranging between 0.0 and 100.0. With the **Operating** tool, double click on the high limit and change it to 100.0 if necessary.
5. Switch to the block diagram by selecting Window≫Show Diagram.
6. Pop up in a free area of the block diagram and choose Functions≫Select a VI... to access the dialog box. Select Temp & Vol.vi in the Learning\Building Blocks directory. Click **Open** in the dialog box to place Temp & Vol.vi on the block diagram.
7. Add the other objects to the block diagram using Figure 4.17 as a guide.
   - Place a Subtract function and a Divide function on the block diagram. These are located in the **Functions** palette on the **Numeric** subpalette. On the Subtract function, add a numeric constant equal to 32. You can add the constant by popping up on the Subtract function and selecting **Create Constant**. Then using the **Labeling** tool change the constant from the default 0.0 to 32.0. Similarly, add a numeric constant of 1.8 on the Divide function. These constants are used to convert from degrees Fahrenheit to Celsius according to the relationship

$$^\circ C = \frac{^\circ F - 32}{1.8}.$$

   - Add the Select function (located on the **Comparison** subpalette on the **Functions** palette). The Select function returns the value wired to the TRUE or FALSE input, depending on the Boolean input value. Use **Show Context Help** for more information on how this function works.
8. Wire the diagram objects as shown in Figure 4.17.

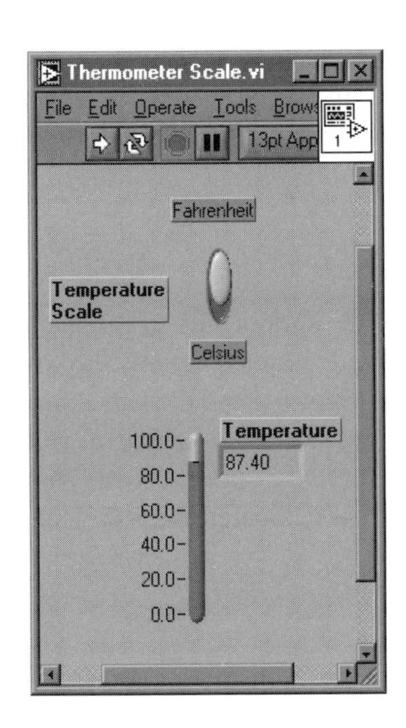

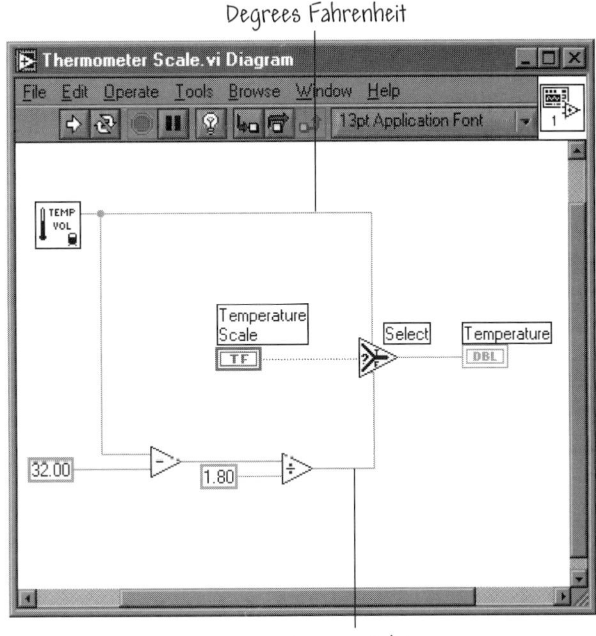

FIGURE 4.17
Calling the subVI Temp & Vol.vi.

9. Switch to the front panel and click the **Continuous Run** button in the toolbar. The thermometer shows the value in degrees Fahrenheit or degrees Celsius, depending on your selection.
10. Switch the scale back and forth (remember to use the **Operating** tool) to select either Fahrenheit or Celsius.
11. When you are finished experimenting with your VI, save it as Thermometer Scale.vi in the Users Stuff directory.

◆

A working version of Thermometer Scale.vi exists in the Chapter 4 folder located in the Learning directory. When you open this VI, notice the use of color on the vertical switch (orange denotes Fahrenheit and blue represents Celsius).

In the previous exercise, the stand-alone Temp & Vol.vi was used in the role of a subVI. Suppose that you have a subVI on your block diagram and you want to examine its contents—to view the code. You can easily open a subVI front panel window by double-clicking on the subVI icon. Once the front panel opens, you can then open the subVI block diagram by selecting **Show Diagram** from the **Window** menu. At that point, any changes you make to the subVI code alter only the version in memory—until you save the subVI. Also note that, even before saving the subVI, the changes affect all calls to the subVI and not just the node you used to open the VI.

4.6 CREATING A SUBVI FROM A SELECTION

The second way to create a subVI is to select components of the main VI and group them into a subVI. You capture and group related parts of VIs by selecting the desired section of the VI with the **Positioning** tool and then choosing **Create SubVI** from the **Edit** pull-down menu. The selection is automatically converted into a subVI, and a default icon replaces the entire section of code. The controls and indicators for the new subVI are automatically created and wired to the existing wires. Using this method of creating subVIs allows you to modularize your block diagram, thereby creating a hierarchical structure.

Building a Sub VI Using the Selection Technique

In this exercise you will modify the Thermometer Scale VI developed in Section 4.5 to create a subVI that converts Fahrenheit temperature to Celsius temperature. The subVI selection process is illustrated in Figure 4.18.

Open Thermometer Scale VI by selecting **Open** from the **File** menu. The VI is located in the Chapter 4 folder of the Learning directory. For reference, the front panel and block diagram are shown in Figure 4.17.

To create a subVI that converts Fahrenheit to Celsius, begin by switching to the block diagram window by choosing **Show Diagram** from the **Window** menu. The goal is to modify the existing block diagram to call a subVI created using the **Create SubVI** option.

Using the **Positioning** tool, select the block diagram elements that comprise the conversion from Fahrenheit to Celsius, as shown in Figure 4.18. A moving dashed line will frame the chosen portion of the block diagram. Now select **Create SubVI** in the **Edit** menu. A default subVI icon will appear in place of the selected group of objects. You can use this selection method of creating subVIs to modularize your VI.

The next step is to modify the icon of the new subVI. Open the new subVI by double-clicking on the default icon called **Untitled 1 (SubVI)**. Two front panel objects should be visible: one numeric control labeled Temp and one unlabeled numeric indicator. Pop up on the numeric indicator and select Visible Items≫Label from the short cut menu. Type Celsius Temp in the text box and then align the two objects on the front panel by their bottom edges.

Use the **Icon Editor** to create an icon similar to the one shown in Figure 4.19. Invoke the editor by popping up in the **Icon Pane** in the front panel of the subVI and selecting **Edit Icon** from the pop-up menu. Erase the default icon by double

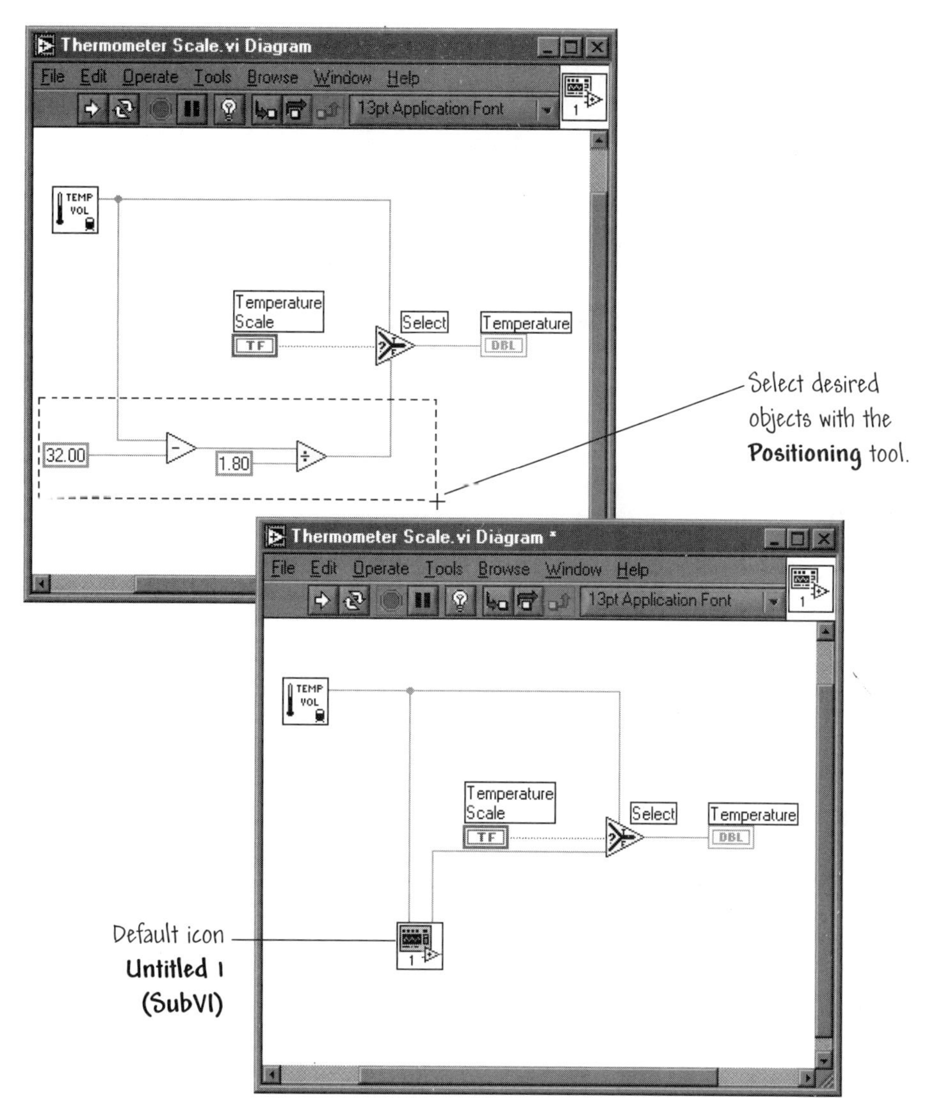

FIGURE 4.18
Selecting the code that converts from Fahrenheit to Celsius.

clicking on the Select tool and pressing <Delete>. Redraw the icon frame by double clicking on the rectangle tool which draws a rectangle around the icon. The easiest way to create the text is using the **Text** tool and the font Small Fonts. The arrow can be created with the **Pencil** tool.

The connector is automatically wired when a subVI is created using the ***Create SubVI*** *option.*

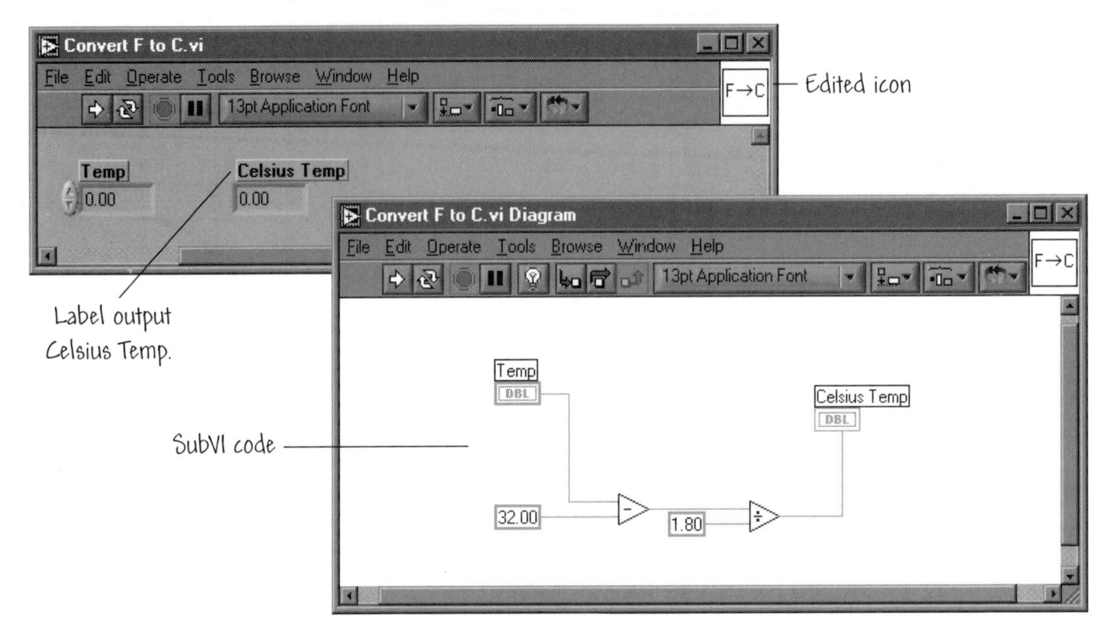

FIGURE 4.19
The new subVI to convert degrees Fahrenheit to degrees Celsius.

When you are finished editing the icon, close the **Icon Editor** by clicking on **OK**. The new icon appears in the upper-right corner of the front panel, as shown in Figure 4.19. Save the subVI by choosing **Save** from the **File** menu. Name the VI Convert F to C and save it in the Users Stuff folder.

You cannot build a subVI from a section of code with more than 28 inputs and outputs because 28 is the maximum number of inputs and outputs allowed on a connector pane.

◆

4.7 SAVING YOUR SUBVI

It is highly recommended that you save your subVIs to a file in a directory rather than in a library. While it is possible to save multiple VIs in a single file called a **VI library**, this is not desirable. Saving VIs as individual files is the most effective storage path because you can copy, rename, and delete files easier than when using a VI library.

VI libraries have the same load, save, and open capabilities as other directories, but they are not hierarchical. That is, you cannot create a VI library inside of another VI library, nor can you create a new directory inside a VI library. Since there is no way to list the VIs in a VI library outside the LabVIEW environment, it would be impossible to locate those VIs by searching the file structure outside of LabVIEW. After you create a VI library, it appears in the **File** dialog box as a file with an icon that is somewhat different from a VI icon.

4.8 THE HIERARCHY WINDOW

When you create an application, you generally start at the top-level VI and define the inputs and outputs for the application. Then you construct the subVIs that you will need to perform the necessary operations on the data as it flows through the block diagram. As discussed in previous sections, if you have a very complex block diagram, you should organize related functions and nodes into subVIs for desired block diagram simplicity. Taking a modular approach to program development creates code that is easier to understand, debug, and maintain.

The **Hierarchy Window** displays a graphical representation of the hierarchical structure of all VIs in memory and shows the dependencies of top-level VIs and subVIs. There are several ways to access the **Hierarchy Window**:

- You can select Browse≫Show VI Hierarchy to open the **Hierarchy Window** with the VI icon of the current **active window** surrounded by a thick red border.
- You can pop up on a subVI and select **Show VI Hierarchy** to open the **Hierarchy Window** with the selected subVI surrounded by a thick red border.
- If the **Hierarchy Window** is already open, you can bring it to the front by selecting it from the list of open windows under the **Window** menu.

The **Hierarchy Window** for the Thermometer Scale subVI.vi is shown in Figure 4.20. The window displays the dependencies of VIs by providing information on VI callers and subVIs. As you move the **Operating** tool over objects in the window, the name of the VI is shown below the VI icon. This window also contains a toolbar, as shown in Figure 4.20, that you can use to configure several types of settings for displayed items.

You can switch the **Hierarchy Window** display mode between horizontal and vertical display by pressing the **Horizontal Layout** or **Vertical Layout** button hierarchy toolbar. In a horizontal display, subVIs are shown to the right of their calling VIs; in a vertical display, they are shown below their calling VIs. In either case, the subVIs are always connected with lines to their calling VIs. The window shown in the Figure 4.20 is displayed vertically.

Arrow buttons and arrows beside nodes indicate what is displayed and what is hidden according to the following rules:

- A red arrow button pointing towards the node indicates some or all subVIs are hidden, and clicking the button will display the hidden subVIs.

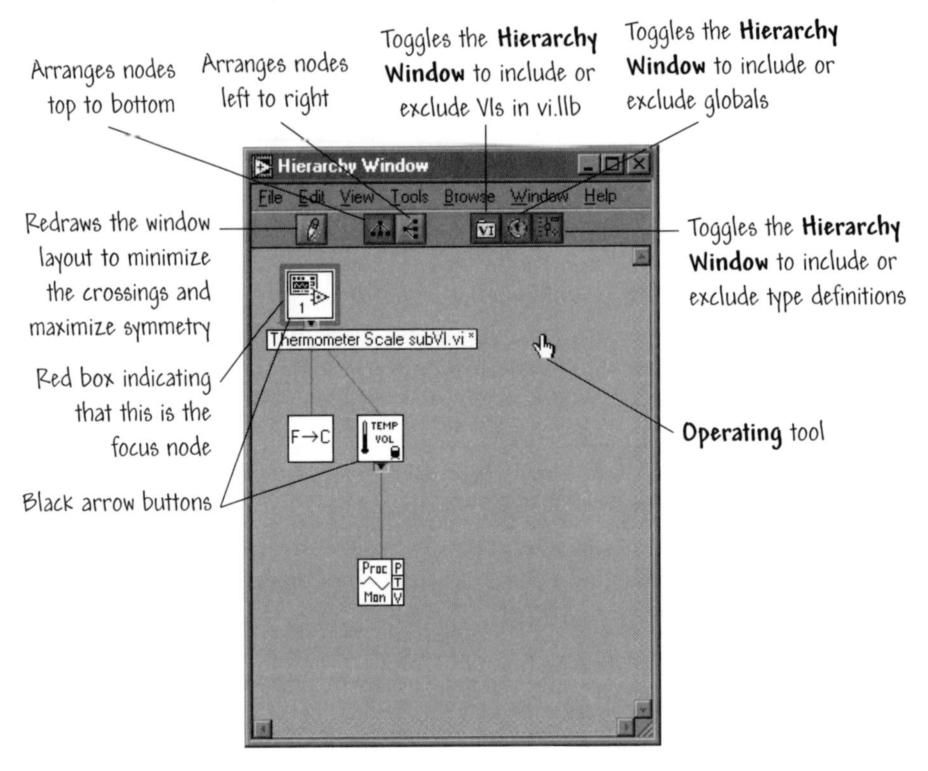

FIGURE 4.20
The **Hierarchy Window** for Thermometer Scale subVI.vi.

- A black arrow button pointing towards the subVIs of the node indicates all immediate subVIs are shown.
- A blue arrow pointing towards the callers of the node indicates the node has additional callers in this VI hierarchy but they are not shown at the present time. If you show all subVIs, the blue arrow will disappear. If a node has no subVIs, no red or black arrow buttons are shown.

Double clicking on a VI or subVI opens the front panel of that node. You also can pop up on a VI or subVI node to access a menu with options, such as showing or hiding all subVIs, opening the VI or subVI front panel, editing the VI icon, and so on.

You can initiate a search for a given VI by simply typing the name of the node directly onto the **Hierarchy Window**. When you begin to type in text from the keyboard a small search window will automatically appear displaying the text that has been typed and allowing you to continue adding text. The search for the desired VI commences immediately. The search window is illustrated in Figure 4.21. If the characters currently displayed in the search window do not match any node names in the search path, the system beeps, and no more char-

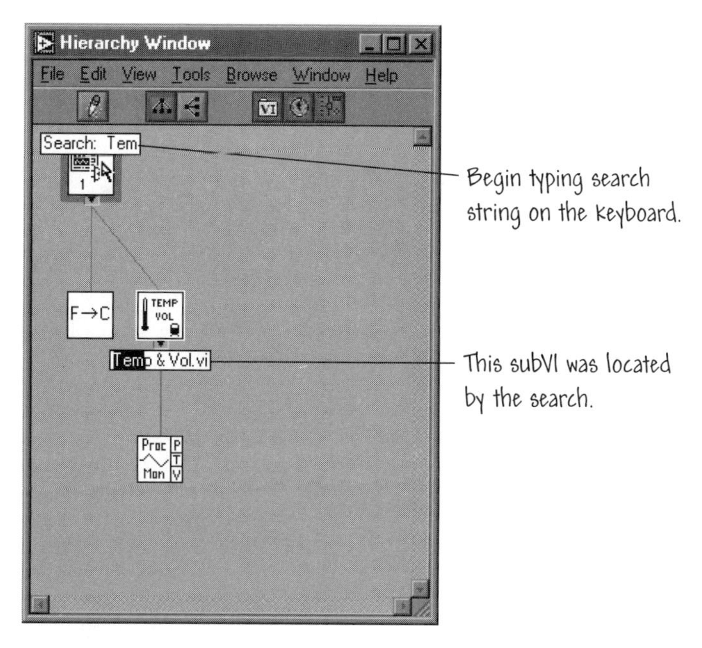

FIGURE 4.21
Searching the **Hierarchy Window** for VIs.

acters can be typed. You can then use the <Backspace> or <Delete> key to delete one or more characters to resume typing. The search window disappears automatically if no keys are pressed for a certain amount of time, or you can press the <Esc> key to remove the search window immediately. When a match is found you can use the right or down arrow key, or the <Enter> key on **Windows** and the <Return> key on **Macintosh** to find the next node that matches the search string. To find the previous matching node, press the left or up arrow key, or the <Shift-Enter> on **Windows** and <Shift-Return> on **Macintosh**.

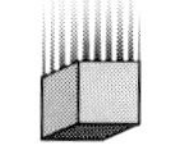

BUILDING BLOCK

4.9 BUILDING BLOCKS: MEASURING VOLUME

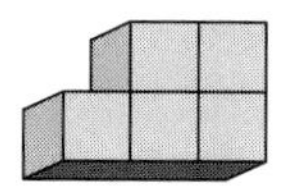

In this exercise you will continue the work on the VI **Volume1** created in Chapter 2. The goal is to make the necessary modifications so that you can use the VI as a subVI. This means that you need to edit the icon and assign the connector terminals. The first step is to create the icon and connector. You can use Figure 4.22 as a guide.

When you have finished readying the subVI, save it in the Users Stuff folder as VolumeSubVI.vi. In many cases you will develop subVIs for general use; hence you will want them accessible from the **Functions** palette. To accomplish

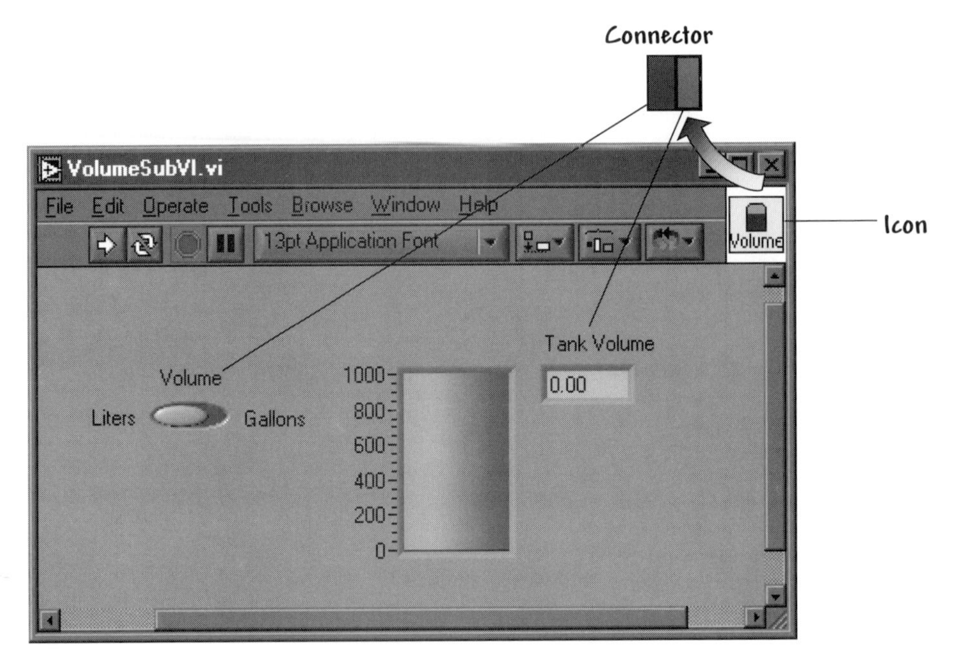

FIGURE 4.22
Wiring the connector of Volume subVI.

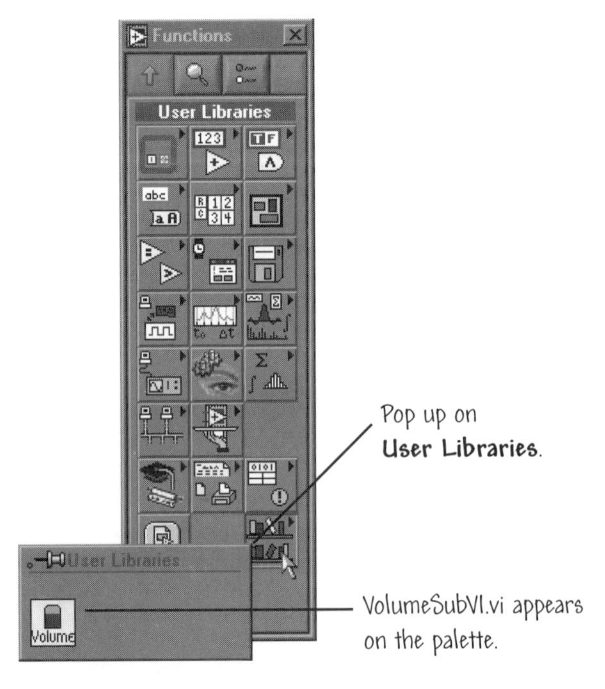

FIGURE 4.23
Locating the Volume subVI on the **User Library** palette.

this task for the VolumeSubVI VI, you need to save it in the folder User.lib. After you have saved the subVI in User.lib, you will need to exit LabVIEW and then open it again. Once this is done, open a new VI, switch to the block diagram, and show the **Functions** palette. Click on the **User Libraries** palette, and you should find the icon for VolumeSubVI, as shown in Figure 4.23. The subVI is now readily accessible to you in your future programming endeavors. You could also have accessed the subVI through Functions≫Select a VI... by navigating to Users Stuff where you previously saved the program.

Using the VolumeSubVI, construct a VI that indicates (using a large LED) whether a volume limit has been exceeded. The volume limit should be a user input parameter. You can use the VI shown in Figure 4.24 as a guide. When you have completed construction of the program and have verified that it is working

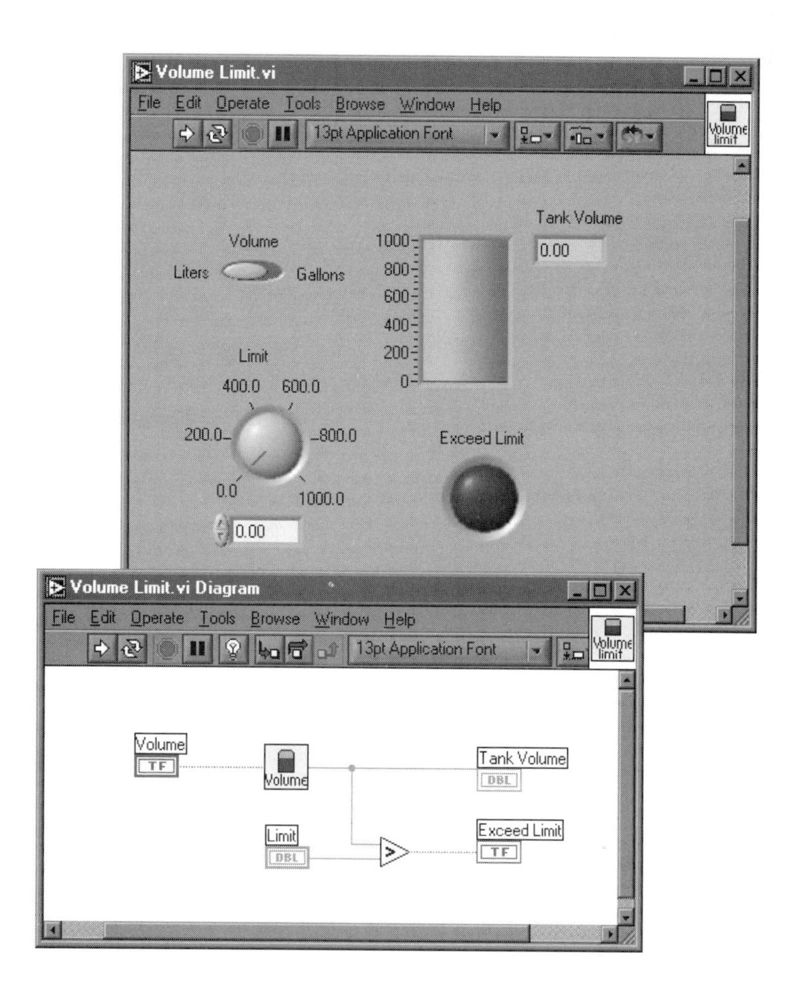

FIGURE 4.24
Using the VolumeSubVI.

properly (using the various debugging tools, such as **Highlight Execution** and probes), you can save the VI as Volume Limit in the Users Stuff folder in the Learning directory.

4.10 RELAXING READING: SCANNING OF ELECTRONIC CIRCUITS

A temperature scanning system was developed at Southern Methodist University for use in the electronic measurements laboratory. The system uses LabVIEW and an NB-MIO-16 data acquisition board as shown in Fig. 4.25 (more on data acquisition [DAQ] in Chapter 8).

A brief description of the system follows. Two types of temperature controllers are available—one type scans temperature (from room temperature to around 150° C) with no required control precision, and the second controls temperature with a higher level of precision. The heat source for the system consists of a power transistor and a buffer transistor. A bipolar transistor provides the temperature sensor. All of these are mounted on an aluminum plate in a small aluminum box, and a 50-pin connector is fixed to the box for connecting to the NB-MIO-16 I/O board.

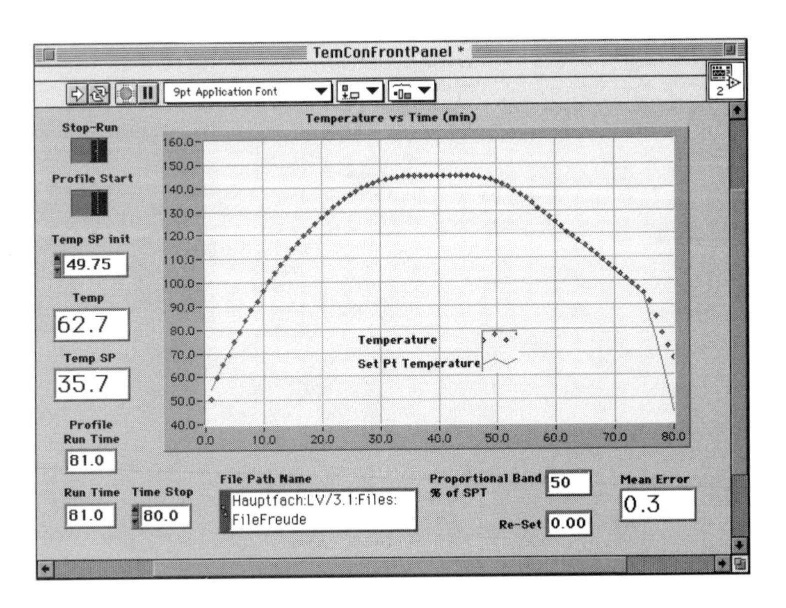

FIGURE 4.25
The front panel for the temperature control system: the profile shown represents a liquid-phase epitaxial growth of gallium arsenide for optoelectronic devices.

As one example of how the system is used in the laboratory, students measure the current gain of the sensor transistor as a function of temperature. Using the

LabVIEW-based system, the students can plot and massage the data to extract things such as the associated SPICE parameters.

Using a LabVIEW program for precision control (proportional-band controller), a complex collection of amplifiers, resistors, and capacitors have now been replaced with the LabVIEW-based software. One feature of the program is its ability to provide any functional temperature profile—there is virtually no limit to the type of profile that can be programmed into the system.

The LabVIEW program for precision temperature control consists of a top-level virtual instrument (VI) that monitors progress of temperature versus time both graphically and with digital indicators. The VI is configured to update the graph for each plotting point, which provides continuous graphical monitoring. The temperature control VI has an extensive subVI hierarchy for converting the measured base-emitter voltage to temperature, computing the base-emitter voltage temperature coefficient, maintaining a constant sensor-transistor bias current, and keeping track of time as well as formulating control functions. At the end of execution, all the desired data is filed under the path named in the VI front panel.

For more information on this application of LabVIEW in the classroom, read the *User Solutions* entitled "Temperature Control for Measuring Electronic Circuits Using LabVIEW," which can be found on the National Instruments website, or contact:

Dr. Ken Ashley
Dept. of Electrical Engineering
Southern Methodist University
P.O. Box 750338
Dallas, TX 75275-0338
e-mail: kla@seas.smu.edu

4.11 SUMMARY

Constructing subVIs was the main topic of this chapter. One of the keys to constructing successful LabVIEW programs is understanding how to build and use subVIs. SubVIs are the primary building blocks of modular programs, which are easy to debug, understand, and maintain. They are analogous to subroutines in text-based programming languages like C or Fortran. Two methods of creating subVIs were discussed—creating subVIs from VIs and creating subVIs from VI selections. A subVI is represented by an icon in the block diagram of a calling VI and must have a connector with terminals properly wired to pass data to and from the calling VI. The **Icon Editor** was discussed as the way to personalize the subVI icon so the information about the function of the subVI is readily apparent from visual inspection of the icon. The editor has a tool set similar to that of most common paint programs. The **Hierarchy Window** was introduced as a helpful tool for managing the hierarchical nature of your programs.

KEY TERMS

Active window: Window that is currently set to accept user input, usually the frontmost window. You make a window active by clicking on it or by selecting it from the **Windows** menu.

Control: Front panel object for entering data to a VI interactively or to a subVI programmatically.

Description box: Online documentation for G objects.

Dialog box: An interactive screen with prompts in which you describe additional information needed to complete a command.

Function: A built-in execution element, comparable to an operator or statement in a conventional programming language.

Hierarchy Window: Window that graphically displays the hierarchy of VIs and subVIs.

Icon Editor: Interface similar to that of a paint program for creating VI icons.

Indicator: Front panel object that displays output.

Marquee: Moving, dashed border surrounding selected objects.

Pseudocode: Simplified, language-independent representation of programming code.

SubVI: A VI used in the block diagram of another VI—comparable to a subroutine.

Top-level VI: The VI at the top of the VI hierarchy. This term distinguishes a VI from its subVIs.

VI library: Special file that contains a collection of related VIs.

EXERCISES

E4.1 Open a new VI and switch to the block diagram window. Select Signal Generator by Duration.vi from the Functions≫Analyze≫Signal Processing≫Signal Generation palette and place it on the block diagram. Click on the **Broken Run** button to access the error list. The one listed error states that the VI has a required input. As you know from the chapter, each input terminal can be categorized as required, recommended, or optional. You must wire values to required inputs before you can run a VI. Recommended and optional inputs do not prevent the VI from running and use the default value for that input. Default values are shown when you open (in this case double-click) on the VI.

Select the error by clicking on it and then choose **Show Error**. Notice that a terminal in the upper left side of the Signal Generator by Duration VI icon is highlighted by a black box, which quickly transitions to a marquee around the icon. Open **Show Context Help** to read about the VI—paying special attention to the inputs listed as **Required**. You should see that the waveform type terminal is a required input. Pop up on the terminal on the icon and and select **Create Control**. Notice that once the waveform type is attached to the icon, the error disappears, and you can run the VI.

E4.2 Create a VI that executes the Quit LabVIEW VI from the Functions≫Application Control palette. Open a new VI and place the Quit LabVIEW VI on the block diagram. To edit the VI properties, select File≫VI Properties... and choose the Execution category. Check the box next to **Run when opened**. Save the VI in Users Stuff. Close the VI and then open it again. What happens? Try to figure out how you can edit the VI. **Hint**: A subVI may be useful.

PROBLEMS

P4.1 Construct a VI that computes the average of three numbers input by the user. One computation in your program should be summing the three input numbers followed by a division by 3. The resulting average should be displayed on the front panel. Also, add a piece of code that multiplies the computed average by a random number in the range $[0, \cdots, 1]$. Create a subVI by grouping the parts of the code that perform the average. Remember to edit the icon so that it represents the function of the subVI, namely, the average of three numbers. Figure P4.1 can be used as a guide.

P4.2 Open While Loop Demo.vi in Learning\Chapter 5—yes, we are looking ahead to the next chapter already! Change the VI so the toolbar, menubar, and scroll

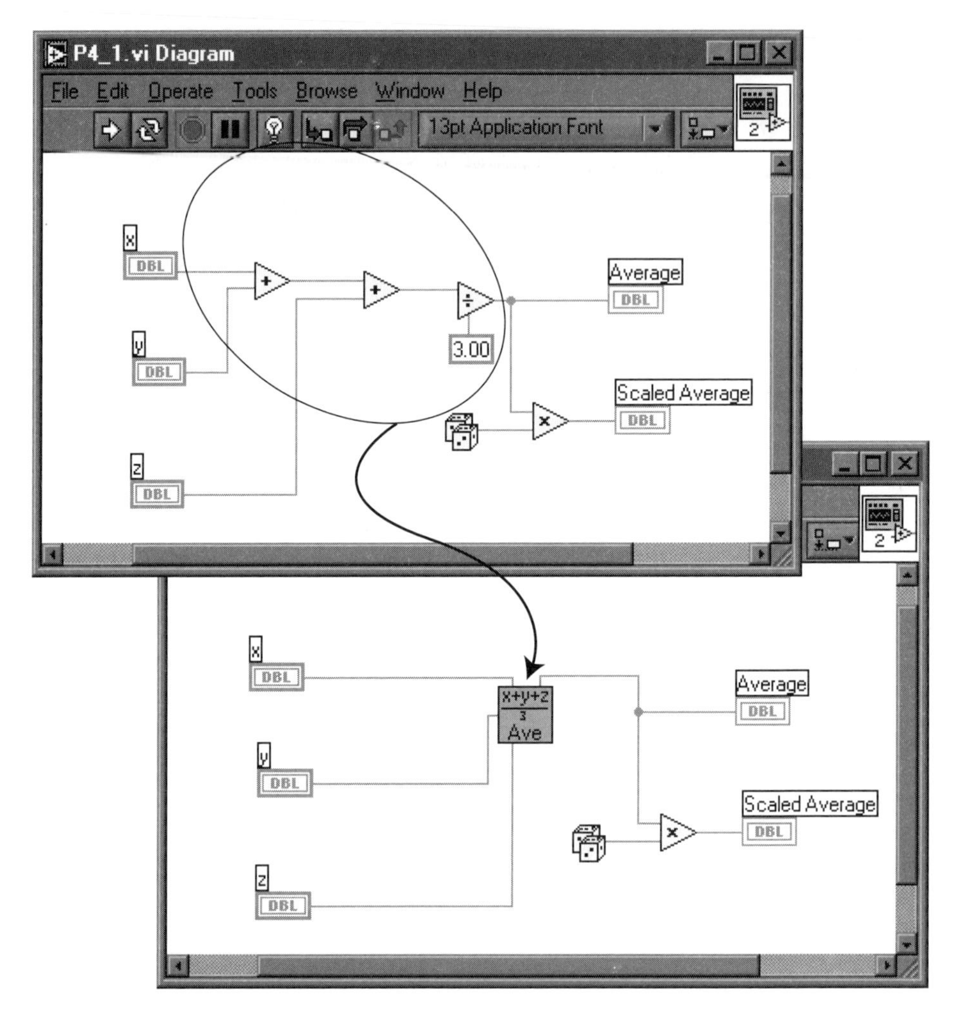

FIGURE P4.1
Computing the average of three numbers using a subVI to compute the average.

bars are not visible when the VI executes. **Hint**: Navigate File≫VI Properties... and choose **Window Appearance**, and then choose **Customize**.

P4.3 Create a subVI that multiplexes four inputs to a single output. The subVI should have four floating point numeric controls (denoted **In1** thru **In4**, one floating point numeric, indicator (denoted by **out**, and one unsigned 8 bit integer. The algorithm to be implemented is as follows: If **Select**=1, then **out=In1**; if **Select**=2, then **out=In2**; if **Select**=3, then **out=In3**; and if **Select** = 4, then **out=In4**. **Hint**: The Select VI from the **Comparison** palette may be useful.

P4.4 Complete the crossword puzzle.

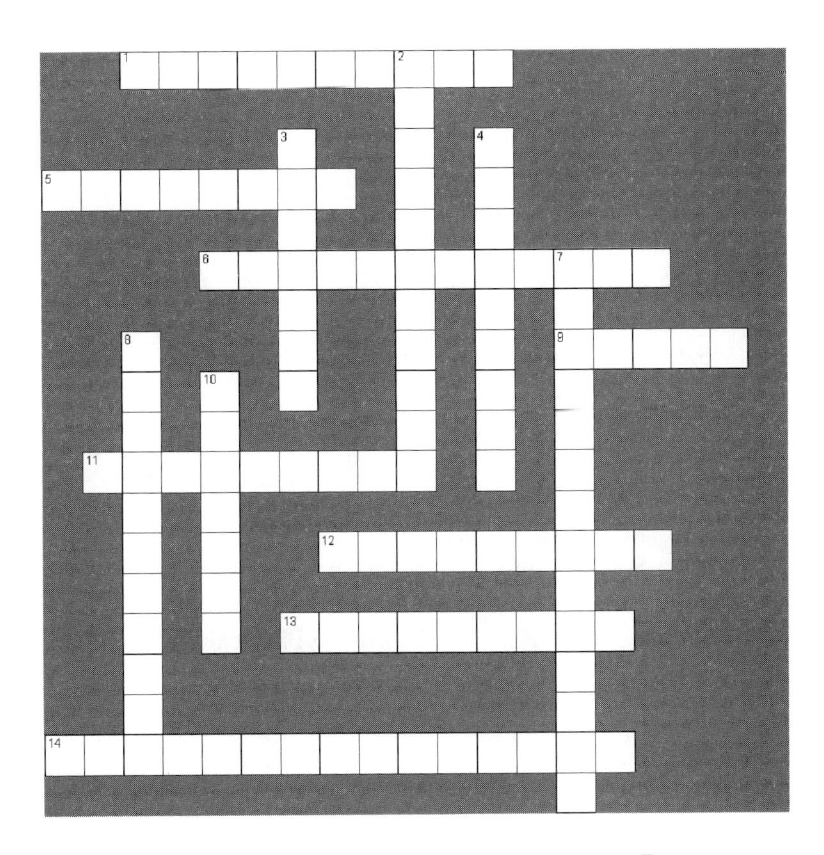

Across

1. Interface similar to that of a paint program for creating VI icons.
5. A built-in execution element.
6. Window that is currently set to accept user input.
9. A VI used in the block diagram of another VI.
11. The VI whose front panel, block diagram, or Icon Editor is the active window.
12. Front panel object that displays output.
13. An interactive screen that prompts you for additional information.
14. Window that graphically displays the hierarchy of VIs and subVIs.

Down

2. The VI at the top of the VI hierarchy. This term distinguishes a VI from subVIs.
3. Front panel object for entering data to a VI interactively or subVI programmatically.
4. Special file that contains a collection of related VIs.
7. Online documentation for G objects.
8. Simplified language-independent representation of programming code.
10. Moving, dashed border surrounding selected objects.

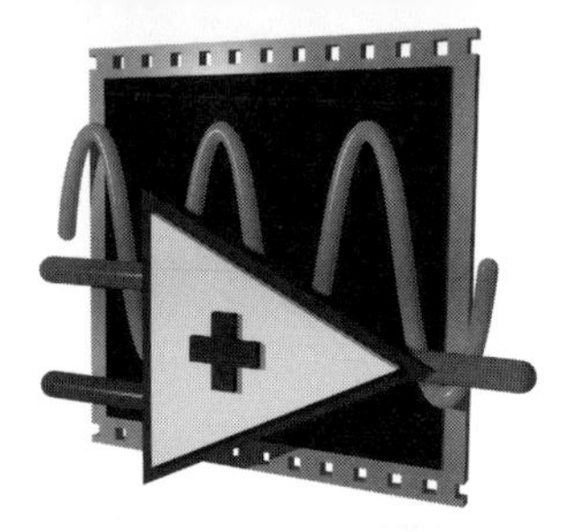

CHAPTER 5

Structures

Structures govern the execution flow in a VI. This chapter introduces you to four structures in LabVIEW: the For Loop, the While Loop, the Case structure, and the Sequence structure. Formula Nodes will be introduced as an effective way to implement mathematical equations. It is shown how script nodes can be used to execute MATLAB code within LabVIEW. We will also discuss the subject of controlling the execution timing of VIs.

GOALS

1. Study For Loops, While Loops, Case structures, and Sequence structures.
2. Understand how to use timing functions in LabVIEW.
3. Understand the use of shift registers.
4. Become familiar with Formula Nodes and MATLAB Script Nodes.
5. Appreciate common wiring errors with structures.

5.1 THE FOR LOOP

For Loops and **While Loops** control repetitive operations in a VI, either until a specified number of iterations completes (i.e., the For Loop) or while a specified condition is true (i.e., the While Loop). A difference between the For Loop and the While Loop is that the For Loop executes a predetermined number of times, and a While Loop executes until a certain conditional input becomes false. For Loops and While Loops are found on the **Structures** palette of the **Functions** menu—see Figure 5.1. In this section, we concentrate on For Loops. While Loops are discussed in the next section.

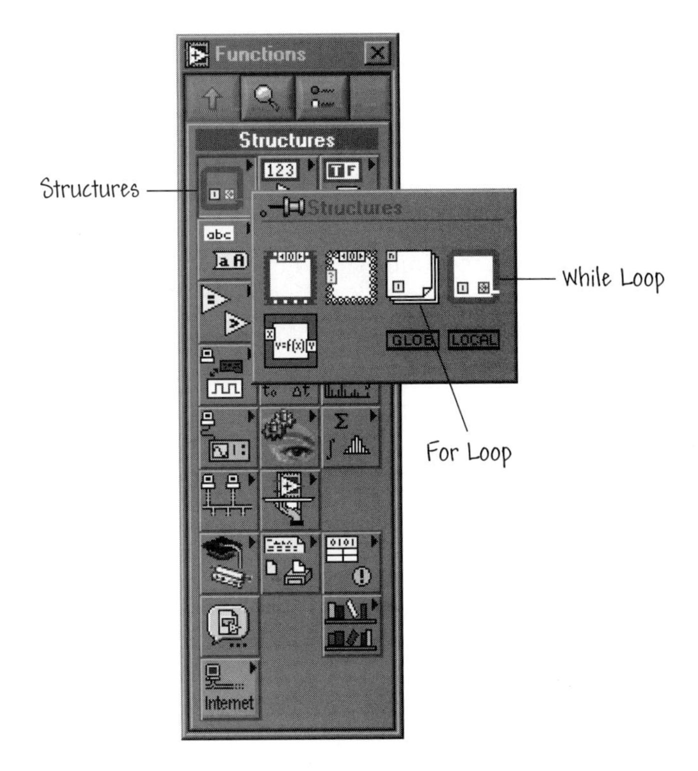

FIGURE 5.1
The For Loop and While Loop are found on the **Structures** palette.

A For Loop executes the code (known as the **subdiagram**) inside its borders a total of N times, where the N equals the value in the count terminal, as illustrated in Figure 5.2. The For Loop has two terminals: the **count terminal** (an input terminal) and the **iteration terminal** (an output terminal). You can set the count explicitly by wiring a value from outside the loop to the count terminal, or you can set the count implicitly with auto-indexing (a topic discussed in Chapter 6). The bottom and right edges of the iteration terminal are exposed to the inside of the loop, allowing you access to the current value of the count. The

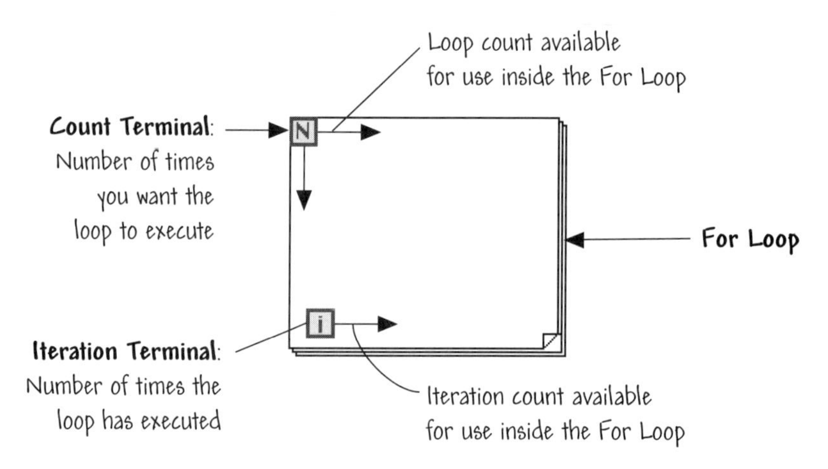

FIGURE 5.2
The For Loop.

iteration terminal (see Figure 5.2) contains the current number of completed loop iterations, where 0 represents the first iteration, 1 represents the second iteration, and continuing up to $N - 1$ to represent the Nth iteration.

Both the count and iteration terminals are long integers with a range of 0 through $2^{31} - 1$. You can wire a floating-point number to the count terminal, but it will be rounded off and **coerced** to lie within the range 0 through $2^{31} - 1$. The For Loop does not execute if you wire 0 to the count terminal!

The For Loop is located on the Functions≫Structures palette. Unlike many other **objects**, the For Loop is not dropped on the block diagram immediately. Instead, a small icon representing the For Loop appears in the block diagram, giving you the opportunity to size and position the loop. To do so, first click in an area above and to the left of all the objects that you want to execute within the For Loop, as illustrated in Figure 5.3. While holding down the mouse button, drag out a rectangle that encompasses the objects you want to place inside the For Loop. A For Loop is created upon release of the mouse button. The For Loop is a resizable box—use the **Positioning** tool for resizing by grabbing a corner of the For Loop and stretching to the desired dimensions. You can add additional block diagram elements to the For Loop by dragging and dropping them inside the loop boundary. The For Loop border will highlight as objects move inside the boundaries of the loop, and the block diagram border will highlight as you drag objects out of the loop.

If you move an existing **structure** (e.g., a For Loop) so that it overlaps another object on the block diagram, the partially covered object will be visible above one edge of the structure. If you drag an existing structure completely over another object, the covered object will display a thick shadow to warn you that the object is underneath.

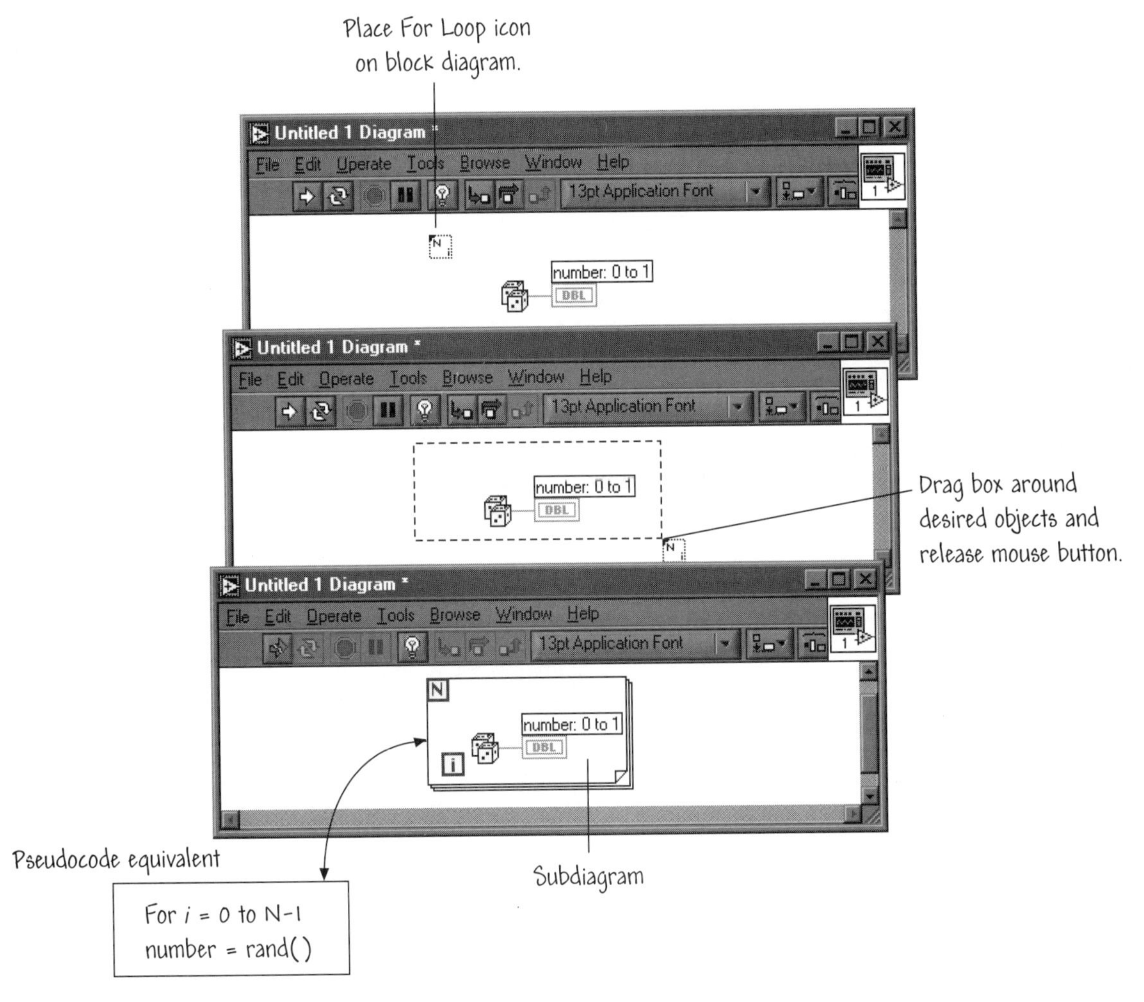

FIGURE 5.3
Placing a For Loop on the block diagram.

5.1.1 Numeric Conversion

Most of the numeric controls and indicators you have used so far have been double precision, floating-point numbers. In G, numbers can be represented as integers (byte [I8], word [I16], or long [I32]) or floating-point numbers (single, double, or extended precision). If you wire together two terminals that are of different data types, LabVIEW converts one of the terminals to the same representation as the other terminal. As a reminder, a **coercion dot** is placed on the terminal where the conversion takes place.

For example, consider the For Loop count terminal shown in Figure 5.4. The terminal representation is long integer. If you wire a double-precision, floating-point number to the count terminal, the number is converted to a long integer.

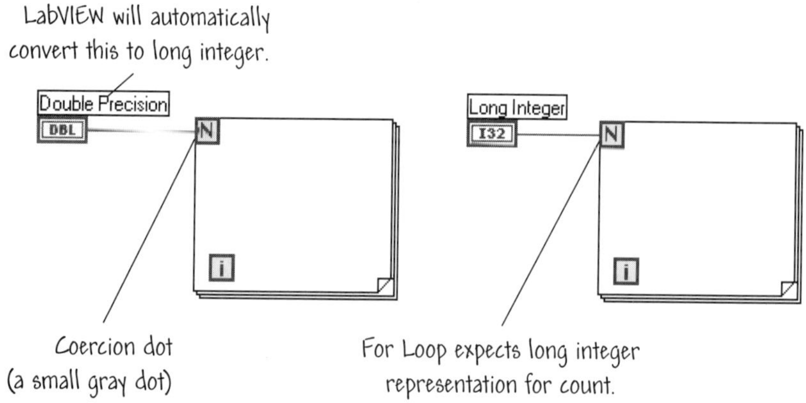

LabVIEW will automatically convert this to long integer.

Double Precision

DBL

N

i

Long Integer

I32

N

i

Coercion dot
(a small gray dot)

For Loop expects long integer representation for count.

FIGURE 5.4
Converting double-precision, floating-point numbers at the count terminal.

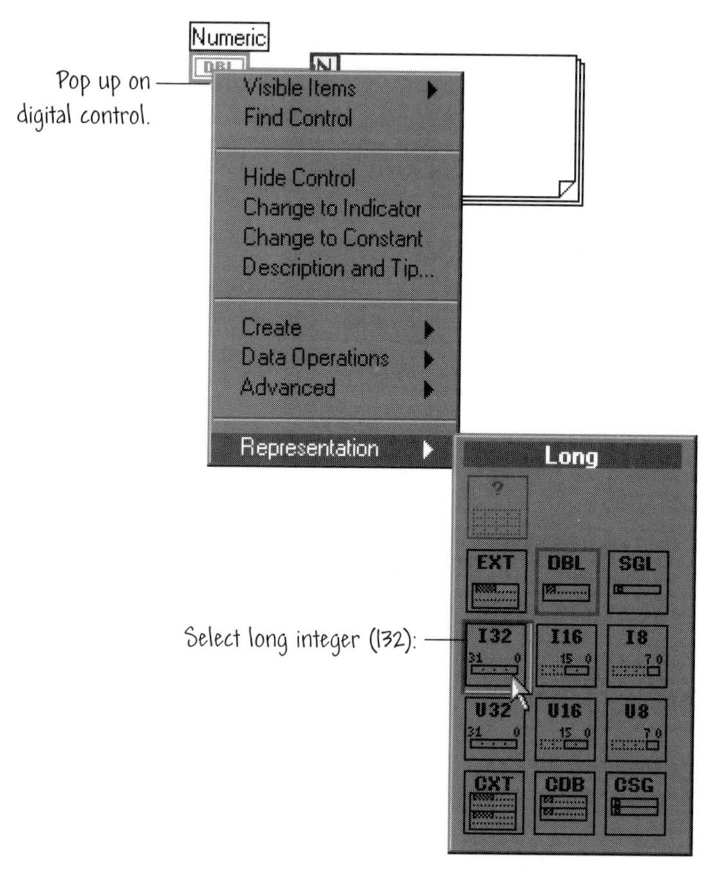

FIGURE 5.5
Changing the representation of a front panel numeric object.

Notice the small gray dot in the count terminal of the first For Loop—that is the coercion dot. To change the representation of the count terminal input, pop up on the terminal and select **Representation**, as illustrated in Figure 5.5. A palette will appear from which you can select the desired representation. In the case of the For Loop, you can change the count terminal input from double-precision, floating-point to long integer.

When the VI converts floating-point numbers to integers, it rounds to the nearest integer. If a number is exactly halfway between two integers, it is rounded to the nearest even integer. For example, the VI rounds 6.5 to 6, but rounds 7.5 to 8. This is an IEEE standard method for rounding numbers.

A For Loop Example

In this exercise we will place a random number object inside a For Loop and display the random numbers and For Loop counter on the front panel. The VI shown in Figure 5.6 can be constructed by following these steps:

1. Place a random number function on the block diagram. The random number function can be found in the Functions≫Numeric palette. Create an indicator on the random number function and label it number: 0 to 1.
2. Place the For Loop on the block diagram so that the random number is enclosed within the loop, as shown in Figure 5.6.
3. Create a constant to the For Loop by popping up on the count terminal and selecting **Constant** from the **Create** menu. Set the value of the constant (which will be zero by default) to 100. This will let the For Loop execute one hundred times.
4. Create an indicator on the iteration terminal and label it Loop number.
5. Debug and run the program. One suggestion is to run the program with **Highlight Execution**, otherwise the program may run too fast to observe the loop execution.
6. On the front panel you will see the loop counter increment from 0 to 99 (that is, 100 iterations), and the random number between 0 and 1 should be displayed each iteration. Notice that the digital indicator counts from 0 to 99, and *not* from 1 to 100!
7. Save the VI as For Loop Demo.vi in the Users Stuff folder in the Learning directory.

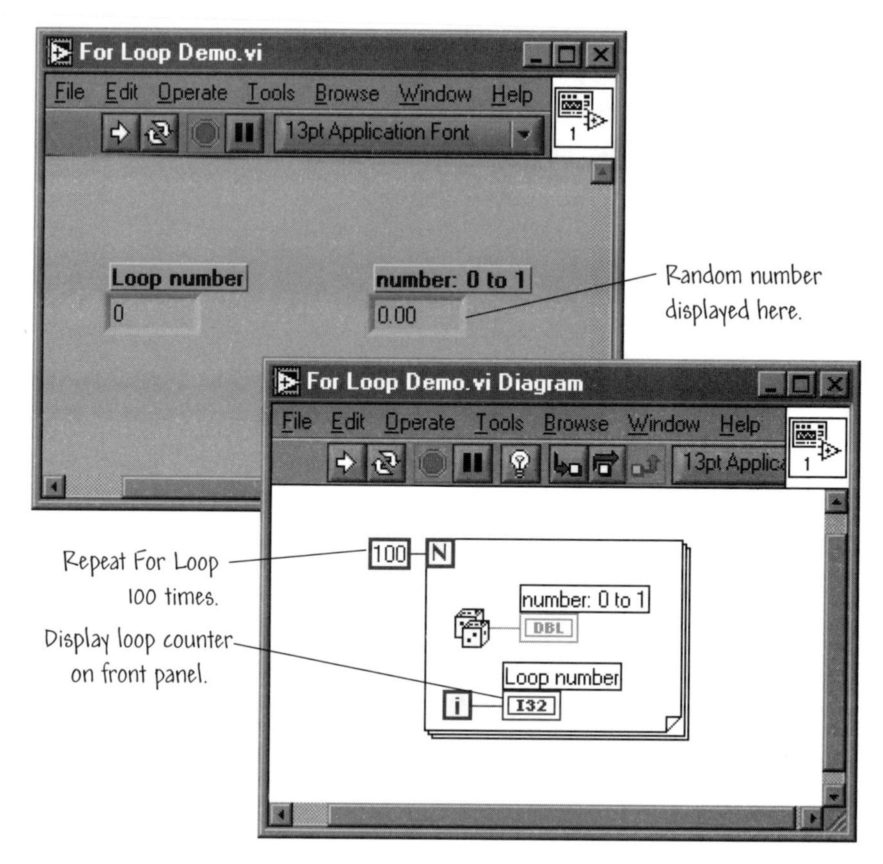

FIGURE 5.6
Displaying a series of random numbers using the For Loop.

A working version of the VI called For Loop Demo.vi *can be found in the* Chapter 5 *folder in the* Learning *directory.* ◆

5.2 THE WHILE LOOP

A While Loop is a structure that repeats a section of code until a certain condition is met. It is comparable to a Do Loop or a Repeat-Until loop in traditional programming languages. The While Loop, shown in Figure 5.7, executes the subdiagram inside its borders until a certain condition is satisfied. The While Loop has two terminals: the **conditional terminal** (an input terminal) and the **iteration terminal** (an output terminal). The iteration terminal of the While Loop behaves exactly like the For Loop iteration terminal. It is an output numeric terminal that outputs the number of times the loop has executed. The conditional terminal input is a Boolean variable: TRUE or FALSE. The While Loop executes until the Boolean value wired to its conditional terminal is FALSE.

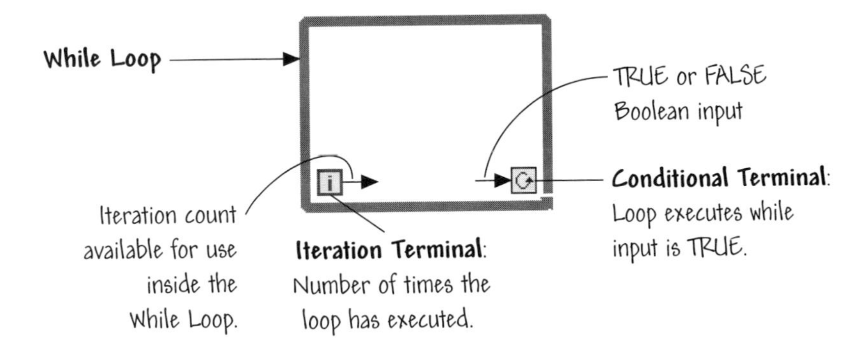

FIGURE 5.7
The While Loop.

The VI checks the conditional terminal at the *end* of each iteration; therefore, the While Loop always executes at least once. If the value at the conditional terminal is TRUE, another iteration is performed; otherwise the loop terminates. The default value of the conditional terminal is FALSE, so it follows that the While Loop iterates only once if you leave the conditional terminal unwired.

You place the While Loop in the block diagram by first selecting it from the **Structures** subpalette of the **Functions** palette, as illustrated in Figure 5.1. Similar to the For Loop, the While Loop is not dropped on the block diagram immediately. Instead, a small icon representing the While Loop appears in the block diagram, giving you the opportunity to size and position the loop, as shown in Figure 5.8. To do so, first click with the While Loop icon in an area above and to the left of all the objects that you want to execute within the While Loop, as shown in Figure 5.8. While holding down the mouse button, drag out a rectangle that encompasses the objects you want to place inside the While Loop. A While Loop is created upon release of the mouse button.

The completed While Loop is a resizable box—once the While Loop is placed on the block diagram, you can use the **Positioning** tool to resize the box by grabbing a corner and stretching the box to the desired dimensions. Additional block diagram elements can be added to the While Loop by dragging and dropping the desired objects inside the While Loop box. As with For Loops, the While Loop border will highlight as the object moves inside, and the block diagram border will highlight when you drag an object out of the While Loop. Once the While Loop (or For Loop) is on the block diagram, you cannot place an object inside the structure by dragging the structure over the object! Doing this will simply cause the structure to be placed over the object (that is, the structure will be the frontmost object). The correct procedure is to drag and drop objects inside existing While Loops (or other structures, such as For Loops).

You can change the way the conditional terminal functions by right clicking on the While Loop and choosing **Stop if True**. Then, rather than having the

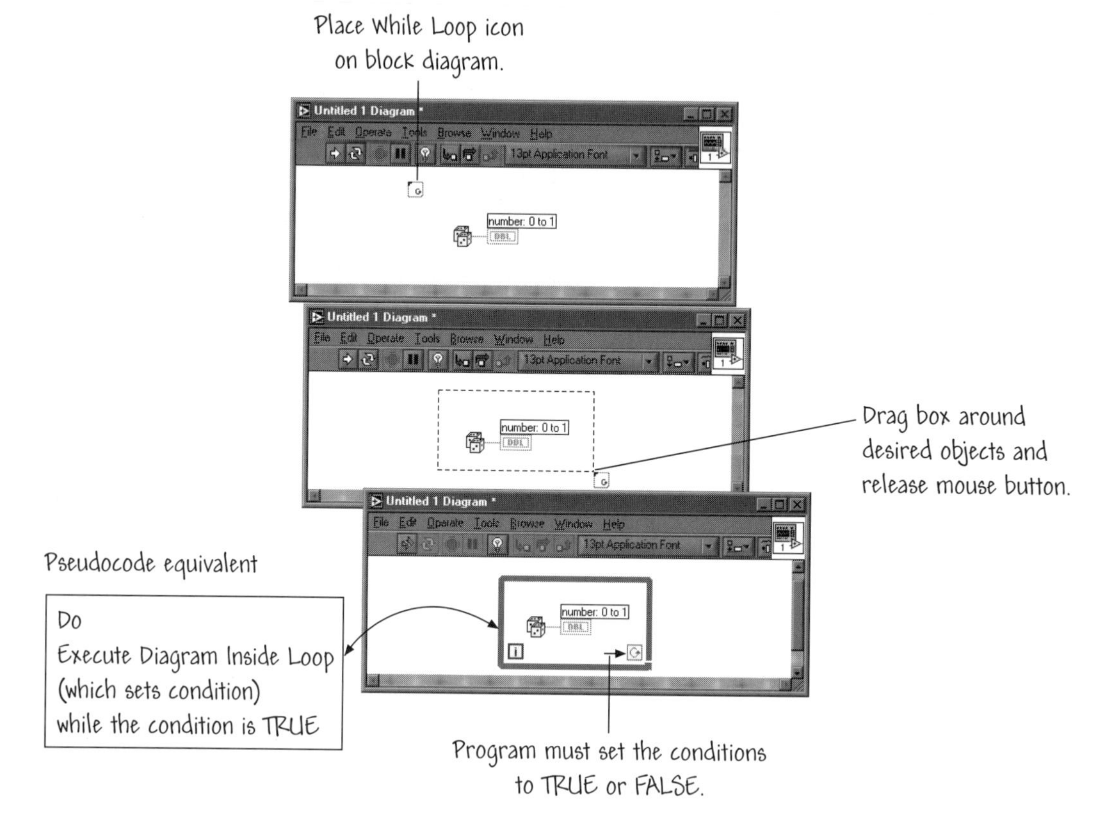

FIGURE 5.8
Placing a While Loop in the block diagram.

While Loop **Continue if True**, it will now **Stop if True**. An example is shown in Figure 5.9. In Fig. 5.9a, the While Loop continues if the value of x is greater than 0.5 and the Enable Boolean is pushed (TRUE). Conversely, in Fig. 5.9b the While Loop will stop if the value of x is greater than 0.5 and the Enable Boolean is pushed (TRUE).

The first time through a For Loop or a While Loop, the iteration count is zero. If you want to register how many times the loop has actually executed, you must add 1 to the count.

A While Loop Example

In this exercise we will place a random number object inside a While Loop and display the random numbers and While Loop counter on the front panel. This is very similar to the For Loop exercise in the previous section. The VI shown in

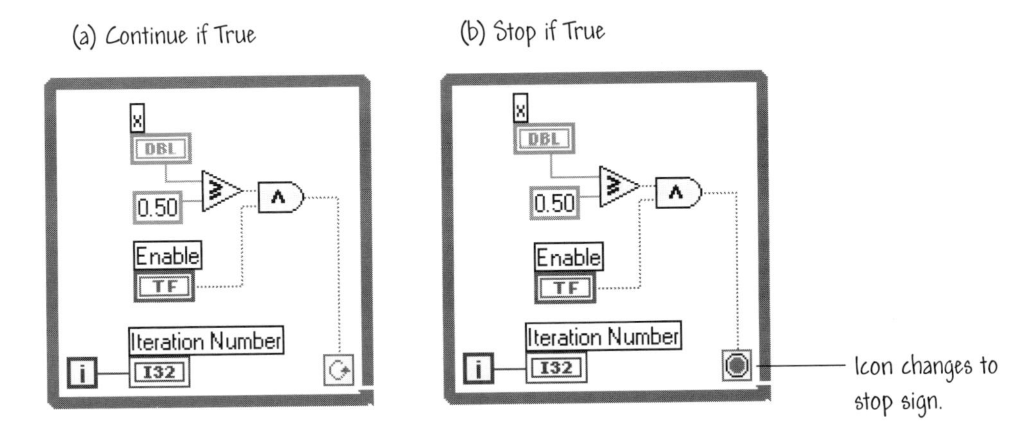

FIGURE 5.9
Continue if True or Stop if True.

Figure 5.10 can be constructed by following these steps:

1. Place a random number function on the block diagram (found in the Functions≫Numeric palette). Create an indicator on the random number function and label it number: 0 to 1.
2. Place the While Loop on the block diagram so that the random number is enclosed within the loop, as shown in Figure 5.10.
3. Create a control to the conditional terminal by popping up on the terminal and selecing **Create Control**. A Boolean variable will appear on the block diagram, and simultaneously an on/off button will appear on the front panel. This will be used to stop the While Loop iterations while in the **Run** mode.
4. Create an indicator on the iteration terminal and label it Loop number.
5. To run the program, use the **Operating** tool and set the button on the front panel to the "on" position. Click on the **Run** button to start the program execution, and run the program with **Highlight Execution** to observe the program data flow.
6. On the front panel you will see the loop counter continue to increment until you push the button to the "off" position. This causes the conditional terminal to change to FALSE, and the While Loop iterations cease.

*If you want to run a VI multiple times, place all the code in a loop rather than using the **Continuous Run** button. It is better to stop a VI with a Boolean control than with the **Abort Execution** button.*

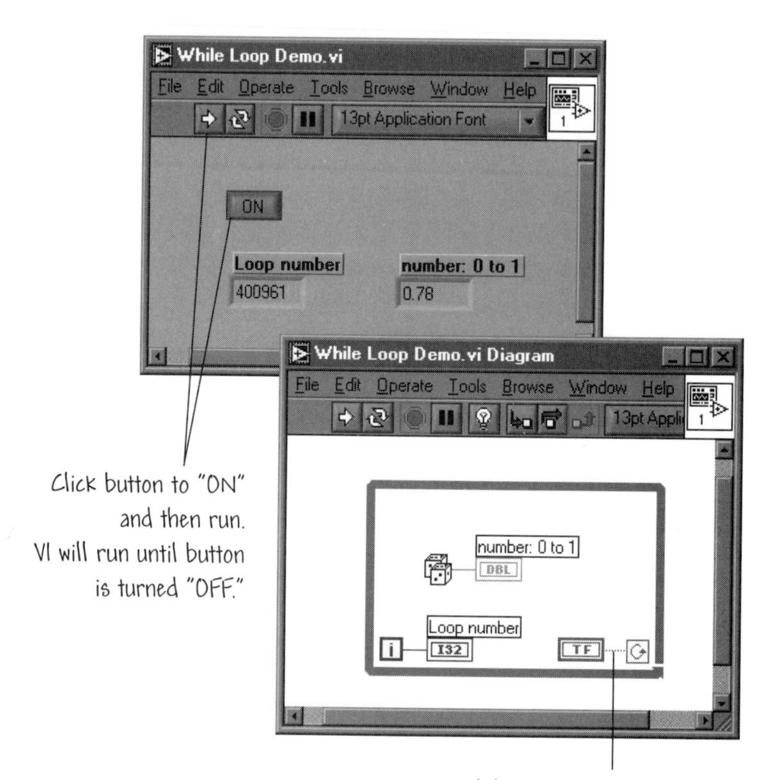

FIGURE 5.10
Displaying a series of random numbers using the While Loop.

7. Save the VI as While Loop Demo.vi in Learning\Users Stuff.

A working version of the VI called While Loop Demo.vi *can be found in the* Chapter 5 *folder in the* Learning *directory.*

If you place the terminal of the Boolean control outside the While Loop, as shown in Figure 5.11, you create either an infinite loop or a loop that executes only once, depending on the initial value. Why? Because in G programming the Boolean input data value is read before it enters the loop—remember data flow programming?—and not within the loop or after completion of the loop.

To experiment with this, modify your VI by placing the Boolean terminal outside of the While Loop. Connect the Boolean terminal to the While Loop

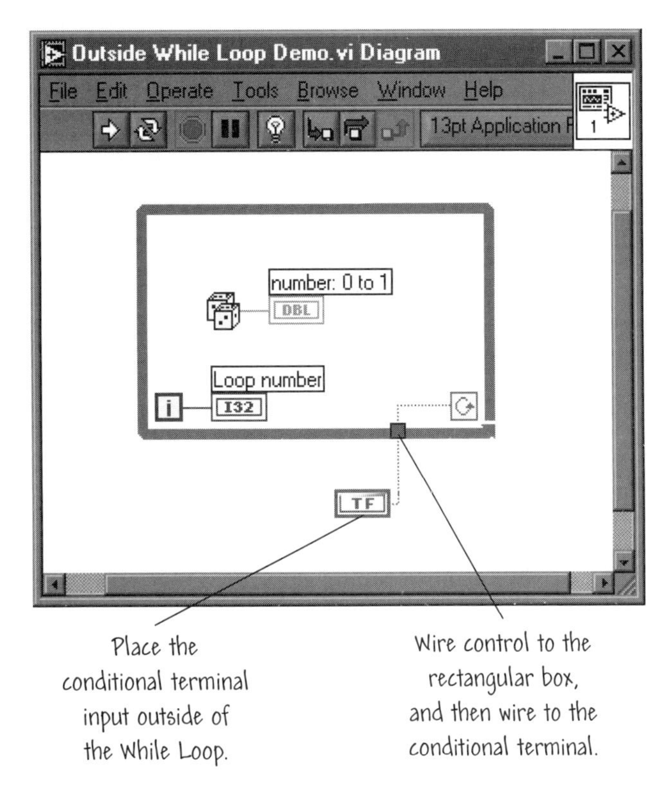

FIGURE 5.11
Placing the conditional terminal input outside of the While Loop.

border, as illustrated in Figure 5.11. A green rectangle will appear on the loop border. Then wire the green rectangle to the conditional terminal.

Conduct the following two numerical experiments: First, set the button state on the front panel to "on" and then run the VI (use **Highlight Execution** to watch the data flow). After a reasonable period of time, press the button to "off." What happens? The VI does not stop running—it is in an infinite loop. Can you explain this behavior using the notion of data flow programming? Since LabVIEW operates under data flow principles, inputs to the While Loop must pass their data before the loop executes. A While Loop passes data out only after the loop completes all iterations. The infinite loop experiment can be stopped by clicking on the **Abort Execution** button on the front panel toolbar.

For a second experiment, set the button state on the front panel to "off" and then run the VI (use **Highlight Execution** to watch the data flow). What happens in this case? You should see that the VI will execute once through the While Loop.

Save the VI with the conditional terminal outside the While Loop as Outside While Loop Demo.vi in the Users Stuff folder in the Learning directory.

A working version of Outside While Loop Demo.vi *can be found in the* Chapter 5 *folder in the* Learning *directory.* ◆

5.3 SHIFT REGISTERS

Shift registers transfer values from one iteration of a For Loop or While Loop to the next. The shift register is comprised of a pair of terminals directly opposite each other on the vertical sides of the loop border, as shown in Figure 5.12. You create a shift register by popping up on the left or right loop border and selecting **Add Shift Register** from the short cut menu.

The right terminal (the rectangle with the up arrow) stores the data as each iteration finishes. The stored data from the previous iteration is shifted and appears at the left terminal (the rectangle with the down arrow) at the beginning of the next iteration, as shown in Figure 5.13. A shift register can hold any data type, including numeric, Boolean, strings (see Chapter 9), and arrays (see Chapter 6), but the data wired to the terminals of each register must be of the same type. The shift register conforms to the data type of the first object that is wired to one of its terminals.

Consider a simple illustrative example. Suppose we have the two situations depicted in Figure 5.14. Both cases look very similar—but the code in (b) contains a shift register. The code in (b) computes a running sum of the iteration

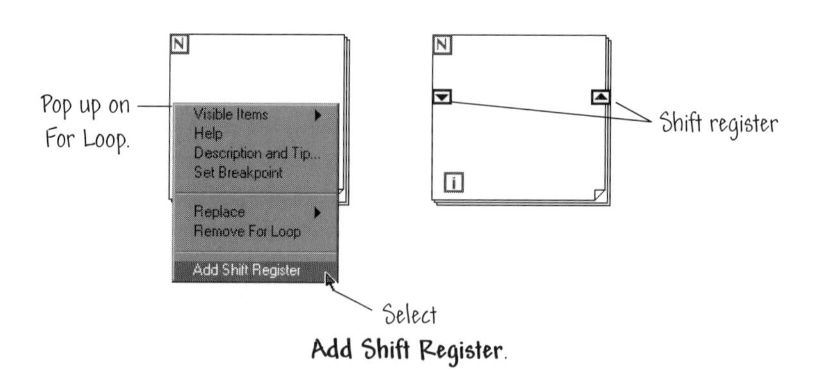

FIGURE 5.12
Shift register containing a pair of terminals.

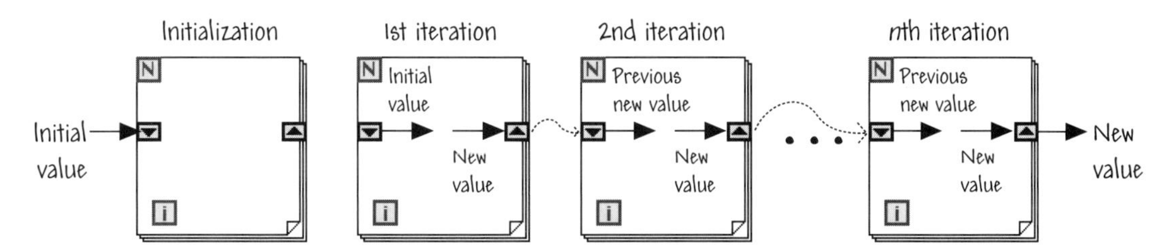

FIGURE 5.13
Passing data from one loop iteration to next using shift registers.

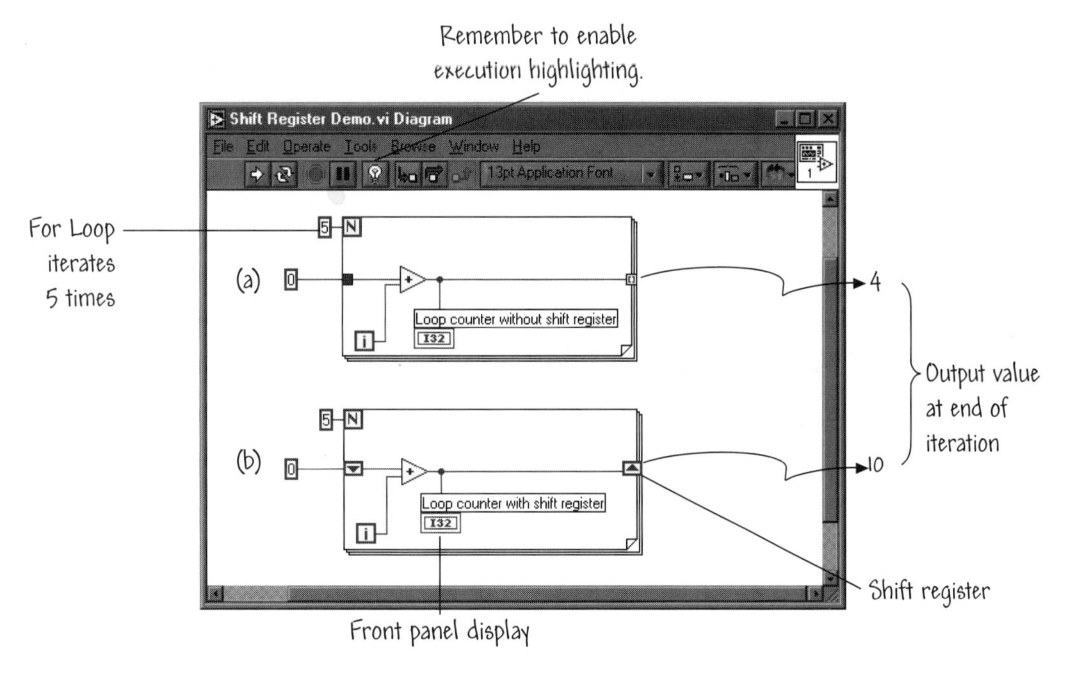

FIGURE 5.14
A simple example showing the effect of adding a shift register to a For Loop.

count within the For Loop. Each time through the loop, the new sum is saved in the shift register. At the end of the loop, the total sum of 10 is passed out to the numeric indicator. Why 10? Because the sum of the iteration count is 10 (= 0+1+2+3+4). On the other hand, the code in (a) that does not contain the shift register does not save values between iterations. Instead, a zero is added to the current iteration count each time, and only the last value of the iteration counter (= 4) will be passed out of the loop.

*You can run the simulation shown in Figure 5.14 by opening the VI titled Shift Register Demo.vi which is in Learning\Chapter 5 directory—enable **Highlight Execution** before running and watch the data flow through the code!*

5.3.1 Using Shift Registers to Remember Data Values from PreviousLoop Iterations

You can configure the shift register to store data values from previous iterations. To prepare the loop to store previous values, first create additional terminals on the loop border, as illustrated in Figure 5.15. If, for example, you add four elements to the left terminal, you can store and access values from the previous four iterations. You add the additional elements by popping up on the left terminal of the shift register and choosing **Add Element** from the short cut menu.

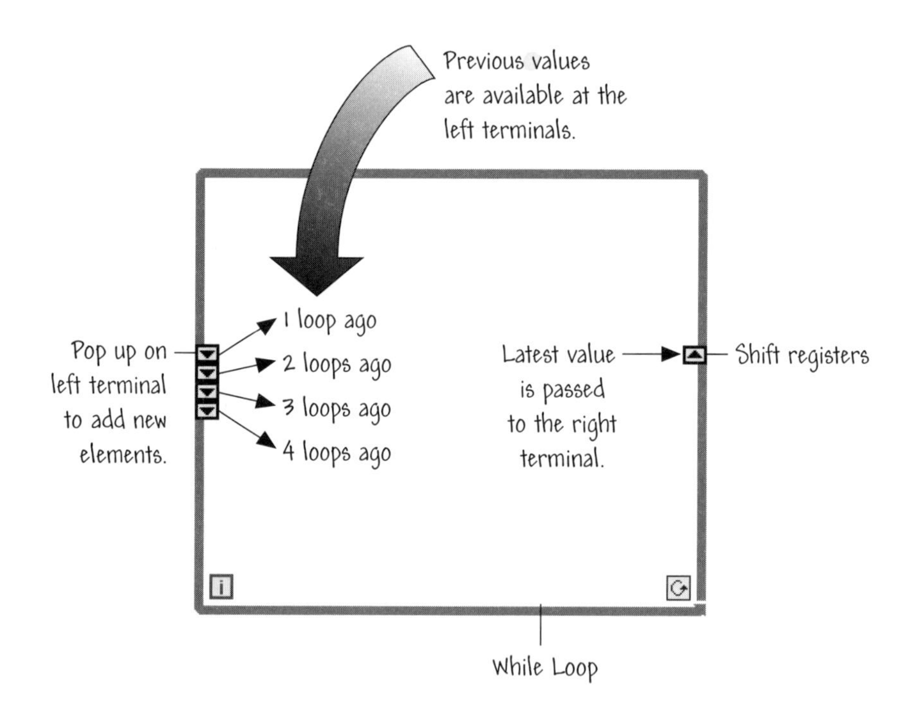

FIGURE 5.15
Adding elements to the shift register to access values from previous loop iterations.

Using Shift Registers

In this example we will open an existing VI and use it to watch the data flow in a While Loop containing shift registers. Begin by opening Viewing Shift Registers.vi located in Learning\Chapter 5.

The front panel has four digital indicators, as shown in Figure 5.16. The $X(i)$ indicator will display the current value, which will shift to the left terminal at the beginning of the next iteration. The $X(i-1)$ indicator will display the value one iteration ago, the $X(i-2)$ indicator will display the value two iterations ago, and the $X(i-3)$ indicator will display the value three iterations ago. The shift register is initialized to zero.

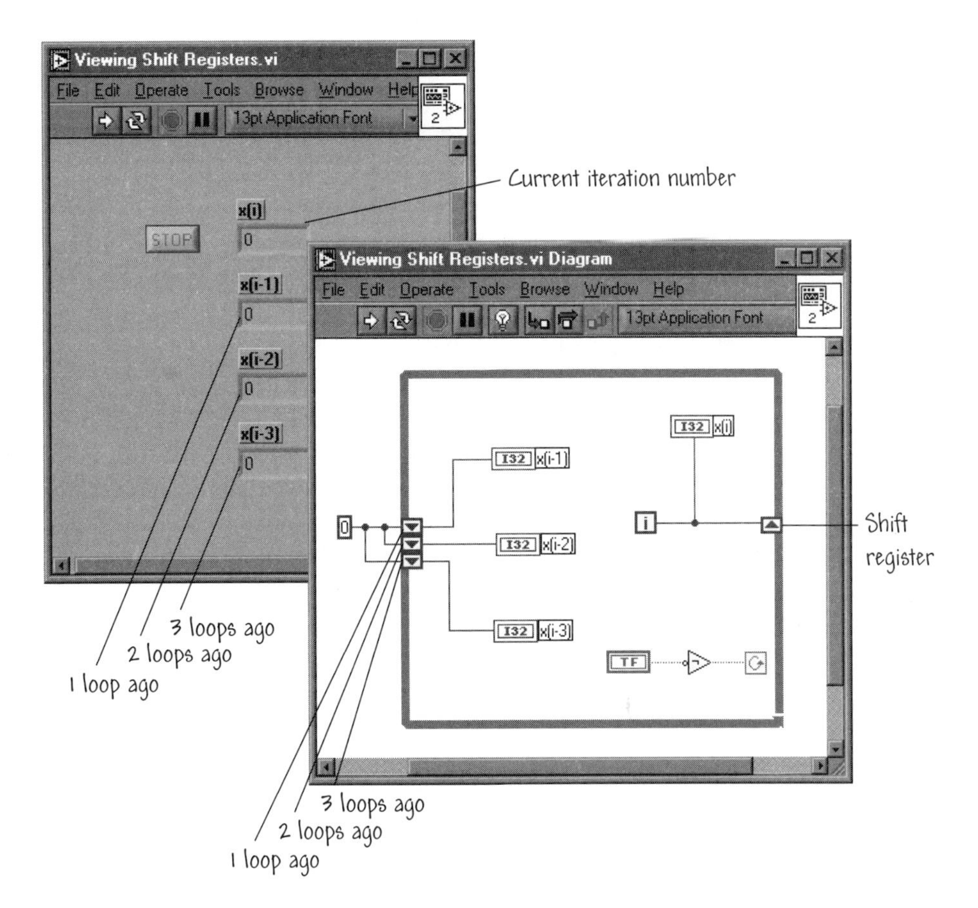

FIGURE 5.16
A demonstration of the use of shift registers to access data from previous iterations of a While Loop.

Before running the program, make sure that **Highlight Execution** is enabled. This will allow you to view the data flow and to watch the shift registers access data from previous iterations of the While Loop. Run the VI and watch the bubbles indicating the data flow. Notice that in each iteration of the While Loop, the VI "funnels" the previous values through the left terminals of the shift register. $X(i)$ shifts to the left terminal, $X(i-1)$, at the beginning of the next iteration. The values at the left terminal funnel downward through the terminals. In this example, the VI retains the last three values. To retain more values, you would need to add more elements to the left terminal of the shift register. When you're finished observing the VI operation, close it and do not save any changes. ◆

5.3.2 Initializing Shift Registers

Shift registers are initialized by wiring a constant or control to the left terminal of the shift register from outside the loop. Shift register initialization is demonstrated in Figure 5.17. On the first execution the final value of the shift register is 12. Why 12? Because the final value is a sum of the iteration count ($= 0 + 1 + 2 + 3 + 4$) plus the initial value ($= 2$). On the second execution of the code that contains the initialized shift register, the result is exactly the same as on the first execution. In fact, the results on all subsequent executions are identical to the first execution: the final value of the shift register is 12. That is, the final value of the shift register from the first execution does not play a role in the second run—this is as it should be!

Unless the shift register is explicitly initialized, the first time the VI is executed the initial value of the shift register will be the default value for the shift register data type—if the shift register data type is Boolean, the initial value will be FALSE. Similarly, if the shift register data type is numeric, the initial value will be zero. The bottom row of Figure 5.17 illustrates what happens if you exe-

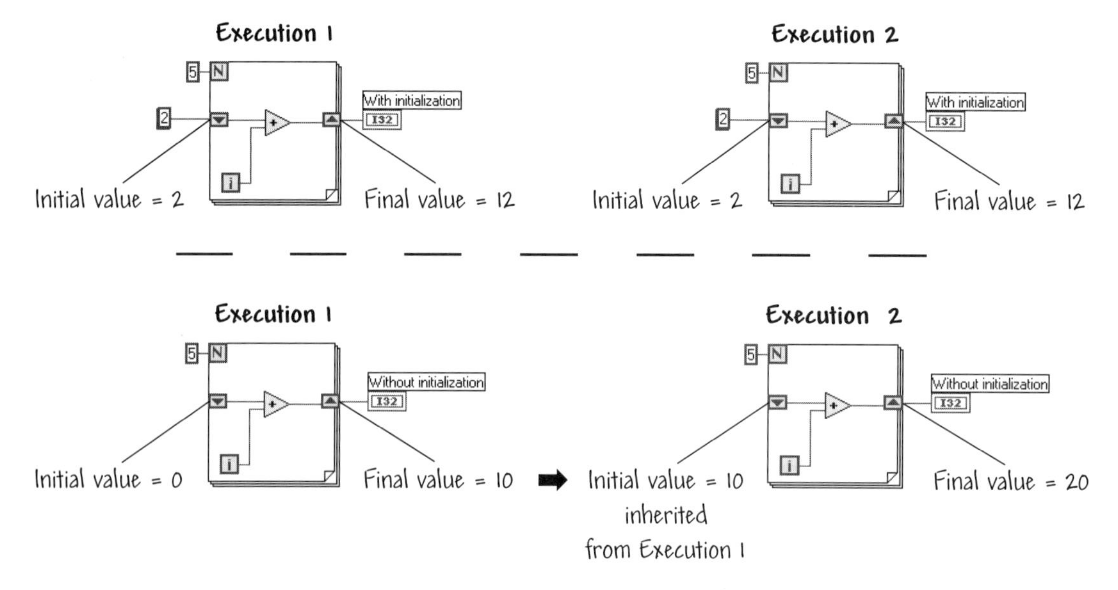

FIGURE 5.17
Initializing shift registers.

cute code twice with uninitialized shift registers. On the first execution the final value of the shift register is 10. Why 10? Can you explain this result? Running the code again, without closing the VI first, results in a final value of the shift register equal to 20. This result is due to the fact that on the second execution the initial value of the shift register is equal to 10 (left over from the previous run), and that is added to the sum of the iteration counter on the second run, which is equal to 10. When the shift register is not explicitly initialized, on the second run the shift register will take on the last value of the first run. Always use intialized shift registers for consistent results!

Data values stored in the shift register are stored until you close the VI and remove it from memory. That is, if you run a VI containing uninitialized shift registers, the initial values taken by the shift registers during subsequent executions will be the last values from previous executions. This can make debugging your VI a difficult process because this situation is hard to detect.

The shift register initialization demonstration depicted in Figure 5.17 can be found in Learning\Chapter 5 *and is called* Shift Register Init Demo.vi.

Computing a Running Average

In this example, we will use shift registers to compute the running average of a sequence of random numbers. Since the random number function in LabVIEW provides random numbers from 0 to 1, we expect that the average will be 0.5. How many random numbers will it take to obtain an average near 0.5?

Figure 5.18 shows a VI that computes the running average of several random numbers. The length of the random number sequence is input via the slide control, as depicted on the front panel in Figure 5.18. Using the block diagram in the figure as a model, develop your own VI to compute the running average of a sequence of random numbers. The formula that is coded to compute the average is

$$Ave_i = \frac{i}{i+1} Ave_{i-1} + \frac{1}{i+1} RN_i \ ,$$

where $i = 0, 1, \cdots, N-1$, Ave_i is the computed average at the ith iteration, and RN_i is the current random number from the random number function. If you aren't familiar with computing a running average, just concentrate on the programming aspects of the VI shown in Figure 5.18 and don't worry about the formula. The main point is to understand how to use the shift registers in conjunction with For Loops and While Loops!

In this example, the shift register is used to pass the value of the variable Ave_{i-1} from one iteration to the next. You should notice as you run the code that for small values of N (e.g., $N = 3$) the average is generally not close to the

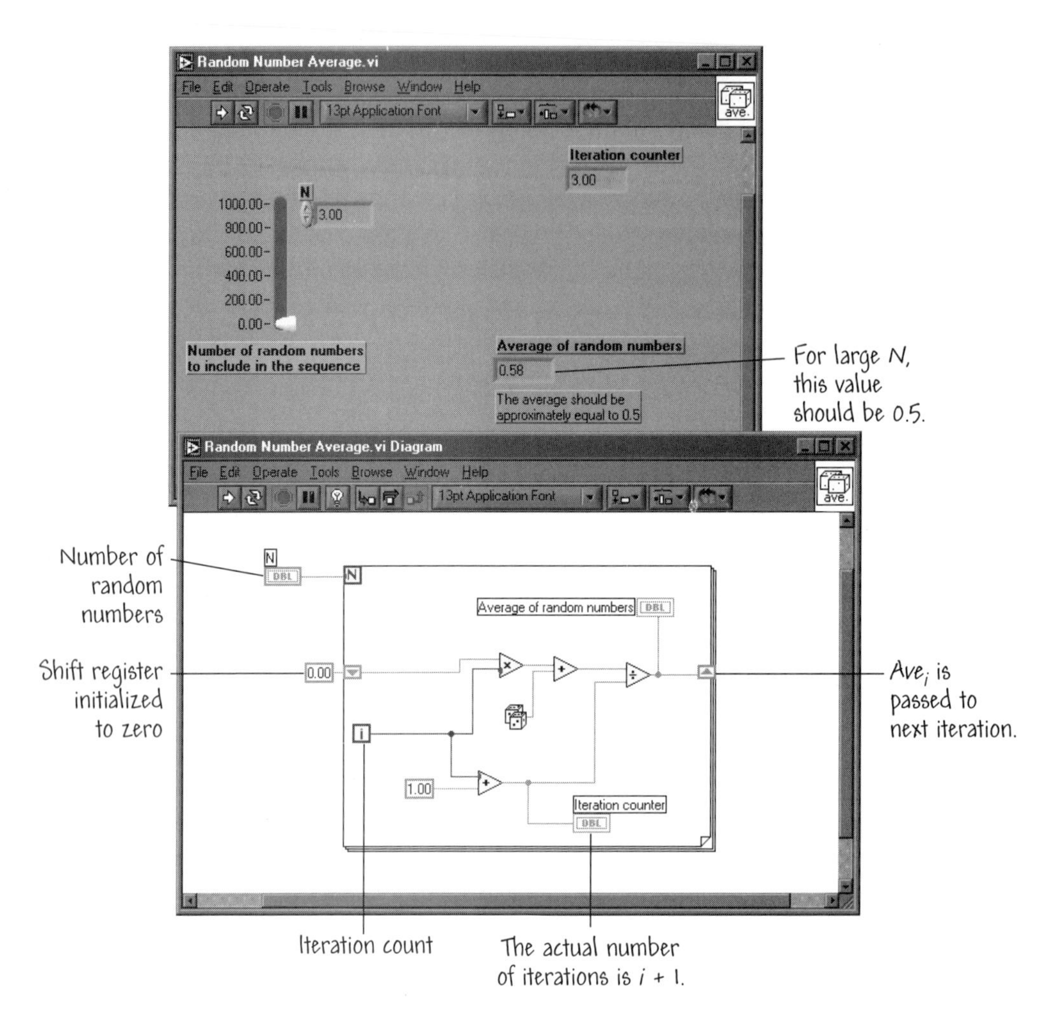

FIGURE 5.18
Computing a running average of a sequence of random numbers.

expected value of $Ave_N = 0.5$; however, as you increase N, the average gets closer and closer to the expected value. Try it out!

The VI depicted in Figure 5.18 can be found in Learning\Chapter 5 *and is called* Random Number Average.vi. *Check it out if you can't get yours to work or if you want to compare your results.* ◆

5.4 CASE STRUCTURES

A **Case structure** is a method of executing conditional text. This is analogous to the common If...Then...Else statements in conventional, text-based programming languages. You place the Case structure on the block diagram by selecting it from the **Structures** subpalette of the **Functions** palette, as shown in Figure 5.19. As with the For Loop and While Loop structures, you can either drag the Case structure icon to the block diagram and enclose the desired objects within its borders, or you can place the Case structure on the block diagram, resize it as necessary, and drag objects inside the structure.

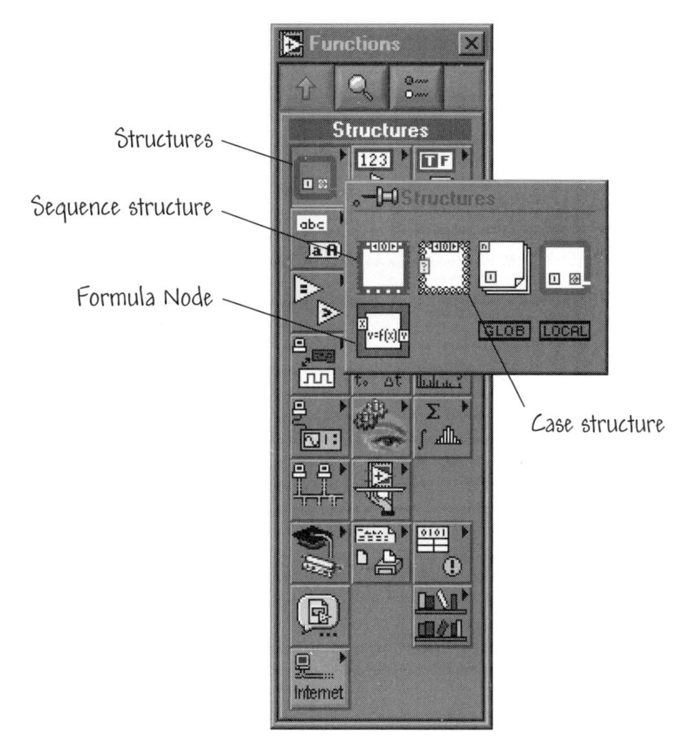

FIGURE 5.19
Selecting the Case structure from the palette.

The Case structure can have multiple subdiagrams. The subdiagrams are configured like a deck of cards of which only one card is visible at a time. At the top of the Case structure border is the diagram identifier. The diagram identifier can be numeric, Boolean, string, or enumerated type control. An enumerated type control is unsigned byte, unsigned word, or unsigned long and is selectable from the decrement and increment buttons, as depicted in Figure 5.20a and b. The diagram identifier displays the values that cause the corresponding subdiagrams to execute. The Sequence structure is similar in form to the Case structure and is depicted in Figure 5.20c and d. If the Sequence structure diagram identifier is numeric, the diagram identifier value is followed by a diagram identifier range, which shows the minimum and maximum values for which the structure contains a subdiagram. To view other subdiagrams (that is, to see the subdiagrams within the "stack of cards"), click the decrement (left) or increment (right) button to display the previous or next subdiagram, respectively. Decrementing from the first subdiagram displays the last, and incrementing from the last subdiagram displays the first subdiagram—it wraps around!

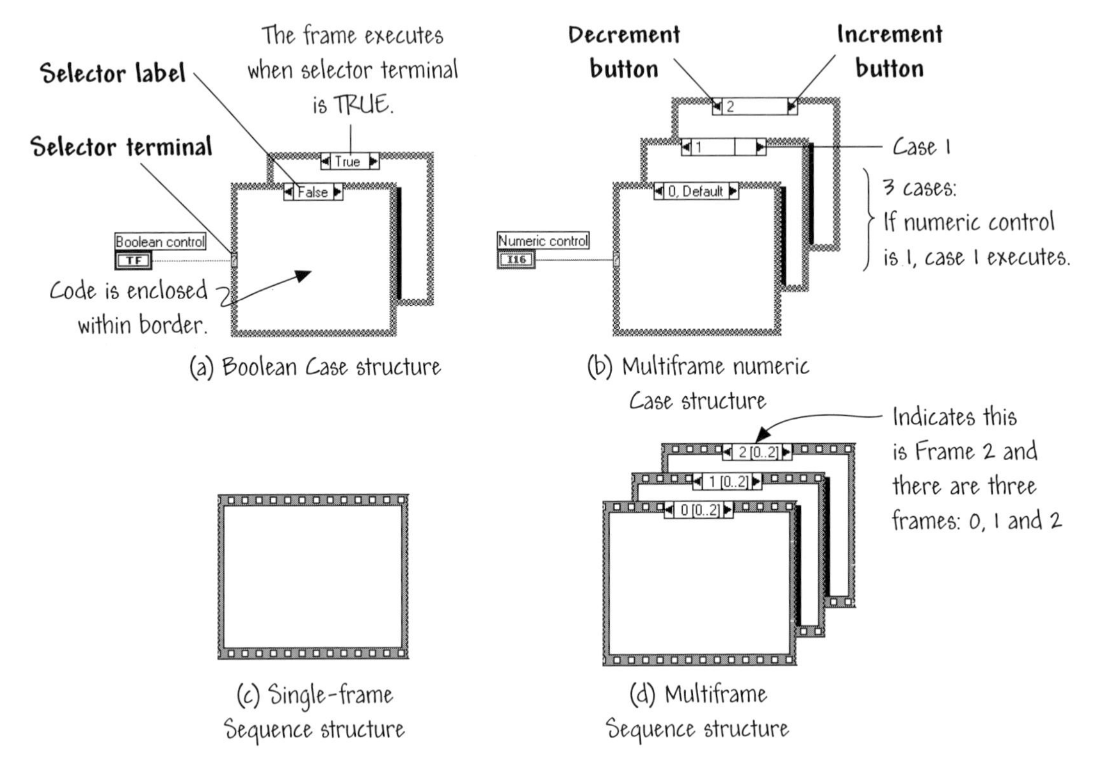

FIGURE 5.20
Overview of Case and Sequence structures.

You can position the selector terminal anywhere on the Case structure along the left border. The selector label automatically adjusts to the input data type. For example, if you change the value wired to the selector from a numeric to a Boolean, then cases 0 and 1 change to FALSE and TRUE, respectively. Here is a point to consider in changing the data type: if the Case structure selector originally received numeric input, then n cases $0, 1, 2, \cdots, n$ may exist in the code. Upon changing the selector input data from numeric to Boolean, cases 0 and 1 change automatically to FALSE and TRUE. However, cases $2, 3, \cdots, n$ are not discarded! You must explicitly delete these extra cases before the Case structure can execute.

You can also type in and edit values directly into the selector label using the **Labeling** tool. The selector values are specified as a single value, as a list or as a range of values. A list of values is separated by commas, such as $-1, 0, 5, 10$. A range is typed in as 10..20, which indicates all numbers from 10 to 20, inclusively. You also can use open-ended ranges. For example, all numbers less than or equal to 0 are represented by ..0. Similarly, 100.. represents all numbers greater than or equal to 100.

The case selector can also usc string values that display in quotes, such as "red," and "green." You don't need to type in the quotes when entering the values unless the string contains a comma or the symbol "..". In a case selector using strings, you can use special backslash codes (such as \r, \n, and \t) for nonalphanumeric characters (carriage return, line feed, tab, respectively).

If you type in a selector value that is not the same type as the object wired to the selector terminal, then the selector value displays in red, and your VI is broken. Also, because of the possible round-off error inherent in floating-point arithmetic, you cannot use floating-point numbers in case selector labels. If you wire a floating-point type to the case terminal, the type is rounded to the nearest integer, and a coercion dot appears. If you try to type in a selector value that is a floating-point number, then the selector value displays in red, and your VI is broken.

5.4.1 Adding and Deleting Cases

If you pop up on the Case structure border, the resulting menu gives you the many options shown in Figure 5.21. You choose **Add Case After** to add a case after

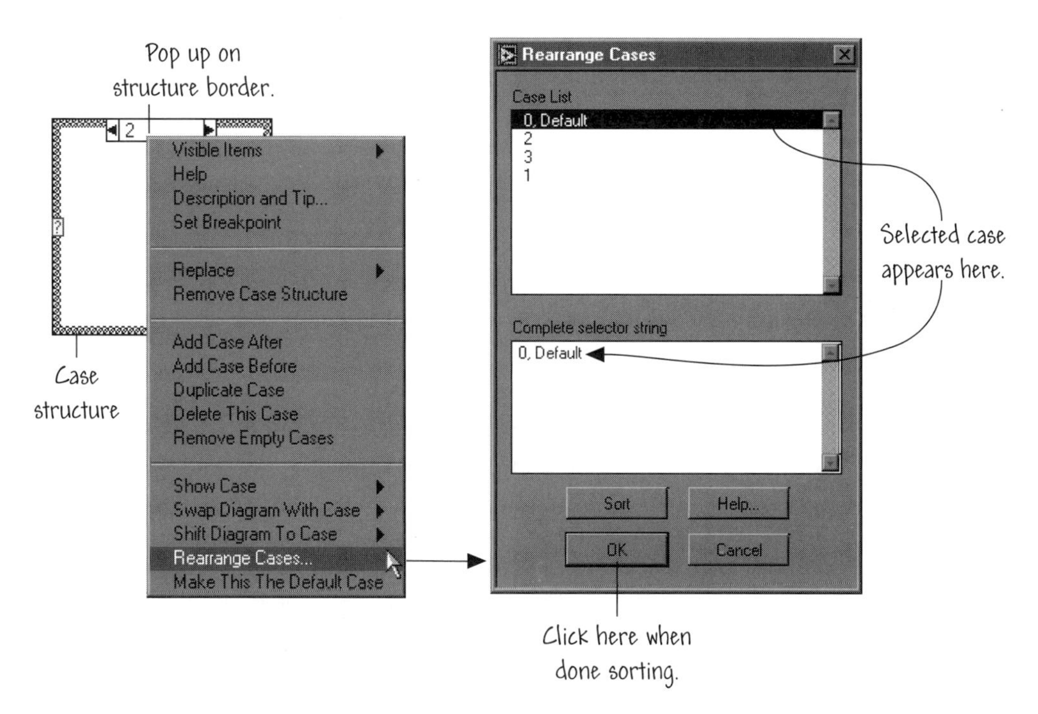

FIGURE 5.21
Adding and deleting cases in Case structures.

the case that is currently visible. **Delete This Case**, or **Add Case Before** to add a case before the currently visible case. You can also choose to copy the currently shown case by selecting **Duplicate Case** and to delete the current visible case by selecting **Delete This Case**. When you add or remove cases (i.e., subdiagrams) in a case structure, the diagram identifiers are automatically updated to reflect the inserted or deleted subdiagrams.

Sometimes you may want to rearrange the listed order of the cases in the structure. For example, rather than have the cases listed $(0, 1, 2, 3, \cdots)$, you might want them listed in the order $(0, 2, 1, 3, \cdots)$. Resorting the order in which the cases appear in the case structure on the block diagram does not affect the run-time behavior of the Case structure! It is merely a matter of programming preference. You can change the order in which the cases are listed in the structure by selecting **Rearrange Cases** from the short cut menu. When you do so, the dialog box shown in Figure 5.21 appears. The **Sort** button sorts the case selector values based on the first selector value. To change the location of a selector, click the selector value you want to move (it will highlight when selected) and drag it to the desired location in the stack. In the **Rearrange Cases** dialog box, the section entitled "Complete selector string" shows the selected case selector name in its entirety in case it is too long to fit in the "Case List" box located in the top portion of the dialog box. Other on-line assistance can be found in the context-sensitive help in the **Help** menu.

The **Make This The Default Case** item in the short cut menu specifies the case to execute if the value that appears at the selector terminal is not listed as one of the possible choices in the selector label. Case statements in other programming languages generally do not execute any case if a case is out of range, but in LabVIEW you must either include a default case that handles out-of-range values or explicitly list every possible input value. For the default case that you define, the word "Default" will be listed in the selector label at the top of the Case structure.

A Simple Case Structure Example

In this exercise you will build a VI that uses a Boolean Case structure, shown in Figure 5.22. The input numbers from the front panel will pass through **tunnels** to the Case structure, and there they are either added or subtracted, depending on the Boolean value wired to the selector terminal.

What is a tunnel? A tunnel is a data entry point or exit point on a structure. You can wire a terminal from outside the Case structure to a terminal within the structure. When you do this a rectanglar box will appear on the structure border—this represents the tunnel. You can also wire an external terminal to the structure border to create the tunnel, and then wire the terminal to an internal terminal in a second step. Tunnels can be found on other structures, such as Sequence structures, While Loops, and For Loops. The data at all input tunnels is available to all cases.

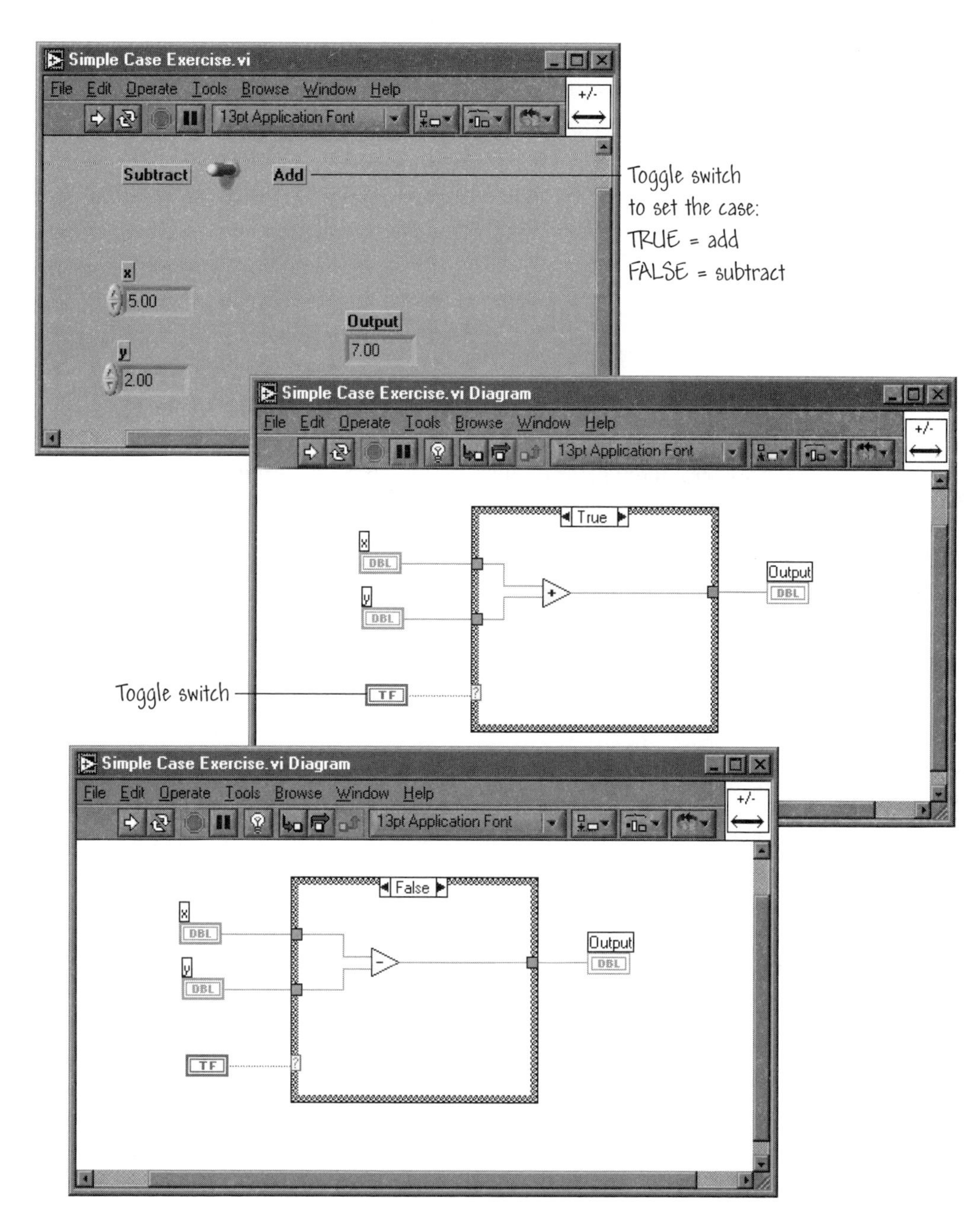

FIGURE 5.22
A Boolean Case structure example to add or subtract two input numbers.

In this exercise you will need to create two tunnels to get data into the Case structure and one tunnel to pass data out of the structure. Using Figure 5.22 as a guide, construct a VI that utilizes the Case structure to perform the following task: if the Boolean wired to the selector terminal is TRUE, the VI will add the numbers; otherwise, the VI will subtract the numbers. The Boolean value of the selector terminal is toggled using a horizontal toggle switch found on the front panel.

The VI depicted in Figure 5.22 can be found in Learning\Chapter 5 *and is called* Simple Case Exercise.vi. ◆

5.4.2 Wiring Inputs and Outputs

As previously mentioned, the data at all input terminals (tunnels and selection terminal) is available to all cases. Cases are not required to use input data or to supply output data, but if any one case supplies output data, all must do so. Forgetting this fact can lead to coding errors. Correct and incorrect wire situations are depicted in Figure 5.23, where for the FALSE case two input numbers are added and sent (through the tunnel) out of the structure, and for the TRUE case the computer system beeps.

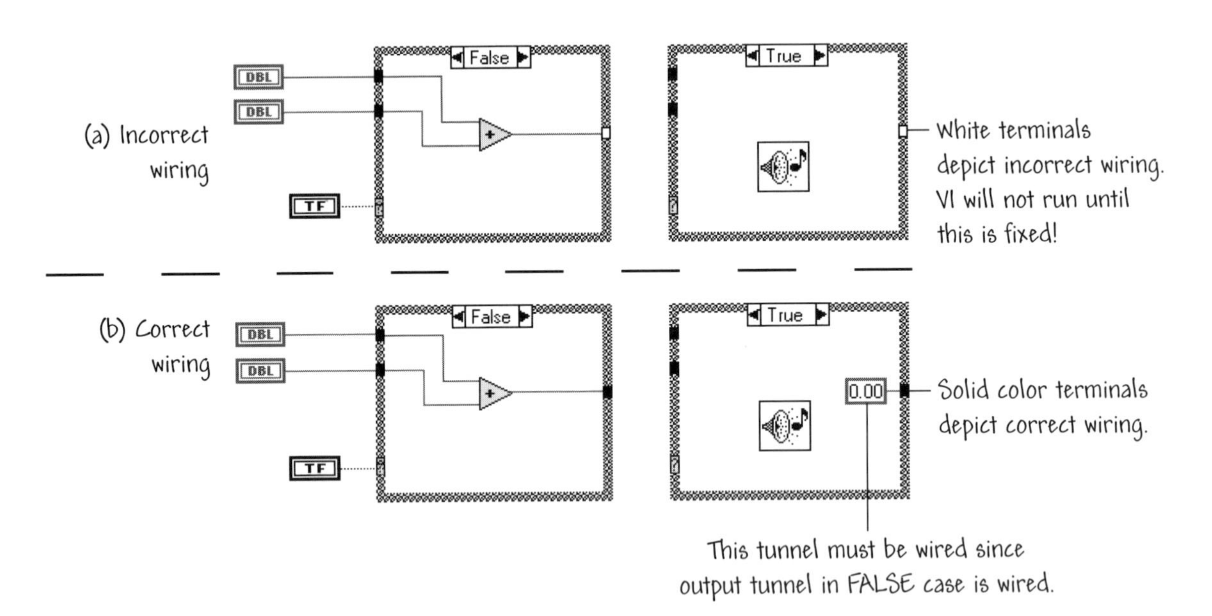

FIGURE 5.23
Wiring Case structure inputs and outputs.

When you create an output tunnel in one case, tunnels appear at the same position on the structure border in the other cases. You must define the output tunnel for each case. Unwired tunnels look like white squares, and when they occur, the

Broken Run button will also appear, so be sure to wire to the output tunnel for each unwired case. You can wire constants or controls to unwired cases by popping up on the white square and selecting **Constant** or **Control** from the **Create** menu. When all cases supply data to the tunnel, it takes on a solid color consistent with the data type of the supplied data, and the **Run** button appears.

Using Case Structures

In this exercise, you will build a VI that computes the ratio of two numbers. If the denominator is zero, the VI outputs ∞ to the front panel and causes the system to make a beep sound. If the denominator is not zero, the ratio is computed and displayed on the front panel.

Begin by opening a new VI. Build the simple front panel shown in Figure 5.24. The digital control x Numerator supplies the numerator x, and the digital control y Denominator supplies the denominator y. The x/y digital indicator displays the ratio of the two input numbers. Switch to the block diagram and place the Case structure in the window.

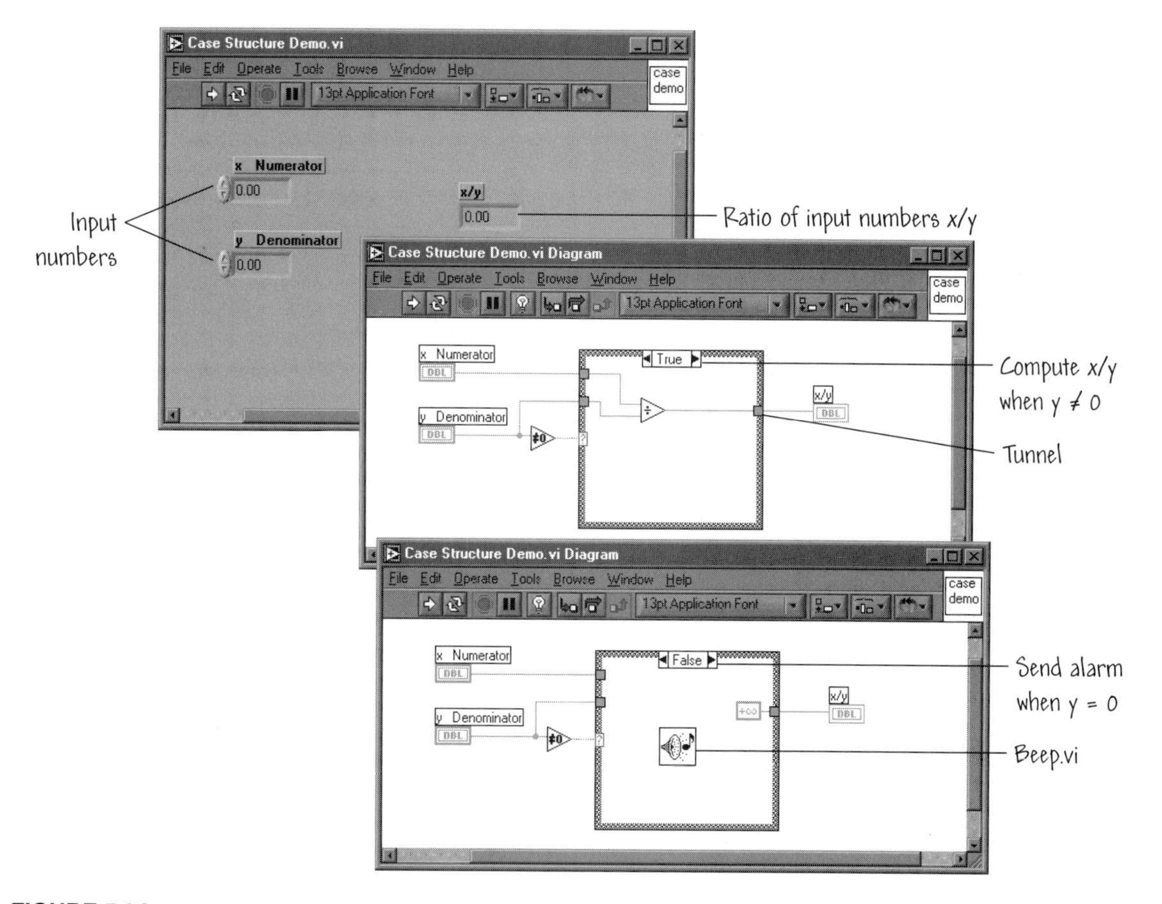

FIGURE 5.24
A VI that computes the ratio of two numbers and uses Case structures to handle division by zero.

By default, the Case structure selector terminal is Boolean—this is the format we desire in this situation. You can display only one case at a time, and we'll start by considering the TRUE case. Remember that to change cases, you need to click on the increment and/or decrement arrows in the top border of the Case structure.

Place the other diagram objects and wire them as shown in Figure 5.24. The main objects are:

1. The Not Equal to 0? function (found in the Functions≫Comparison menu) checks whether the denominator is zero. The function returns a TRUE if the denominator is not equal to 0.
2. The Divide function (found in the Functions≫Numeric menu) returns the ratio of the numerator and denominator numbers.
3. The Positive Infinity constant indicates that a divide by zero has been attempted (see the Functions≫Numeric≫Additional Numeric Constants menu).
4. The Beep.vi function (found in the library **vi.lib\platform\system.lib**) causes the system to issue an audible tone when division by zero has been attempted. On **Windows** platforms, all input parameters to the beep function are ignored, while on the **Macintosh**, you can specify the tone frequency in Hertz, the duration in milliseconds, and the intensity as a value from 0 to 255, with 255 being the loudest. Double click on the icon to see this VI's front panel.

You must define the output tunnel for each case. When you create an output tunnel in the TRUE case, an output tunnel will appear at the same position in the FALSE case. Unwired tunnels look like white squares. In this and all other VI development, be sure to wire to the output tunnel for each unwired case. In Figure 5.24, the constant ∞ is wired to the output tunnel in the FALSE case.

The VI will execute either the TRUE case or the FALSE case. If the denominator number is not equal to zero, the VI will execute the TRUE case and return the ratio of the two input numbers. If the denominator number is equal to zero, the VI will execute the FALSE case and output ∞ to the digital indicator and make the system beep.

Once all objects are in place and wired properly, return to the front panel and experiment with running the VI. Change the input numbers and compute the ratio. Make the denominator input equal to zero and listen for the system beep. When you are finished, save the VI as Case Structure Demo.vi in the Users Stuff folder in the Learning directory. Close the VI.

A working version of this VI called Case Structure Demo.vi can be found in Learning\Chapter 5. ◆

5.5 SEQUENCE STRUCTURES

The **Sequence structure** executes subdiagrams sequentially. The subdiagrams look like a frame of film; hence they are known as **frames**. The Sequence structure is shown in Figure 5.20. Determining the execution order of a program by arranging its elements in a certain sequence is called **control flow**. Most text-based programming languages (such as Fortran and C) have inherent control flow because statements execute in the order in which they appear in the program. In data flow programming, a node executes when data is available at all of the node inputs (this is known as **data dependency**), but sometimes you cannot connect one node to another. When data dependencies are not sufficient to control the data flow, the Sequence structure is a way of controlling the order in which nodes execute. You use the Sequence structure to control the order of execution of nodes that are not connected with wires. Within each frame, as in the rest of the block diagram, data dependency determines the execution order of nodes.

The code that you want to execute first is placed inside the border of the Sequence structure frame 0(0..x), the code to be executed second is placed inside the border of frame 1(0..x), and so on. The interval (0..x) represents the range of frames in the Sequence structure. Only when the last frame completes does data leave the structure.

The Sequence structure is found in the **Structures** palette of the **Functions** menu, as shown in Figure 5.19. As with the Case structure, only one frame is visible at a time. Clicking on the arrows at the top of the structure allows you to flip through the other frames. New frames are created by popping up on the structure border and selecting **Add Frame After** or **Add Frame Before**, as illustrated in Figure 5.25.

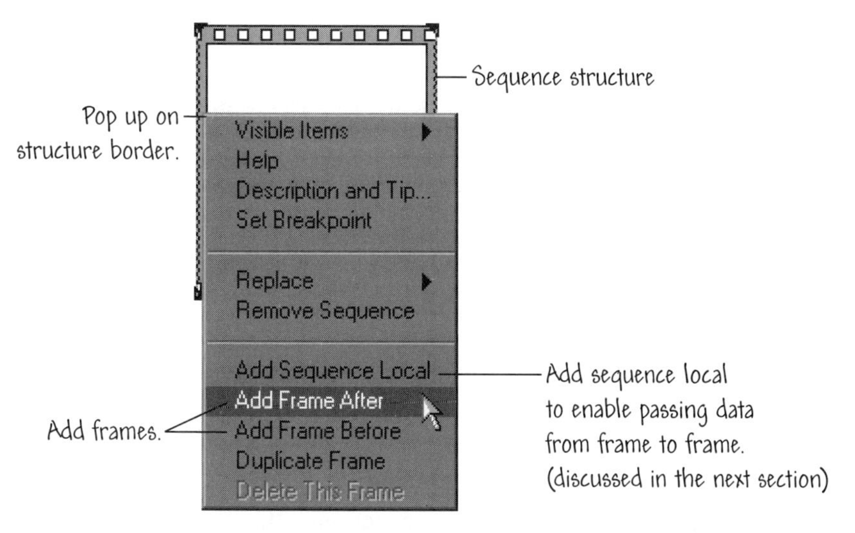

FIGURE 5.25
Removing and adding frames to the Sequence structure.

Recall that the outputs on Case structures can have one data source *per case*. In contrast, on Sequence structures, output tunnels can have only one data source. The output can originate from any frame, but data leaves the Sequence structure only when it completes execution entirely, not when the individual frames finish. Data at input tunnels is available to all frames, just as with Case structures.

5.5.1 Sequence Locals

Sequence locals are variables that pass data between frames of a Sequence structure. You create sequence locals on the border of a frame—the data wired to a sequence local is then available in subsequent frames. The data, however, is not available in frames preceding the frame where the sequence was created.

To obtain a sequence local, choose **Add Sequence Local** from the structure border short cut menu (see Figure 5.25). You cannot add another sequence local if you pop up too close to an existing sequence local or over the subdiagram display window. Once a sequence local terminal is placed on the border, you can drag it to any other unoccupied location on the border. At first, a sequence local terminal is just a small yellow box, but once a data source is wired to the terminal, an outward-pointing arrow appears in the frame of the terminal. The sequence local terminals in subsequent frames contain inward-pointing arrows to show that they are a data source for that frame. You cannot use the sequence local in frames preceding the source frame since it hasn't been assigned a value yet! To remind you that the sequence local does not contain a value in preceding frames (hence is not available for use), it appears as a dimmed rectangle. To remove a sequence local terminal, pop up on the terminal and select **Remove**.

The example in Figure 5.26 shows a four-frame Sequence structure. A sequence local in frame 1 passes the value of a random number function to subsequent frames. The random number value is available in frame 2—as indicated by the arrow pointing into frame 2. In this simple example, the random number in frame 2 is being displayed by a digital indicator. In frame 3 we attempt to wire the random number to a digital control. The result is a broken wire since we cannot wire a digital control to a sequence local that is a source of data. Also, notice that the random number value is not available in frame 0, as indicated by the dimmed yellow square. Remember that data is not available in frames preceding the frame in which the sequence local is created!

5.5.2 Evaluate and Control Timing in a Sequence Structure

It is useful to be able to control and time the execution of VIs. Using Wait (ms) and Tick Count (ms) functions (located in the **Time & Dialog** palette of the **Functions** menu) in conjunction with a Sequence structure lets you accomplish these tasks.

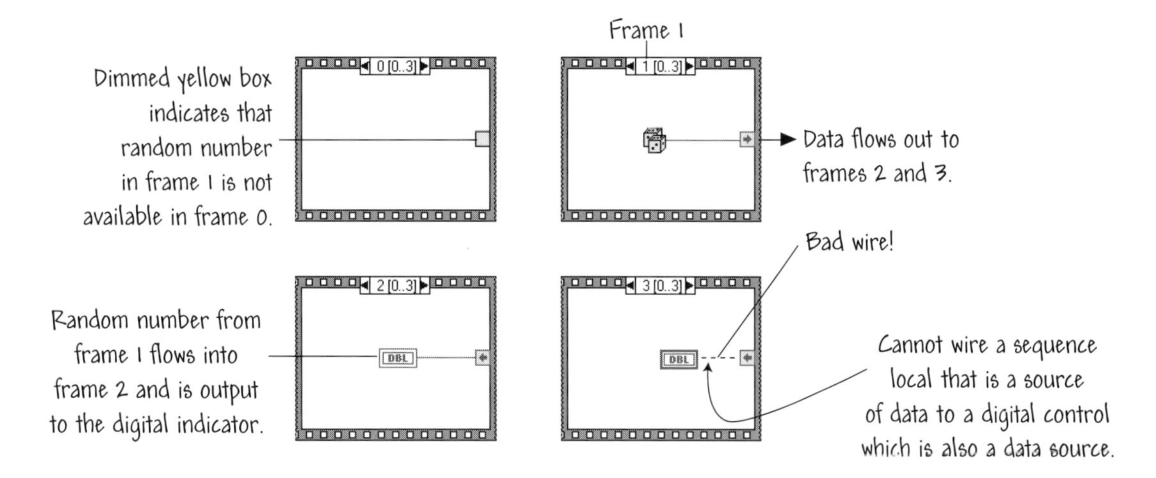

FIGURE 5.26
Many forms of the sequence local.

› The Wait (ms) function causes your VI to wait a specified number of milliseconds before it continues execution. The function waits the specified number of milliseconds and then returns the millisecond timer's end value.

› The Tick Count (ms) function returns the value of the millisecond timer and is commonly used to calculate elapsed time. The base reference time (that is, zero milliseconds) is undefined. Therefore, you cannot convert the millisecond timer output value to a real-world time or date. Note that the value of the millisecond timer wraps from $2^{32} - 1$ to 0.

The internal clock does not have high resolution—about 1 ms on ***Windows 2000/NT/9X*** *and* ***Macintosh****. The resolutions are driven by operating system limitations and not by LabVIEW.*

A simple example illustrating the use of the Wait (ms) and Tick Count (ms) functions is shown in Figure 5.27. Open the VI Timing with Sound Demo located in Learning\Chapter 5. Set the time between beeps on the slider control and run the VI in **Continuous Run** mode. You should hear a sound approximately every *n* seconds according to the setting on the slider control. The actual time between beeps (to the resolution of the clock) is displayed on the front panel to verify that the desired timing has been achieved. On **Windows 2000/NT/9X** and **Macintosh**, if you reduce the time between beeps to 0.001 second (using the

Labeling tool to type in the value in the digital display of the slider control), you will find that the actual time between beeps is about 1 ms. However, if you continue to reduce the time below 0.001, the actual time between beeps will not reduce accordingly—you have reached the clock resolution of your system.

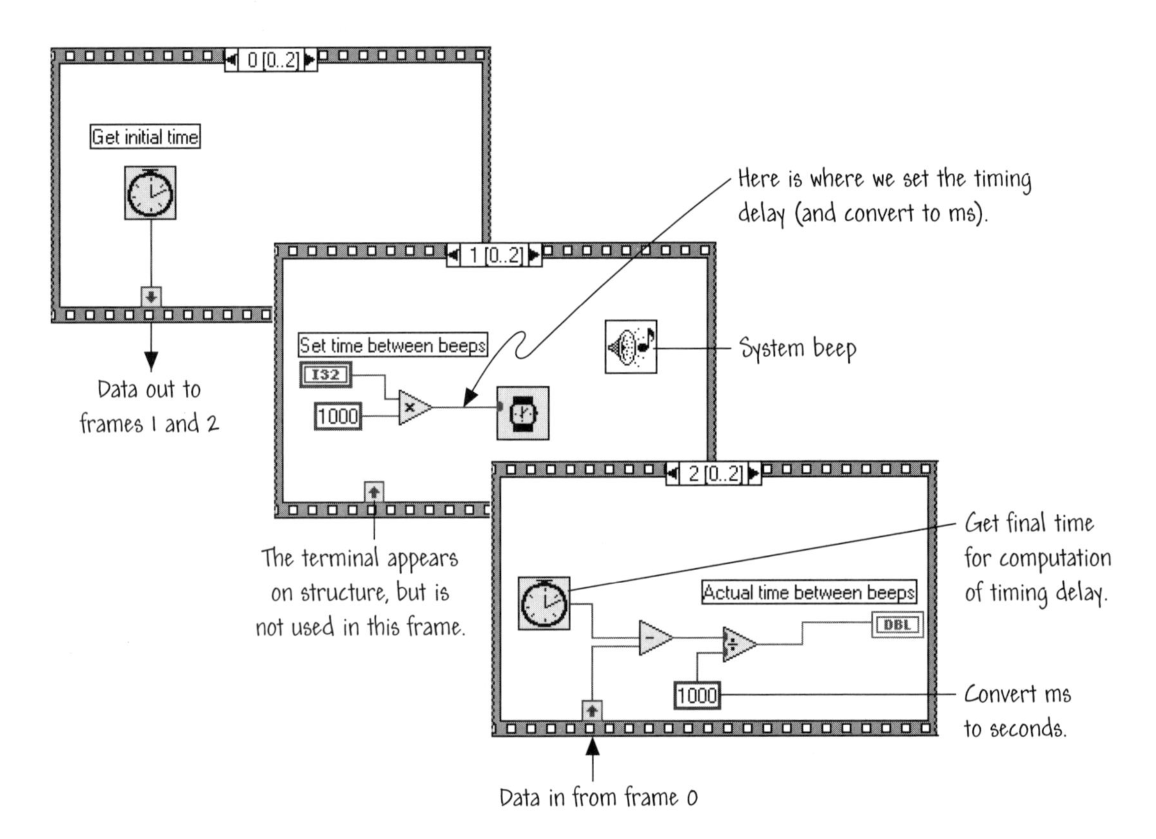

FIGURE 5.27
Controlling the timing of a VI using the Sequence structure.

5.5.3 Avoid the Overuse of Sequence Structures

In general, VIs can operate with a great deal of inherent parallelism. Sequence structures tend to hide parts of the program and to interfere with the natural data flow. The use of Sequence structures prohibits parallel operations, but does guarantee the order of execution. Asynchronous tasks that use I/O devices can run concurrently with other operations if Sequence structures do not prevent them from doing so.

We will be discussing I/O devices (such as GPIB, serial ports, and data acquisition boards) more in subsequent chapters (see Chapters 8 and 10), and you will understand better how the use of Sequence structures can inhibit the performance of such devices. The objective here is to alert you to the idea that you should avoid the overuse of Sequence structures. The use of Sequence structures does not negatively impact the computational performance of your program, but it does interrupt the data flow. You should write programs that take advantage of the concept of data flow programming!

5.6 THE FORMULA NODE

The **Formula Node** is a structure that allows you to program one or more algebraic formulas using a syntax similar to most text-based programming languages. It is useful when the equations have many variables or otherwise would require a complex block diagram model for implementation. The Formula Node itself is a resizable box (similar to the Sequence structure, Case structure, For Loop, and While Loop) in which you enter formulas directly in the code, in lieu of creating block diagram subsections.

Consider the equation

$$y = x - e \sin x,$$

where $0 \leq e \leq 1$. This is a famous equation in astrodynamics, known as Kepler's equation. If you implement this equation using regular LabVIEW arithmetic functions, the block diagram looks like the one in Figure 5.28. You can implement the same equation using a Formula Node as shown in the same figure.

5.6.1 Formula Node Input and Output Variables

You place the Formula Node on the block diagram by selecting it from the **Structures** subpalette of the **Functions** palette (see Figure 5.19). You create the input and output terminals of the Formula Node by popping up on the border of the node and choosing **Add Input** or **Add Output** from the short cut menu, as shown in Figure 5.29. Output variables have a thicker border than input variables. You can change an input to an output by selecting **Change to Output** from the short cut menu, and you can change an output to an input by selecting **Change to Input** from the short cut menu.

Once the necessary input and/or output terminals are on the Formula Node, enter the input and output variable names in their respective boxes using the **Labeling** tool. Every variable used in the Formula Node must be declared as an input or an output with no two inputs and no two outputs possessing the same name. An output can, however, have the same name as an input. Intermediate variables

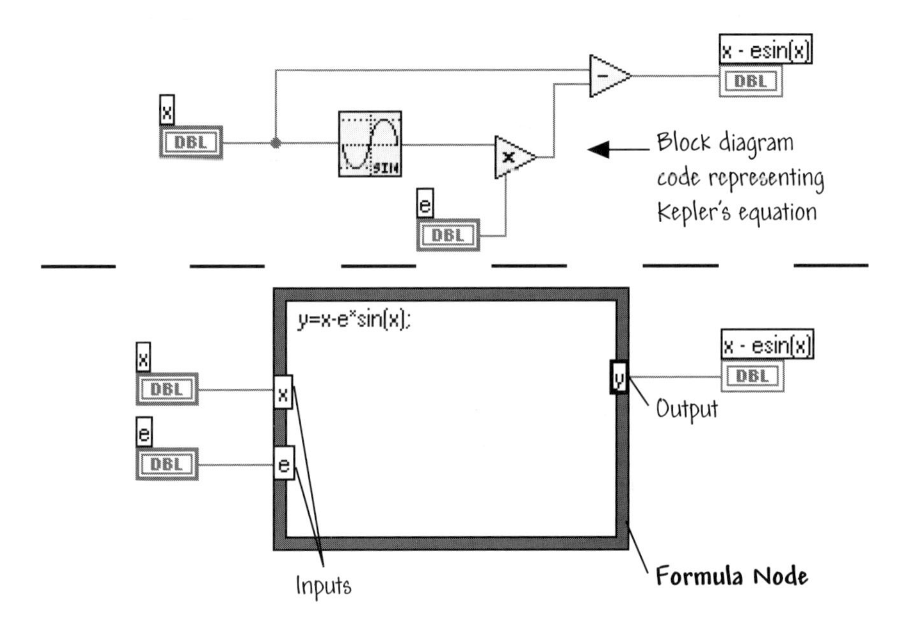

FIGURE 5.28
Implementing formulas in a Formula Node can often simplify the coding.

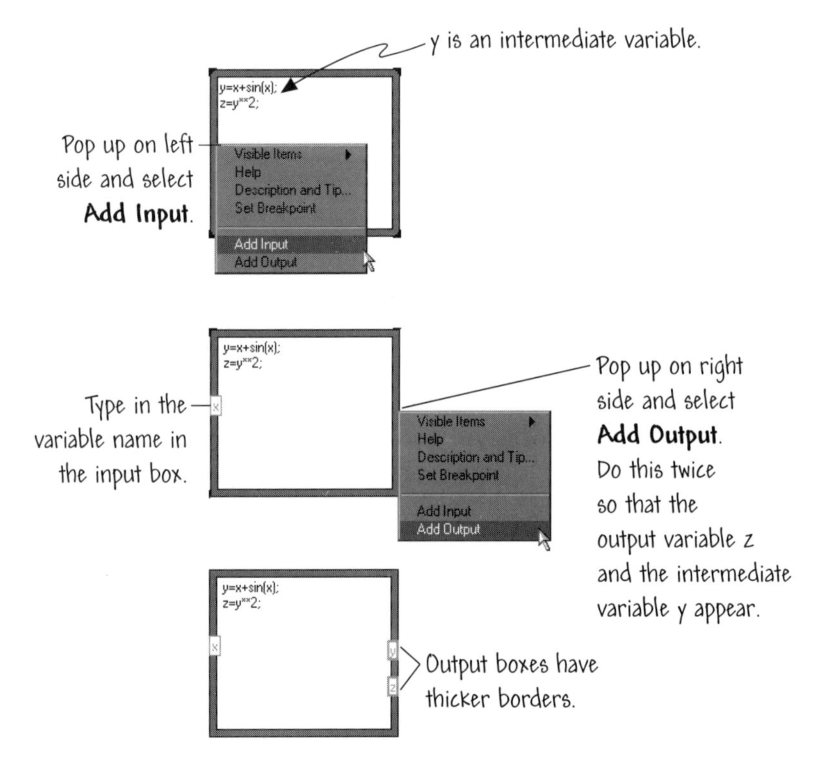

FIGURE 5.29
Formula Node input and output variables.

(that is, variables used in internal calculations) must be declared as outputs, although it is not necessary for them to be wired to external nodes. In Figure 5.29, the y variable and the z variable are both declared as outputs, although the intermediate variable y does not have to be wired to an external node.

5.6.2 Formula Statements

Formula statements use a syntax similar to most text-based programming languages for arithmetic expressions. You can add comments by enclosing them inside a slash-asterisk pair (/*comment*/). Figure 5.30 shows the operators and functions that are available inside the Formula Node. You can access the same list of available operators and functions by choosing Help≫Show Context Help and moving the cursor over the Formula Node in the block diagram.

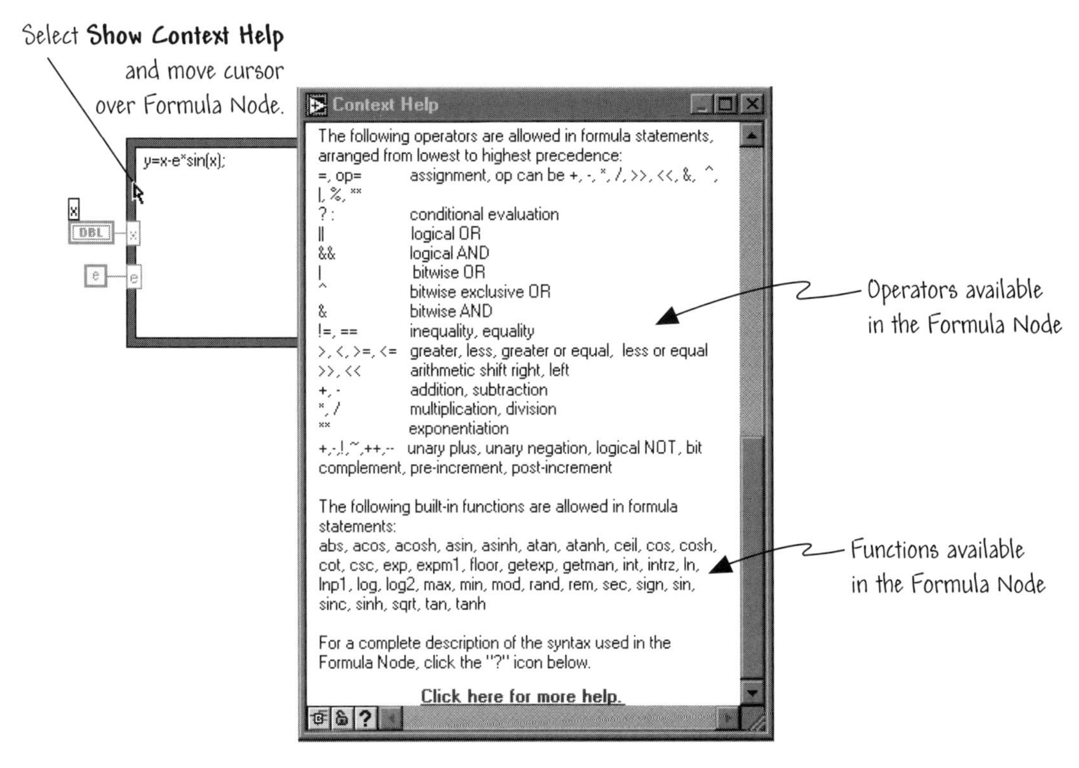

FIGURE 5.30
Operators and functions available in Formula Nodes.

You enter the formulas inside the Formula Node using the **Labeling** tool. Each formula statement must terminate with a semicolon, and variable names are case-sensitive. There is no limit to the number of variables or formulas in a Formula Node. If you have a large number of formulas, you can either enlarge the Formula Node using the **Positioning** tool, or you can pop up in the Formula Node (not on the border) and choose **Scrollbar**. The latter method will put a scrollbar in the Formula Node, and with the **Operating** tool you can scroll down through the list of formulas for viewing.

The following example shows how you can perform conditional branching inside a Formula Node. Consider the following code fragment that computes the ratio of two numbers, x/y:

if $(y \neq 0)$ then
$z = x/y$
else
$z = +\infty$
end if

When $y = 0$, the result is set to ∞. You can implement the code fragment given above using a Formula Node, as illustrated in Figure 5.31.

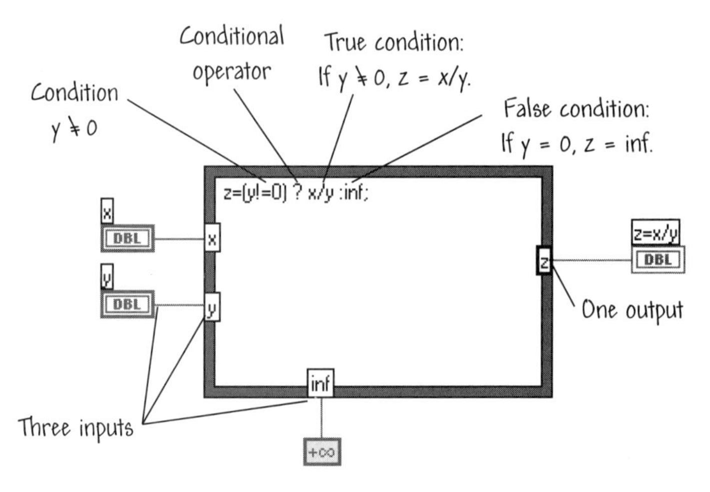

FIGURE 5.31
Implementing the formula $z = x/y$ in a Formula Node.

The Formula Node demonstration depicted in Figure 5.31 can be found in Learning\Chapter 5 *and is called* Formula Code.vi. *Open the VI and try it out!*

5.7 MATLAB SCRIPT NODES

LabVIEW has script nodes that allow you to execute external scripts. In particular, you can execute both HiQ and MATLAB scripts using a script node. HiQ [1] is a high-performance, interactive, problem-solving environment contained as part

1. HiQ is a registered trademark of the National Instruments Corporation

of the *LabVIEW Student Edition* installation. MATLAB [2] is a popular interactive program for scientific and engineering calculations and visualization. Since many students have developed and used MATLAB scripts (known as m-files), in this section we focus on using script nodes in LabVIEW to execute MATLAB scripts. More information on HiQ script nodes can be found in the LabVIEW on-line help utilities.

LabVIEW uses ActiveX technology to implement MATLAB script nodes, therefore MATLAB script nodes are available only on Windows platforms.

5.7.1 Accessing the MATLAB Script Node

The MATLAB script node is accessed on the **Functions** palette, as illustrated in Figure 5.32. Notice that the HiQ script node is also accessible from the **Functions** palette on the Mathematics≫Formula subpalette. The process of placing

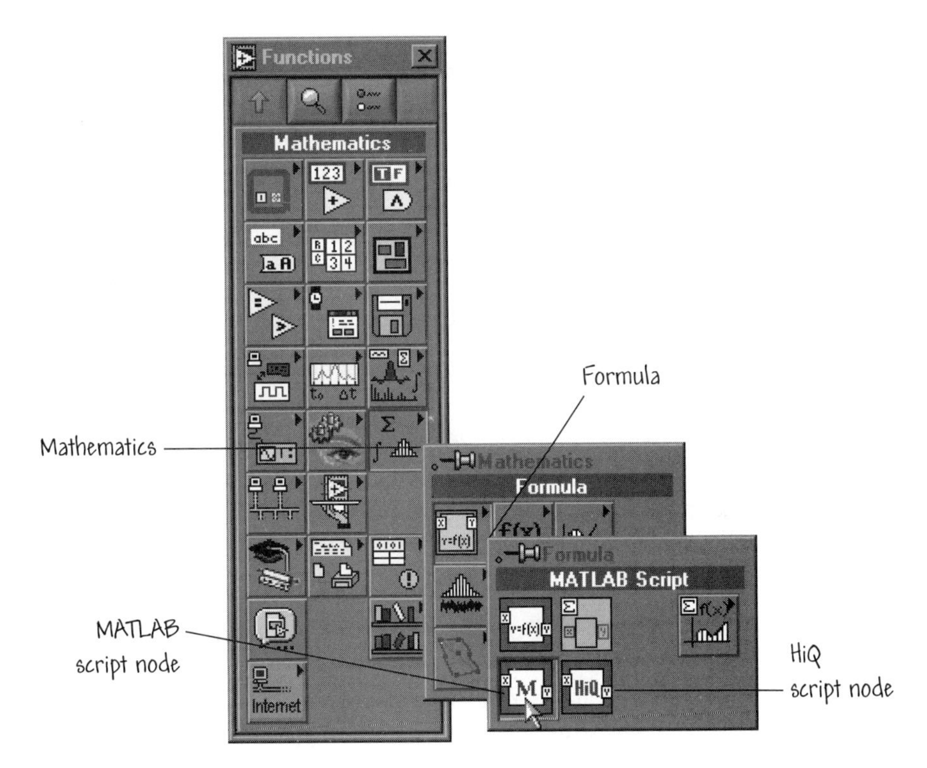

FIGURE 5.32
Accessing the MATLAB script node.

2. MATLAB is a registered trademark of The Mathworks, Inc.

the script node on the block diagram is similar to **For Loops** and **While Loops**. First select the MATLAB script node from the palette and place it on the block diagram. Using the **Positioning** tool, the script node is extended to the desired size, as illustrated in Figure 5.33.

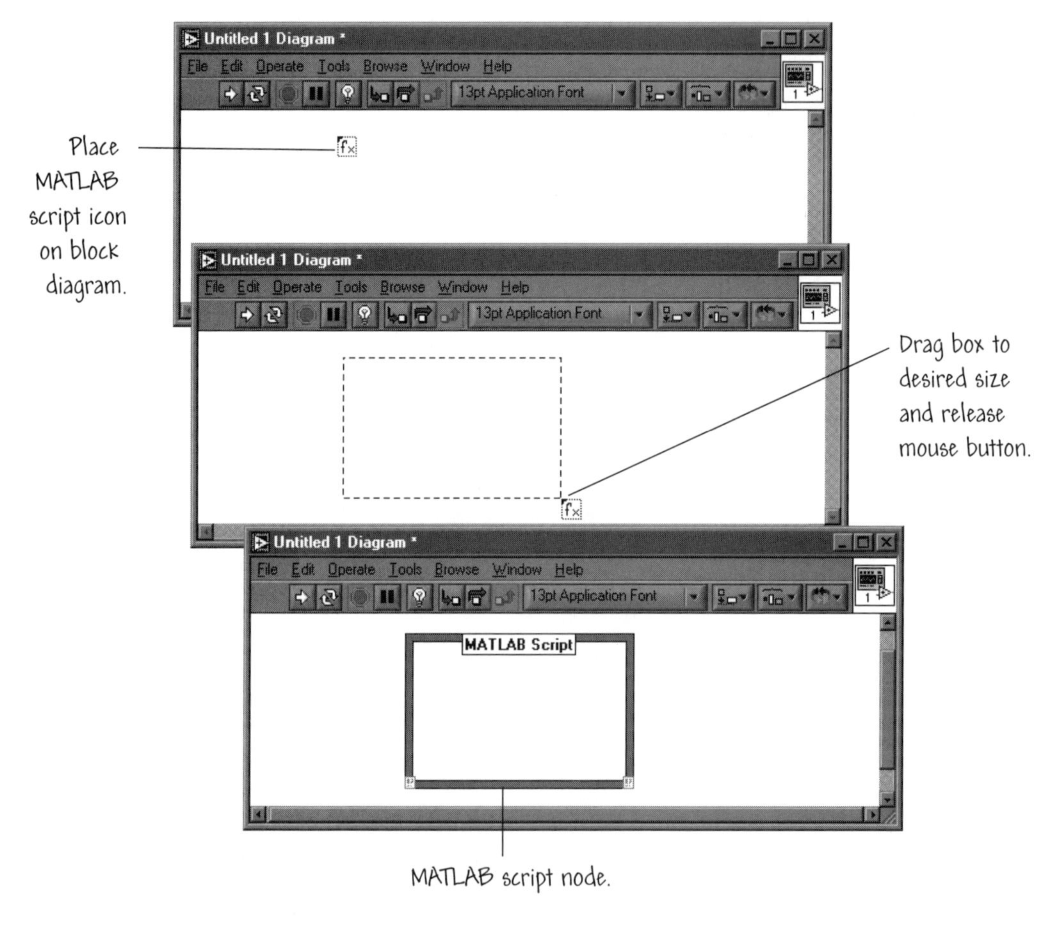

FIGURE 5.33
Placing a MATLAB script node on the block diagram.

5.7.2 Entering Scripts into the MATLAB Script Node

There are two ways to place a MATLAB script in the script node. You can use the **Operating** or **Labeling** tool to enter the script in the MATLAB script node, or if you already have a script written, you can import it. To import a script, right-click the MATLAB script node and select **Import** from the short cut menu to display the **Choose a script** dialog box, as shown in Figure 5.34. When the dialog box appears, select the file you want to import and click **Open**. The MATLAB script text will then appear in the script node. It is suggested that you write your script and run it within the MATLAB environment for testing and debugging purposes before you import it into LabVIEW.

You must have MATLAB installed on your computer to use MATLAB script nodes because script nodes invoke the MATLAB script server to execute MATLAB scripts.

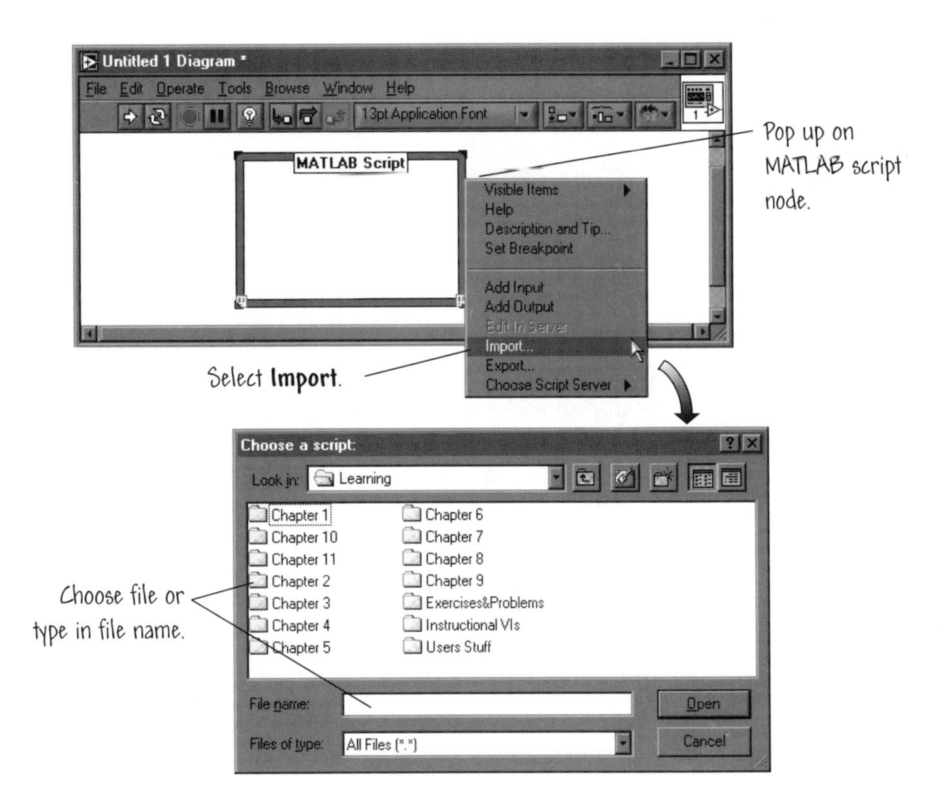

FIGURE 5.34
Importing a script into a MATLAB script node.

5.7.3 Input and Output Variables

As with Formula Nodes, you need to add inputs and outputs for variables to the MATLAB script node. To add an output variable, right-click the MATLAB script node frame and select **Add Output** from the short cut menu, as shown in Figure 5.35. Similarly, to add an input variable, right-click the MATLAB script node frame and select **Add Input** from the short cut menu. When the input and output variables appear on the node, you can add their names. You can also use the

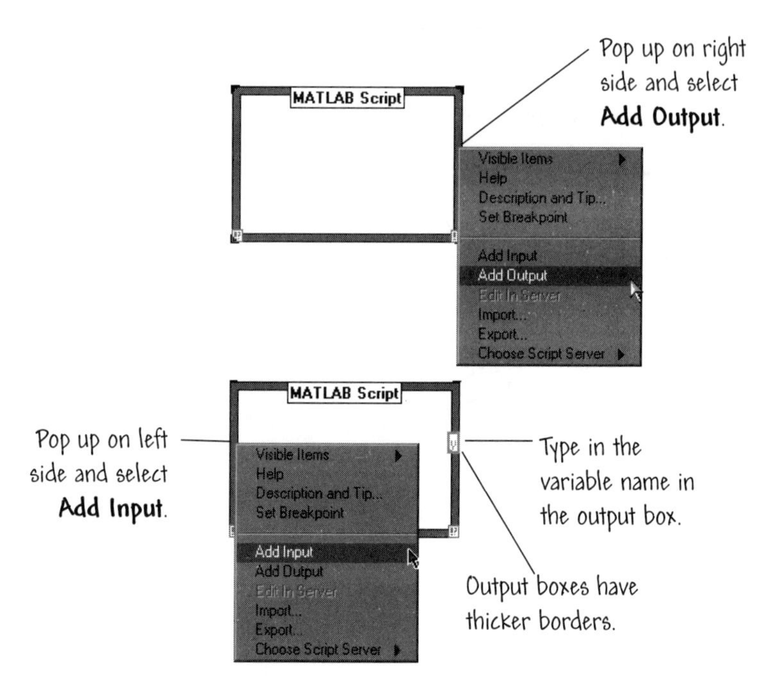

FIGURE 5.35
Adding input and output variables to the MATLAB script node.

Labeling tool to edit the variable names at any time. By default the MATLAB script node includes one input and one output for the error in and error out parameters. To take advantage of the error-checking parameters for debugging information, it is suggested that you create an indicator for the error out terminal on the MATLAB script node before you run the VI. This allows you to view the error information generated at runtime.

Just as with a regular formula node, you can display a scrollbar within your script node by popping up on the node and selecting Visible Items≫Scrollbar.

You can create controls and indicators for each input and output on the MATLAB script node. For example, to create an indicator on an output terminal, right-click the output terminal select Create≫Indicator from the short cut menu. LabVIEW will create an indicator on the front panel and wire terminals to the output on the block diagram, as shown in Figure 5.36.

5.7.4 Saving MATLAB Scripts

You may want to save your MATLAB script to a text file. In this way, you can later open this text file in LabVIEW, thus importing the MATLAB script into LabVIEW. To save a MATLAB script, right-click the MATLAB script node and

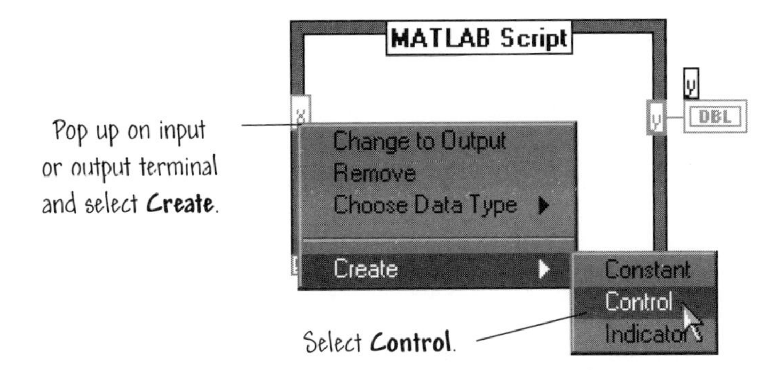

FIGURE 5.36
Adding control and indicators to the input and output variables.

select **Export** from the short cut menu to display the **Name the script** dialog box, as illustrated in Figure 5.37. Enter the desired new file name or select the file you want to overwrite and click **Save**. MATLAB script files are text files, and although text files usually have a **.txt** extension, MATLAB files have a **.m** extension. This is consistent with MATLAB m-file naming conventions.

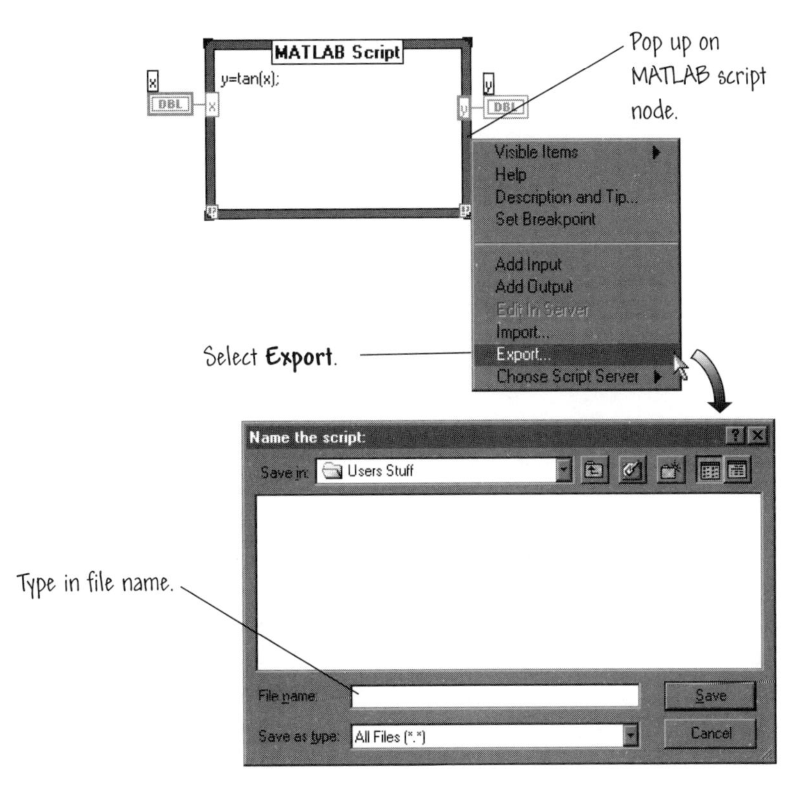

FIGURE 5.37
Saving MATLAB scripts to a text file.

5.7.5 MATLAB Data Types in LabVIEW

LabVIEW recognizes the MATLAB datatypes. Table 5.1 shows the LabVIEW data types and the corresponding data types in MATLAB. You can change the data type of an input or output terminal on a script node. MATLAB is a loosely

TABLE 5.1 LabVIEW and MATLAB Data Types.

	LabVIEW Data Types	MATLAB Data Types
I32	Signed 32-bit integer numeric	N/A
	Double-precision floating-point numeric	Real
abc	String	N/A
[I32]	1D array signed 32-bit integer numeric	N/A
[DBL]	1D array double-precision floating-point numeric	Real Vector
[I32]	Multidimensional array signed 32-bit integer numeric	N/A
[DBL]	Multidimensional array double-precision floating point numeric	Real Matrix
[CDB]	Complex double	Complex
CDB	1D array complex double	Complex Vector
[CDB]	Multidimensional array complex double	Complex Matrix

typed script language, hence the datatype of a variable is not determined until after the script executes. Therefore, LabVIEW cannot determine the variable type in Edit mode. LabVIEW queries the script server to find out possible datatypes, and lets you choose which LabVIEW datatype each terminal should be.

If you incorrectly configure a variable's datatype, LabVIEW produces either an error or incorrect information at runtime.

You should always verify the data type of the script node inputs and outputs. In MATLAB, the default data type for any new input or output in Real. To change the datatype of an input or output terminal on a script node, first right-click the terminal of the input or output and select **Choose Data Type** on the short cut menu. A list of the available datatypes appears from which you can choose the preferred data type.

A MATLAB Example

In this exercise you will construct a VI that utilizes a MATLAB script node. The purpose of the VI is to generate and plot a given number of random numbers in MATLAB. The script will also compute the average of the random numbers for output. The VI is shown in Figure 5.38. Open a new VI and begin by placing a MATLAB script node on the block diagram. Size it large enough to allow for the required lines of code (refer to the block diagram in Figure 5.38).

Add an input and an output to the script node. The input variable is the number of random numbers to generate, and the output variable is the average of the random numbers. Make the input and output variables double-precision floating-point numerics. Also, place a string indicator on the script node to show an error

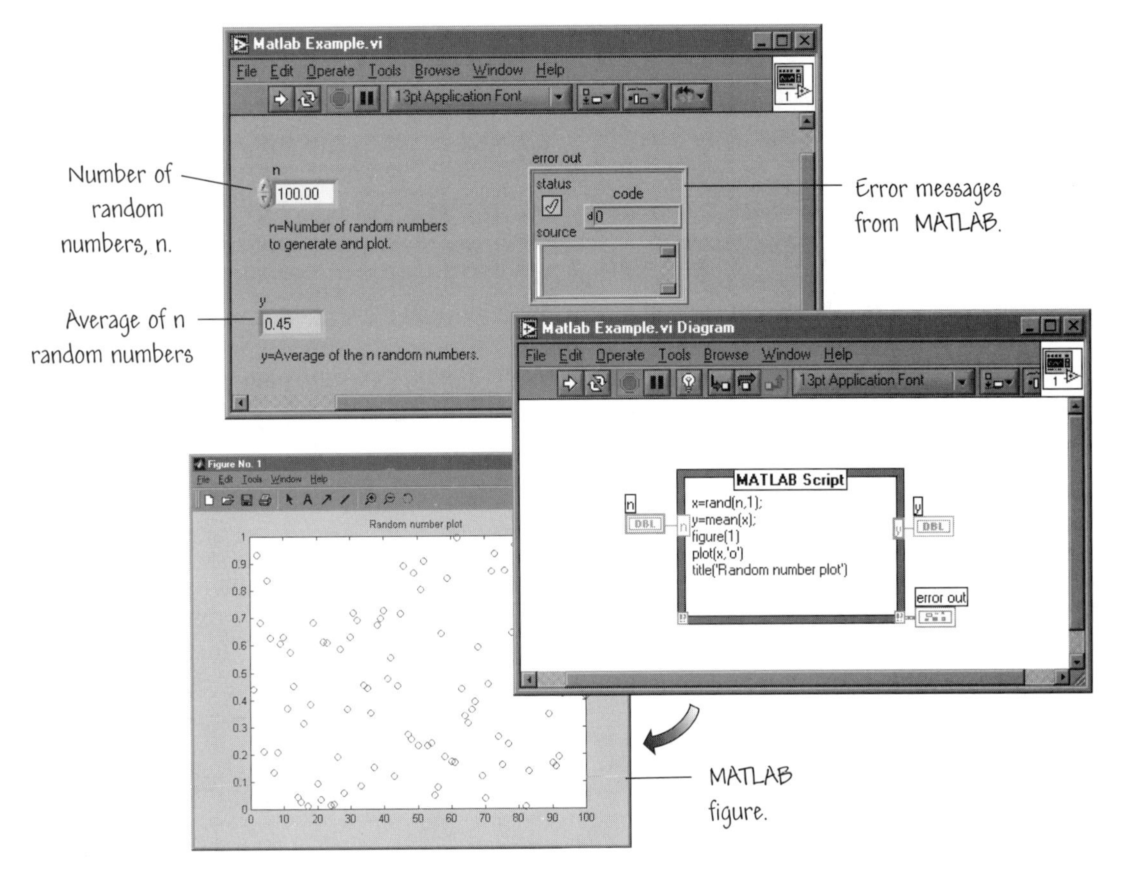

FIGURE 5.38
Using a script node to generate a MATLAB plot.

out messages generated in MATLAB. Run the VI when ready. LabVIEW will launch the MATLAB application if it is not already opened. Vary the number of random numbers in the set and see how the average varies correspondingly. The MATLAB script node will generate a plot of the random numbers using MATLAB graphics. When you are finished experimenting, save the VI as Matlab Example.vi in the Users Stuff folder.

A working version of Matlab Example.vi can be found in the Chapter 5 folder in the Learning directory. ◆

5.8 SOME COMMON PROBLEMS IN WIRING STRUCTURES

When wiring the structures presented in this chapter, you may encounter wiring problems. In this section, we discuss some of the more common wiring errors and present suggestions on how to avoid them. Five common problems with wiring structures are:

- Assigning more than one value to a Sequence local.
- Wiring from multiple frames of a Sequence structure.
- Failing to wire a tunnel in all cases of a Case structure.
- Overlapping tunnels.
- Wiring underneath rather than through a structure.

The first two problems are actually variations of the multiple sources error.

5.8.1 Assigning More Than One Value to a Sequence Local

A sequence local can be assigned a value in only one frame. Figure 5.39 shows the value π assigned to the sequence local in frame 0 and then another attempt to assign a value to this same local variable in frame 1—this results in a bad wire. You can use the value of the sequence local in all frames that follow the frame in which the assignment took place.

5.8.2 Wiring from Multiple Frames of a Sequence Structure

Figure 5.40 depicts two Sequence structure frames attempting to assign values to the same tunnel. The tunnel turns white to signal this error, which is just another variation of the multiple sources error.

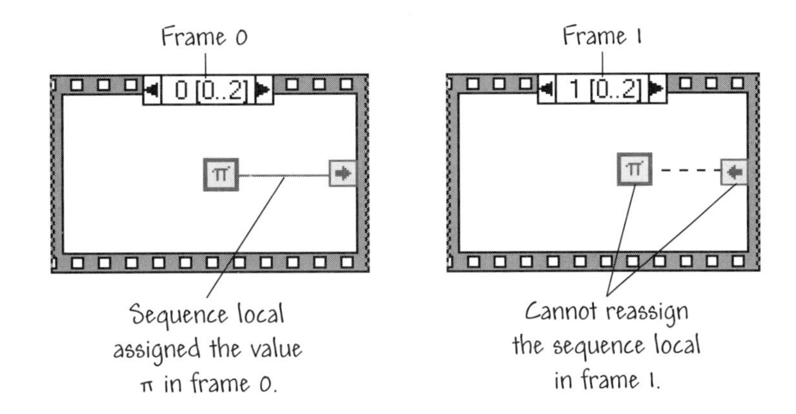

FIGURE 5.39
Local variables in a Sequence structure can be assigned a value in only one frame.

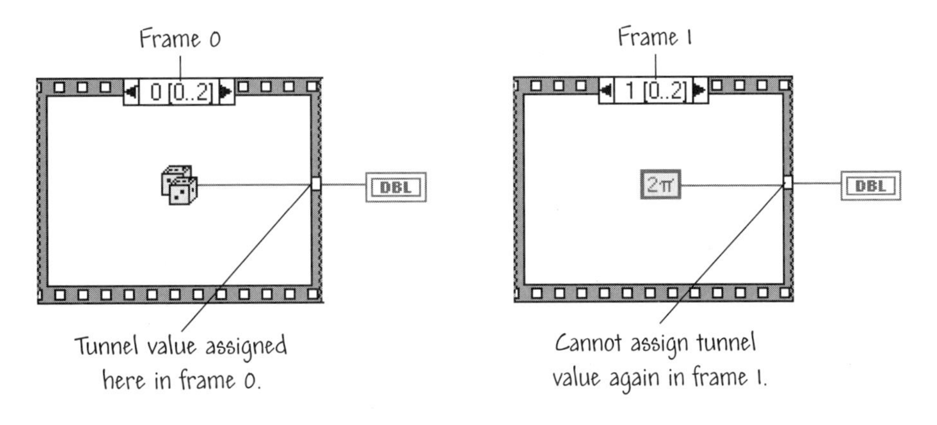

FIGURE 5.40
Two Sequence structure frames cannot assign values to the same tunnel.

5.8.3 Failing to Wire a Tunnel in All Cases of a Case Structure

When you wire from a Case structure to an object outside the structure, you must connect output data from all cases to the object. Failure to do so will result in a bad tunnel, as illustrated in Figure 5.41. This problem is a variation of the no source error. Why? Because at least one case would not provide data to the object outside the structure when that case executed. The problem is easily solved by wiring to the tunnel in all cases. Can you explain why this is not a multiple sources violation? It seems like one object is receiving data from multiple sources. The answer is that only one case executes at a time and produces only one output value per execution of the Case structure. If each Case did not output a value, then data flow execution would stop on the cases that did not output a value.

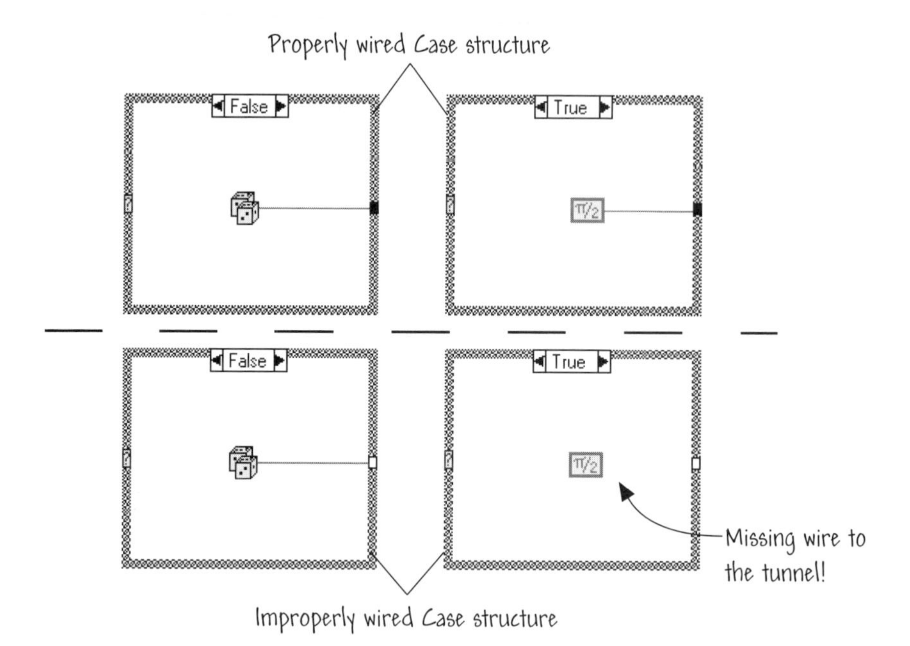

FIGURE 5.41
Failing to wire a tunnel in all cases of a Case structure leads to problems.

5.8.4 Overlapping Tunnels

Tunnels are created as you wire, resulting occasionally in tunnels that overlap each other. Overlapping tunnels do not affect the execution of the diagram, so this is not really an error condition. But overlapping tunnels make editing and debugging of the VI more difficult. You should avoid creating overlapping tunnels! The fix is easy—drag one tunnel away with the **Positioning** tool to expose the other tunnels. A problematic situation is illustrated in Figure 5.42.

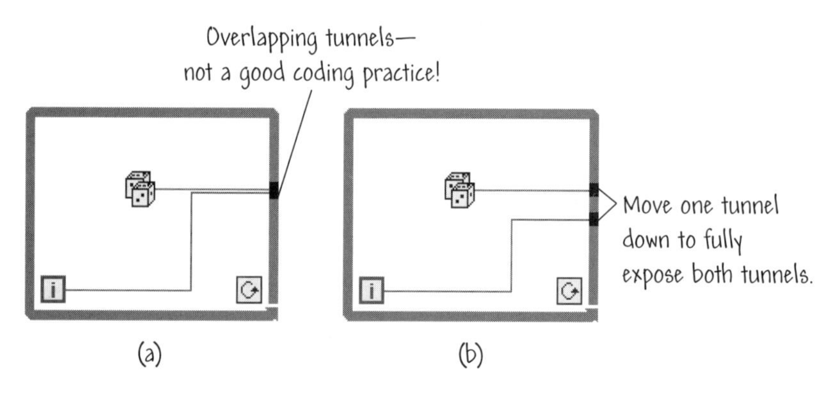

FIGURE 5.42
Sometimes overlapping tunnels occur when wiring structures.

5.8.5 Wiring Underneath Rather Than Through a Structure

Suppose that you want to pass a variable through a structure. You accomplish this by clicking either in the interior or on the border of the structure as you are wiring. This process is illustrated in Figure 5.43. If you fail to click in the interior or on the border, the wire will pass underneath the structure, and some segments of the wire may be hidden. This condition is not an error per se, but hidden wires are visually confusing and should be avoided.

As you wire through the structure, you are provided with visual cues for guidance during the wiring process. For example, when the **Wiring** tool crosses the left border of the structure, a highlighted tunnel appears. This lets you know that a tunnel will be created at that location as soon as you click the mouse button. You should click the mouse button! However, if you continue to drag the tool through the structure (without clicking the mouse on the left hand-side border) until the tool touches the right border of the structure, a second highlighted tunnel appears. If you continue to drag the **Wiring** tool past the right border of the structure without clicking, both tunnels disappear. When this occurs, the wire passes underneath the structure rather than through it, as depicted in Figure 5.43.

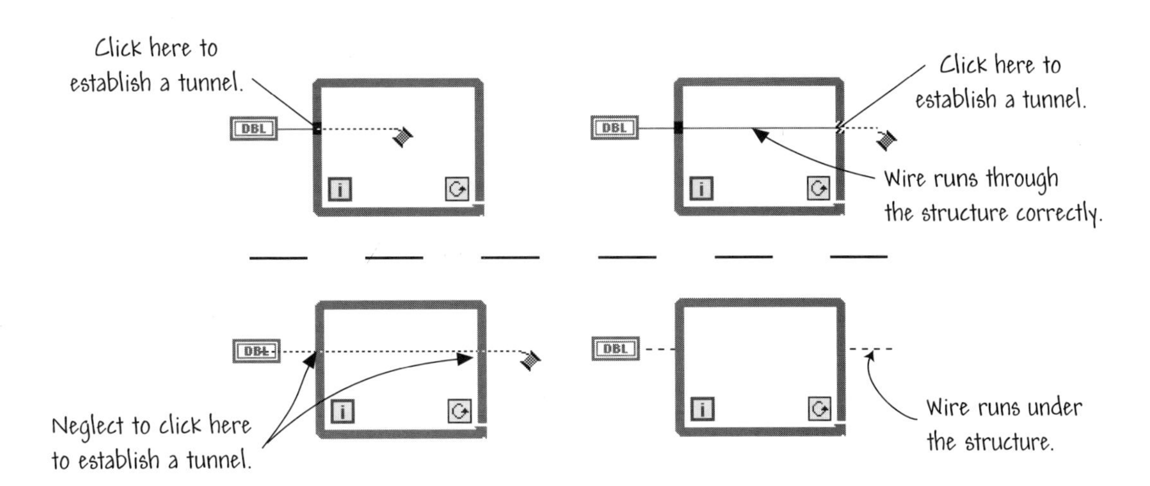

FIGURE 5.43
Wiring underneath rather than through a structure is a common problem.

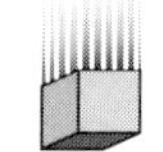

BUILDING BLOCK

5.9 BUILDING BLOCKS: MEASURING VOLUME

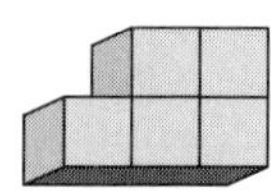

In this exercise you will continue the work on the building blocks VI Volume Limit created in Chapter 4. The goal is to make the necessary modifications so that you can acquire multiple volume measurements using a For Loop and control the timing of the loop. You should use Figure 5.44 as a guide.

When you have finished editing and debugging the VI, save it in the Users Stuff folder as Multiple Volume Points.vi.

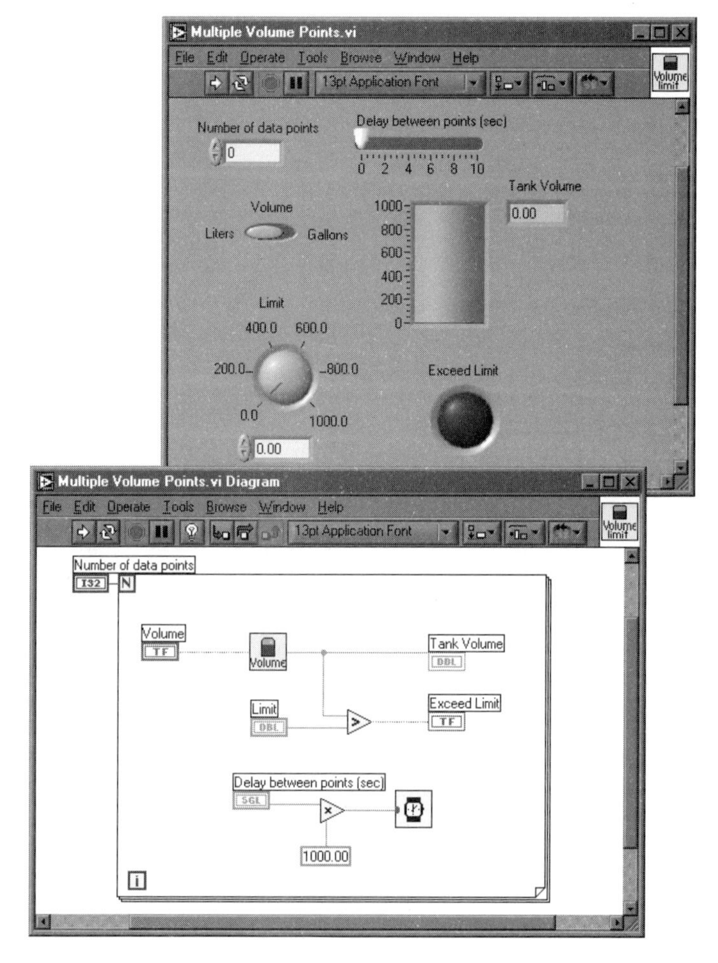

FIGURE 5.44
Wiring the connector of the Volume subVI.

Run the VI for various delays between data points—try 5 sec and count the interval to see if the delay really is 5 seconds (actually, the timing will only be approximate!). You will see in Chapter 7 that a better way to control the For Loop timing is to use the Wait Until Next ms Multiple function.

5.10 RELAXING READING: IMPROVING PHYSICAL UNDERSTANDING

The teaching laboratory plays an important role in the education process. For many students it represents their only opportunity (while in school) to explore real-world applications of fundamental theories learned in the classroom. The learning experience is often compromised by the lack of time on the part of the faculty to prepare the laboratory and a lack of new hardware and support staff to maintain the equipment. Faculty at the University of Wisconsin—Madison have examined ways of automating their undergraduate laboratory on electrical and electromechanical power conversion—and LabVIEW is an integral part of their solution. Figure 5.45 shows a typical LabVIEW equipped laboratory.

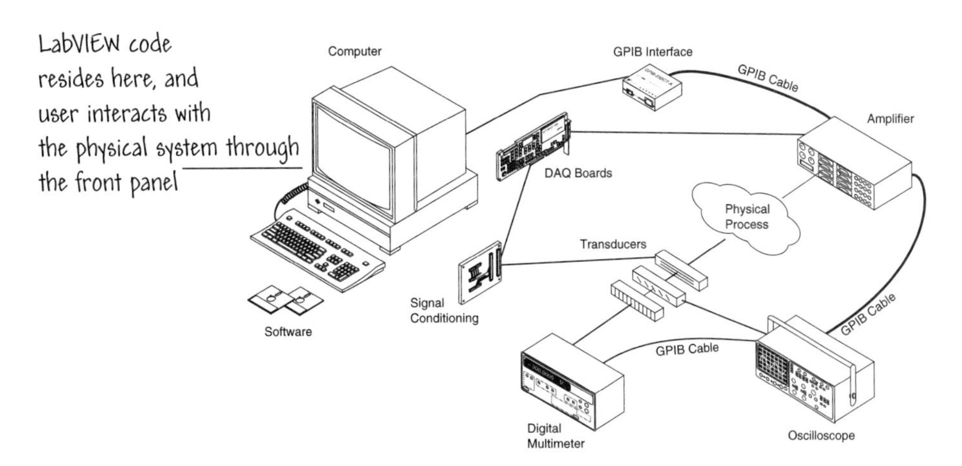

FIGURE 5.45
A typical equipment setup in an instrument laboratory equipped with LabVIEW.

The four laboratory automation goals of the faculty were:

- To reduce the instrumentation learning time, so that most of the lab time was used for "core" material.
- To automate the data acquisition process.
- To expedite all forms of feedback and analysis so that the lab session results in a meaningful discussion of the experimental results.
- To encourage discussion and preparation of report documents within the laboratory itself.

Most automated laboratories use a computer workstation with high-speed communication interfaces; instrumentation under computer control; analog and digital hardware interface boards on the computer bus; software for numerical analysis, graphics and word processing; software for bidirectional communication between hardware and the workstation; and a programming environment for providing custom graphical user interfaces. The communication and user interface software used in the University of Wisconsin—Madison study was LabVIEW.

One of the most effective examples of the improvement in productivity was found in an introductory DC machine experiment. The students determined the armature resistance and the back-emf constant of the motor. Using these values in the steady-state model, the students then applied the model to independently predict the armature voltage necessary to attain speeds at given loads. In effect, the students were playing the role of a cruise control in a fictional electric vehicle. Once the actual motor speed reached ± 1 percent of the target speed, the load requirement was changed. Following this, the same set of operating points was achieved by closing a speed control loop on the motor. After completing the experiment, the students compared the armature voltages they computed with those resulting from the speed controller. This interactive use of the model within the experiment not only illustrates the usefulness of the model in a real-world setting, but also exposes the limitations of the model, which did not incorporate friction, drag, and other second-order effects.

Another means of employing the theoretical model during the experiment is by plotting the actual and the theoretical results simultaneously. This practice offers an excellent opportunity for the instructor to interact with students by asking them to explain any differences. This was very successfully employed in both a magnetic actuator experiment and in a power electronics experiment in which a simple DC-DC buck converter circuit was investigated.

Laboratory automation can significantly improve teaching and student productivity by including analytical models in comparison to experimental results, online analysis of experimental vs. theoretical discrepancies, online plotting, and a rapid means for data recording to files. The interaction between instructor and student at the University of Wisconsin—Madison focused less on procedural matters, such as debugging the experimental setup, and more on the discussion of the key concepts covered in the experiment. The introduction of laboratory automation incorporating LabVIEW is powerful tool in emphasizing the understanding and exploration of the fundamental physical principles presented in electrical engineering teaching laboratories.

For more information on this application of LabVIEW in the classroom, read the *User Solutions* entitled "Improving Physical Understanding Through Laboratory Automation" by Steven A. Orth and Robert D. Lorenz at The University of Wisconsin—Madison. This *User Solution* is located on the NI website. More information can be found at http://www.engr.wisc.edu/ece/courses/ece377.html.

5.11 SUMMARY

In this chapter we studied four structures (For Loops, While Loops, Sequence and Case structures). The While Loop and the For Loop are used to repeat exccution of a subdiagram placed inside the border of the resizable loop structure. The While Loop executes as long as the value at the conditional terminal is TRUE. The For Loop executes a specified number of times. Shift registers are variables that transfer values from previous iterations to the beginning of the next iteration.

The Case structure and the Sequence structure are utilized to control data flow. Case structures are used to branch to different subdiagrams depending on the values of the selection terminal of the Case structure. Sequence structures are used to execute diagram functions in a specific order. The portion of the diagram to be executed first is placed in the first frame of the Sequence structure, the diagram to be executed second is placed in the second frame, and so on. Sequence locals pass values between Sequence structure frames.

You can directly enter formulas on the block diagram in the Formula Node. The Formula Node is useful when the function equation has many variables or otherwise would require a complex block diagram model for implementation.

LabVIEW has script nodes that allow you to execute external scripts. In this chapter, we discussed the MATLAB script node, and showed how to input external MATLAB code for execution in LabVIEW.

We also discussed the matter of controlling and timing the execution of structures. The **Time & Dialog** palette of the **Functions** menu provides functions to pop up dialog boxes and control or monitor VI timing. The Wait (ms) function pauses your VI for the specified number of milliseconds, and the Tick Count (ms) returns the value of the internal clock. Both of these functions can be used in conjunction with the various different structures to control and monitor the execution speed of the code.

KEY TERMS

Case structure: A conditional branching control structure that executes one and only one of its subdiagrams, based on specific inputs. It is similar to If-Then-Else and Case statements in conventional programming languages.

Coercion: The automatic conversion performed in LabVIEW to change the numeric representation of a data element.

Coercion dot: Dot that appears where LabVIEW is forced to convert a numeric representation of one terminal to match the numeric representation of another terminal.

Conditional terminal: The terminal of a While Loop that contains a Boolean value that determines whether the loop performs another iteration.

Count terminal: The terminal of a For Loop whose value determines the number of times the For Loop executes its subdiagram.

Control flow: Determining the execution order of a program by arranging its elements in a certain sequence.

Data dependency: The concept that block diagram nodes do not execute until data is available at all the node inputs.

For Loop: Iterative loop structure that executes its subdiagram a set number of times.

Formula Node: Node that executes formulas that you enter as text. Especially useful for lengthy formulas too cumbersome to build in the block diagram.

Frames: Diagrams associated with Sequence structures that look like frames of film and control the execution order of the program.

Iteration terminal: The terminal of a For Loop or While Loop that contains the current number of completed iterations.

Object: Generic term for any item on the front panel or block diagram, including controls, nodes, and wires.

Sequence local: Terminal that passes data between the frames of a Sequence structure.

Sequence structure: Program control structure that executes its subdiagram in numeric order. Commonly used to force nodes that are not data-dependent to execute in a desired order.

Shift register: Optional mechanism in loop structures used to pass the value of a variable from one iteration of a loop to a subsequent iteration.

Structure: Program control element, such as a Sequence, Case, For Loop, or While Loop.

Subdiagram: Block diagram within the border of a structure.

Tunnels: Relocatable connection points for wires from outside and inside the structures.

While Loop: Loop structure that repeats a section of code until a Boolean condition is False.

EXERCISES

E5.1 Open Rearrange Cases.vi in the Exercises&Problems\Chapter 5 folder in the Learning directory. Open the block diagram and analyze all the possible temperature values listed in the Case structure. Notice that the case values are not in ascending numerical order. In order to change the order of case frames, pop up on the Case border and select **Rearrange Cases....** Then click on the **Sort** button. The cases are now listed in ascending numerical order. Click **OK** if you want to leave them in ascending order. Otherwise, if you want the case frames to display in any other order, click and drag the values in **Case List**.

E5.2 Open Gray Dots.vi in the Exercises&Problems\Chapter 5 folder in the Learning directory. Switch to the block diagram and notice there are gray dots to signify numeric conversion. In general, you should try to eliminate as many gray dots in your VIs as possible because they slow down execution and use more memory for mathematical operations. To eliminate gray dots, pop up on the terminal with the gray dot and select **Representation....** In this VI, pop up on the blue indicator terminal outside the For Loop (on the right side) and choose the same representation as the input value, which is DBL. To eliminate the other gray dot on the N terminal of the For Loop, change the representation on the input value to match the N terminal, which is I32. Both gray dots should be gone!

E5.3 Construct a VI that displays a random number between 0 and 1 once every second. Also, compute and display the average of the last four random numbers generated. Only display the average after 4 numbers have been generated, otherwise display a 0. Each time the random number exceeds 0.5, generate a beep sound using the **beep.vi**.

E5.4 Using a single **While Loop**, construct a VI that executes a loop N times or until the user presses a stop button. Place a **Wait Until Next ms Multiple** VI in the loop so that the user has time to press the stop button.

E5.5 Use a **For Loop** to generate 100 random numbers and determine the most current maximum and minimum number as the random numbers are being generated. This is sometimes referred to as a "running" maximum and minimum. Display the running maximum and minimum random number on the front panel.

E5.6 Use the MATLAB script node to obtain a plot of a sine wave on a MATLAB figure. The sine wave frequency should be an input to the script node.

PROBLEMS

P5.1 Complete the crossword puzzle.

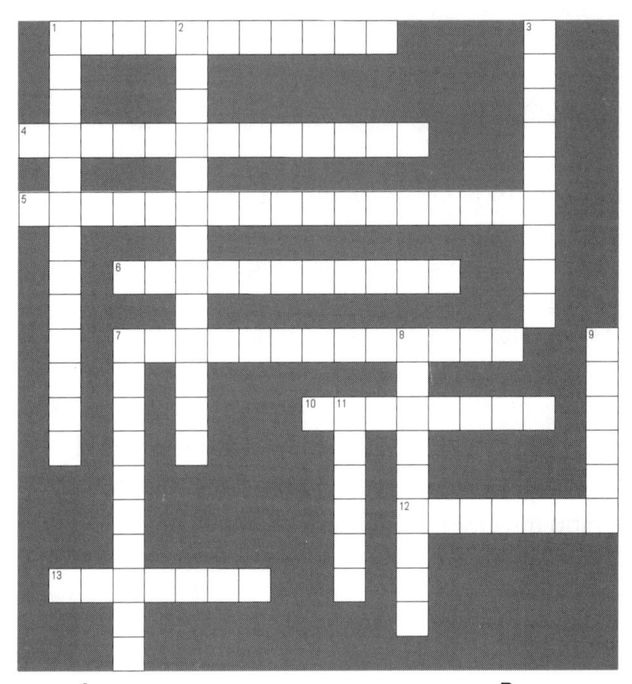

Across

1. Dot that appears where LabVIEW converts numeric representations of terminals to match.
4. Terminal that passes data between the frames of a Sequence structure.
5. The terminal of a For Loop or While Loop that contains the current number of completed iterations.
6. Node that executes formulas that you enter as text.
7. Optional mechanism in loop structures used to pass the value of a variable from iteration to iteration.
10. The automatic conversion performed to change the numeric representation of a data element.
12. Relocatable connection points for wires from outside and inside the structures.
13. Iterative loop structure that executes its subdiagram a set number of times.

Down

1. A conditional branching structure that executes one of its subdiagrams based on specific inputs.
2. The terminal of a For Loop whose value determines the number of times the For Loop executes.
3. Loop structure that repeats a section of code until a Boolean condition is False.
7. Block diagram within the border of a structure.
8. Program control element, such as a Sequence, Case, For Loop, or While Loop.
9. Diagrams associated with Sequence structures that control the execution order of the program.
11. Generic term for any item on the front panel or block diagram, including controls, nodes, and wires.

P5.2 Open Case Errors.vi in the Exercises&Problems\Chapter 5 folder in the Learning directory. Click on the **Broken Run** button and use the error list to determine how to fix this VI so you can run it.

P5.3 Open Avoid Sequence.vi in the Exercises&Problems\Chapter 5 folder in the Learning directory. Rewrite this VI without a Sequence structure. The objects in this VI can execute in the correct order without the Sequence structure. Remember that it is important to use the Sequence structure only when passing data from one object to another does not create the proper execution order.

P5.4 Construct a VI that has three LEDs. When the VI executes, the first LED should turn on for one second, then the second LED for two seconds, and finally the third LED should turn on for three seconds.

P5.5 Create two subVIs to calculate the function $Y = \sin(X^2/4 + 2X + 1)$. In the first VI, use a **Formula Node**, and in the second VI, use functions from the **Numeric** subpalette. Compute the execution time (in ms) required for each VI to evaluate the function on the interval $X = 0.0, 0.01, \ldots, 999.99, 1000.00$.

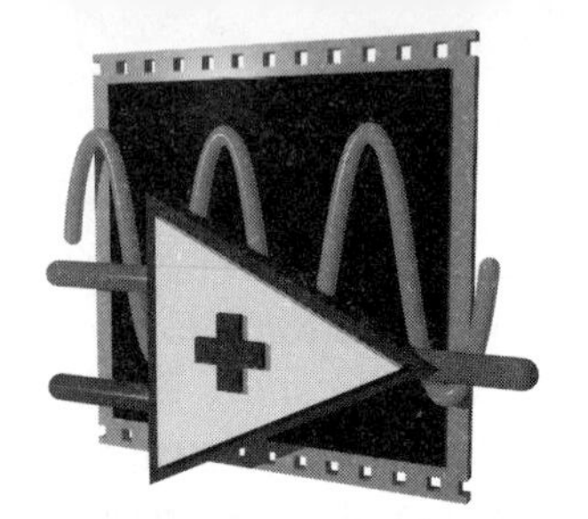

CHAPTER 6

Arrays and Clusters

The array data type and the cluster data type are presented in this chapter. An array is a variable-sized collection (or grouping) of data elements that are all the same type, such as a group of floating-point numbers or a group of strings. A cluster is a fixed-sized collection of data elements of mixed types, such as a group containing floating-point numbers and strings. You will learn how to use built-in functions to manipulate arrays and clusters. The important concept of polymorphism is introduced. Polymorphism is the ability of a function to adjust to input data of different types, dimensions, or representations.

GOALS

1. Understand how to create and use arrays.
2. Learn to use a variety of built-in array functions.
3. Understand the concept of polymorphism.
4. Become familiar with creating and using clusters.
5. Learn to manipulate clusters using built-in functions.

6.1 ARRAYS

An **array** is a *variable-sized* collection of data elements that are all the *same type*. In contrast, a **cluster** is a *fixed-sized* collection of data elements of *mixed types*. In what situations might you use arrays? One scenario that benefits from the use of arrays and which you will encounter frequently involves working with a collection of data for plotting purposes. Arrays are quite useful as a mechanism for organizing data for plotting. Arrays are also helpful when you perform repetitive computations or when you are solving problems that are naturally formulated in matrix-vector notation, such as solving systems of linear equations. Using arrays in your VIs leads to compact block diagram codes that are easier to develop because of the large number of built-in array functions and VIs.

Arrays can have one or more dimensions with up to 2^{31} elements per dimension. The maximum number of elements depends on the available memory. The individual elements of an array can be any type with the exceptions that you cannot have an array of arrays, an array of charts, or an array of graphs. You access an individual array element by its index. The index is zero-based, implying that the array index is in the range 0 to $n-1$, where n is the number of elements in the array. A one-dimensional (1D) array is shown in Figure 6.1. The first element of the 1D array has index 0, the second element has index 1, and so on.

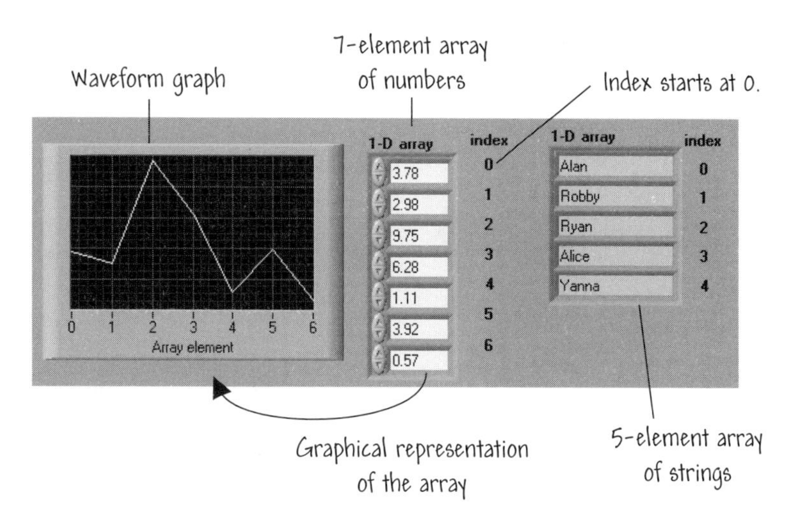

FIGURE 6.1
One-dimensional array examples.

Another way to display an array is with a waveform graph. In Figure 6.1, the waveform graph is used to display the numeric array in which each successive element of the array is plotted on the graph. To generate an x versus y graph, you could use a two-dimensional array (2D) with one column containing the x data points and the other column containing the y data points. To learn more about graphs, refer to Chapter 7.

6.1.1 Creating Array Controls and Indicators

It takes two steps to create an array control or indicator. The two steps involve combining an **array shell** from the **Array & Cluster** subpalette of the **Controls** palette, as shown in Figure 6.2, with a valid element, which can be a numeric, Boolean, or cluster. Actually, valid elements also include strings—we will discuss strings in Chapter 9. In any case, charts, graphs, or other arrays are not valid elements to combine with the array control or indicator.

The two steps that lead to the creation of an array control or indicator follow:

- Select an empty array shell from the **Array & Cluster** subpalette of the **Controls** palette and drop it onto the front panel, as illustrated in Figure 6.3.
- Drag a valid data object (such as a numeric, Boolean, or string) into the array shell, as shown in Figure 6.4. The array shell resizes to accommodate its new type. You can also place the valid data object directly into the shell using the array shell's short cut menu.

To display more elements of the array, use the **Positioning** tool to grab a resizing handle on the corner of the array window and stretch the object to the desired number of visible array elements. The index value shown in the box on the left side of the array corresponds to the first visible element in the array. You can move through the array by clicking on the up and down arrows in the index display.

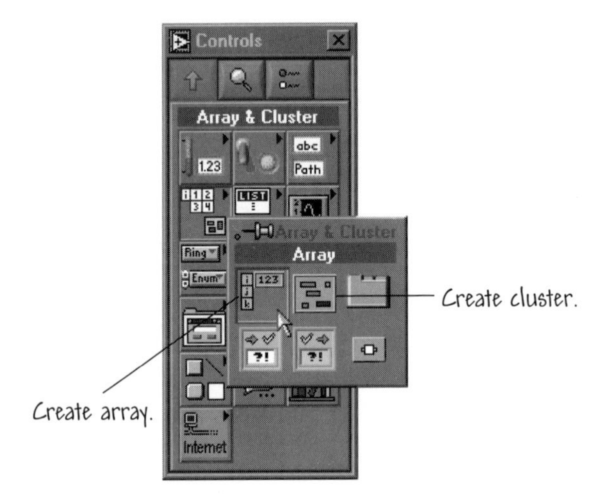

FIGURE 6.2
Creating an array control from the **Array & Cluster** subpalette of the **Controls** palette.

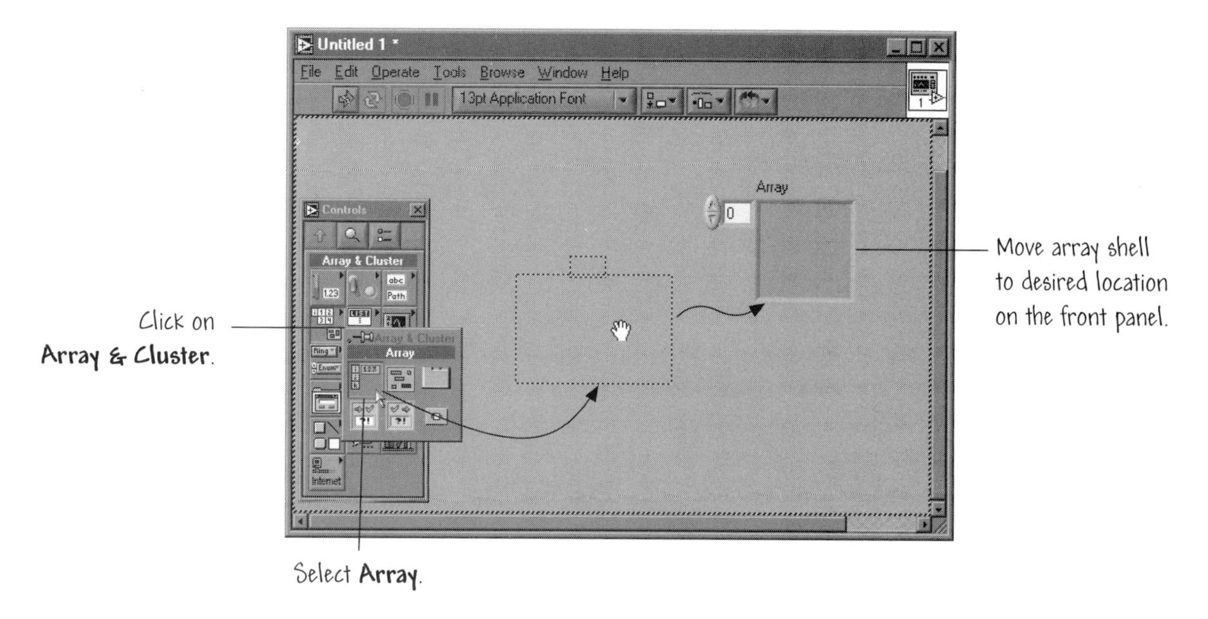

FIGURE 6.3
Placing an empty array shell on the front panel.

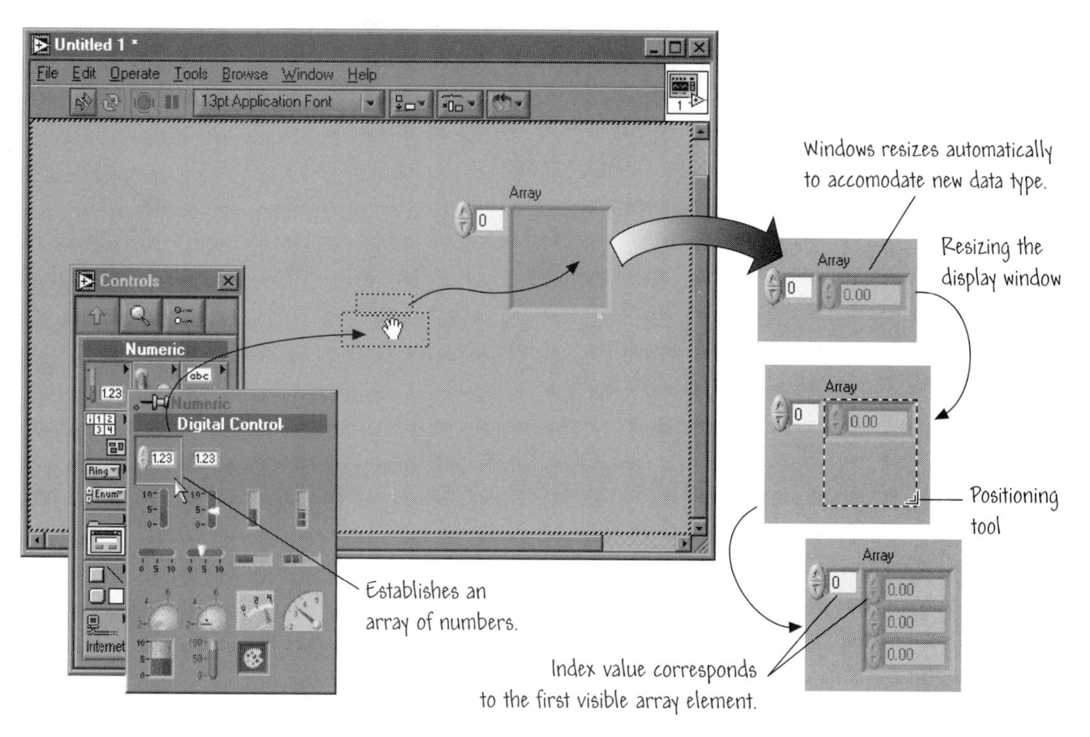

FIGURE 6.4
Drag a valid data object into the array shell to establish the array type.

The block diagram terminal of an array shell is black when first dropped on the front panel. This indicates that the data type is undefined. The terminal contains brackets that denote array structures, as illustrated in Figure 6.5. When you drop a valid data object (such as a numeric, Boolean, or string) into the array shell, the array terminal on the block diagram turns from black to a color reflecting the data type. In Figure 6.5 the array block diagram terminal is pink, indicating that the array contains strings. When you wire arrays in the block diagram, you will find that array wires are thicker than wires carrying a single value.

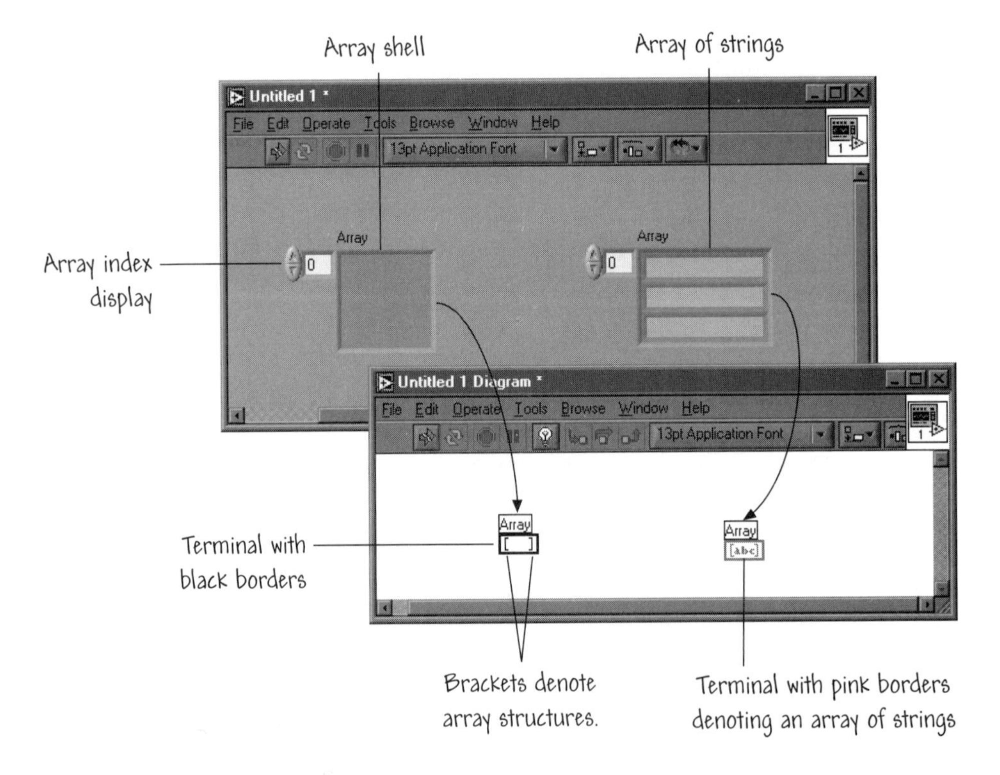

FIGURE 6.5
The terminal of an array shell is black, denoting an undefined data type.

Remember that you must assign a data object to the empty array shell before using the array on the block diagram. If you do not assign a data type, the array terminal will appear black with an empty bracket.

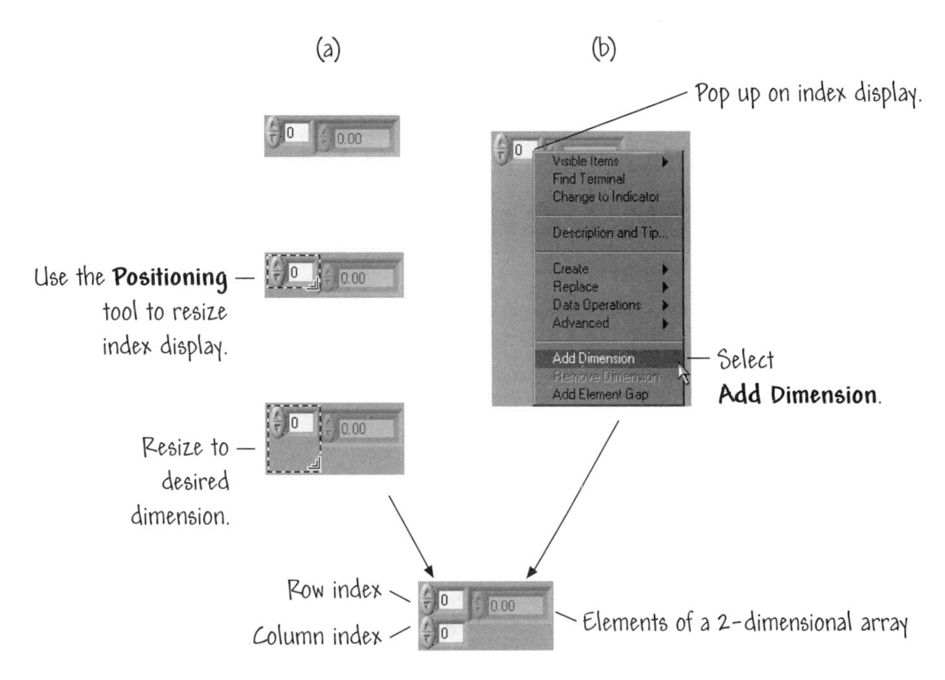

FIGURE 6.6
Adding dimensions to an array.

6.1.2 Multidimensional Arrays

A two-dimensional (2D) array requires two indices—a row index and a column index—to locate an element. A three-dimensional array requires three indices, and in general, a n-dimensional array requires n indices. You add dimensions to the array in one of two ways: (a) by using the **Positioning** tool to resize the index display, or (b) by popping up on the array index display and choosing **Add Dimension** from the short cut menu (see Figure 6.6). An additional index display appears for each dimension you add. The example in Figure 6.6 shows a 2D digital control array.

You can reduce the dimension of an array by resizing the index display appropriately or by selecting **Remove Dimension** from the pop-up menu.

6.2 CREATING ARRAYS WITH LOOPS

With the For Loop and the While Loop you can create arrays automatically with a process known as **auto-indexing**. Figure 6.7a shows a For Loop creating a 10-element array using auto-indexing. On each iteration of the For Loop, the next element of the array is created. In this case the loop counter is set to 10; hence a 10-element array is created. If the loop counter was set to 20, then a 20-element array would be created. The array passes out of the loop to the indicator after the loop iterations are complete.

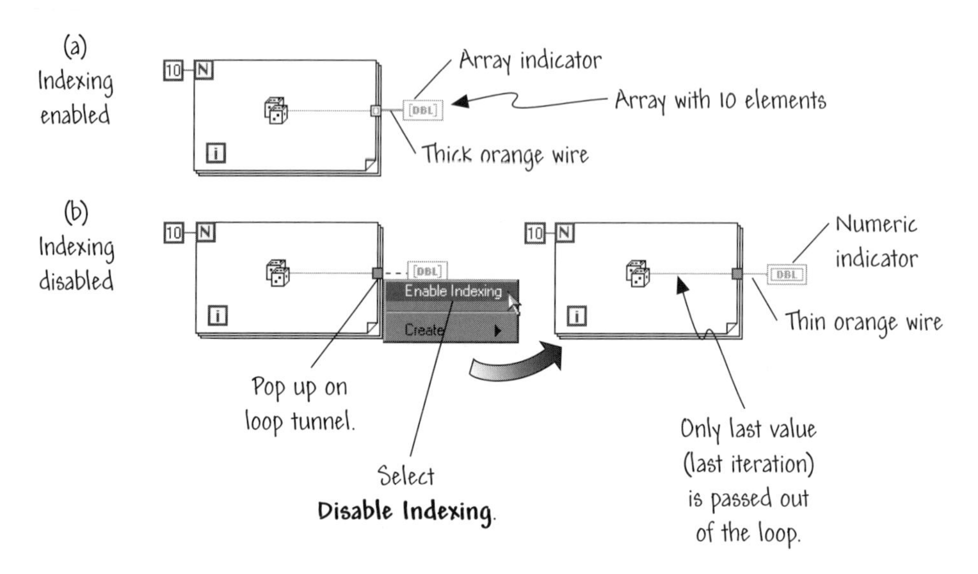

FIGURE 6.7
Auto-indexing is the ability of For Loops and While Loops to automatically index and accumulate arrays at their boundaries.

It is possible to pass a value out of a loop without creating an array. This requires that you disable auto-indexing by popping up on the tunnel (the square on the loop border through which the data passes) and selecting **Disable Indexing** from the short cut menu. In Figure 6.7b, auto-indexing is disabled; thus when the VI is executed, only the last value returned from the Random Number function passes out of the loop.

Open and run Array Auto Index Demo.vi *shown in Figure 6.7. It can be found in the* Chapter 6 *folder in the* Learning *directory. After running the program, you should find that with auto-indexing enabled, the indicator array will have 10 elements (indexed* $0, 1, \cdots, 9$*), and with auto-indexing disabled only the last value of the random number function is passed out of the For Loop.*

You can also pass arrays into a loop one element at a time or the entire array at once. With auto-indexing enabled, when you wire an array (of any dimension) from an external node to an input tunnel on the loop border, then elements of the array enter the loop one at a time, starting with the first component. This is illustrated in Figure 6.8a. With auto-indexing disabled, the entire array passes into the loop at once. This is illustrated in Figure 6.8b.

With For Loops auto-indexing is enabled by default. In contrast, with While Loops auto-indexing is disabled by default. If you desire auto-indexing, you need

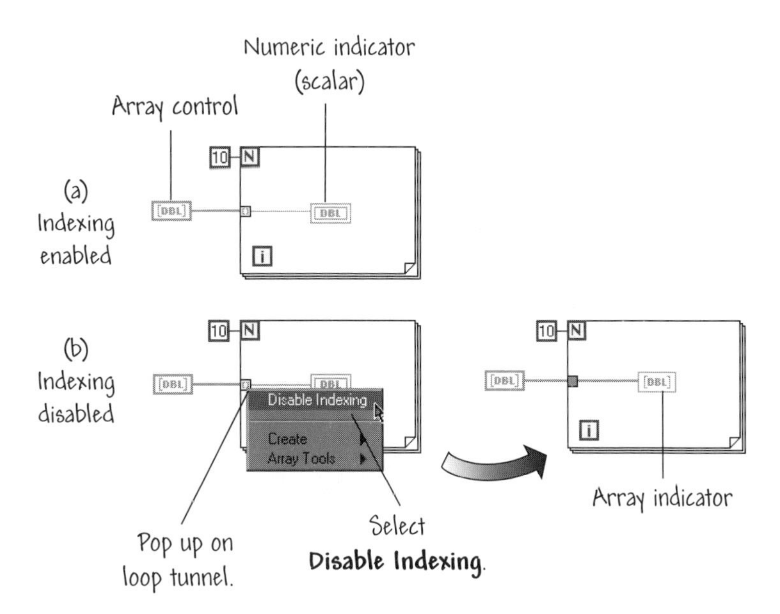

FIGURE 6.8
Auto-indexing applies when you are wiring arrays into loops.

*to pop up on the While Loop tunnel and choose **Enable Indexing** from the short cut menu.*

With a For Loop with auto-indexing enabled, an array entering the loop automatically sets the loop count to the number of elements in the array, thereby eliminating the need to wire a value to the loop count, N. What happens if you explicitly wire a value to the loop count that is different than an array size entering the loop? Or what if you wire two arrays with a different number of elements to a For Loop? The answer is that if you enable auto-indexing for more than one array entering a For Loop, or if you set the loop count by wiring a value to N with auto-indexing enabled, the actual loop count becomes the smaller of the two. For example, in Figure 6.9, the array size (= 4), and not the loop count N (= 5), sets the For Loop count—because the array size is the smaller of the two. The Array Size function used in the VI shown in Figure 6.9 will be discussed in the next section along with other common built-in array functions provided.

Open and run Array Auto Count Set.vi shown in Figure 6.9. It can be found in the Chapter 6 folder in the Learning directory. After running the program, you should find that only four random numbers are generated, despite the fact that the loop count N has been set to five!

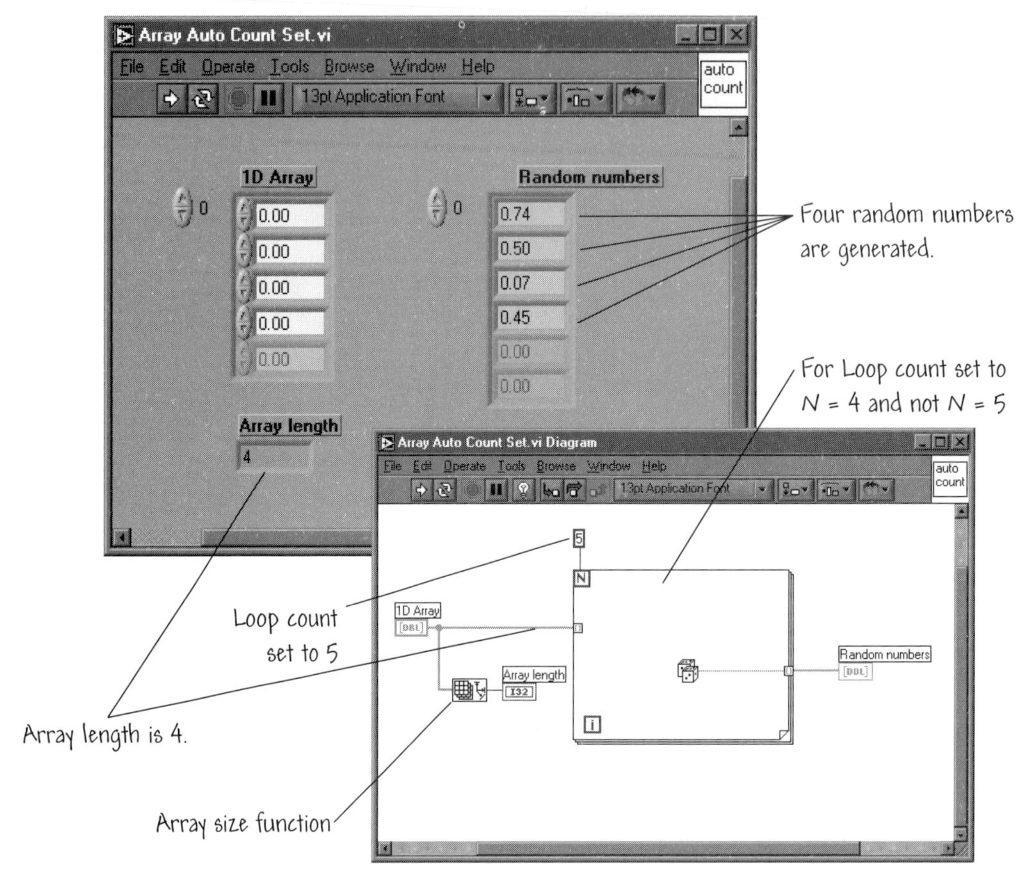

FIGURE 6.9
Automatically setting the For Loop count to the array size.

6.2.1 Creating Two-Dimensional Arrays

You can use two nested For Loops (that is, one loop inside the other) to create a 2D array. The outer For Loop creates the row elements, and the inner For Loop creates the columns of the 2D array. Figure 6.10 shows two For Loops creating a 2D array of random numbers using auto-indexing.

Open and run Two-Dimensional Array Demo.vi *shown in Figure 6.10. It can be found in the* Chapter 6 *folder in the* Learning *directory. After running the program, you should find that a 2D array has been created.*

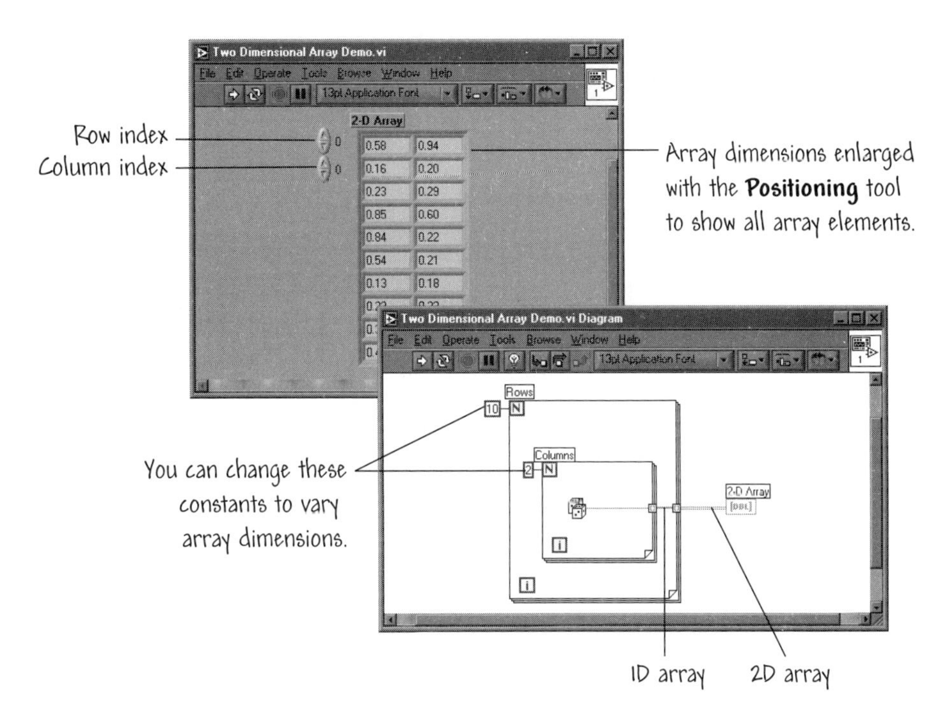

FIGURE 6.10
Creating a 2D array using two For Loops and auto-indexing.

6.3 ARRAY FUNCTIONS

Many built-in functions are available to manipulate arrays. Most of the common array functions are found in the **Array** subpalette of the **Functions** palette, as illustrated in Figure 6.11. Several of the heavily used functions are pointed out in Figure 6.11 and discussed in this section.

6.3.1 Array Size

The function Array Size returns the number of elements in the input array, as illustrated in Figure 6.12. If the input is an n-dimensional array, the output of the array function Array Size is a one-dimensional array with n elements—each element containing a dimension size.

You can run the array size demonstration shown in Figure 6.12 by opening Array Size Demo.vi in the Chapter 6 folder in the Learning directory. After executing the VI once to verify that the 1D array has length 4 and the 2D array has 2 rows and 4 columns, add an element to the 1D array and run the VI again. What happens? The 1D Array size value increments by 1.

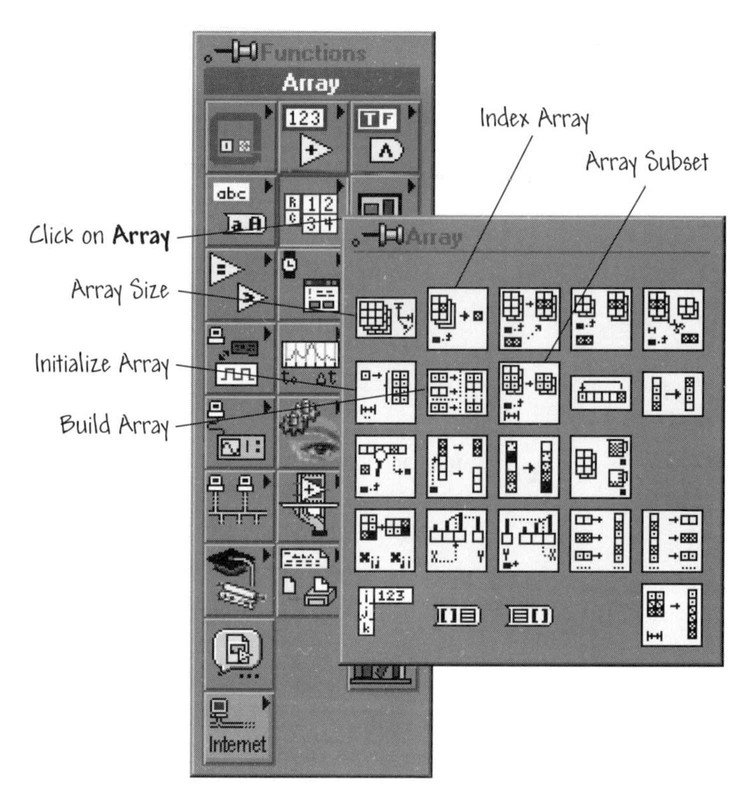

FIGURE 6.11
The array functions palette.

6.3.2 Initialize Array

The function Initialize Array creates an n-dimensional array with elements containing the values that you specify. All elements of the array are initialized to the same value. To create and initialize an array that has more than one dimension, pop up on the lower-left side of the Initialize Array node and select **Add Dimension** or use the **Positioning** tool to grab a resizing handle to enlarge the node. You can remove dimensions by selecting **Remove Dimension** from the function short cut menu or with the **Resizing** cursor. The Initialize Array is useful for allocating memory for arrays. For instance, if you are using shift registers to pass an array from one iteration to another, you can initialize the shift register using the Initialize Array function.

The input determines the data type and the value of each element of the initialized array. The dimension size input determines the length of the array (see Figure 6.13a). For example, if the input value **element** is a double-precision floating-point number with the value of 5.7, and dimension size has a value of 100, the result is a 1D array of 100 double-precision floating-point numbers all set to 5.7, as illustrated in Figure 6.13b. You can wire inputs to the Initialize Array function from front panel control terminals, from block diagram constants,

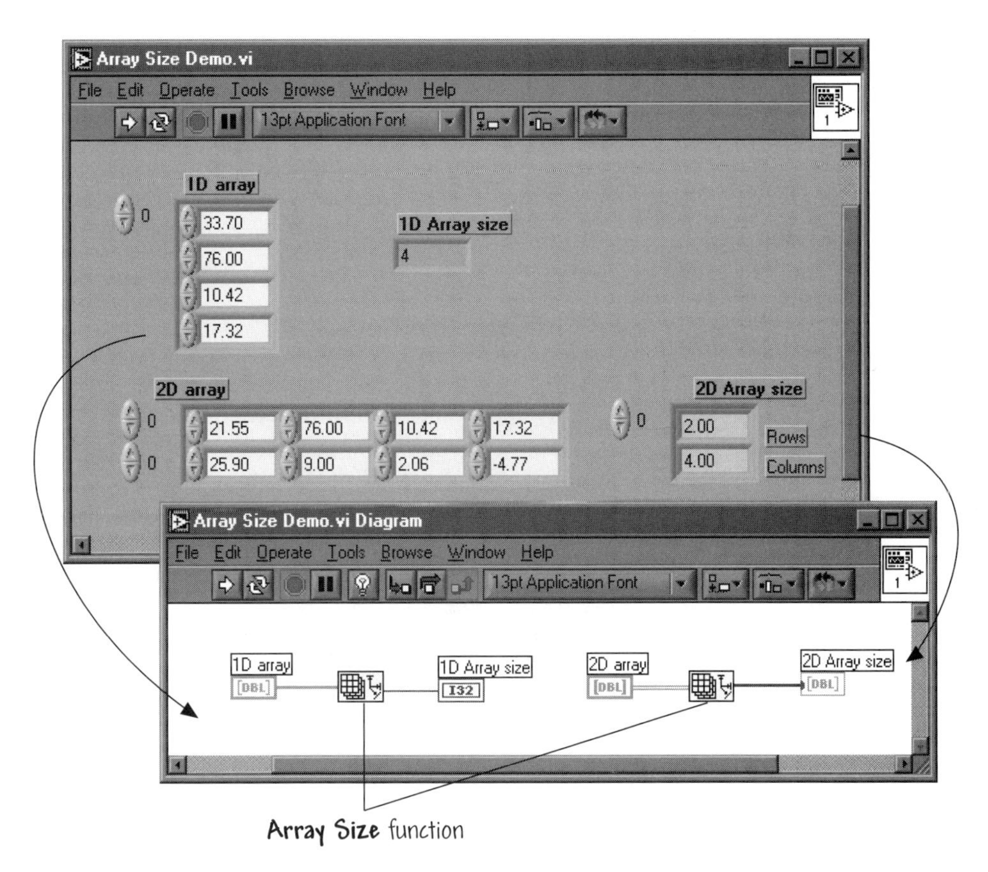

FIGURE 6.12
The function Array Size returns the number of elements in the input array.

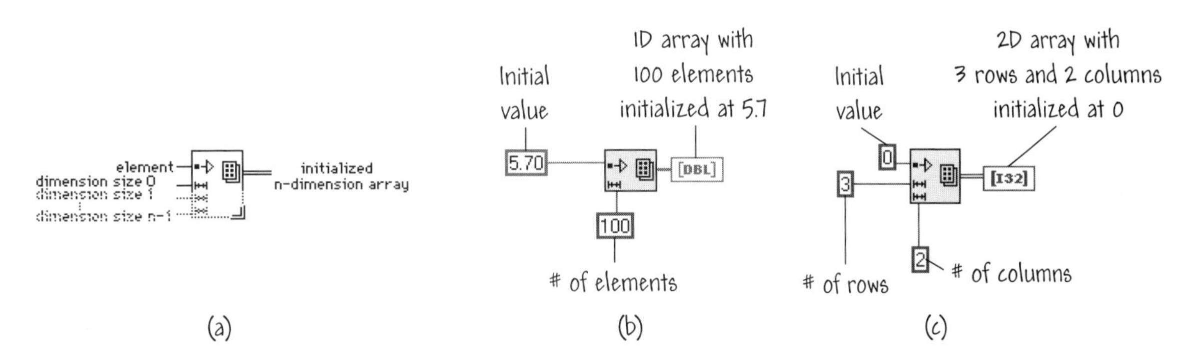

FIGURE 6.13
The function Initialize Array creates an array of a certain dimension containing the same value for each element.

or from calculations in other parts of the block diagram. One method of obtaining the output array is to pop up on the Initialize Array function and choose Create≫ Indicator. Also, you can pop up and choose Create≫ Control and Create≫ Constant to create inputs for the element and dimension size. Figure 6.13c depicts a 2D array with 3 rows and 2 columns initialized with the long integer value of 0.

You can run the array initialization demonstration shown in Figure 6.13 by opening Array Initialization Demo.vi *located in the* Chapter 6 *folder in the* Learning *directory. Change the initial value of the two arrays on the block diagram, run the VI and examine the result.*

6.3.3 Build Array

The function Build Array concatenates multiple arrays or adds extra elements to an array. The function has two types of inputs—scalars and arrays—therefore it can accommodate both arrays and single-valued elements.

The Build Array function appears with one scalar input when initially placed in the block diagram window, as shown in Figure 6.14.

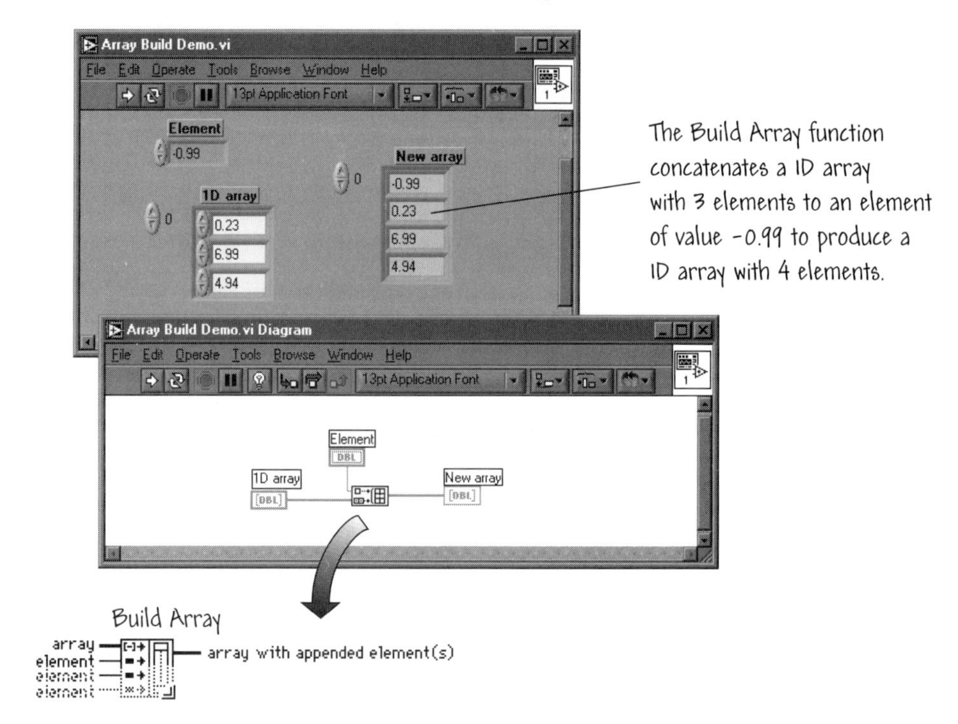

FIGURE 6.14
Build Array concatenates multiple arrays or appends elements to an array.

Array inputs have brackets, while element inputs do not. Array inputs and scalar inputs are not interchangeable! Pay special attention to the inputs of a Build Array function, otherwise you may generate wiring errors that are difficult to detect.

You can add as many inputs as you need to the Build Array function. Each input to the function can be either a scalar or an array. To add more inputs, pop up on the left side of the function and select **Add Input**. You also can enlarge the Build Array node by placing the **Positioning** tool at one corner of an object to grab and drag the resizing handle. You can remove inputs by shrinking the node with the resizing handle or by selecting **Remove Input** from the short cut window.

The input type (element or array) is automatically configured when wired to the Build Array function. The Build Array function shown in Figure 6.14 is configured to concatenate an array and one scalar element into a new array.

The Build Array function concatenates the elements or arrays in the order in which they appear in the function, top to bottom. If the inputs are of the same dimension, you can right-click on the function and select **Concatenate Inputs** to concatenate the inputs into a longer array of the same dimension. If you leave the **Concatenate Inputs** option unselected, you will add a dimension to the array.

The Array Build demonstration shown in Figure 6.14 is located in the Chapter 6 *folder in the* Learning *directory and is called* Array Build Demo.vi. *Open, run and experiment with the VI.*

6.3.4 Array Subset

The function Array Subset returns a portion of an array starting at **index** and containing **length** elements. In the example shown in Figure 6.15 you'll find that the array index begins with 0.

The array subset demonstration shown in Figure 6.15 can be found in the Chapter 6 *folder in the* Learning *directory and is called* Array Subset Demo.vi. *Open and run the VI in the* ***Continuous Run*** *mode. Using the* ***Operating*** *tool, change the value of* ***Index Number*** *and watch the Array Subset function return a portion of the array, starting at the index that you specify and including the three subsequent elements.*

6.3.5 Index Array

The function Index Array accesses an element of an array. The example shown in Figure 6.16 uses the input **Index number** to specify which element of the array to access. Remember that the index number of the first element is zero.

The index array demonstration shown in Figure 6.16 can be found in the Chapter 6 *folder of the* Learning *directory and is called* Array Index Demo.vi. *Open and run the VI in* ***Continuous Run*** *mode. Using the* ***Operating*** *tool, change*

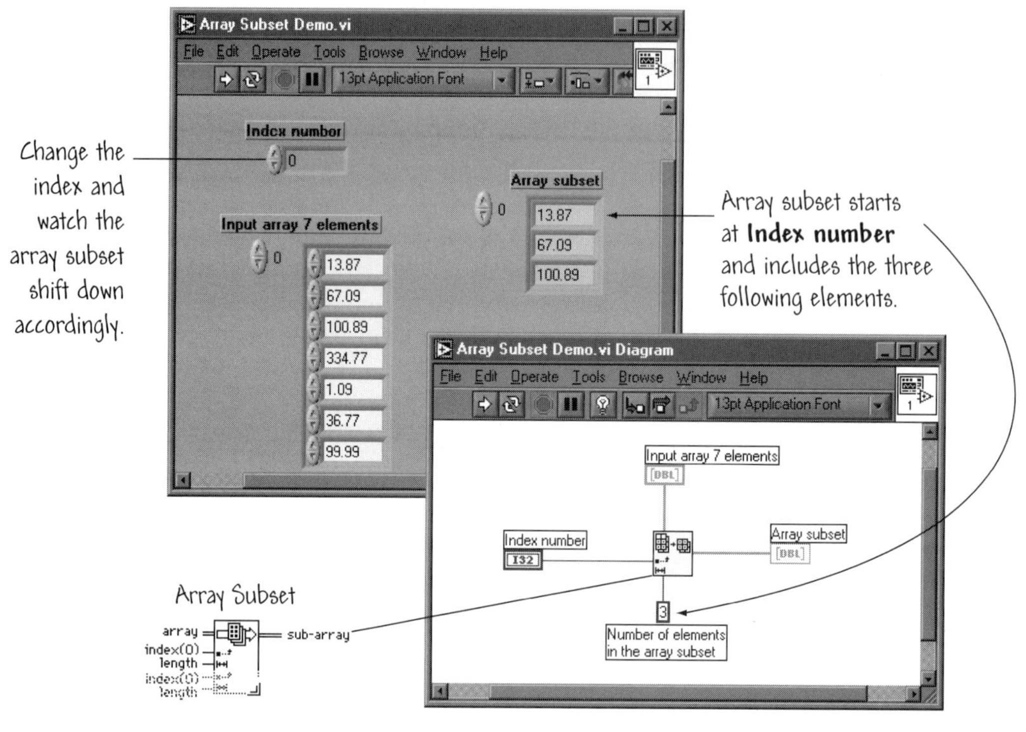

FIGURE 6.15
The function Array Subset returns a portion of an array starting at **index** and containing **length** elements.

*the value of **Index number** and watch the Index Array function return a different element based on the index value.*

The Index Array function automatically resizes to match the dimensions of the wired input array. For example, if you wire a 1D array to the Index Array function, a single index input will show. Similarly, if you wire a 2D array the Index Array function, two index inputs will show—one for the row and one for the column.

Once you have wired an input array to the Index Array function, you can access more than one element, or subarray (e.g., a row or a column) using the **Positioning** tool to manually resize the function once placed on the block diagram. When you expand the Index Array function it will expand in increments determined by the dimensions of the array wired to the function. In Figure 6.17, the Index Array function is expanded so that three subarrays can be extracted: a row, a column, and a single element. The index inputs you wire determine the shape of the subarray you want to access or modify. For example, if the input to an Index Array function is a 2D array and you wire only the row input, you extract a complete 1D row of the array. If you wire only the column input, you extract a

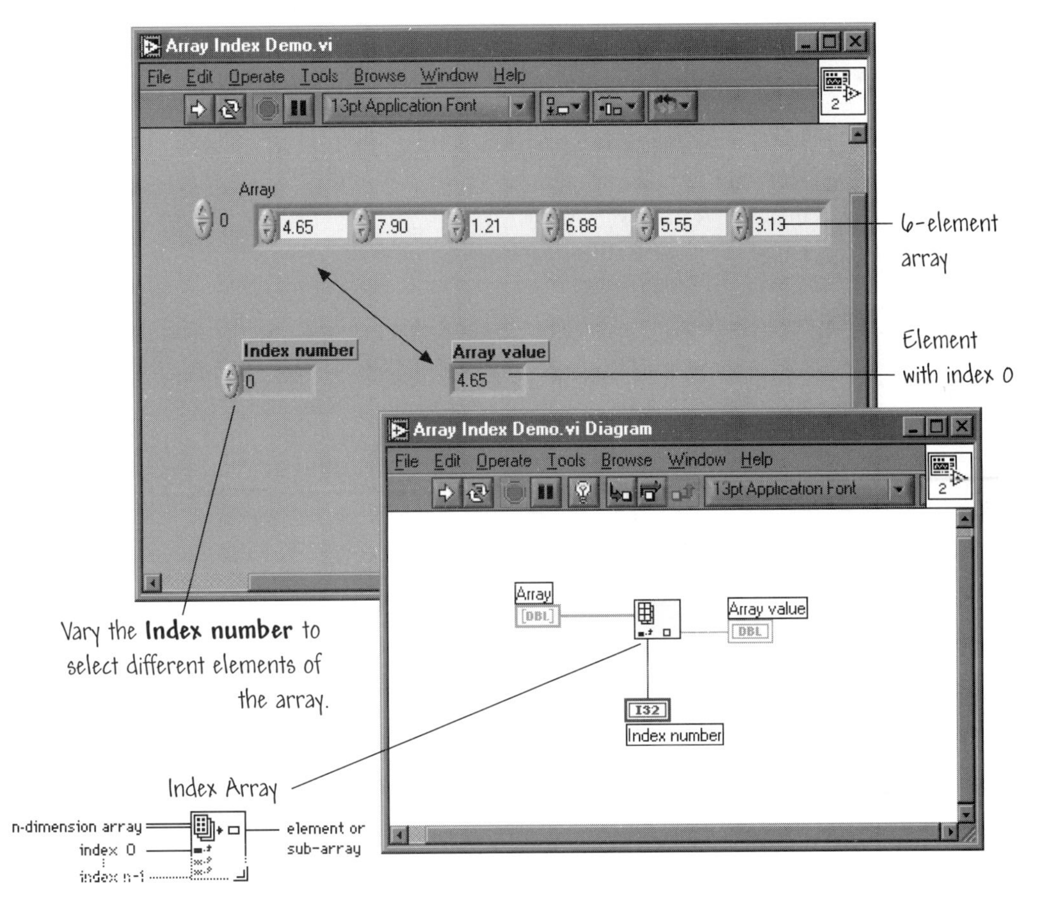

FIGURE 6.16
The function Index Array accesses an element of an array.

complete 1D column of the array. If you wire the row input and the column input, you extract a single element of the array. These three cases are illustrated in Figure 6.17. Each input group is independent and can access any portion of any dimension of the array.

The 2D index array demonstration shown in Figure 6.17 can be found in the Chapter 6 *folder of the* Learning *directory and is called* 2D Array Index Demo.vi*. Open and run the VI in* ***Continuous Run*** *mode. Using the* ***Operating*** *tool, change the* ***Column number*** *and* ***Row number*** *values and verify the numbers returned in* ***Column*** *and* ***Row****.*

Practice with Arrays

In this exercise you will get more practice using array functions. Open Practice with Arrays.vi located in the Chapter 6 folder of the Learning directory.

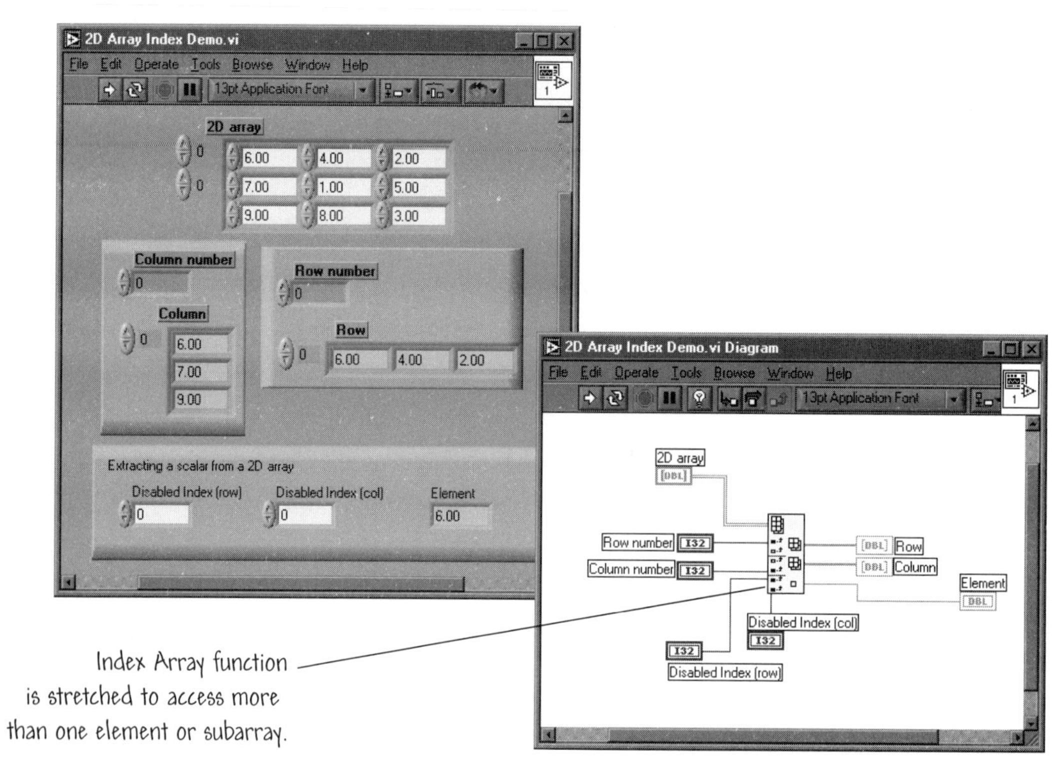

FIGURE 6.17
Using Index Array to extract a row or column of an array.

The front panel and block diagram of the incomplete VI are shown in Figure 6.18. This VI is not completed—you will finish wiring and debugging the VI for practice.

The front panel contains four arrays and one digital control. The completed VI concatenates the input arrays and the digital control values to form a new array. Using the Array Size function and the Array Initialize function, the VI creates a new array of appropriate dimension and with all elements of the array initialized to 1. In the final computation of the VI, the difference between the two new arrays is computed, and the result is displayed on the front panel. When you compute the difference of two arrays with the same number of elements, the differencing operation subtracts the array values element by element.

The completed block diagram should resemble the one shown in Figure 6.19. Remember to save and close the VI in the Users Stuff folder.

The completed VI for this exercise can be found in the Chapter 6 *folder of the* Learning *directory and is called* Practice with Arrays Done.vi.

◆

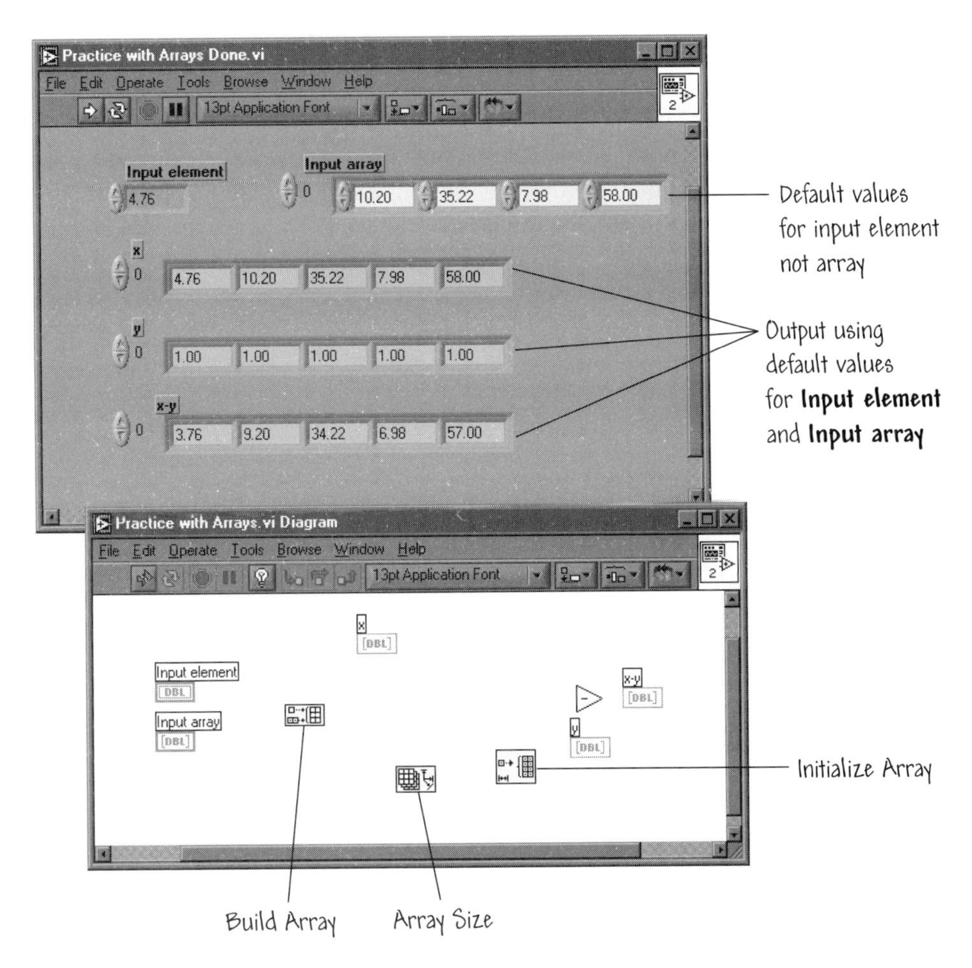

FIGURE 6.18
Practice with array functions.

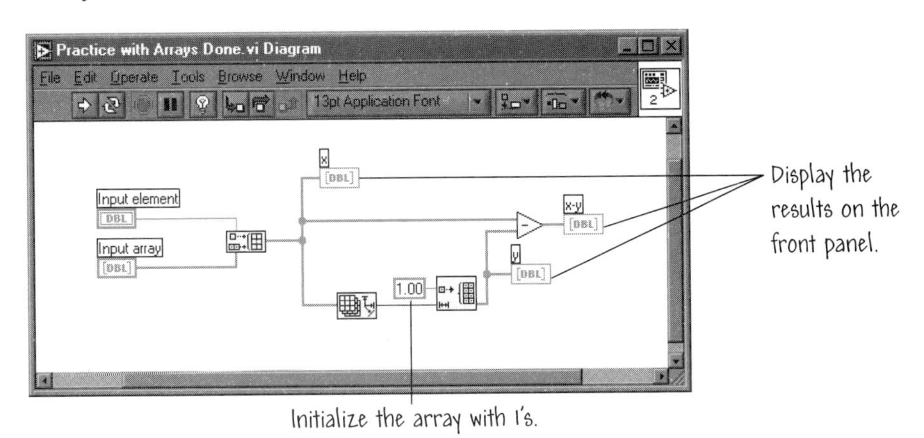

FIGURE 6.19
The completed block diagram for the array practice exercise.

6.4 POLYMORPHISM

Polymorphism is the ability of certain LabVIEW functions (such as Add, Multiply, and Divide) to accept as inputs of different dimensions and representations. Arithmetic functions that possess this capability are polymorphic functions. For example, you can add a scalar to an array or add together two arrays of different lengths. Figure 6.20 shows some of the polymorphic combinations of the Add function.

In the first combination shown in Figure 6.20a, the result of adding a scalar and a scalar is another scalar. In the second combination shown in 6.20b, the

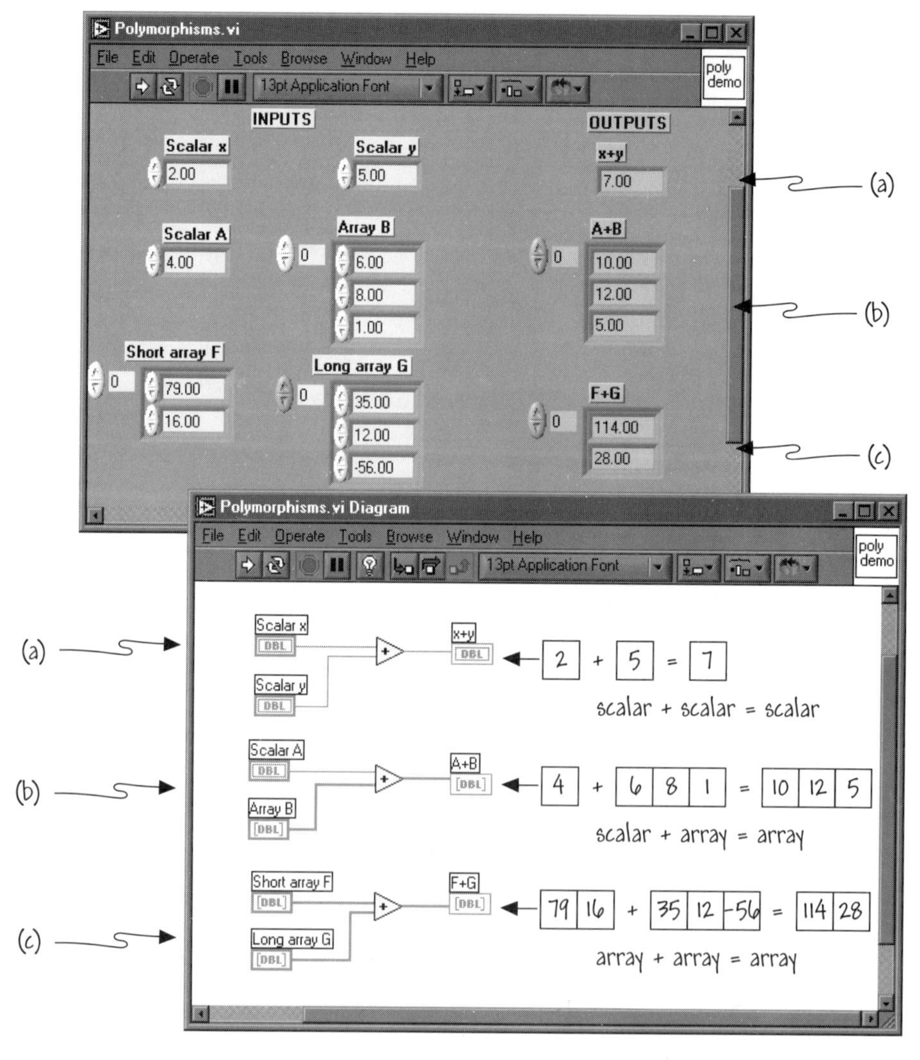

FIGURE 6.20
Polymorphic combinations of the Add function.

result of adding a scalar to an array is another array. In this situation, the scalar input is added to each element of the input array. In the third combination depicted in 6.20c, an array of length 2 is added to an array of length 3, resulting in an array of length 2 (the length of the shorter of the two input arrays). Array addition is performed component-wise; that is, each element of one array is added to the corresponding element of the other array.

When two input arrays have different lengths, the output array resulting from some arithmetic operation (such as adding the two arrays) will be the same size as the smaller of the two input arrays. The arithmetic operation applies to corresponding elements in the two input arrays until the shorter array runs out of elements—then the remaining elements in the longer array are ignored!

Consider the VI depicted in Figure 6.21. Each iteration of the For Loop generates a random number that is stored in the array and readied for output once the loop iterations are finished. After the loop finishes execution (after 10 iterations), the Multiply function multiplies each element in the array by the scaling factor 2π. Notice that the Multiply function has two inputs: an array and a scalar. The polymorphic capability of the Multiply function allows the function to take inputs of differing dimension (in this instance, an array of length 10 and a scalar) and to produce a sensible output. What happens when you have two array inputs for a Multiply function? In that case, corresponding elements of each array are multiplied.

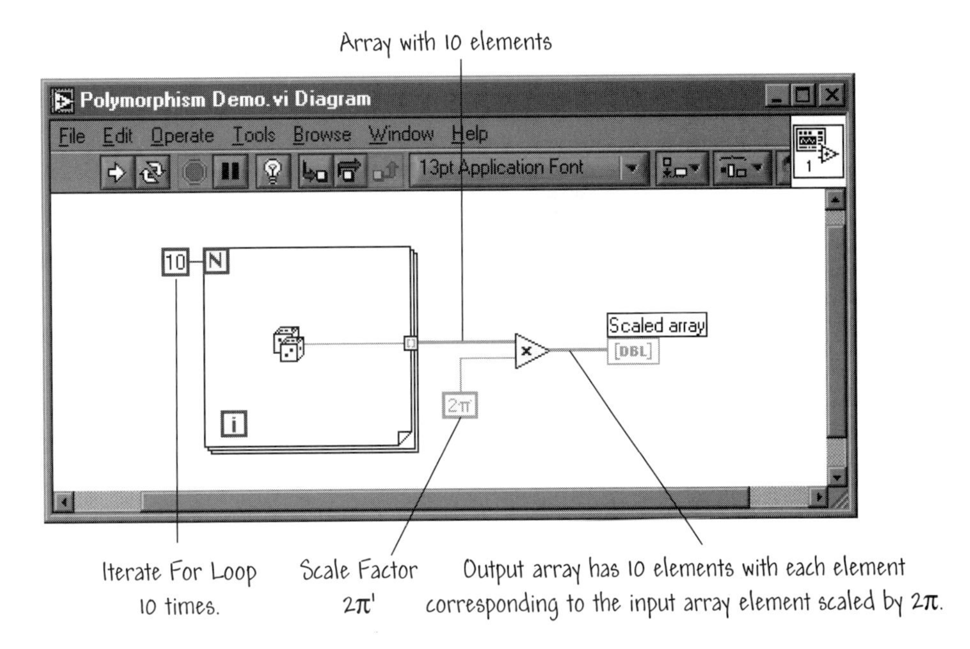

FIGURE 6.21
Demonstrating the polymorphic capability of the Multiply function.

The two demonstrations of polymorphism shown in Figures 6.20 and 6.21 can be found in the Chapter 6 *folder in the* Learning *directory and are called* Polymorphisms.vi *and* Polymorphism Demo.vi, *respectively. Open and run each VI. Vary the values of the array elements and examine the results. You can also edit the array lengths and test your knowledge of polymorphisms by first predicting the result and then verifying your prediction by running the VI.*

Practice Using Polymorphism

Open a new VI and recreate the front panel and block diagram shown in Figure 6.22. When you have finished the construction, experiment with the VI. A suggested way to run the VI is in **Continuous Run** mode. Increment and decrement the variable **Array length** and watch the array length change.

This exercise demonstrates two applications of polymorphism. Referring to Figure 6.22, it can be seen that in one case the Multiply function operates on two inputs of different dimension: an array and a scalar $\pi/2$. In the second

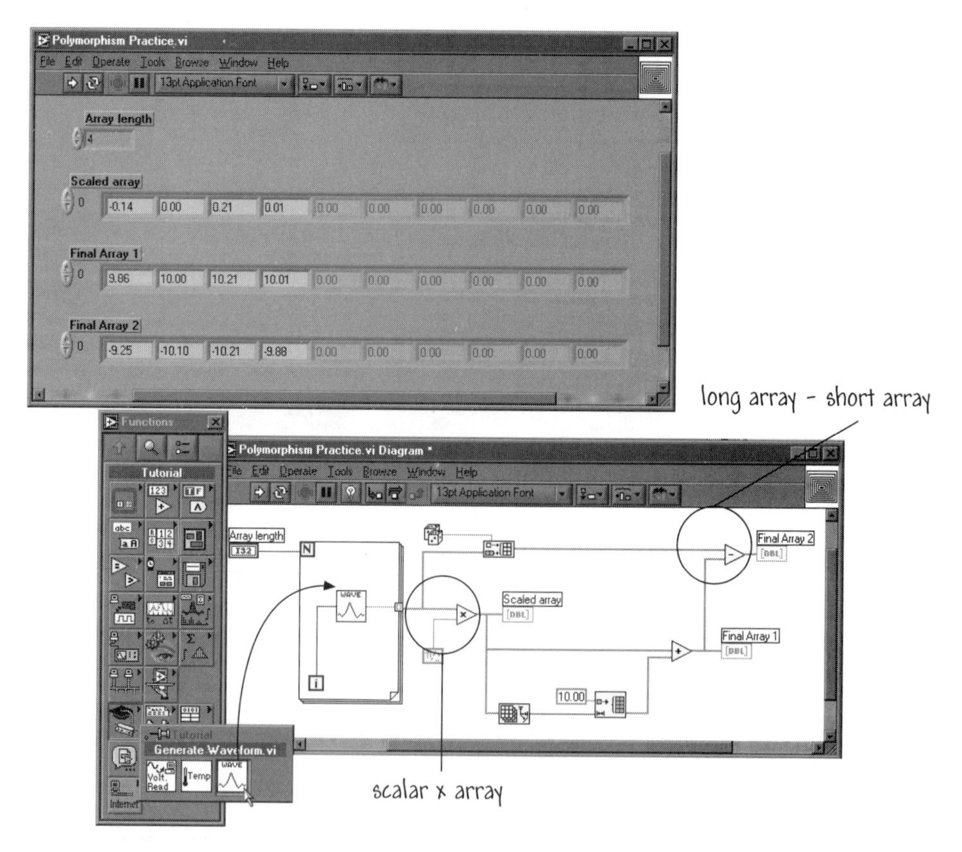

FIGURE 6.22
Practicing using polymorphism.

illustration of polymorphism, the Subtract function has two array inputs of differing lengths. The resulting array **Final Array 2** has length equal to the length of the shorter array (the length is equal to length of **Scaled array**). Three array functions are also utilized: Build Array, Array Size, and Initialize Array. Save the VI as Polymorphism Practice.vi and place it in the Users Stuff folder.

The completed polymorphism practice VI shown in Figure 6.22 can be found in the Chapter 6 folder in the Learning directory and is called Polymorphism Practice.vi.

◆

6.5 CLUSTERS

A **cluster** is a data structure that, like arrays, groups data. However, clusters and arrays have important differences. One important difference is that clusters can group data of different types, whereas arrays can group only like data types. For example, an array might contain ten digital indicators, whereas a cluster might contain a digital control, a toggle switch and string control. And although cluster and array elements are both ordered, you access cluster elements by **unbundling** some or all the elements at once rather than indexing one element at a time. Clusters are also different from arrays in that they are of a fixed size.

One similarity between arrays and clusters is that both are made up of either controls or indicators. In other words, clusters cannot contain a mixture of controls and indicators. An example of a cluster is shown in Figure 6.23. The cluster shown has four elements: a digital control, a horizontal toggle switch, a string control (more on strings in Chapter 9), and a knob. The cluster data type appears frequently when graphing data (as we will see in the next chapter).

Clusters are generally used to group related data elements that appear in multiple places on the block diagram. Because a cluster is represented by only one wire in the block diagram, its use has the positive effect of reducing wire clutter and the number of connector terminals needed by subVIs. A cluster may be thought of as a bundle of wires wherein each wire in the cable represents a different element of the cluster. On the block diagram, you can wire cluster terminals only if they have the same type, the same number of elements, and the same order of elements. Polymorphism applies to clusters as long as the clusters have the same number of elements in the same order.

6.6 CREATING CLUSTER CONTROLS AND INDICATORS

Cluster controls and indicators are created by placing a **cluster shell** on the front panel, as shown in Figure 6.24. A new cluster shell has a resizable border and an (optional) label. When you pop up in the empty element area of the cluster shell, the **Controls** palette appears, as illustrated in Figure 6.25. You create a cluster by

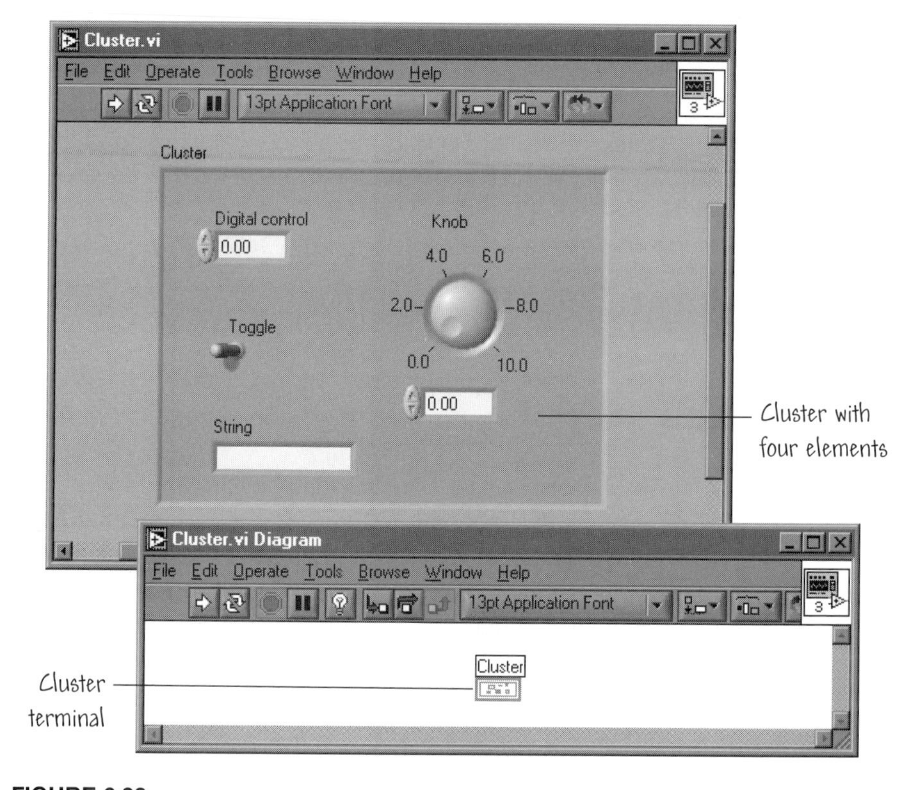

FIGURE 6.23
A cluster with four elements: digital control, toggle switch, string control, and a knob.

placing any combination of numerics, Booleans, strings, charts, graphs, arrays, or even other clusters from the **Controls** palette into the cluster shell. Remember that a cluster can contain controls *or* indicators, but not both. The cluster becomes a control cluster or indicator cluster depending on the first element you place in the cluster. For example, if the first element placed in a cluster shell is a digital control, then the cluster becomes a control cluster. Any objects added to the cluster afterwards become control objects. Selecting **Change To Control** or **Change To Indicator** from the short cut menu of any cluster element changes the cluster and all its elements from indicators to controls or from controls to indicators, respectively. You can also drag existing objects from the front panel into the cluster shell.

6.6.1 Cluster Order

The elements of a cluster are ordered according to when they were placed in the cluster rather than according to their physical position within the cluster shell. The first object placed in the cluster shell is labeled element 0, the second

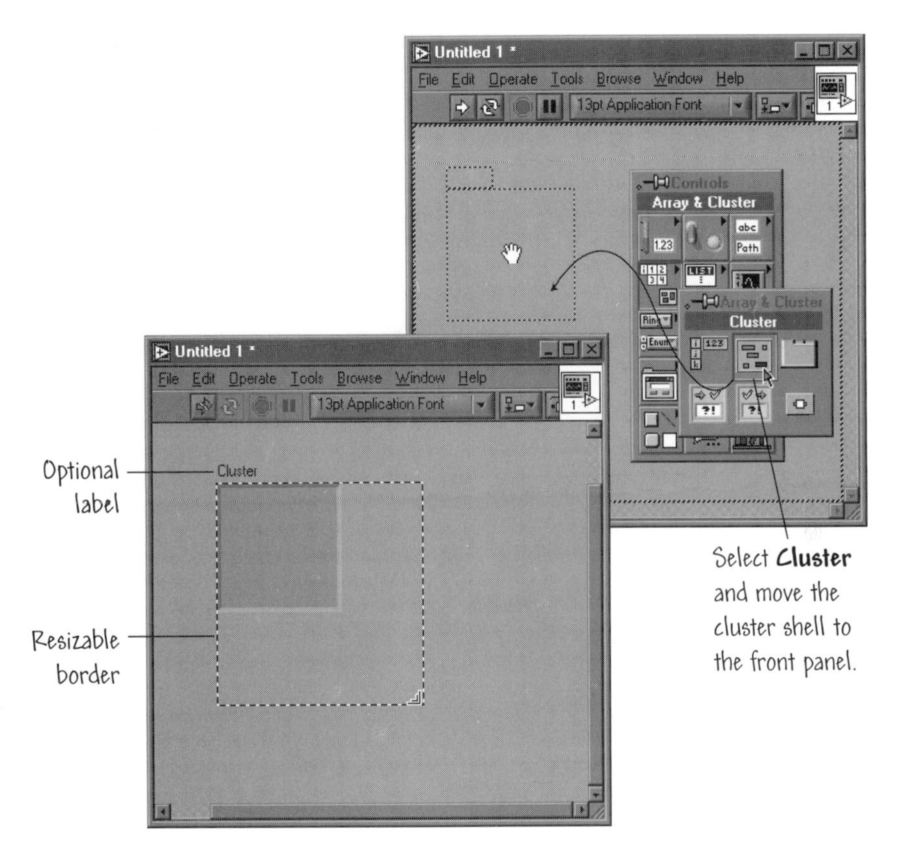

FIGURE 6.24
Creating and resizing a cluster.

object inserted is element 1, and so on. The order of the remaining elements is automatically adjusted when an element is removed from the cluster. The cluster order determines the order in which the elements appear as terminals on the Bundle and Unbundle functions in the block diagram (more on these cluster functions in the next sections). You must keep track of your cluster order if you want to access individual elements in the cluster, because individual elements in the cluster are accessed by order—not by name!

You can examine and change the order of the elements within a cluster by popping up on the cluster border on the front panel and choosing the item **Reorder Controls in Cluster...** from the short cut menu, as shown in Figure 6.26. Notice in the figure that a new set of buttons replaces the toolbar, and the cluster appearance changes noticeably. Even the cursor changes to a hand with a # sign above—this is the cluster order cursor.

Two boxes appear side-by-side in the bottom right corner of each element in the cluster. The white boxes indicate the current place of that element in the

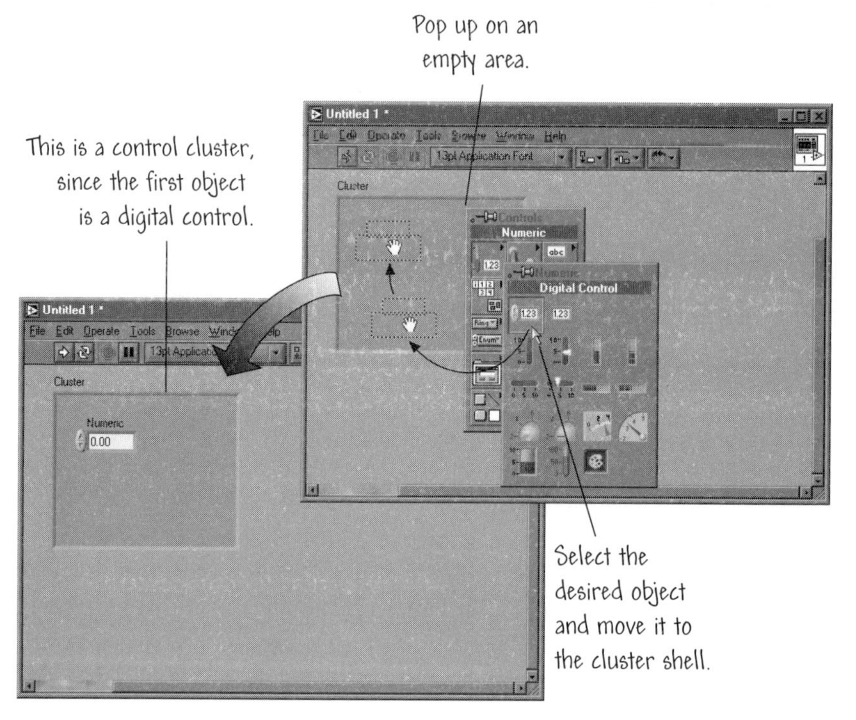

FIGURE 6.25
Creating a cluster by placing objects in the cluster shell.

cluster order. The black boxes indicates the new location in the order, in case you have changed the order. The numbers in the white and black boxes are identical until you make a change in the order. Clicking on an element with the cluster order cursor changes the current place of the element in the cluster order to the number displayed in the toolbar at the top of the front panel. When you have completed arranging the elements within the cluster to your satisfaction, click on the OK button, or you can revert to the old order by clicking on the X button.

A simple example shows the importance of the cluster order. Consider the situation depicted in Figure 6.27a, where the front panel contains two clusters. In the first cluster, the first component is a numeric control, while in the second cluster, the first component is a numeric indicator. In the associated block diagram, the cluster control properly wires to the cluster indicator. Now, in Figure 6.27b, the cluster order has been changed where the string control is the first component of the cluster control. The numeric indicator is still the first component of the cluster indicator. The wire connecting is now broken, and if you attempted to run the VI you would get an error message stating that there is a type conflict.

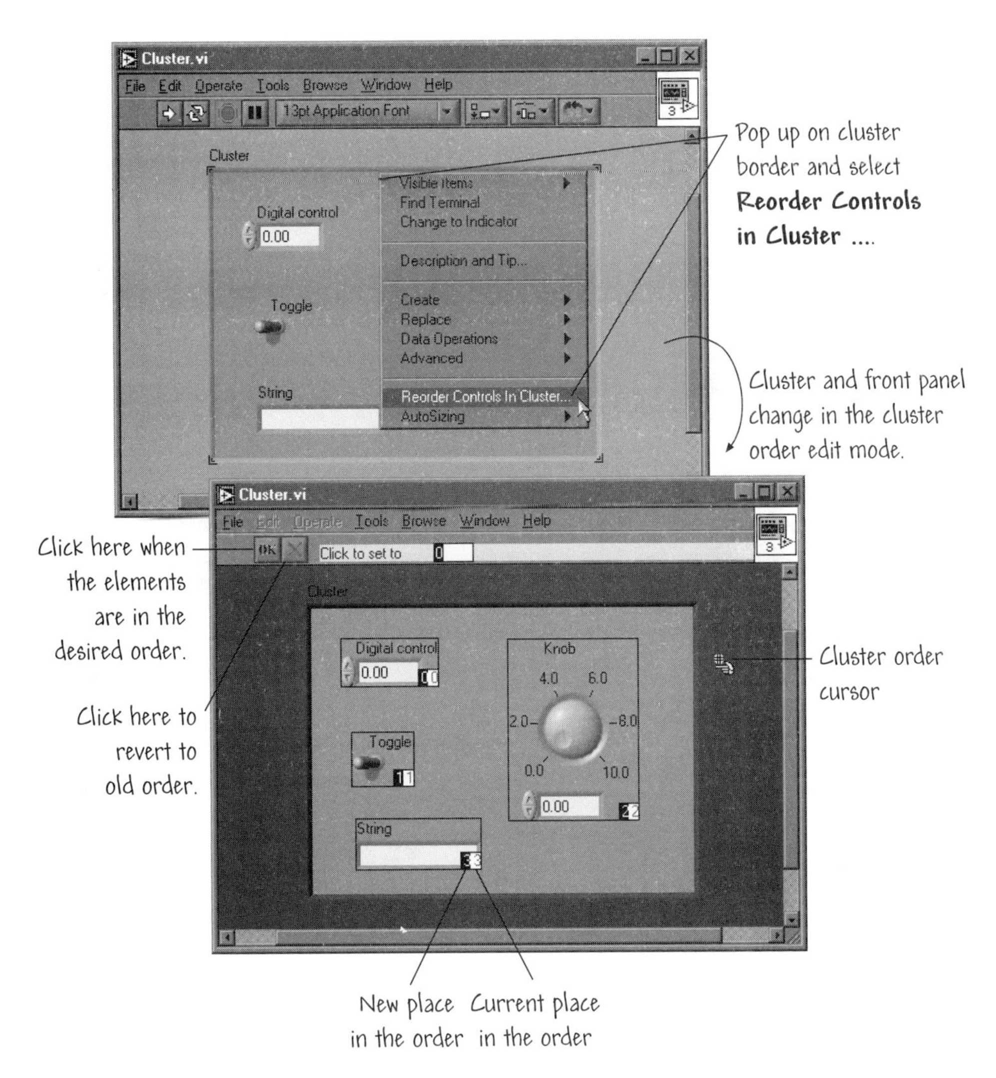

FIGURE 6.26
Examining and changing the cluster order.

6.6.2 Using Clusters to Pass Data to and from SubVIs

A VI connector panel can have a maximum of 28 terminals (as discussed in Section 4.3). In general, you do not want to pass information individually to all 28 terminals when calling a VI or a subVI. A good rule-of-thumb is to use connector passes with 14 or fewer terminals. Otherwise, the terminals are very small and may be difficult to wire. Using clusters, you can group related controls together. One cluster control uses one terminal on the connector, but it can contain may controls. Similarly, you can use a cluster to group indicators.

This allows the subVI to pass multiple outputs using only one terminal. Figure 6.28, illustrates the benefit of using clusters to pass data to and from subVIs.

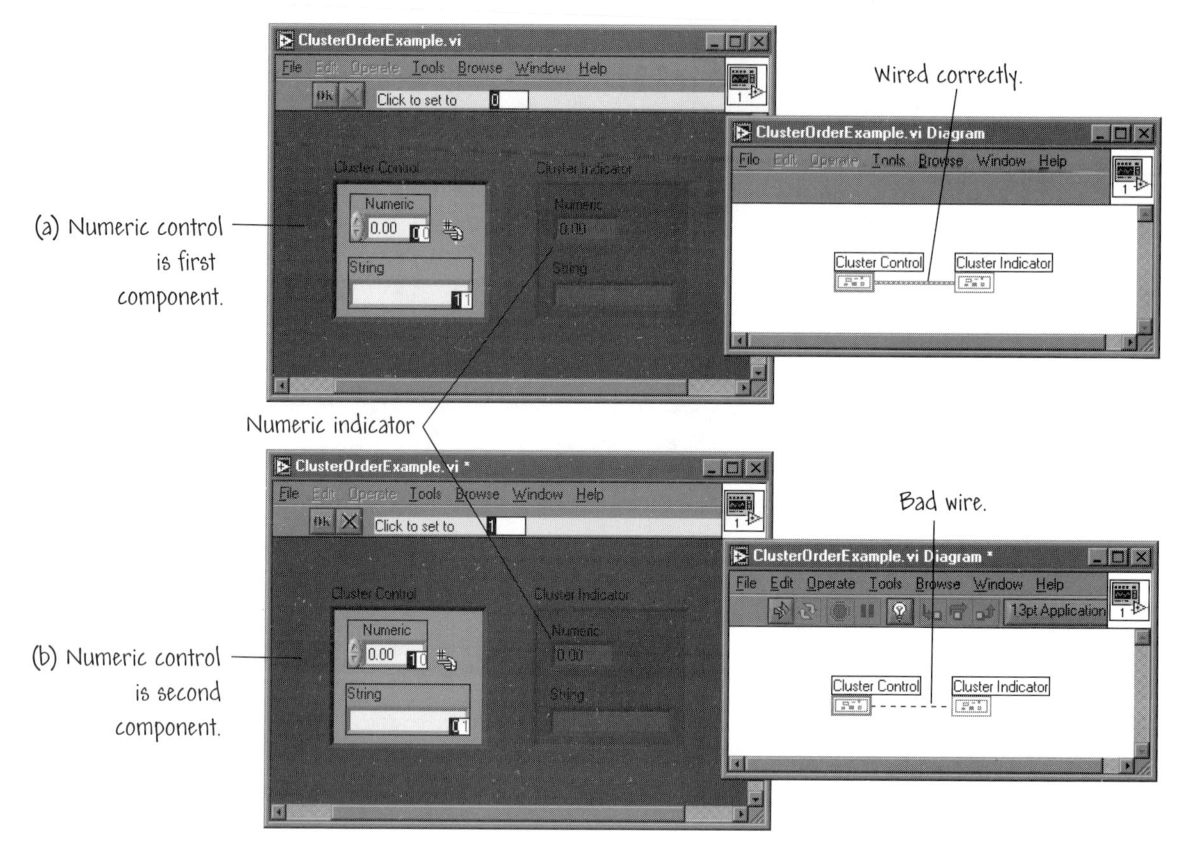

FIGURE 6.27
Example showing the importance of cluster order.

6.7 CLUSTER FUNCTIONS

Two of the more important cluster functions are the Bundle and Unbundle functions. These functions (and others) are found in the **Cluster** subpalette of the **Functions** palette, as illustrated in Figure 6.29.

6.7.1 The Bundle Function

The Bundle function is used to assemble individual elements into a single new cluster or to replace elements in an existing cluster. When placed on the block diagram, the Bundle function appears with two element input terminals on the left side. You can increase the number of inputs by enlarging the function icon (with the resizing handles) vertically to create as many terminals as you need. You can also increase the number of inputs by popping up on the left side of the function and choosing **Add input** from the menu. Since you must wire all the inputs you create, only enlarge the icon to show the exact desired number of element inputs. When you wire to each input terminal, a symbol representing the

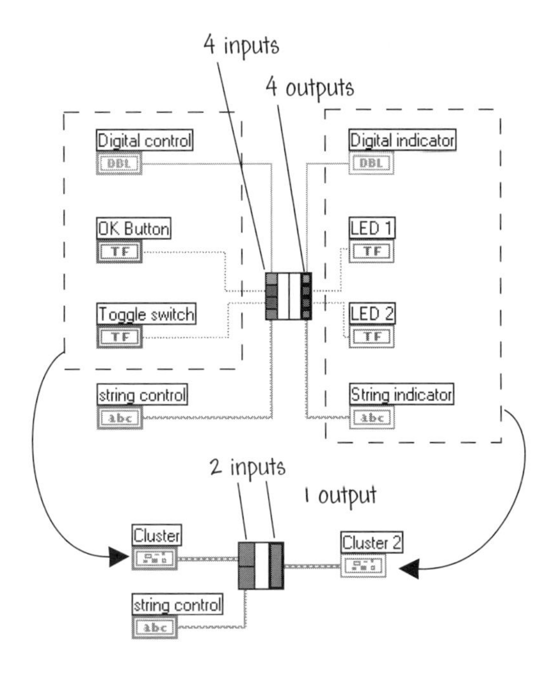

FIGURE 6.28
Using clusters to pass data to and from subVIs.

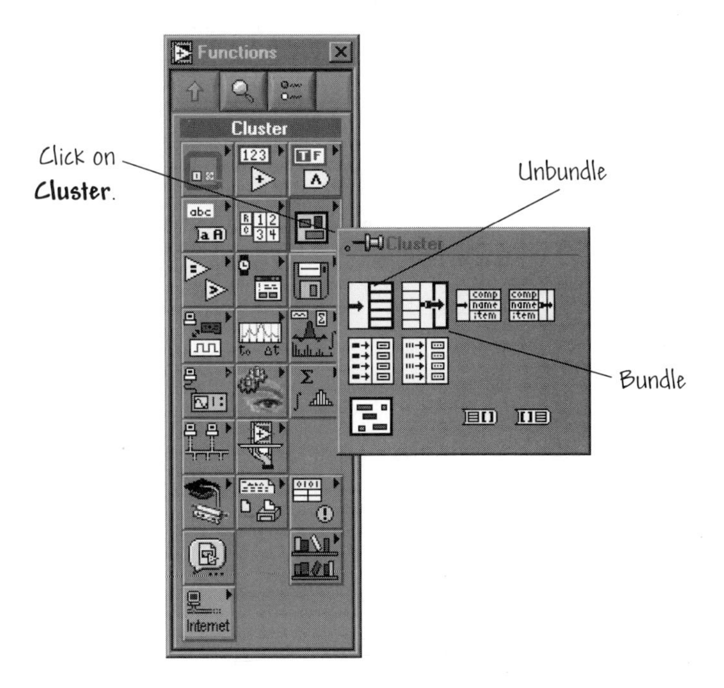

FIGURE 6.29
The cluster functions palette.

data type of the wired element appears in the originally empty terminal, as illustrated in Figure 6.30. The order of the elements in the cluster is equal to the order of inputs to the Bundle function. The order is defined top to bottom, which means that the element you wire to the top terminal becomes element 0, the element you wire to the second terminal becomes element 1, and so on.

The example depicted in Figure 6.30 shows the Bundle function creating a cluster from three inputs: a floating-point real number, an integer, and an array of numbers generated by the For Loop. The output of the new cluster is wired to a waveform graph that displays the random numbers on a graph. Graphs (such as the waveform graph) are covered in Chapter 7.

The example shown in Figure 6.30 can be found in the Chapter 6 *folder in the* Learning *directory and is called* Cluster Bundle Demo.vi. *When you open and run the VI you will see that varying the input x0 changes the x-axis origin.*

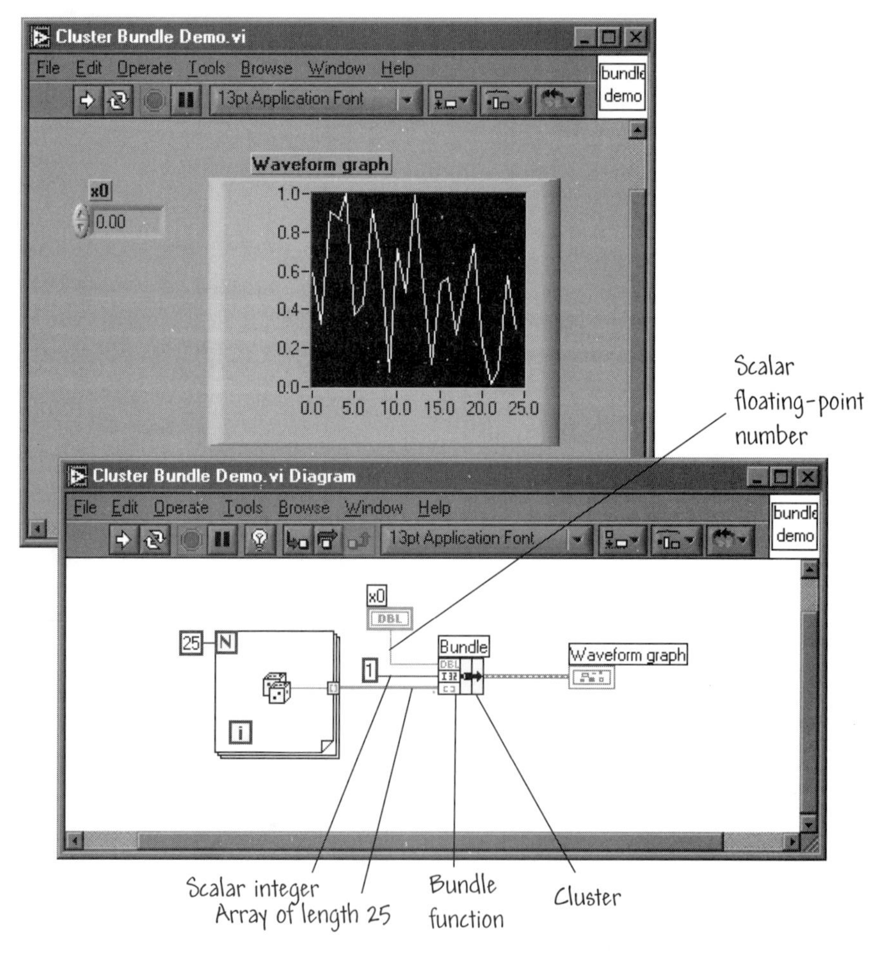

FIGURE 6.30
The Bundle function.

In addition to input terminals for elements on the left side, the Bundle function also has an input terminal for clusters in the middle, as shown in Figure 6.31. Sometimes you want to replace or change the value of one or two elements of a cluster without affecting the others. The example in Figure 6.31 shows a convenient way to change the value of two elements of a cluster. Because the cluster input terminal (that is, the middle terminal) of the Bundle function is wired to an existing cluster named **Cluster**, the only element input terminals that must be wired are those that are associated with cluster elements that you want to replace—in the illustration in Figure 6.31 the second and fourth element values are being replaced. To replace an element in the cluster using the Bundle function, you first place the Bundle function on the block diagram and then resize the function to show exactly the same number of input terminals as there are elements in the existing cluster that you want to modify. The next step is to wire

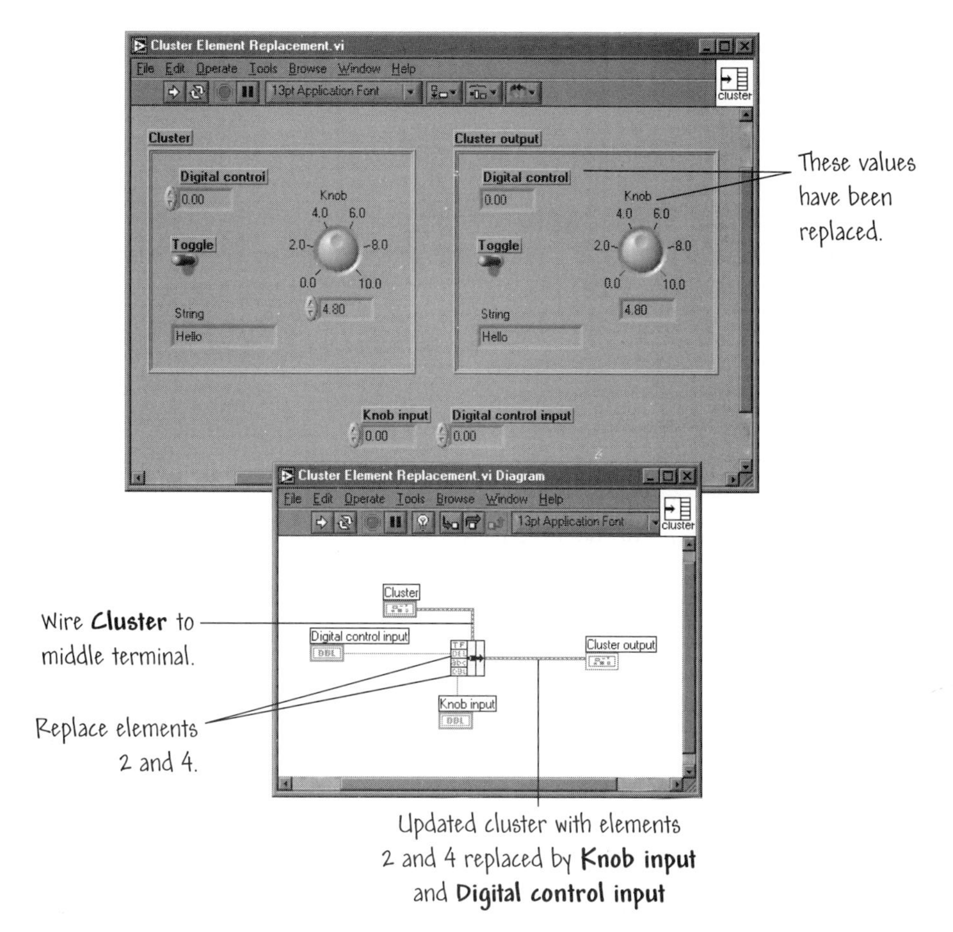

FIGURE 6.31
Using the Bundle function to replace elements of an existing cluster.

the existing cluster to the middle input terminal of the Bundle function. Afterwards, any other input terminals on the left side that are wired will replace the corresponding elements in the existing cluster. Remember that if the objective is to create a new cluster rather than modify an existing one, you do not need to wire an input to the center cluster input of the Bundle function.

The example depicted in Figure 6.31 shows the Bundle function being used to change the value of the digital control and the knob. The corresponding VI can be found in the Chapter 6 folder in the Learning directory and is called Cluster Element Replacement.vi. Open and run the VI in **Continuous Run** mode. **Cluster** is a control cluster and **Cluster Output** is an indicator cluster. Toggle the horizontal toggle switch (using the **Operating** tool). You should see that the toggle switch in **Cluster Output** also switches. Now vary the **Knob** value away from the default value of 4.8. Notice that the **Knob** in **Cluster Output** does not change. This phenomenom occurs because the **Knob** is the fourth element of the cluster and it's value is being replaced by the value set in the digital control **Knob input**. To change the value of the **Knob** in **Cluster Output** you change the value of **Knob input**. Try it out!

6.7.2 The Unbundle Function

The Unbundle function extracts the individual components of a cluster. The output components are arranged from top to bottom in the same order as in the cluster. When placed on the block diagram, the Unbundle function appears with two element output terminals on the right side. You adjust the number of terminals with the resizing handles following the same method as with the Bundle function, or you can pop up on the right side of the function and choose **Add Output** from the menu. Element 0 in the cluster order passes to the top output terminal, element 1 passes to the second terminal, and on down the line. The cluster wire remains broken until you create the correct number of output terminals, at which point the wire becomes solid. When you wire an input cluster to the correctly sized Unbundle, the previously blank output terminals will assume the symbols of the data types in the cluster.

*While all elements are unbundled using the **Unbundle** function, the **Unbundle By Name** function can access one or more elements in a cluster.*

The example depicted in Figure 6.32 shows the Unbundle function being used to unpack the elements of the data structure **Cluster**. The cluster has four elements, and each element is split from the cluster and wired to its own individual indicator for viewing on the front panel.

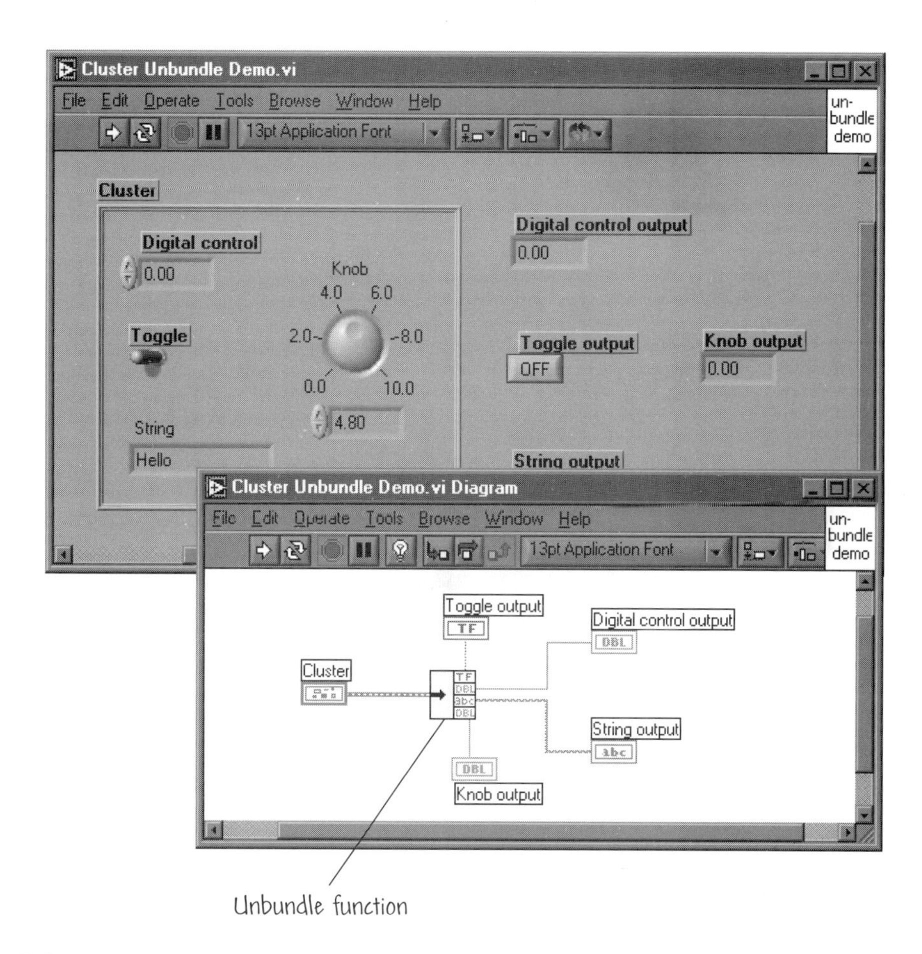

FIGURE 6.32
The Unbundle function.

The example shown in Figure 6.32 can be found in the Chapter 6 *folder in the* Learning *directory and is called* Cluster Unbundle Demo.vi. *Open and run the VI in* ***Continuous Run*** *mode and notice that varying any values in the control cluster immediately changes the values of the various indicators.*

6.7.3 Creating Cluster Constants on the Block Diagram

You can use the same technique you used on the front panel to create a cluster constant on the block diagram. On the block diagram, choose Cluster≫Cluster Constant from the **Functions** palette to create the cluster shell as illustrated in Figure 6.33. Click in the block diagram to place the cluster shell, and place other constants of the appropriate data type within the cluster shell.

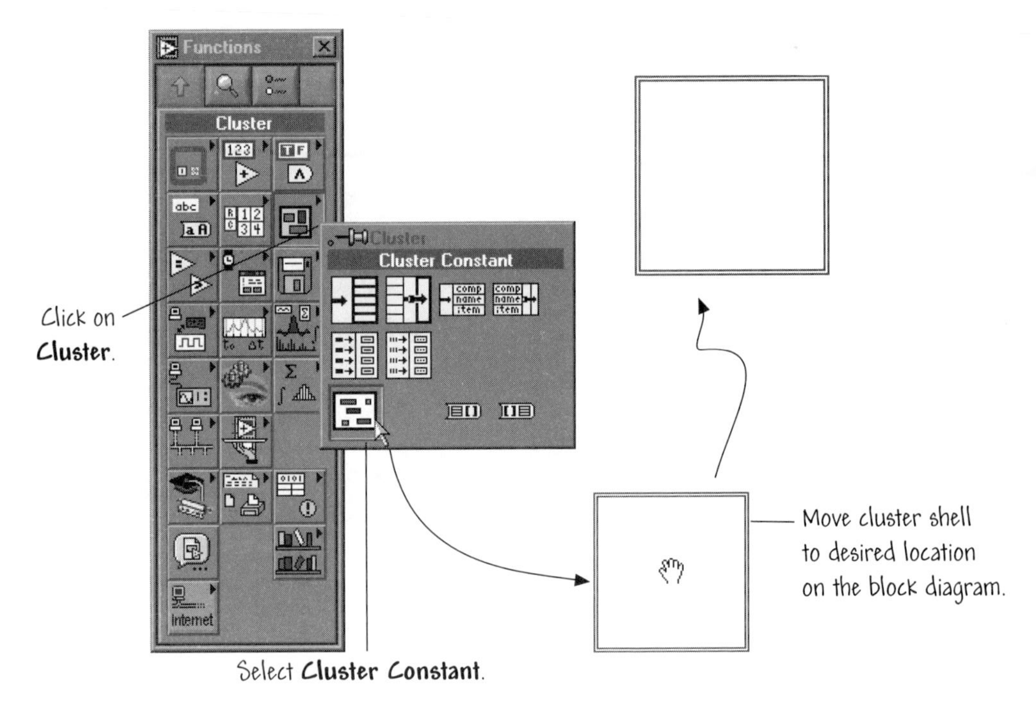

FIGURE 6.33
Creating a cluster constant on the block diagram.

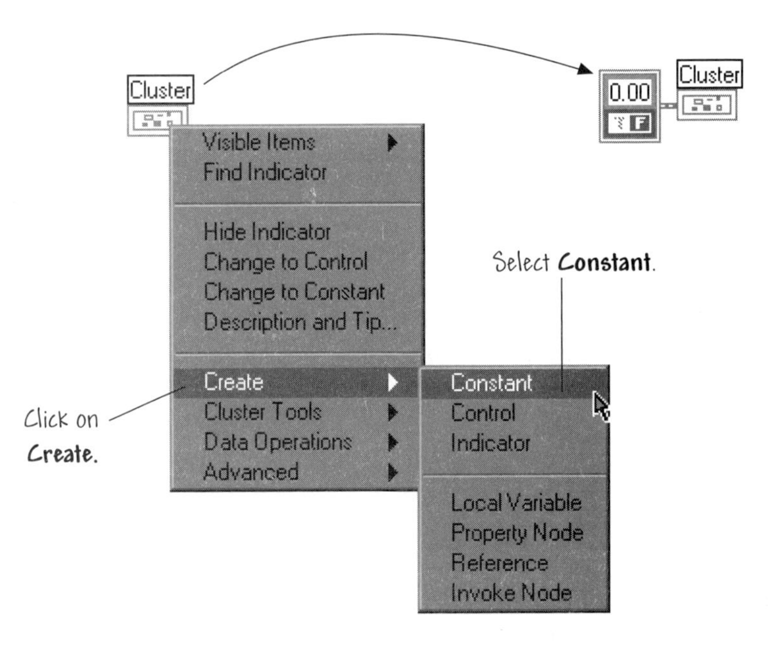

FIGURE 6.34
Create a cluster constant using a cluster on the front panel.

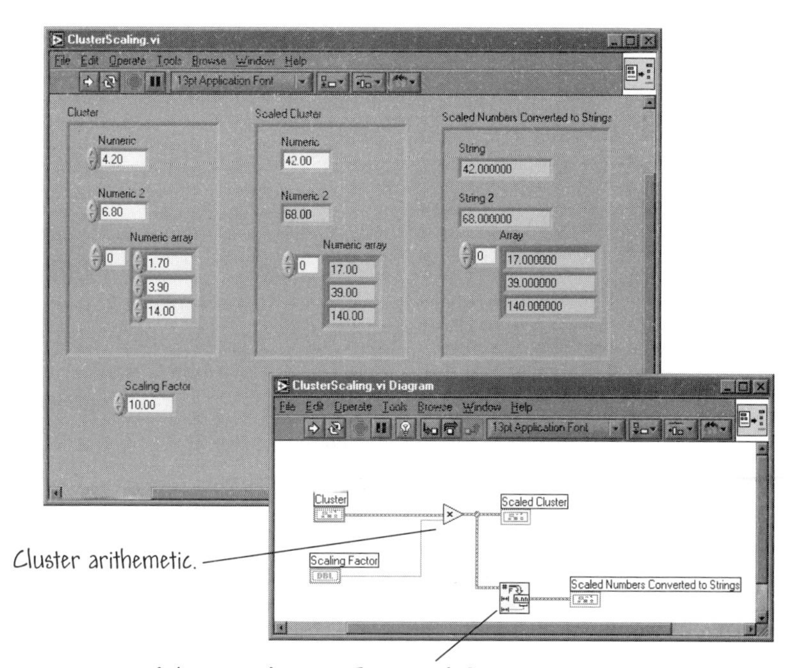

Select **Number To Fractional String** on the **String/Number Conversions** subpalette found on the **Strings** palette.

FIGURE 6.35
Using polymorphism with clusters.

If you have a cluster control or indicator on the front panel and want to create a cluster constant containing the same components in the block diagram, you can either drag that cluster from the panel to the diagram or select Create≫Constant from the short cut menu, as depicted in Figure 6.34.

6.7.4 Using Polymorphism with Clusters

Since the arithmetic functions are polymorphic, they can be used to perform computations on clusters of numbers. As shown in Figure 6.35, you use the arithmetic functions with clusters in the same way you use them with arrays of numerics. You can also use the string-to-number functions to convert a cluster of numerics to a cluster of strings (strings are covered in Chapter 9).

The VI in Figure 6.35 can be found in Chapter 6 of the Learning directory. The VI is called ClusterScaling.

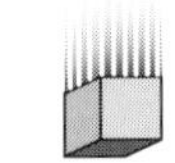

BUILDING BLOCK

6.8 BUILDING BLOCKS: MEASURING VOLUME

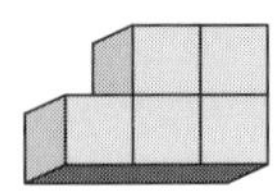

In the Chapter 5 Building Block exercise, you constructed a VI that could acquire multiple volume measurements using a For Loop, and then you controlled the timing of the loop. The output was displayed on a tank indicator along with an LED to indicate when a volume limit had been exceeded. In this exercise you will modify the VI Multiple Volume Points that you developed in Chapter 5 and saved in the Users Stuff folder.

The objective is to use the Build Array function to organize the volume data in a manner suitable for graphing. The target front panel and block diagram are shown in Figure 6.36. Open Multiple Volume Points.vi and edit it using Figure 6.36 as a guide.

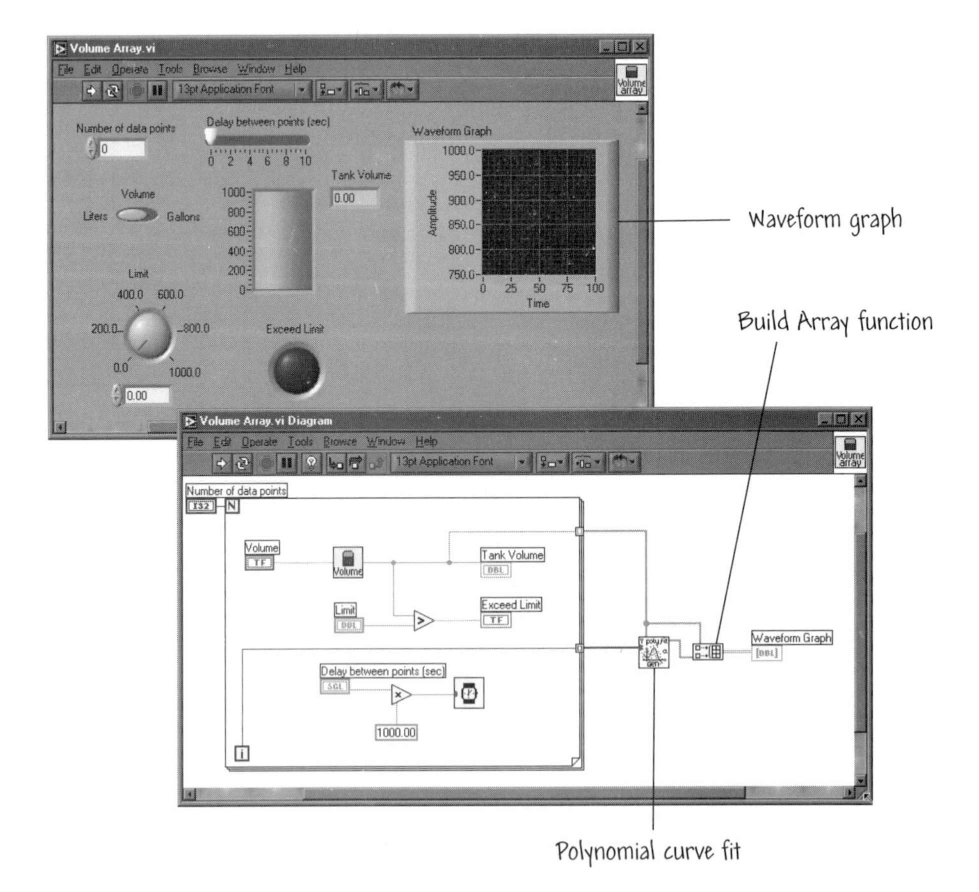

FIGURE 6.36
Using the Build Array function to organize volume data for graphing purposes.

There are two new objects used in this VI: the waveform graph and the General Polynomial Fit function. Waveform graphs are the subject of Chapter 7 and are used here as a way to motivate the upcoming discussions on graphs and charts. The subject of curve fits is covered in Chapter 11 where various data analysis techniques are presented. Figure 6.37 shows where to look to find the waveform graph and the General Polynomial Fit function. It is not important at this point to know the details about waveform graphs or polynomial curve fits—that information will come later!

When you have finished editing and debugging of the VI, save it in the Users Stuff folder as Volume Array.vi.

Experiment with your new VI. Set the variable Number of data points to 10 and run the VI with no delay between data points. The VI will display the volume measurements and curve fit the data with a smooth line. Run the VI several times and observe the results—the data points should be easily fitted with the smooth curve fit. Experiment with the VI by switching between **Liters** and **Gallons** and increasing the number of data points.

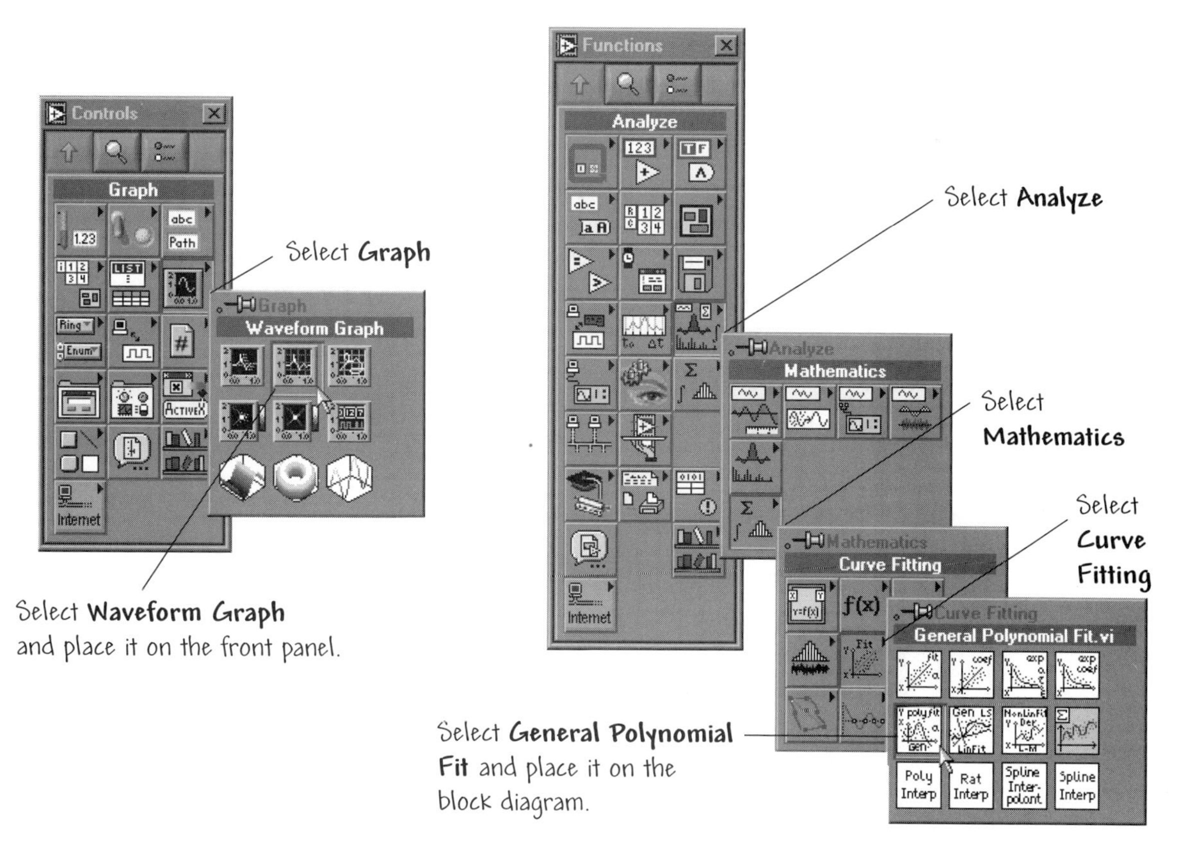

FIGURE 6.37
Finding waveform graphs and polynomial curve-fitting functions.

6.9 RELAXING READING: PLASMA ETCHING CHAMBERS

At this point in our study of LabVIEW, we should begin to consider how we might use LabVIEW in a serious, real-world application requiring data acquisition and instrument control in real time. Although DAQ and GPIB are subjects of subsequent chapters, it is time to look forward and begin to consider how we might use LabVIEW to solve a difficult and challenging problem.

Researchers at the University of Michigan Center for Display Technology and Manufacturing have considered the problem of real-time control of plasma etching chambers for semiconductor and flat panel display production. Their challenge was to achieve improved etching performance by monitoring and controlling plasma parameters beyond the built-in capabilities of the etching tool. Their solution is to use a PC-based virtual instrumentation system with DAQ and GPIB boards controlled by LabVIEW.

Current state-of-the-art plasma processing methods used fixed values of important plasma generation input parameters (such as pressure, power, and gas flows) without regard for plasma condition variations. Researchers at Michigan chose to adjust these parameters in real time using closed-loop feedback control techniques. Their goal was to maintain specific plasma parameters (such as bias voltage and fluorine concentration) to better control the final wafer characteristics. The manufacturing challenge was to reduce variations in etch depth, angle, and rate, both within a run and between runs of the etching process for semiconductors and flat panel displays. This need for repeatable, high-quality manufacturing of products is common to many manufacturing processes.

An instrumentation system was required for controlling etch plasmas in a real-time environment with *in situ* data monitoring, chart display, and data logging. A PC-based system running LabVIEW and data acquisition (DAQ) and GPIB boards from National Instruments was selected for the following reasons:

- flexibility,
- ease of programming,
- the performance needs were satisfied at the lowest cost, and
- the system possessed the desired control rate performance.

A block diagram schematic of the plasma etching system is shown in Figure 6.38. The system front panel is depicted in Figure 6.39. The objective was to develop a safe process for transferring some of the actuator control from the tool controller to the LabVIEW system. This was needed to produce a reliable and industrially viable system.

An AT-MIO-64E3 board was configured to accept up to 32 analog input signals from both the tool controller and additional external sensors, and to trigger

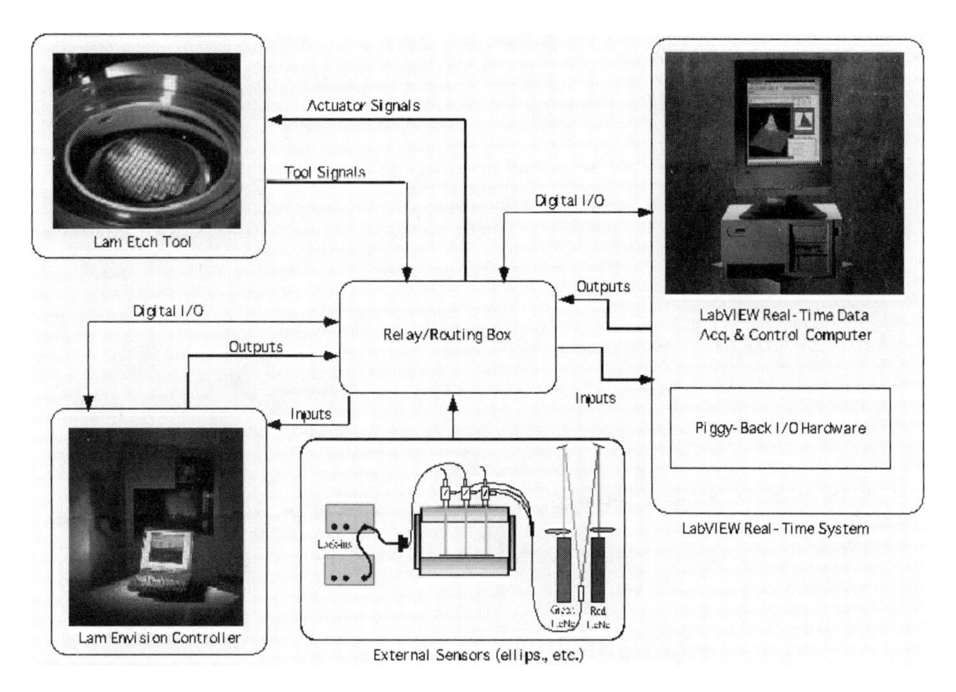

FIGURE 6.38
Block diagram of the plasma etching system.

the control hand-off mechanism with the on-board digital lines. The AT-MIO-64E3 is a high-speed, 12-bit, E-Series DAQ board available from National Instruments. The National Instruments AT-AO-10 analog output board was used to control various tool input signals, including plasma power, pressure, throttle position, and gas flow rates. Both analog and digital signals were routed from the tool controller to a custom-built relay switching box, then through anti-aliasing filters, and then back to both the tool controller and the LabVIEW system. Additional plasma sensing signals (such as fluorine concentration and etch rate) were routed directly through the switching box to signal-conditioning cards and then to the LabVIEW system. Actuator signals were routed through a bank of 10 relays for toggling between normal control of actuators and LabVIEW actuator control, dependent on the proper sequence of software-controlled digital handshaking between the tool controller and LabVIEW system.

If the preceding paragraph contained technical jargon beyond your current technical know-how—don't worry! The main point being made is that LabVIEW was selected by a research team looking for a practical solution to an important manufacturing problem. LabVIEW was an integral component of a very complex manufacturing process. Your application may not be as complex (or it may be more complex!), but it is comforting to know that LabVIEW (while an important

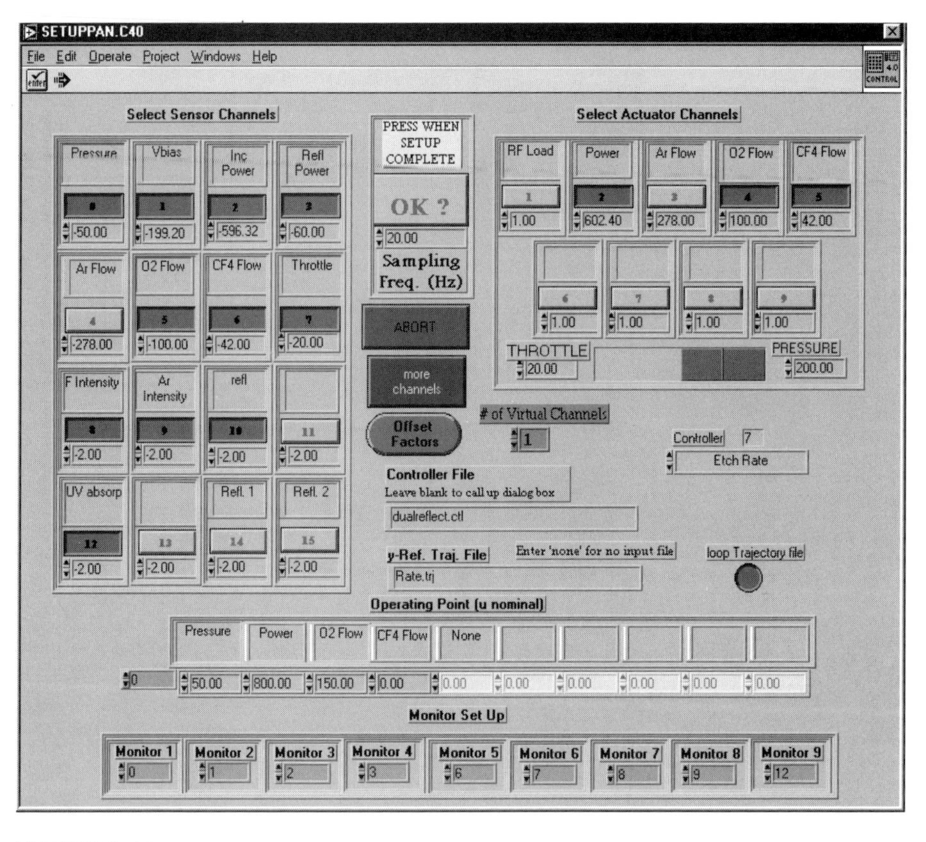

FIGURE 6.39
On-the-fly programming of input and output channels, data and header files, selectable control laws, signal trajectory tracking, and a host of nonreal-time initialization routines are available from the set-up screen.

resource for academic laboratories) can provide viable real-world solutions to common problems facing engineers in industry.

What about the safety features of the LabVIEW controlled plasma etching system? With the LabVIEW subsystem in control, could the etch tool now be operated under conditions it was never intended for, or even under potentially hazardous conditions? The most important safety feature designed into the system was that the original tool controller remained in control of all safety features of the etch process at all times. The tool controller continued to monitor critical system parameters for compliance with established safety rules. If any system parameter fell out of compliance, the tool controller reestablished direct control from the LabVIEW controller. Therefore, the LabVIEW system did not override or impinge upon any of the original safety features of the manufacturing tool.

The LabVIEW system performs a multitude of other handy functions. Before each etch, an AT-GPIB/TNT (PnP) board is used to remotely calibrate and set up eight separate plasma sensor components located outside the clean room. The AT-GPIB/TNT is a low-cost, high-performance IEEE 488 interface developed by National Instruments for the IBM PC AT and compatible computers equipped with 16-bit ISA slots. The user can interactively select input and output channels, all with corresponding multiplicative and additive scale factors. The user can assign channels to nine real-time strip charts (more on graphics in the next chapter), and select various control algorithms, sampling rate, screen update rate, signal conditioning gain, and cutoff frequency. A header file containing channel, scaling, date, time, name, and a host of other relevant information is automatically generated for each experiment. LabVIEW was found to be very flexible—new features were regularly and easily incorporated into the LabVIEW framework.

The main conclusion resulting from the experience of the researchers at The University of Michigan is that users can implement LabVIEW as a real-time controller for semiconductor processing at a low cost. Real-time data acquisition features were implemented with LabVIEW using a nonreal-time operating system (Win95) just by using some clever triggering schemes!

For more information on this application of LabVIEW, read the *User Solutions* entitled "Real-Time Feedback Control of Plasma Etching Chambers Using LabVIEW," which can be found on the National Instruments website, or contact:

Pete Klimecky
University of Michigan
Electronics Manufacturing Lab
3300 Plymouth Rd.
Ann Arbor, MI 48105-2551
email: crown@eecs.umich.edu

6.10 SUMMARY

The main topics of the chapter were arrays and clusters. Arrays are a variable-sized collection (or grouping) of data elements that are all the same type, such as a group of floating-point numbers or a group of strings. LabVIEW offers many functions to help you manipulate arrays. In this chapter we discussed the following array functions:

- Array Size
- Initialize Array

- Build Array
- Array Subset
- Index Array

A cluster is a fixed-sized collection of data elements of mixed types, such as a group containing floating-point numbers and strings. As with arrays, LabVIEW offers built-in functions to help you manipulate clusters. In this chapter we discussed the following cluster functions:

- Bundle
- Unbundle

The important concept of polymorphism was introduced. Polymorphism is the ability of a function to adjust to input data of different types, dimensions, or representations. For example, you can add a scalar to an array or add two arrays of differing lengths together.

Array: Ordered and indexed list of elements of the same type.

Array shell: Front panel object that contains the elements of an array. It consists or an index display, a data object window, and an optional label.

Array Size: Array function that returns the number of elements in the input array.

Array Subset: Array function that returns a portion of an array.

Auto-indexing: Capability of loop structures to assemble and disassemble arrays at their borders.

Build Array: Array function that concatenates multiple arrays or adds extra elements to an array.

Bundle: A cluster function that assembles all the individual input components into a single cluster.

Cluster: A set of ordered, unindexed data elements of any data type, including numeric, Boolean, string, array, or cluster.

Index Array: Array function accessing elements of arrays.

Initialize Array: Array function that creates an n-dimensional array with elements containing the values that you specify.

Polymorphism: The ability of a function to adjust to input data of different dimension or representation.

Unbundle: A cluster function that splits a cluster into each of its individual components.

EXERCISES

E6.1 Open While Loop Indexing.vi in the Exercises&Problems\Chapter 6 folder in the Learning directory. Click on the **Broken Run** button to access the error list. Highlight the error and press the **Show Error** button. This error is a result of an array wired to a terminal that is a scalar value. Recall that While loops disable indexing on tunnels by default. To eliminate the error, pop up on the loop tunnel and enable indexing. Run the VI and after a few moments click on the **Stop** push button on the front panel. An array will appear in the digital indicator **Temperature measurements**.

E6.2 Open Cluster Order.vi in the Exercises&Problems\Chapter 6 folder in the Learning directory. Click on the **Broken Run** button. Then click on the error and press the **Show Error** button. The error is a result of the cluster elements not being in the same order as the elements that are bundled. To eliminate the error, pop up on the cluster border in the front panel and select **Reorder Controls in Cluster...**. Change the order so **Data Points** is the first element, the **Limit** is the second element, and **Over Limit?** is the third element. Now you can run the VI. Try it out!

E6.3 Open Unbundle By Name.vi in the Exercises&Problems\Chapter 6 folder in the Learning directory. Switch to the block diagram. This VI is the same as the Cluster Unbundle Demo.vi found in the Chapter 6 folder in the Learning directory and discussed earlier in this chapter. However, in this VI the Unbundle By Name function is used instead of Unbundle. The advantage of using the Unbundle By Name is that you can easily identify the cluster values that are unbundled. You can only use Unbundle By Name if every object in the cluster has a label associated with it. Open Cluster Unbundle Demo.vi and compare the block diagram with the Unbundle By Name.vi block diagram. Using the on-line help as a source of information, write a short description of the cluster function Unbundle By Name.

E6.4 Create a subVI that calculates the dot (inner) product of two n-dimensional vectors. If the two vectors are denoted by v_1 and v_2, where

$$v_1 = \begin{bmatrix} v_1(0) \\ v_1(1) \\ \vdots \\ v_1(n) \end{bmatrix} \quad \text{and} \quad v_2 = \begin{bmatrix} v_2(0) \\ v_2(1) \\ \vdots \\ v_2(n) \end{bmatrix},$$

then the dot product is given by

$$v_1 \cdot v_2 = v_1(0)v_2(0) + v_1(1)v_2(1) + \cdots + v_1(n)v_2(n).$$

E6.5 Create a subVI that calculates the cross product of two 3-dimensional vectors. If the two vectors are denoted by v_1 and v_2, where

$$v_1 = \begin{bmatrix} v_1(0) \\ v_1(1) \\ v_1(2) \end{bmatrix} \quad \text{and} \quad v_2 = \begin{bmatrix} v_2(0) \\ v_2(1) \\ v_2(2) \end{bmatrix},$$

then the cross product is given by

$$v_1 \times v_2 = \begin{bmatrix} v_1(1)v_2(2) - v_1(2)v_2(1) \\ v_1(2)v_2(0) - v_1(0)v_2(2) \\ v_1(0)v_2(1) - v_1(1)v_2(0) \end{bmatrix}.$$

E6.6 Create a VI that performs matrix multiplication for two input matrices A and B. The matrix A is an $n \times m$ matrix, and the matrix B is an $m \times p$ matrix. The resulting matrix C is an $n \times p$ matrix, where

$$C = AB.$$

PROBLEMS

P6.1 Complete the crossword puzzle.

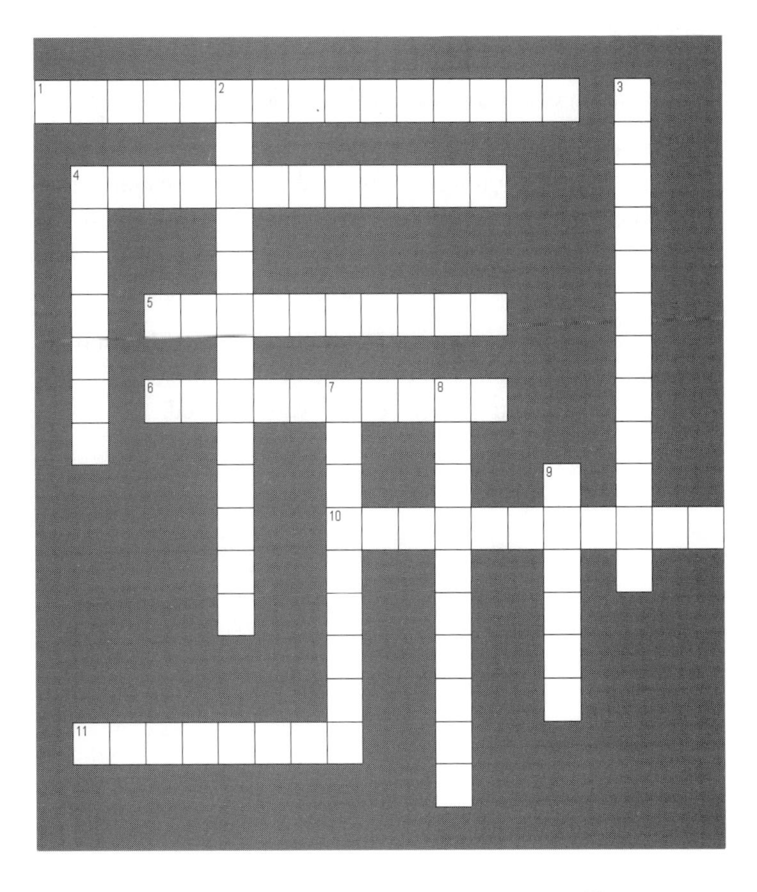

Across

1. Array function that creates an n-dimensional array with elements containing the value that you specify.
4. Front panel object that contains the elements of a cluster.
5. Array function that concatenates multiple arrays or adds extra elements to an array.
6. Array function accessing elements of arrays.
10. Array function that returns a portion of an array.
11. A cluster function that splits a cluster into each of its individual components.

Down

2. Capability of loop structures to assemble and disassemble arrays at their borders.
3. The ability to adjust to input data of different dimension or representation.
4. A set of ordered, unindexed data elements of any data type.
7. Array function that returns the number of elements in the input array.
8. Ordered and indexed list of elements of the same type.
8. Front panel object that contains the elements of an array.
9. A function that assembles all the individual input componentsinto a single cluster.

P6.2 Create a VI that reads 100 temperature measurements by using the Digital Thermometer.vi in the Activity directory. With each temperature measurement, bundle the time and date of the measurement.

P6.3 Build a VI that generates and plots 100 random numbers on a waveform graph. Compute the average of the random numbers and display the result on the front panel. Use the Add Array Elements function found on the **Numeric** palette in the computation of the average of the random numbers.

P6.4 Create a VI that computes and plots the second-order polynomial $y = Ax^2 + Bx + C$. The VI should use controls on the front panel to input the coefficients A, B, and C, and also should use front panel controls to enter the number of points N to evaluate the polynomial over the interval x_0 to x_f. Plot y versus x on a waveform graph.

P6.5 Create a VI with a cluster of six buttons labeled **Option1...Option6**. When executing, the VI should wait for one of the buttons to be pressed, then display a dialog box indicating which option was selected.

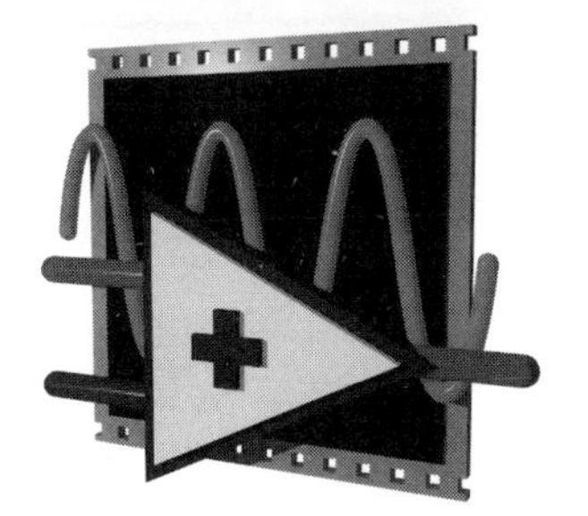

CHAPTER 7

Charts and Graphs

Graphs and charts are used to display data in a graphical form. Three types of charts are discussed—scope, strip, and sweep charts—and two types of graphs—waveform and XY graphs. Charts and graphs are different! Data is presented on charts by appending new data as it becomes available to the existing plot, much in the same manner as on a strip chart you might find in a laboratory. On the other hand, graphs are used to display pregenerated arrays of data in a more traditional fashion, such as the typical $x - y$ graph. In this chapter, you will learn about charts and graphs, the type of data required by each, and several ways to utilize graphics. Since data presentation is an important component of communication, it is necessary to be able to customize the appearance of charts and graphs. The subject of customizing charts and graphs using the palette and the legend (and other tools) is discussed in the chapter.

GOALS

1. Learn about charts and graphs and recognize their similarities and differences.
2. Understand the three modes of a chart: strip, scope, and sweep.
3. Learn to customize the appearance of charts and graphs.

7.1 WAVEFORM CHARTS

A **waveform chart** is a special kind of indicator—located in the **Graph** subpalette of the **Controls** palette, as illustrated in Figure 7.1.

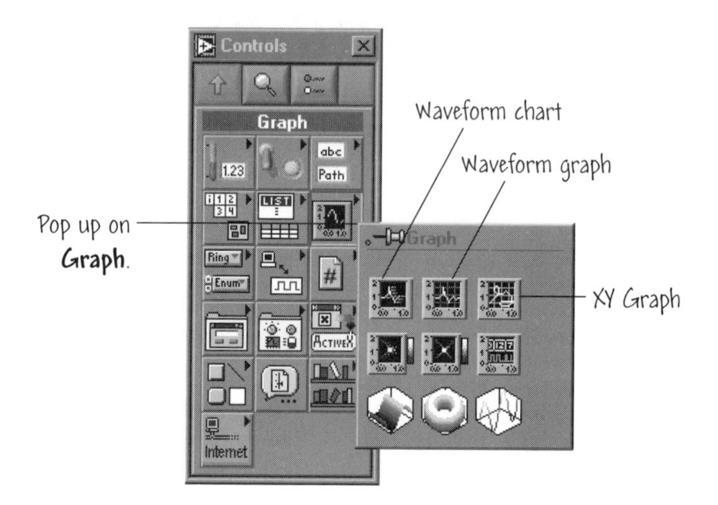

FIGURE 7.1
The graphs and chart are located in the **Graph** subpalette of the **Controls** palette.

There is only one type of waveform chart, but it has three different update modes for interactive data display—**strip chart**, **scope chart**, and **sweep chart**, as shown in Figure 7.2. You select the update mode that you want to use by popping up on the waveform chart and choosing one of the options from the

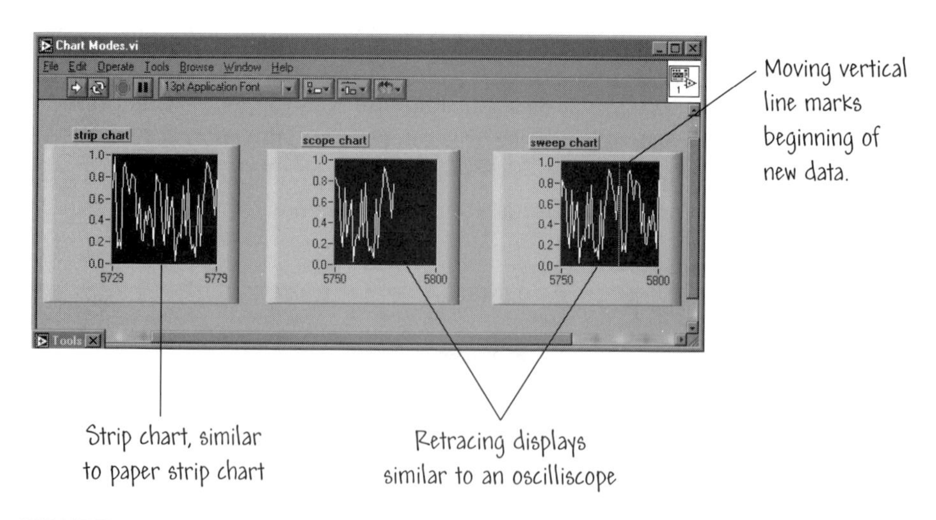

FIGURE 7.2
The waveform chart has three update modes.

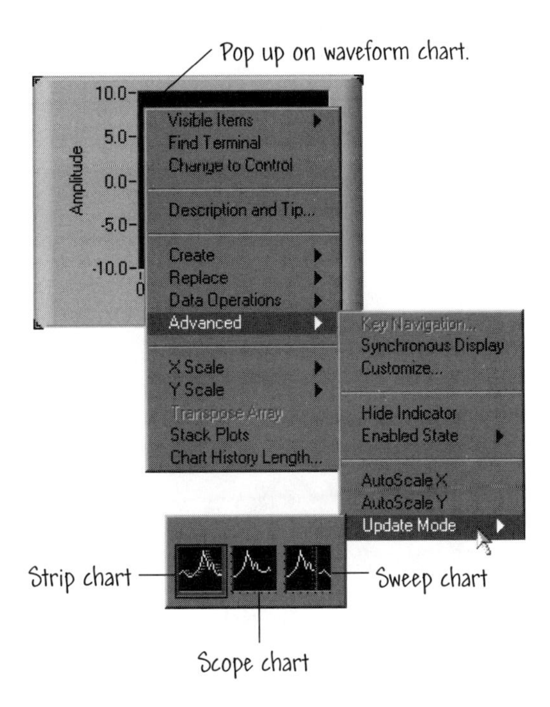

FIGURE 7.3
Changing waveform chart modes: strip chart, scope chart, and sweep chart.

Advanced≫Update Mode menu. This process is illustrated in Figure 7.3. The strip chart, scope chart, and sweep chart all handle the incoming data a little differently. The strip chart has a display that scrolls so that as each new data point arrives, the entire plot moves to the left—the oldest data point falls off the chart, and the latest data point is added at the rightmost part of the plot. This action is very similar to a paper strip chart commonly found in laboratories.

The scope and sweep charts more closely resemble the action of an oscilloscope. When the number of data points is sufficient so that the plot reaches the right border of the plotting area of the scope chart, the entire plot is erased, and the plotting begins again from the left side. The sweep chart acts much like the scope chart except that, rather than erasing the plot when the data reaches the right border, a moving vertical line marks the beginning of new data and moves across the display as new data is added. The scope chart and the sweep chart run faster than the strip chart.

You can run the waveform chart demonstration shown in Figure 7.2 by opening Chart Modes.vi *located in* Chapter 7 *of the* Learning *directory. Run the VI in* ***Continuous Run*** *mode and examine the three different waveform charts: the strip chart, the scope chart, and the sweep chart. Select* ***Highlight Execution*** *if you want to observe the data as it plots in slow motion.*

Waveform charts may display single or multiple traces. An example of a multiple plot waveform chart is shown in Figure 7.4.

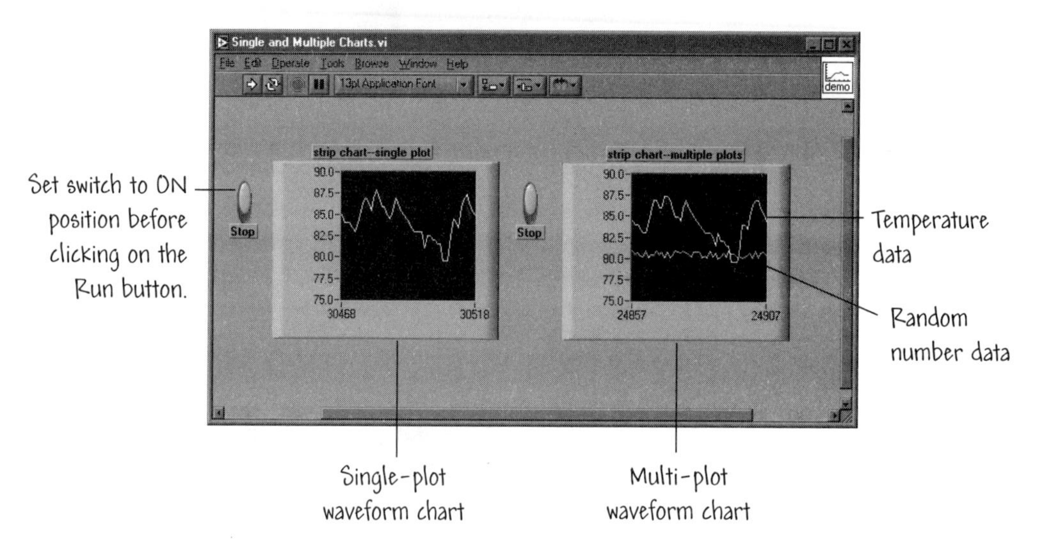

FIGURE 7.4
Examples of single- and multiple-plot waveform charts.

To generate a single-plot waveform chart, you wire a scalar output directly to the waveform chart. The data type displayed in the waveform chart will match the input. In the example shown in Figure 7.5, a new temperature value is plotted on the chart each time the While Loop iterates.

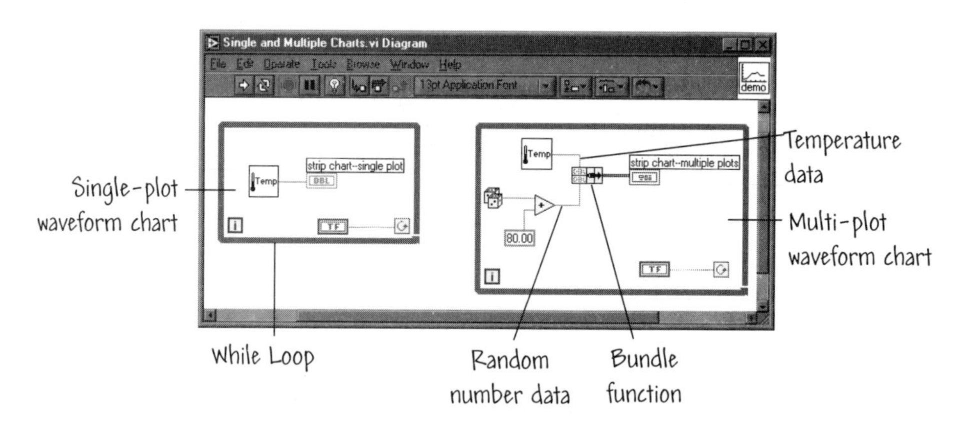

FIGURE 7.5
Block diagrams associated with single- and multiple-plot waveform charts.

As illustrated in Figure 7.4, waveform charts can display multiple plots. To generate a multi-plot waveform chart, you bundle the data together using the Bundle function (see Chapter 6 for a review). In the example in Figure 7.5, the Bundle function groups the output of the two different data sources—Digital Thermometer.vi and Random Number (0–1) function—that acquire temperature and random number data for plotting on the waveform chart. The random number is biased by the constant 80 so that the plots are easily displayed on the same y-axis scale. Remember that Digital Thermometer.vi simulates acquiring temperature data presented in degrees Fahrenheit in the general range of 80°. You can add more plots to the waveform chart by increasing the number of Bundle function input terminals—that is, by resizing the Bundle function.

You can run the single- and multi-plot waveform chart demonstration shown in Figure 7.4 by opening Single and Multiple Charts.vi *located in* Chapter 7 *of the* Learning *directory. Run the VI and examine the single- and multi-plot waveform chart. You can stop the plotting of each chart using the vertical switch button. After stopping the plotting with the vertical switch, remember to reset the switch to the up position before the next execution of the VI.*

Practice with Waveform Charts

In this exercise you will construct a VI to compute and display a baseball batting average on a waveform strip chart. This VI utilizes the Batting Average subVI.vi you developed in Chapter 4. Figure 7.6 shows the front panel that you can use as a guide as you construct your VI.

The front panel has six items:

- Four digital controls
- One vertical switch
- One waveform strip chart

Open a new VI front panel and place the six objects listed above in the window. Label the four digital controls **H**, **K**, **O**, and **E**. Add a free label under the vertical switch indicating the switch position for stopping the execution. Place a waveform chart in the front panel window and locate the front panel objects in approximately the same locations as seen in Figure 7.6.

Since a baseball batting average always has a value between 0 and 1000 (with 1000 representing a perfect batting average), we need to rescale the waveform chart y-axis to insure the scaling is always 0 to 1000. Using the **Labeling** tool, double-click on the 10.0 on the waveform chart y-axis scale, type 1000, and click on the **Enter** button. This changes the maximum value of the y-axis to 1000. The minimum value is 0 by default, so no additional adjustment is required.

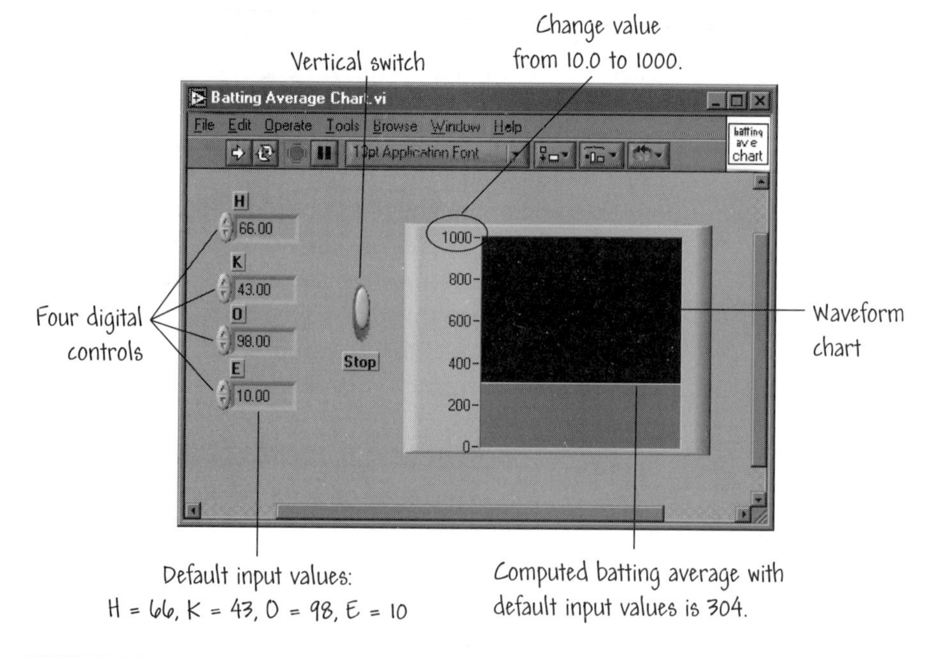

FIGURE 7.6
The front panel for a waveform chart to plot a baseball batting average.

Input default values for the number of hits (**H**), strike-outs (**K**), put-outs (**O**), and fielding errors (**E**). Enter the following data: **H** = 66, **K** = 43, **O** = 98, and **E** = 10. From the pull-down menu **Operate**, select **Make Current Values Default** so that it is not necessary to input the default values each time the VI is opened again.

Figure 7.7 shows the block diagram that you can use as a guide for building your VI. The digital control terminals and waveform chart and vertical switch terminals will automatically appear on the block diagram after you have placed their associated objects on the front panel.

The block diagram has two additional items:

- While Loop
- Batting Average subVI.vi

Switch to the block diagram window of your VI. First enclose the four digital controls, waveform chart, and vertical switch terminals inside a While Loop. The only other object that is needed is the subVI to compute the batting average. The Batting Average subVI.vi is located in Chapter 4 in the Learning directory. You can access the subVI by popping up on the block diagram and from the **Functions** palette choosing **Select a VI...**, as illustrated in Figure 7.8. You will need to navigate through the file structure to reach the Chapter 4 directory. Once there, you select the desired subVI—this is demonstrated in Figure 7.8.

Batting Average Chart.vi Diagram
File Edit Operate Tools Browse Window Help
13pt Application Font
batting ave chart
H
DBL
K
DBL
O
DBL
E
DBL
ave
I32
TF
Add the While Loop.
Wire the vertical switch to the Conditional Terminal.
Add the Batting Average subVI VI and wire the terminals.

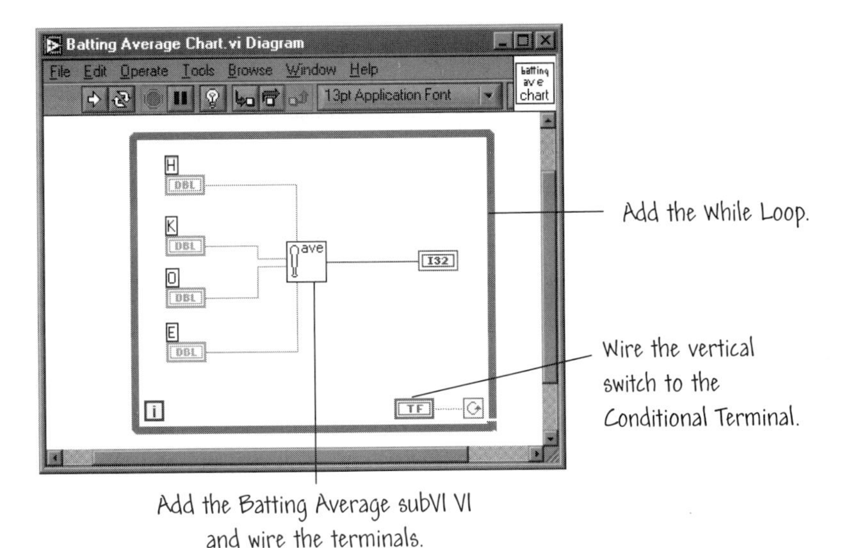

FIGURE 7.7
The block diagram for a waveform chart to plot a baseball batting average.

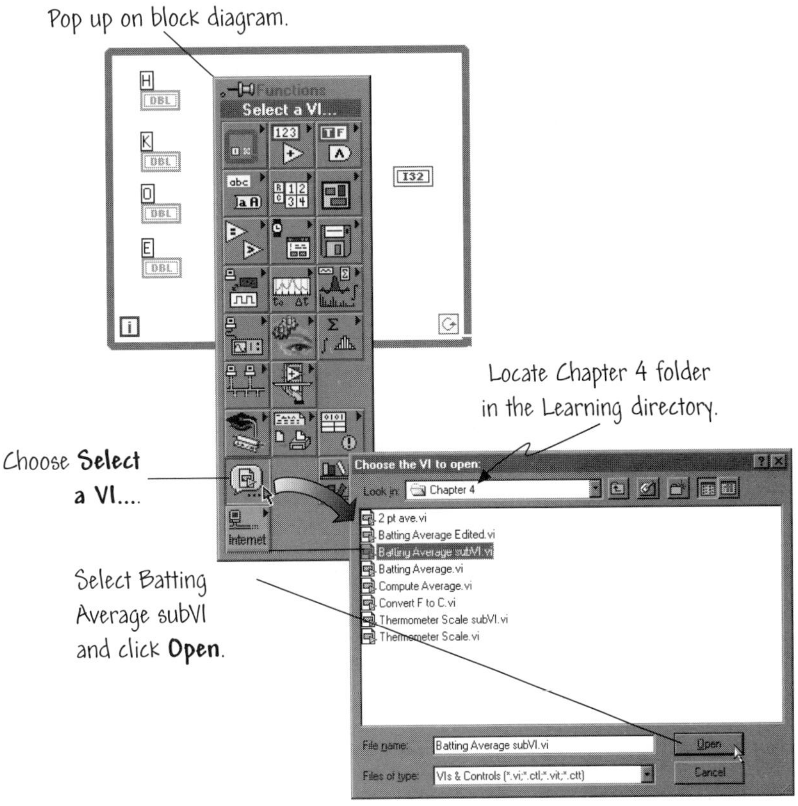

Pop up on block diagram.
Functions
Select a VI...
H
DBL
K
DBL
O
DBL
E
DBL
I32
Internet
Locate Chapter 4 folder in the Learning directory.
Choose **Select a VI...**
Choose the VI to open:
Look in: Chapter 4
2 pt ave.vi
Batting Average Edited.vi
Batting Average subVI.vi
Batting Average.vi
Compute Average.vi
Convert F to C.vi
Thermometer Scale subVI.vi
Thermometer Scale.vi
Select Batting Average subVI and click **Open**.
File name: Batting Average subVI.vi
Files of type: VIs & Controls (*.vi;*.ctl;*.vit;*.ctt)
Open
Cancel

FIGURE 7.8
Placing the Batting Average subVI.vi on the block diagram.

Once all the required objects are on the block diagram, wire the diagram as shown in Figure 7.7. As an optional step, you can remove the x-axis scale. Pop up on the waveform chart on the front panel and under the Visible Items category choose X Scale to deselect the X scale. In this example, the value of the batting average is being computed and plotted each iteration of the While Loop—the X scale does not have any physical meaning, other than as a count of the total loop iterations.

Once everything is ready for execution, run the VI. With the default values you should find that the computed batting average is 304. Using the **Operating** tool, increase the number of hits or the number of strike-outs and observe how it affects the batting average. As you vary the various input parameters (that is, the number of hits, strike-outs, put-outs, and errors) you should see a line across the waveform chart move up or down, depending on which inputs you are varying (hits make the batting average increase, while increasing any of the other three parameters will reduce the batting average!). Save your VI in the Users Stuff directory and name it Batting Average Chart.vi.

You can find a completed VI for the waveform chart demonstration shown in Figure 7.6 in Chapter 7 *of the* Learning *directory. The VI is called* Batting Average Chart.vi.

◆

In the previous example, the While Loop executed as quickly as the computer system would allow. In many situations you may want to acquire and plot data at specified intervals. This can be accomplished by controlling the loop timing using the Wait Until Next ms Multiple function. This timing control function, located in the **Time & Dialog** subpalette, as shown in Figure 7.9, waits until the

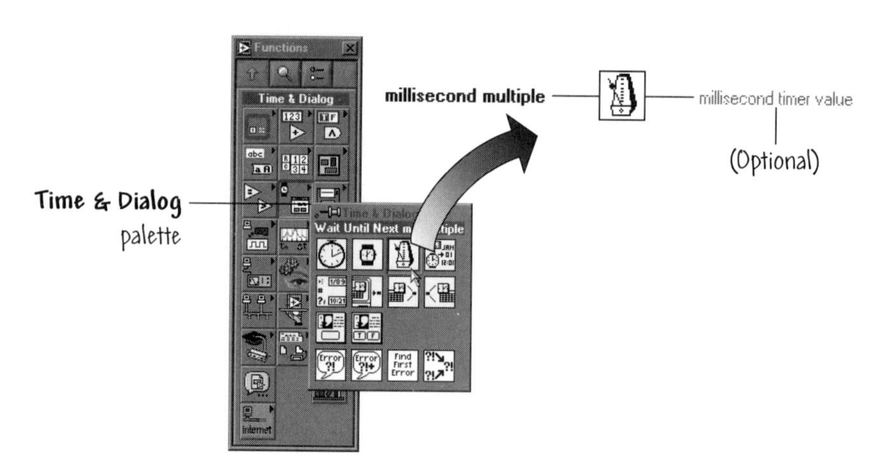

FIGURE 7.9
The Wait Until Next ms Multiple function.

millisecond timer is a multiple of the specified input value before returning the (optional) millisecond timer value (the output of the function need not be wired!). Therefore, you can control the loop iteration to execute only once every specified number of milliseconds. As with the Wait (ms) function, the timer resolution is system dependent and may be less accurate than 1 millisecond.

The example shown in Figure 7.10 shows how to control the timing of a For Loop to execute once every second (that is, once per 1000 ms). Placing the Wait Until Next ms Multiple function within the For Loop, each iteration will take approximately 1 second, and the random number will be plotted at 1-second intervals. Can you predict how much time it will take to complete the execution of the For Loop shown in Figure 7.10? It should take around 10 seconds. You can test it out by opening and running the VI shown in Figure 7.10. It is called Loop Timing Demo.vi and can be found in Chapter 7 of the Learning directory.

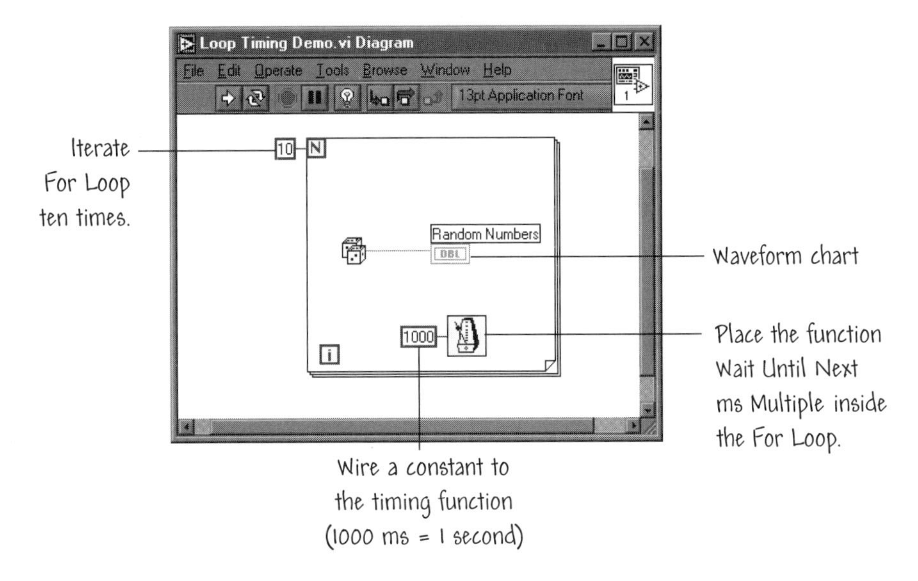

FIGURE 7.10
Controlling the loop timing and rate at which data is plotted on a waveform chart.

Practice with Timing

Open the VI called Batting Average Chart.vi that you constructed in the previous example (it should be located in the Users Stuff directory). Add the capability to control the loop timing by introducing a Wait Until Next ms Multiple function within the While Loop, as shown in Figure 7.11.

Set the loop timing for 25 ms. When the VI is now executing the batting average is computed and the waveform chart is updated at 25 millisecond intervals. The effect is that you should be able to watch the computed batting average trends

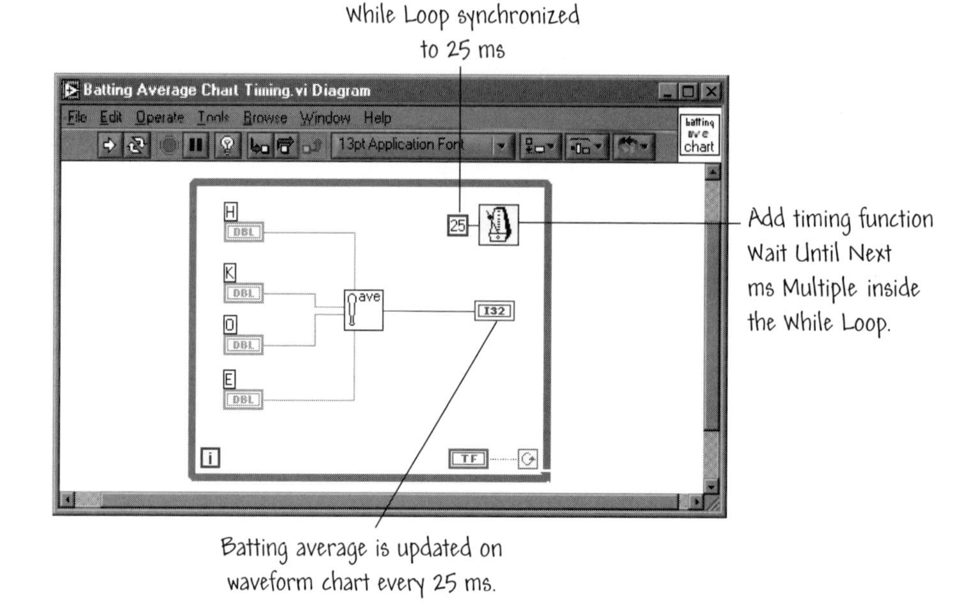

FIGURE 7.11
Controlling the rate on the waveform chart displaying a computed baseball batting average.

change more clearly than when the While Loop was executing as fast as possible. Vary the loop-timing parameter and observe the effect on the waveform chart.

Save your updated VI in the Users Stuff directory and name it Batting Average Chart Timing.vi.

You can find a completed VI for the waveform chart timing demonstration in Chapter 7 of the Learning directory. The VI is called Batting Average Chart Timing.vi.

◆

7.2 WAVEFORM GRAPHS

A **waveform graph** is an indicator that displays one or more data arrays. This is equivalent to the familiar 2D plot with horizontal and vertical axes. There are two types of graphs: waveform graphs and XY graphs. In this section the focus is on waveform graphs—XY graphs are the subject of the next section.

Waveform graphs and XY graphs are functionally different, but they look the same on the front panel. Both waveform graphs and XY graphs plot existing arrays of data all at once, unlike waveform charts, which plot new data as they become available. An example of a waveform graph is shown in Figure 7.12. But the waveform graph plots only single-valued functions with uniformly spaced

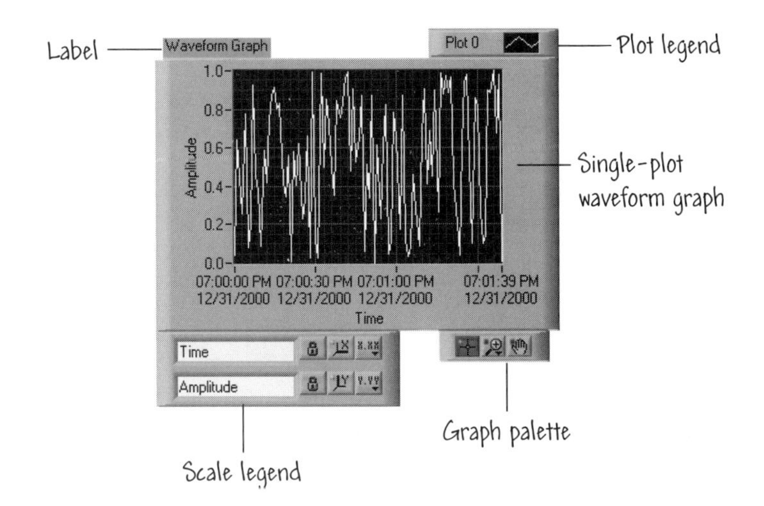

FIGURE 7.12
Example of a waveform graph.

points and is ideal for plotting arrays of data in which the points are evenly distributed. Conversely, XY graphs are general-purpose Cartesian graphs suitable for plotting data available at irregular intervals or plotting two dependent variables against each other. The waveform graphs and XY graphs accept different types of input data!

Waveform graphs are located on the **Graph** subpalette of the **Controls** palette (see Figure 7.1). For basic, single-plot graphs, an array of Y values (along the vertical axis) can pass directly to a waveform graph. This method assumes the initial X value (along the horizontal axis) is $X_0 = 0$ and the $\Delta X = 1$. The value of ΔX determines the X marker spacing. When passing only the Y values to the waveform graph, the graph icon will appear as an array indicator, as seen in Figure 7.13a.

If you want to start plotting other than $X_0 = 0$, or if your data points are spaced differently than $\Delta X = 1$, you can wire a cluster consisting of the initial value X_0, ΔX, and a data array to the waveform graph. The graph terminal will then appear as a cluster indicator, as seen in Figure 7.13b.

If you want to plot more than one curve on a single graph, you can pass a 2D array of data to create a multiple-plot waveform graph. Two methods for wiring multiple-plot waveform graphs are illustrated in Figure 7.14. The graph icon assumes the data type to which it is wired.

Graphs always plot the *rows* of a two-dimensional array. The two-dimensional array in Figure 7.14 has two rows with 10 columns per row—a 2×10 array. If your data is given in columns, you must transpose the array before graphing. Once you have wired your 2D array of data to the graph terminal, go to the front panel, pop up on the graph, and select **Transpose Array**.

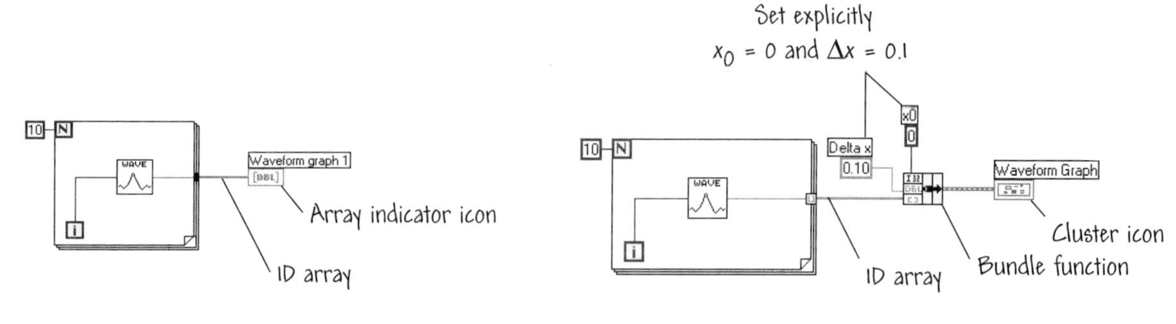

(a) Default values for initial X and ΔX (b) User defined initial X and ΔX

FIGURE 7.13
Single-plot waveform graphs.

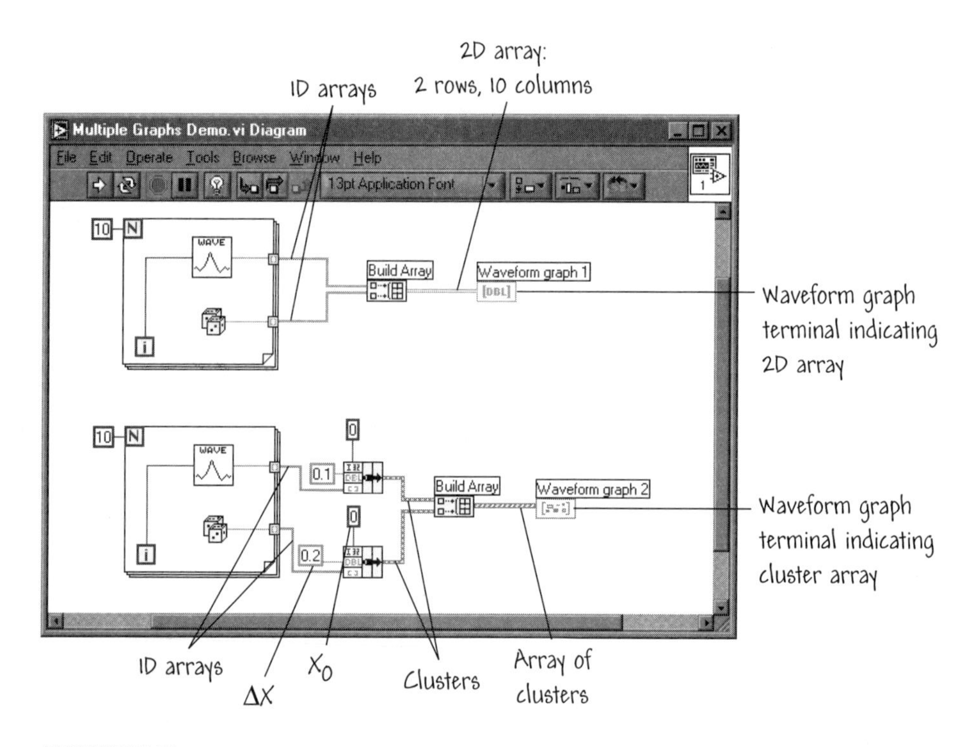

FIGURE 7.14
Two ways to wire multiple-plot waveform graphs.

The waveform, single-plot graph example shown in Figure 7.13 is called Waveform Graphs demo.vi. *The waveform multiple-plot graph example shown in Figure 7.14 is called* Multiple graphs demo.vi. *Both VIs can be found in* Chapter 7 *of the* Learning *directory.*

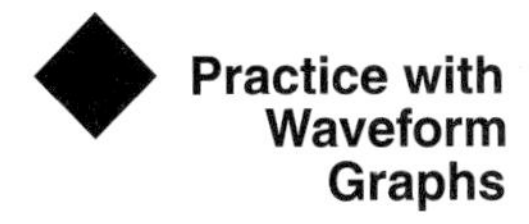

Practice with Waveform Graphs

Open a new VI and build the front panel shown in Figure 7.15. Place three objects on the front panel:

- Digital control labeled No. of points
- Digital control labeled Rate of growth, r
- Waveform graph labeled Population

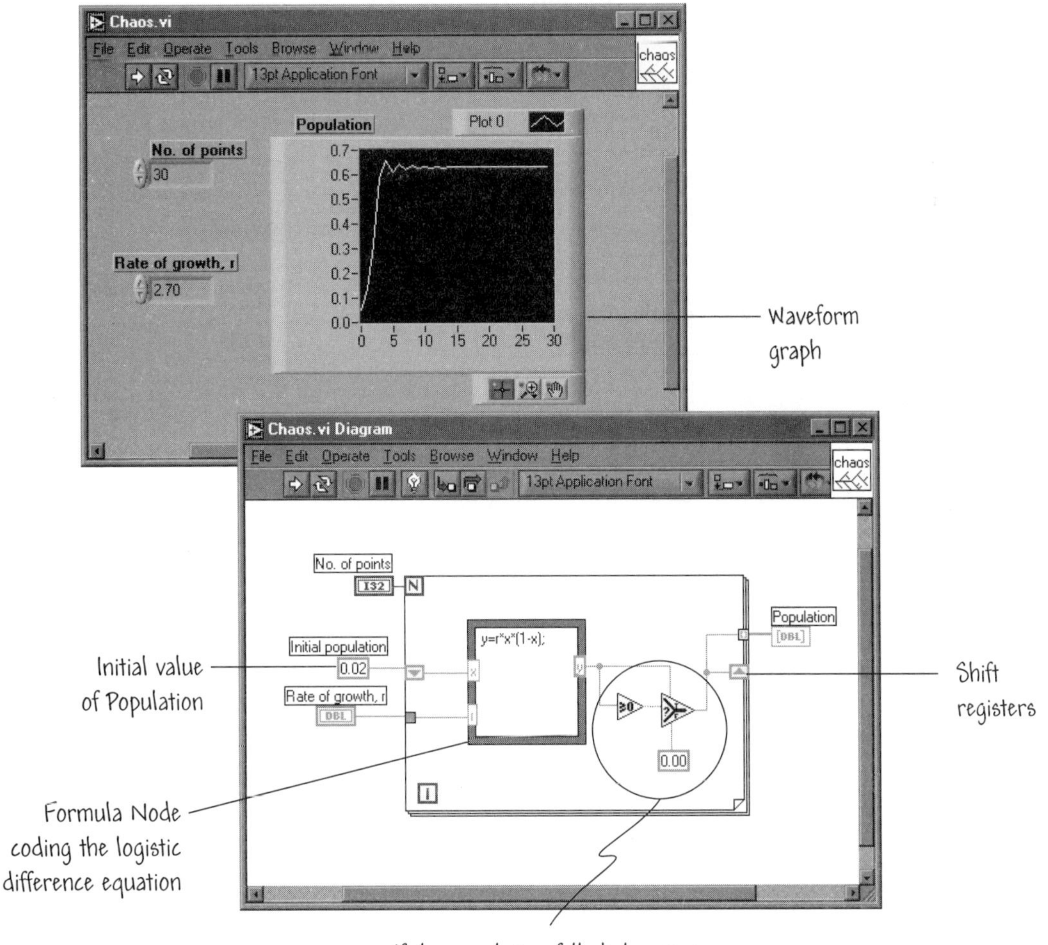

FIGURE 7.15
Front panel and block diagram arrangement to experiment with chaos.

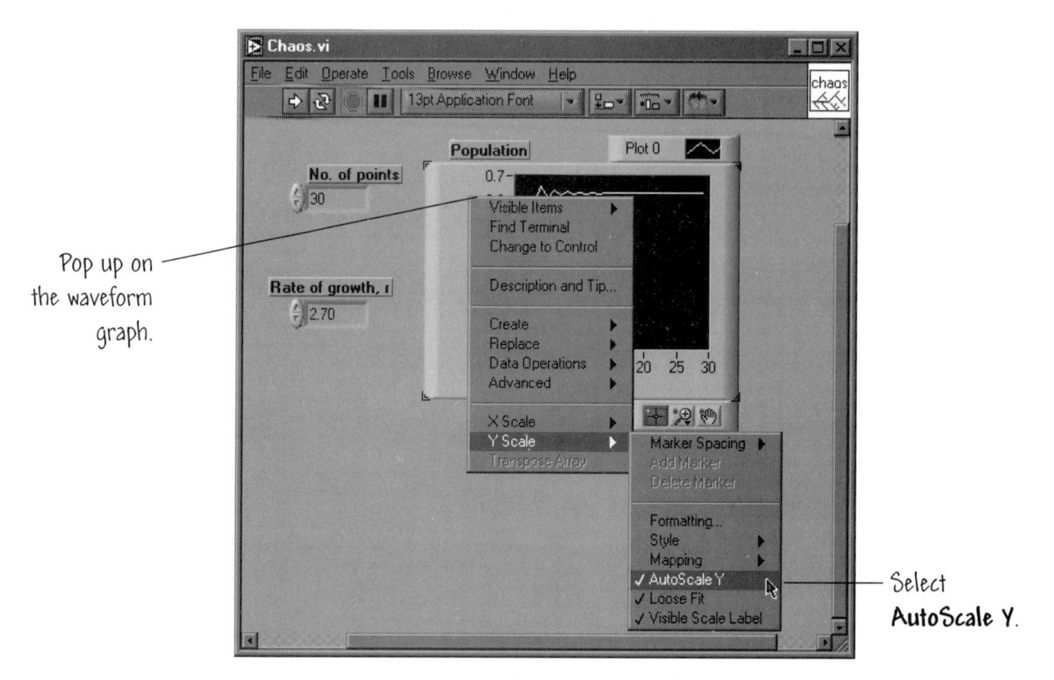

FIGURE 7.16
Changing the waveform graph y-axis limits to allow autoscaling.

We would like the waveform graph y-axis scale to be driven by the data values, rather than remaining at some predetermined fixed values. To modify the y-axis limits of the waveform graph, pop up on the waveform graph and change the **Y Scale** to **Autoscale Y**, as illustrated in Figure 7.16.

The block diagram that you need to construct is shown in Figure 7.15. You will need to add the following objects:

- For Loop with the loop counter wired to the digital control **No. of points**
- Shift registers on the For Loop to transfer data from one iteration to the next
- A Formula Node containing the equation

$$y = rx(1 - x) ,$$

where the source of the parameter r is the digital control **Rate of growth**, **r**.

The equation in the Formula Node is known as the *logistic difference equation*, and is given more formally as

$$x_{k+1} = rx_k(1 - x_k)$$

where $k = 0, 1, 2, \cdots$, and x_0 is a given value. In the block diagram (see Figure 7.15), the initial value is wired as $x_0 = 0.02$. The logistic difference equation has been used as a model to study population growth patterns. The model has been scaled so that the values of the population vary between 0 and 1, where 0 represents extinction and 1 represents the maximum conceivable population.

Wire the block diagram as shown in Figure 7.15 and prepare the VI for execution. Set the variable **No. of points** initially to 30 and the variable **Rate of growth, r** to 2.7. Execute the VI by clicking on the **Run** button. The value of the population is shown graphically in the waveform graph. You should observe the population reach a steady-state value around 0.63.

Experiment with the VI by changing the values of the **Rate of growth, r**. For values of $1 < r < 3$, the population will reach a steady-state value. You will find that as $r \to 3$, the population begins to oscillate, and in fact, the "steady-state" value oscillates between two values. When the parameter $r > 4$, the behavior appears erratic. As the parameter increases, the behavior becomes chaotic.[1] Verify that when the parameter $r = 4.1$, the population exceeds the maximum conceivable value (that is, it exceeds 1.0 on the waveform graph) and subsequently becomes extinct! Save your updated VI in the Users Stuff directory and name it Chaos.vi.

The working version of the chaos example shown in Figure 7.15 is called Chaos.vi *and is located in* Chapter 7 *of the* Learning *directory.*

◆

7.3 XY GRAPHS

As previously mentioned, waveform graphs are ideal for plotting evenly sampled waveforms. But what if your data is available at irregular intervals, or what if you want to plot two dependent variables against each other (e.g., x versus y)? The **XY graph** is well-suited for use in situations where you want specify points using their (x, y) coordinates. The XY graph is a general-purpose Cartesian graph that can also plot multivalued functions—such as circles and ellipsoids. XY graphs are located on the **Graph** subpalette of the **Controls** palette.

A single-plot XY graph and its corresponding block diagram are shown in Figure 7.17. The Bundle function is used to combine the X and Y arrays into a

1. For more on chaos, see *Chaos: Making a New Science*, by James Gleick, Penguin Books (New York, 1987).

cluster that is then wired to the XY graph. For the XY graph, the components are a bundled X array (the top input) and a Y array (the bottom input). The XY graph terminal will appear on the block diagram as a cluster indicator. For a multiple-plot XY graph, you need to build an array of the clusters of X and Y values, as shown in Figure 7.17.

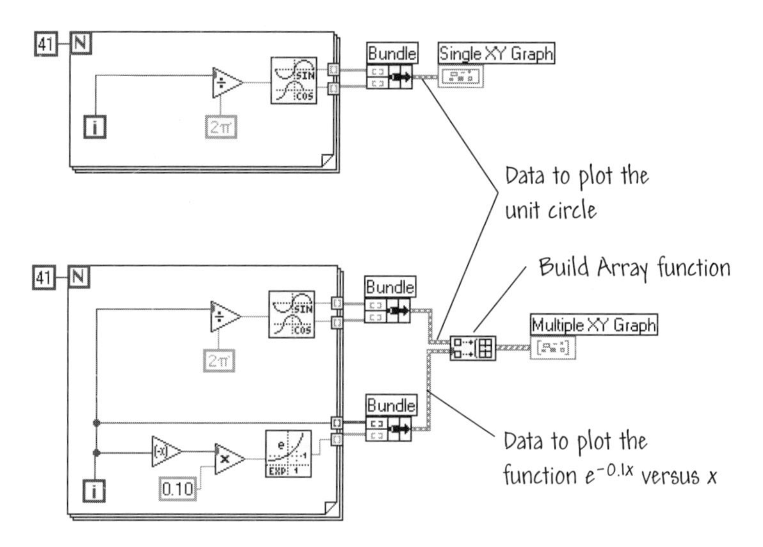

FIGURE 7.17
Examples of the XY graphs: single and multiple plots.

The XY graph example shown in Figure 7.17 is called XY Graphs demo.vi *and is located in* Chapter 7 *of the* Learning *directory. The single XY graph produces a unit circle, and the multiple XY graph produces a unit circle and a plot of $e^{-0.1x}$ for $0 \leq x \leq 41$.*

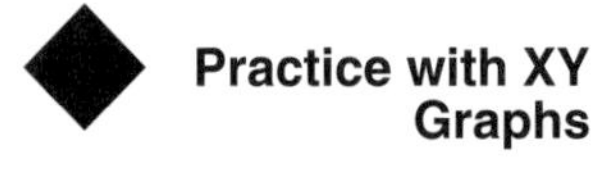

Practice with XY Graphs

Suppose that you need to borrow \$1000 from a bank and are given the option of choosing between simple interest and compound interest. The annual interest rate is 10% and, in the case of compound interest, the interest accrues annually. In either case, you pay the entire loan off in one lump sum at the end of the loan period. Construct the VI shown in Figure 7.18 to compute and graph the amount due at the end of each loan period, where the loan period varies from $N = 1$ year to $N = 20$ years in increments of 1 year.

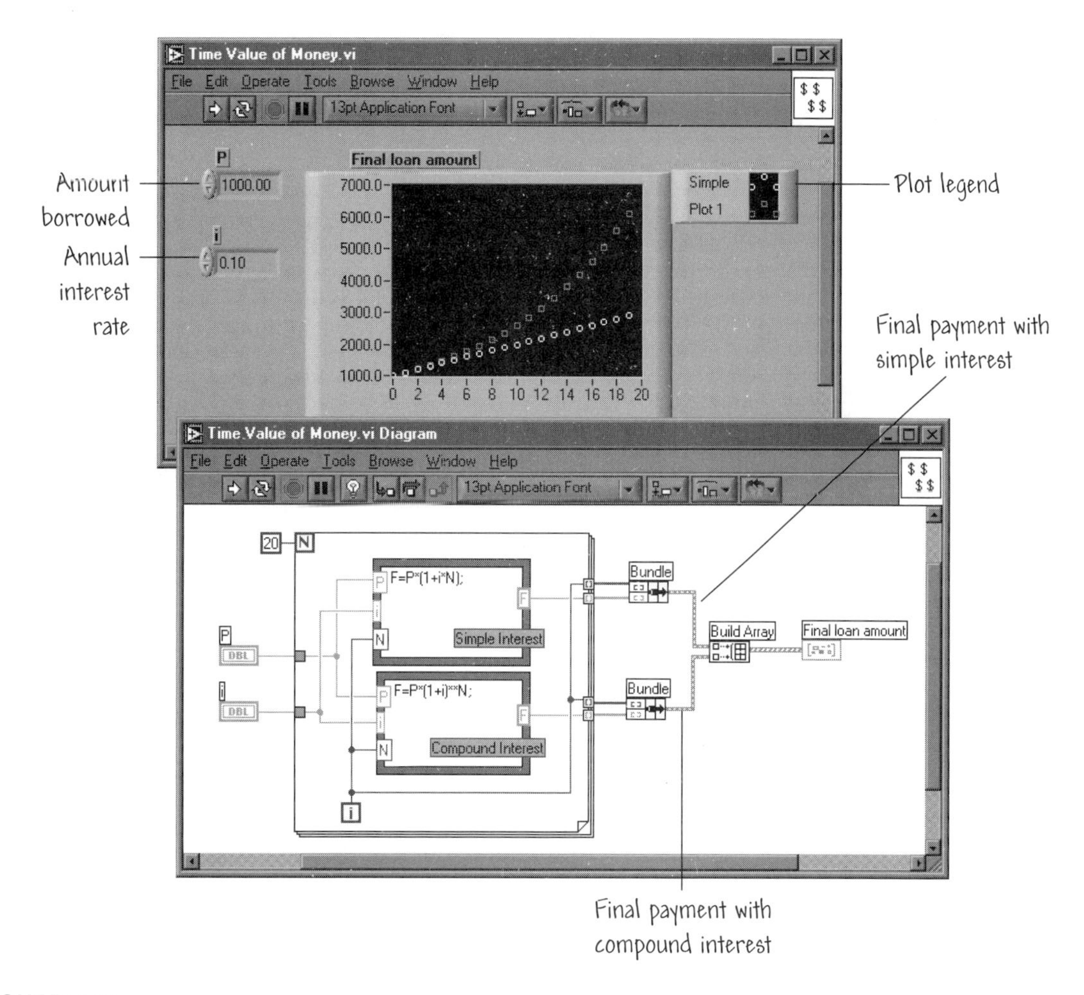

FIGURE 7.18
A VI to compute and plot the time value of money.

Create a multiple-plot XY graph showing the final payments due for both the simple interest and the compound interest situations.

Define the following variables: F = final payment due, P = amount borrowed, i = interest rate, N = number of interest periods. The relevant formulas are:

- Simple interest

$$F = P(1 + iN)$$

- Compound interest

$$F = P(1 + i)^N$$

Open a new front panel and place the following three items:

- Digital control labeled **P**

- Digital control labeled **i**
- XY Graph

The XY graph in Figure 7.18 has been edited somewhat. Pop up on the first plot in the legend and select **Common Plots** and choose the scatter plot (top row, center). Then pop up in the same place again and select a **Point Style**. Repeat this process for the second plot listed in the legend.

One solution to coding the VI is shown in Figure 7.18. In this block diagram, the formulas are coded using Fomula Nodes. Can you think of other ways to code the same equations?

Once your VI is working, you can use it to experiment with different values of annual interest and different initial loan amounts. Based on your investigations, would you prefer to have simple interest or compound interest?

The XY graph example shown in Figure 7.18 is called Time Value of Money.vi *and is located in* Chapter 7 *of the* Learning *directory.* ◆

7.4 CUSTOMIZING CHARTS AND GRAPHS

Charts and graphs have editing features that allow you to customize your plots. This section covers how to configure some of the more important customization features. In particular, the following items are discussed:

- Autoscaling the chart and graph x- and y-axes
- Using the Plot Legend
- Using the Graph Palette
- Using the Scale Legend
- Chart customizing features, including the scrollbar and the digital display

By default, charts and graphs have the **plot legend** showing when first placed on the front panel. Using the **Positioning** tool, you can move the scale and plot legends and the graph palette anywhere relative to the chart or graph. Figure 7.19 shows some of the more important components of the chart and graph customization objects.

7.4.1 Axes Scaling

The x- and y-axes of both charts and graphs can be set to automatically adjust their scales to reflect the range of the plot data. The **autoscaling** feature is enabled or disabled using the **AutoScale X** and **AutoScale Y** options from the **Advanced**

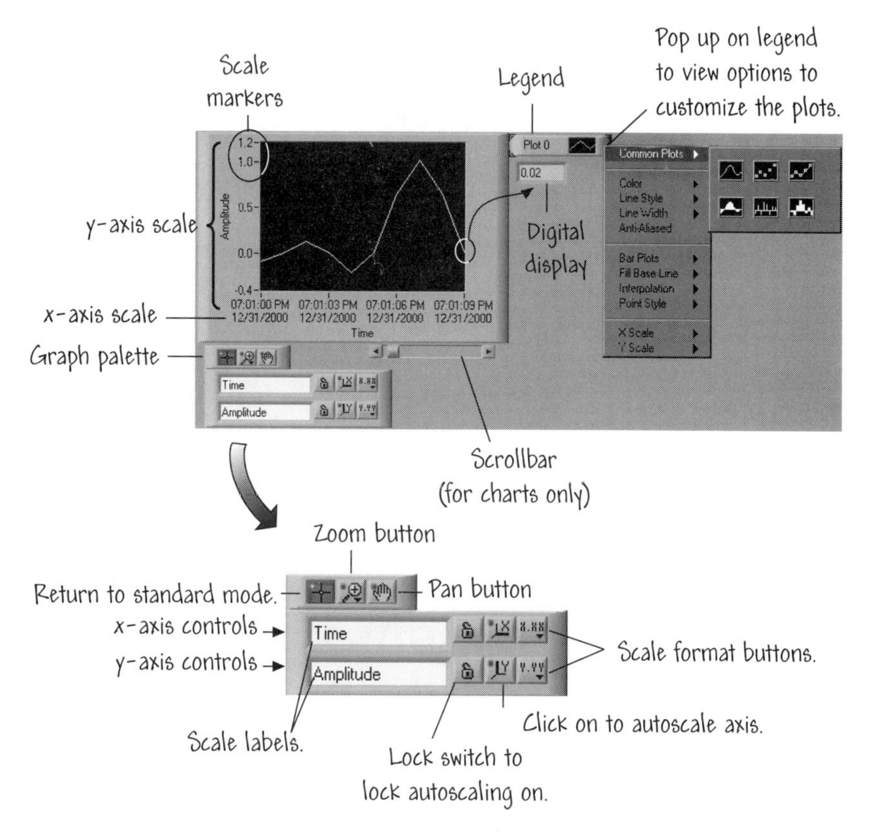

FIGURE 7.19
The chart and graph customization objects.

menu or the **X Scale** or **Y Scale** submenus of the short cut menu, as shown in Figure 7.20.

The use of autoscaling may cause the chart or graph to update more slowly.

The X and Y scales can be varied manually using the **Operating** or **Labeling** tool—just in case you don't want to use the autoscale feature. For instance, if the graph x-axis end marker has the value 10, you can use the **Operating** tool to change that value to 1 (or whatever other value you want!). The graph x-axis marker spacing will then change to reflect the new maximum value. In typical situations, the scales are set to fit the exact range of the data. You can use the **Loose Fit** option (see Figure 7.20) if you want to round the scale to multiple values of the scale increment—in other words, with a loose fit the scale marker numbers are rounded to a multiple of the increment used for the scale. For example, if the

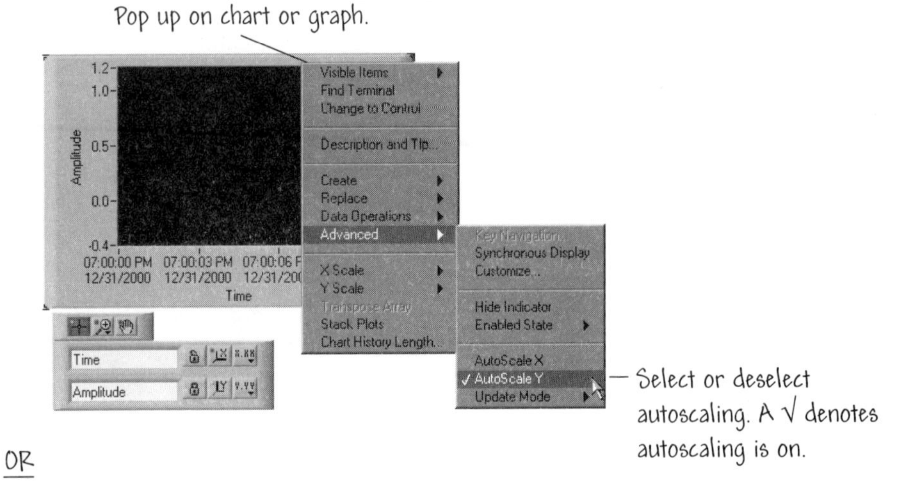

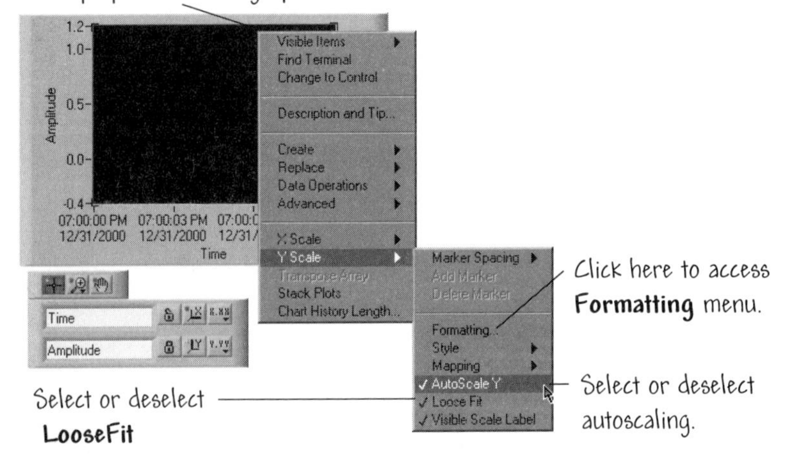

FIGURE 7.20
Using the **Autoscaling** feature of charts and graphs.

scale markers increment by 5, then with a loose fit, the minimum and maximum scale values are set to a multiple of 5.

On both the **X Scale** or **Y Scale** submenus, you will find the **Formatting...** option. Choosing the option opens up a dialog box as shown in Figure 7.21 that allows you to configure many components of the chart or graph. You can modify individually the x- and y-axis characteristics. The following choices are available:

- The **Scale Style** menu lets you select major and minor tick marks for the scale, or none at all. A major tick mark corresponds to a scale label, while a minor tick mark denotes an interior point between labels. You have to click on the **Scale Style** icon to view the choices. With this menu you can make the axes markers invisible.

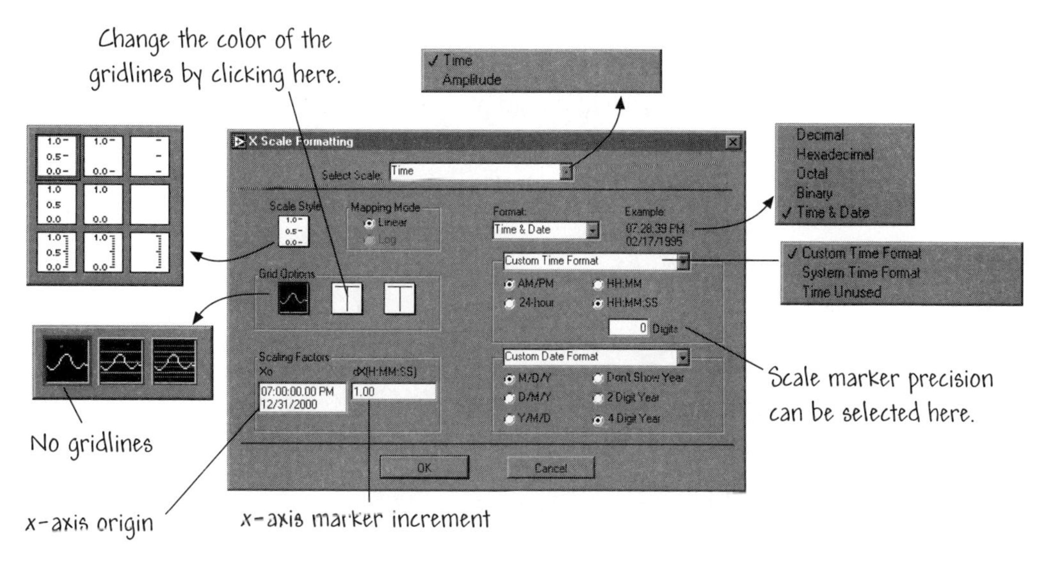

FIGURE 7.21
Scale formatting.

- The **Mapping Mode** provides the choice of either a linear or a logarithmic scale.
- The **Grid Options** provides control over the gridlines: no gridlines, gridlines only at major tick mark locations, or gridlines at both major and minor tick marks. This option also allows you to change the color of the grid lines.
- Under **Scaling Factors**, you can set the axis origin X_0 (or Y_0) and the axis marker increment ΔX (or ΔY).
- Within the **Precision** area, you can choose the number of digits of precision for the scale marker numbers, as well as the notation (**Floating Point**, **Scientific** or **Time**). With LabVIEW you can specify a time and date format for graphs, as illustrated in Figure 7.21. In the **Format** pull-down menu you can choose among Decimal, Hexadecimal, Octal, Binary, or Time & Date. The choices under the Scaling Factors and Precision Change depend on the format choice.

7.4.2 The Legend

The legend provides a way to customize the plots on your charts and graphs. This is where you choose the data point style, the plot style, the line style, width and color, and other characteristics of the appearance of the chart or graph. For example, on multiple-plot graphs you may want one curve to be solid line and the other curve to be a dashed line or one curve to be red and the other blue. An example of a legend is shown in Figure 7.22. When you move the chart or graph around on the front panel, the legend moves with it. You can change the position of the legend relative to the graph by dragging the legend with the **Positioning** tool to the desired location.

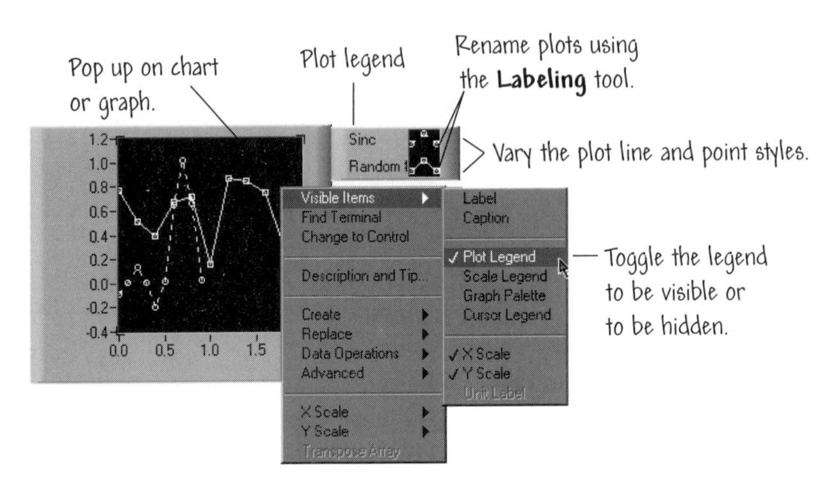

FIGURE 7.22
The legend for charts and graphs.

The legend can be visible or hidden—and you use the **Visible Items** submenu of the chart or graph short cut menu (see Figure 7.22) to choose. After you customize the plot characteristics, the plot retains those settings regardless of whether the legend is visible or not. If your chart or graph receives more plots than are defined in the legend, they will have the default characteristics.

When the legend first appears, it is sized to accomodate only a single plot (named Plot 0, by default). If you have a multiple-plot chart or graph, you will need to show more plot labels by dragging down a corner of the legend with the **Resizing** cursor to accomodate the total number of curves. Each curve in a multiple-plot chart or graph is labeled as Plot 0, Plot 1, Plot 2, and so forth. You can modify the default label by assigning a name to each plot in the legend with the **Labeling** tool. Choosing names that reflect the physical value of the data depicted in the chart makes good sense—for instance, if the curve represents the velocity of an automobile you might use the label Automobile velocity in km/hr. The legend can be resized on the left to give the plot labels more room or on the right to give the plot samples more room. This is useful when you assign long names to the various curves on a multiple-plot chart or graph. You can pop up on each plot in the legend and change the plot style, line, color, and point styles of the plot. The short cut menu is shown in Figure 7.19.

7.4.3 The Graph Palette and Scale Legend

The graph palette and scale legend allows you access to various aspects of the chart and graph appearances while the VI is executing. You can autoscale the axes and control the display format of the axes scales. To aid in the analysis of the data presented on the charts and graphs, buttons allow you to pan and zoom in and out. The graph palette and scale legend shown in Figure 7.23 can be displayed using the **Visible Items** submenu of the chart or graph short cut menu.

On the left of the scale legend are three buttons that control the autoscaling of the axes. Clicking on the autoscaling button (either the x- or y-axis buttons) will cause the chart or graph to autoscale the axis. When you press the lock button, the chart or graph will autoscale either of the axes scales continuously.

The x-axis and y-axis labels can also be modified by entering the desired labels (using the **Labeling** tool) into the text area provided on the left side of the scale legend. The upper text is for the x-axis.

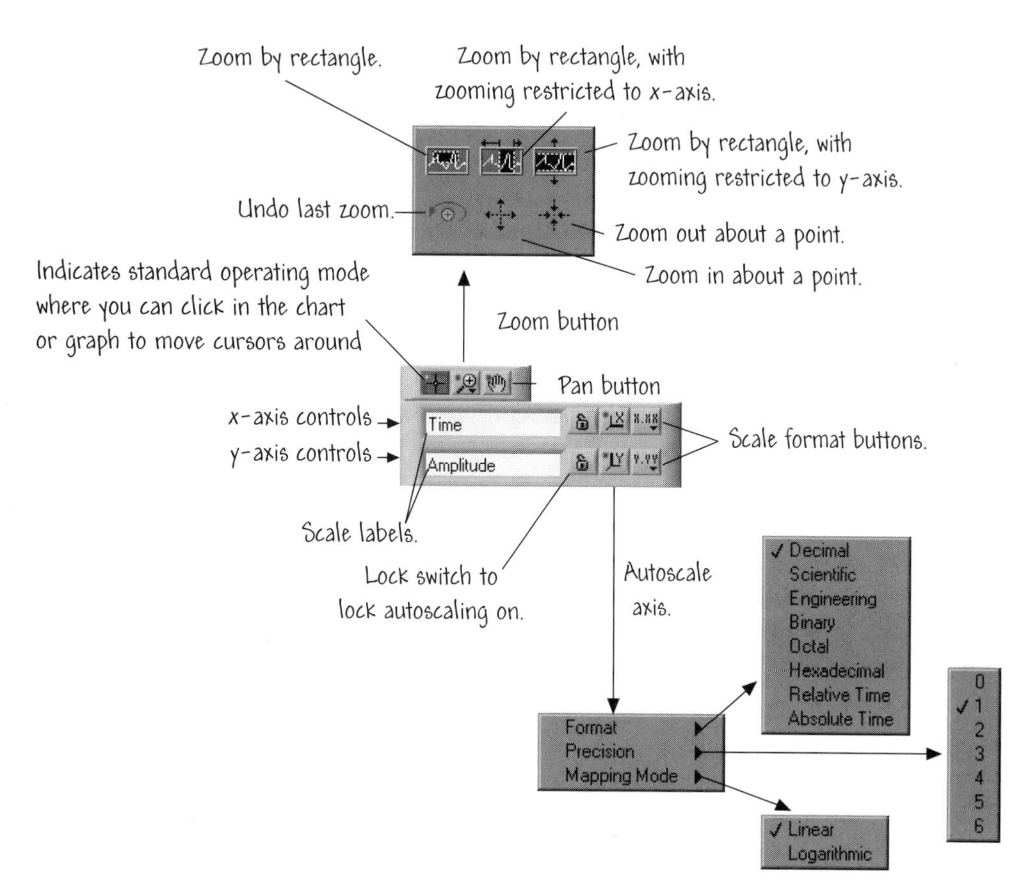

FIGURE 7.23
The chart and graph palette.

The scale format buttons (on the right of the scale legend) provide run-time control over the format and precision of the *x* and *y* scale markers, respectively. As shown in Figure 7.23, you can control the **Format** of the axes markers, the **Precision** of the marker numbers (from 0 to 6 digits after the decimal point), and the **Mapping Mode**. Remember that in the edit mode, you can control the format, precision, and mapping mode through short cut menus on the chart or graph. The chart or graph scale legend gives you the added feature of being able to control the format and precision of the *x* and *y* scale markers while the VI is executing.

The three buttons on the graph palette allow you to control the operational mode of the chart or graph. You are in the standard mode under normal circumstances where the button with the cross-hair (left side of the palette) is selected. This indicates that you can click in the chart or graph area to move cursors around. Other operational modes include panning and zooming. When you press the pan button (depicted in Figure 7.23), you can scroll through the visible data by clicking and dragging sections of the graph with the pan cursor. Selecting the zoom button gives you access to a short cut menu that lets you choose from several methods of zooming.

As depicted in Figure 7.23, you have several zoom options:

- **Zoom by Rectangle**. In this mode, you use the cursor to draw a rectangle around the area you want to zoom in and when you release the mouse button, the axes will rescale to zoom in on the desired area.
- **Zoom by Rectangle**—with zooming restricted to X data. This is similar in function to the **Zoom by Rectangle** mode, but the Y scale remains unchanged.
- **Zoom by Rectangle**—with zooming restricted to Y data. This is similar in function to the **Zoom by Rectangle** mode, but the X scale remains unchanged.
- **Undo Last Zoom**. This resets the axes scales to their previous setting.
- **Zoom In about a Point**. With this option you hold down the mouse on a specific point on the chart or graph to continuously zoom in until you release the mouse button.
- **Zoom Out about a Point**. With this option you hold down the mouse on a specific point and the graph will continuously zoom out until you release the mouse button.

For the zoom in and zoom out about a point modes, <shift>-clicking will zoom in the other direction.

7.4.4 Special Chart-Customizing Features

Charts have the same customizing features as graphs—with two additional options: **Scrollbars** and **Digital Displays**. The **Visible Items** submenu (of the chart short cut menu) is used to show or hide the optional digital display(s) and a scrollbar.

Charts have scrollbars that can be used to display older data that has scrolled off the chart. The chart scrollbar is depicted in Figure 7.19. The scrollbar can be made visible by popping up on the chart and selecting the scrollbar from the **Visible Items** submenu, as illustrated in Figure 7.24.

There is one digital display per plot that displays the latest value being plotted on the chart. You can place a digital display on the chart by selecting the **Digital Display** from the **Visible Items** submenu, as illustrated in Figure 7.24. The digital display can be moved around relative to the chart using the **Positioning** tool, but by default it will initially be placed on the legend. The last value

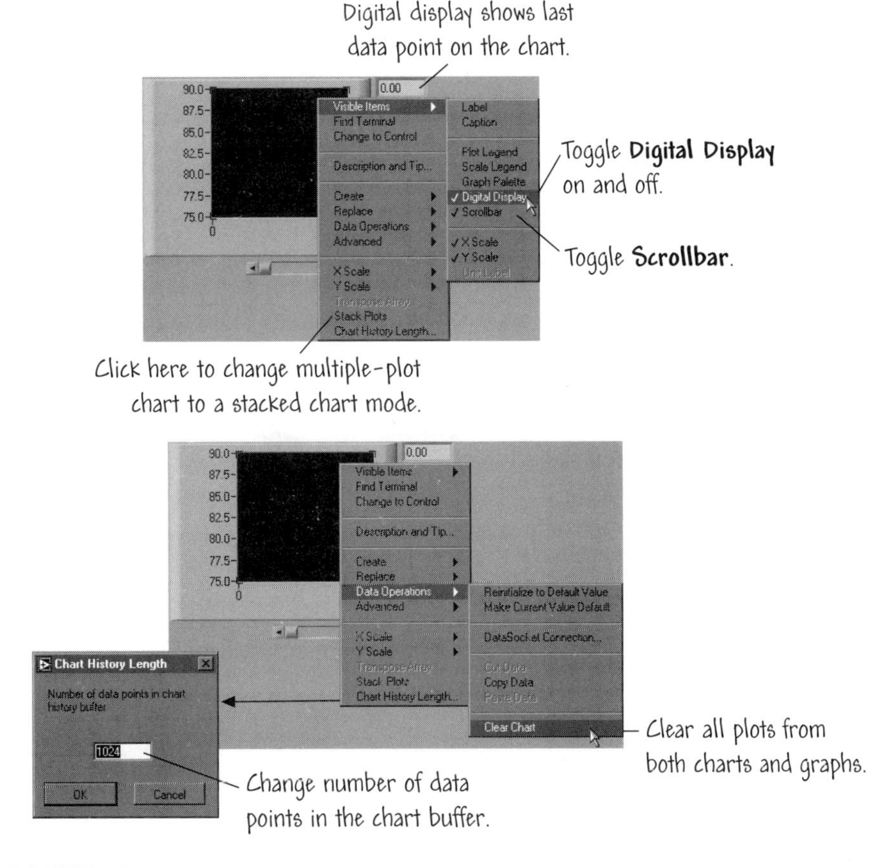

FIGURE 7.24
The scrollbar and the digital display for charts.

passed to the chart from the block diagram is the latest value for the chart and the value that is displayed on the digital display. If you want to view past values contained on the chart data you can navigate through the older data using the scrollbar. There is a limit to the amount of data the chart can hold in the buffer. By default, a chart can store up to 1024 data points. When the chart reaches the limit, the oldest point(s) are discarded to make room for new data. If you want to change the size of the chart data buffer, select **Chart History Length...** from the short cut menu and specify a new value of up to 100,000 points.

When you are not running the VI, you can clear the chart by popping up on the chart and selecting **Clear Chart** from the **Data Operations** menu, as shown in Figure 7.24. However, if you are running the VI (you are in the run mode), then **Clear Chart** becomes a short cut menu option and it will automatically appear in the chart short cut menu after clicking on the **Run** button. This allows you to clear the chart while the VI is executing.

If you have a multiple-plot chart, you can choose between overlaid plot or stacked plot. When you display all plots on the same set of scales, you have an overlaid plot. When you give each plot its own scale, you have a stacked plot. You can select **Stack Plots** or **Overlay Plots** from the chart short cut menu (see Figure 7.24) to toggle between stacked and overlaid modes. Figure 7.25 illustrates the difference between stacked and overlaid plots.

The multiple-plot chart example shown in Figure 7.25 is called Chart Customizing.vi *and is located in* Chapter 7 *of the* Learning *directory. The two charts on the block diagram display the same type of data, except that the chart on the left uses the stacked chart, and the chart on the right uses the overlaid chart.*

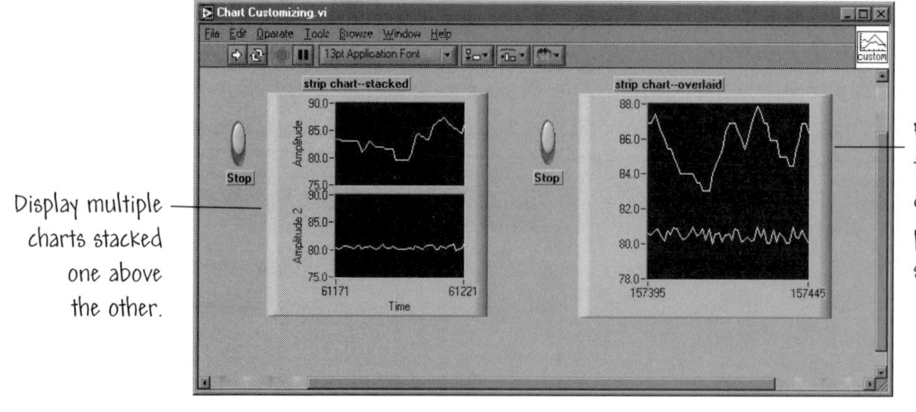

FIGURE 7.25
Stacked and overlaid charts.

7.4.5 Special Graph-Customizing Features: Cursor Legend

To add cursors to a graph, you first bring the **Cursor Legend** to the front panel graph by popping up on the graph and selecting Visible Items≫Cursor Legend from the short cut menu, and clicking anywhere in a cursor legend row to activate a cursor, as illustrated in Figure 7.26. To add multiple cursors to the graph, the cursor legend displays the exact value of that point. The cursor legend is shown in Figure 7.27.

You can place and label cursors on all graphs. To lock a cursor onto a plot, select the lock icon associated with that icon on the **Cursor Legend**. A graph can

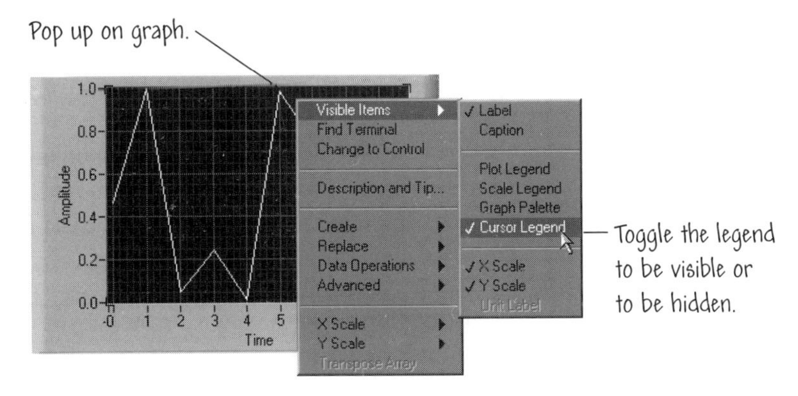

FIGURE 7.26
Placing the **Cursor Legend** on the front panel.

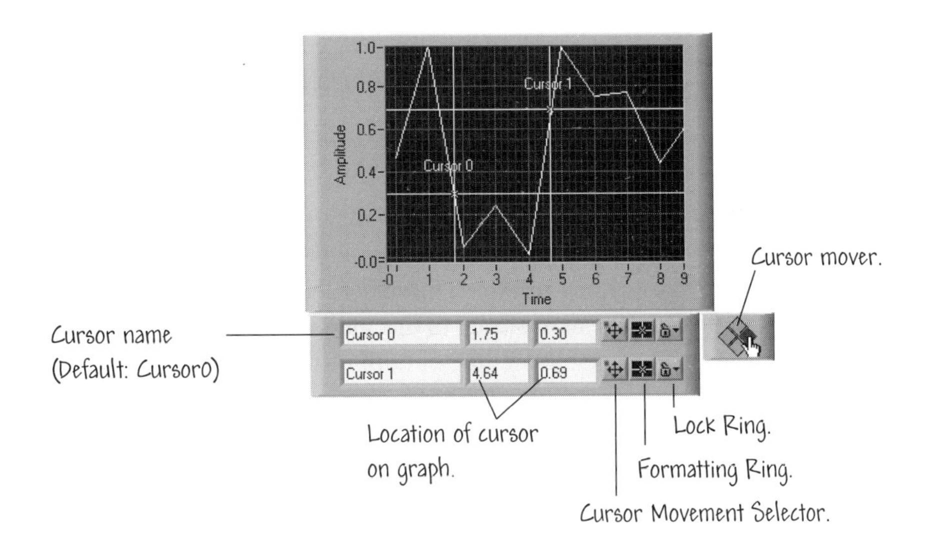

FIGURE 7.27
The **Cursor Legend** is associated with graphs.

have any number of cursors and multiple cursors can be moved at the same time. The **Operating** tool is used to click any of the following buttons on the right of the cursor legend to customize your cursor:

- **Cursor Movement Selector**: Click this button to move the cursor using the cursor mover (see Figure 7.27). If you click the Cursor Movement Selector for more than one cursor, the cursors move in parallel.
- **Formatting Ring**: Customize the appearance of each cursor.
- **Lock Ring**: Customize the action of each cursor or associate a particular cursor with a particular plot.

On the **Formatting Ring** you can choose to customize the appearance of your cursor. The following options are available:

- **Color**: Lets you choose the color of your cursor from the Color Picker.
- **Cursor Style**: Provides various cursor styles.
- **Point Style**: Provides various point styles for the intersection of your cursor.
- **Line Style**: Provides various solid and dotted line styles.
- **Line Width**: Provides various line widths.
- **Show Name**: Displays the name of the cursor on the graph. You can use the **Positioning** tool to move the name in relation to the cursor.
- **Bring to Center**: Centers the cursor on the graph without changing the X and Y scales.
- **Go to Cursor**: Changes the X and Y scales to show the cursor at the center of the graph.

The **Lock Ring** provides three options to customize the action of your cursor:

- **Free**: Click the cursor on the graph or enter values in the X and Y coordinates in the cursor legend to move the cursor.
- **Snap to Point**: The cursor moves to the closest plot point. The cursor can switch to another plot in this mode.
- **Lock to Plot**: The cursor locks to a particular plot. The cursor cannot switch to another plot in this mode. If you have more than one plot, LabVIEW lists the plots at the bottom of the Lock Ring menu. Click the plot you want to associate with each cursor.

7.4.6 Using Context Help

For students new to LabVIEW, it can often be confusing when you try to wire data to charts and graphs. Do you use a **Build Array** function, a **Bundle** function, or both? What order do the input terminals use? You can find valuable information in the Context Help window. For example, if you select **Show Context**

Help from the **Help** menu and put your cursor over a Waveform Graph terminal in the diagram, you will see the information shown in Figure 7.28. The Context

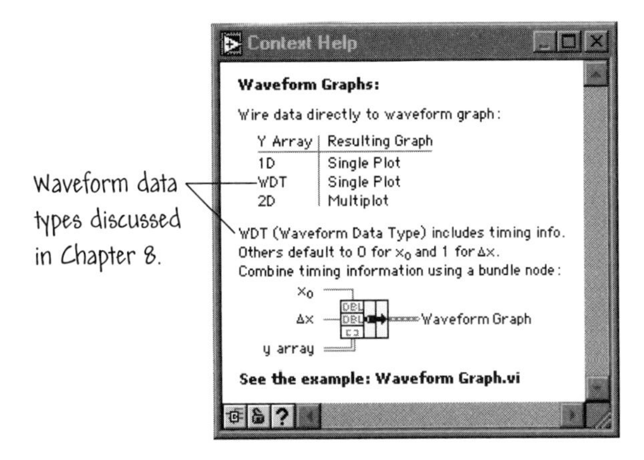

FIGURE 7.28
Using Context Help to determine which data types to wire to the Waveform Graph.

Help window shows you what data types to wire to the Waveform Graph, how to specify point spacing with the **Bundle** function, and which example to use when you want to see the different ways you can use a Waveform Graph. The Context Help window also shows similar information for XY Graphs and Waveform Charts.

BUILDING BLOCK

7.5 BUILDING BLOCKS: MEASURING VOLUME

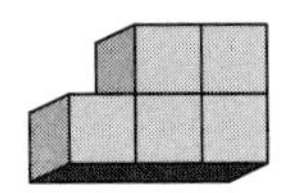

The work on the VI called Volume Array.vi created in Chapter 6 will be used as the basis for a new VI that displays the volume data on a chart. The goal is to make the necessary modifications to the existing Volume Array.vi so you can plot the data on a strip chart. You will also control the timing of the loop using the Wait Until Next ms Multiple function. Use Figure 7.29 as a guide for your VI construction.

The **Limit** input found on the front panel of the VI existed on Multiple Volume Points.vi, which you constructed in the building blocks exercise of Chapter 5. You may want to revisit that during your work on this exercise. When you have finished constructing the VI, save it in the Users Stuff folder as Volume Chart.vi.

Run the VI and observe the operation of the strip chart. The parameters that you can play with are controlled by the Limit knob, the Volume Units horizontal switch, and the Delay between points (sec) horizontal slide. You can easily change the chart mode to scope or sweep mode (even while the VI is running)

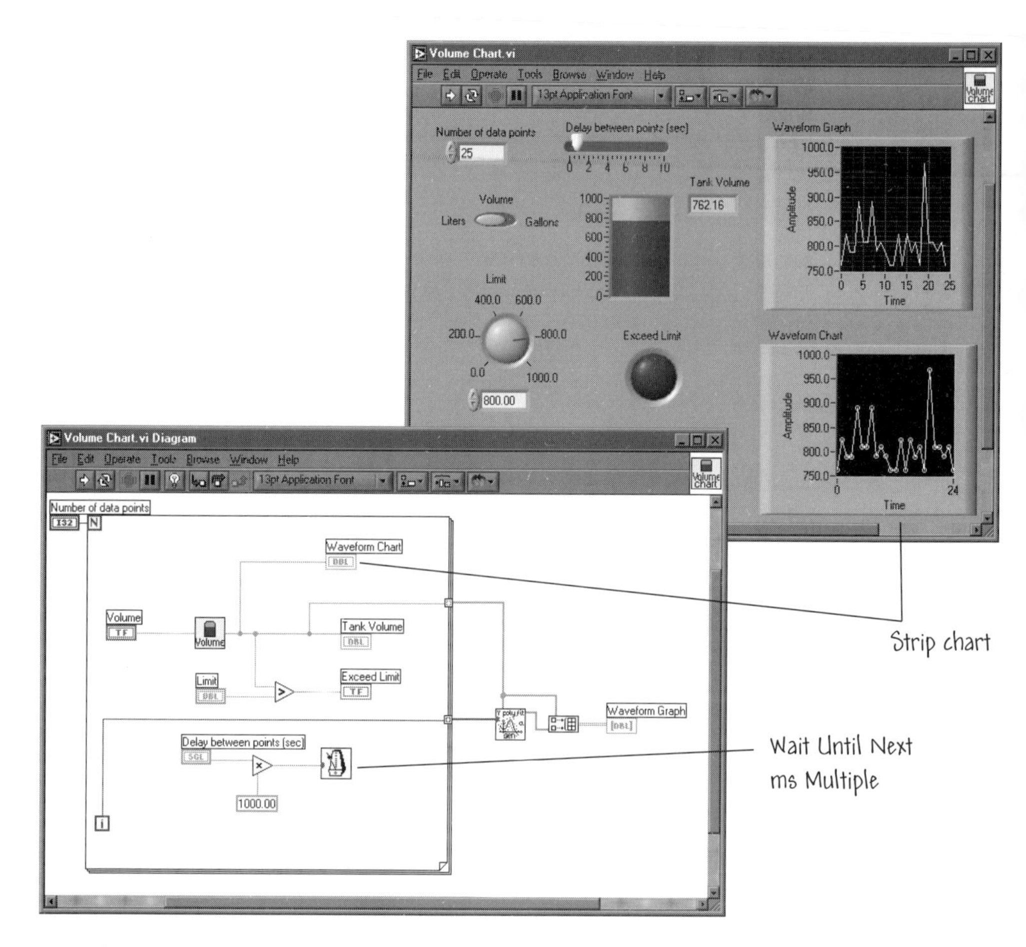

FIGURE 7.29
The Volume Chart.vi block diagram and front panel.

by popping up on the chart and selecting the desired chart mode in the **Update Mode** menu.

7.6 RELAXING READING: LEARNING ANALYTICAL CHEMISTRY

Universities and colleges often present analytical chemistry in two courses—"quantitative analysis" and "instrumental analysis." Many educators (and students!) believe that the chemistry courses are in need of major overhauls. The quantitative analysis course at St. Olaf, for instance, was once called "dry as dust and heavy as a Russian dinner." Faculty members at St. Olaf have worked to change both the quantitative analysis and instrumental analysis courses to reflect the way that industry and academic analytical chemistry are evolving into analytical sciences. A key feature of these two courses at St. Olaf is their organization into small group class and lab activities, with LabVIEW virtual instruments at their core.

The learning method is based on the concept of "role playing" with four roles:

- Manager—responsible for orchestrating the activity mission.
- Hardware—responsible for connection and operation of the instruments.
- Chemist—responsible for preparation and delivery of the samples and reagents.
- Software—responsible for linking and operating the computer resources that run the instruments, acquire the data, and make the Manager's plan work using the Chemist's solutions and Hardware's connections.

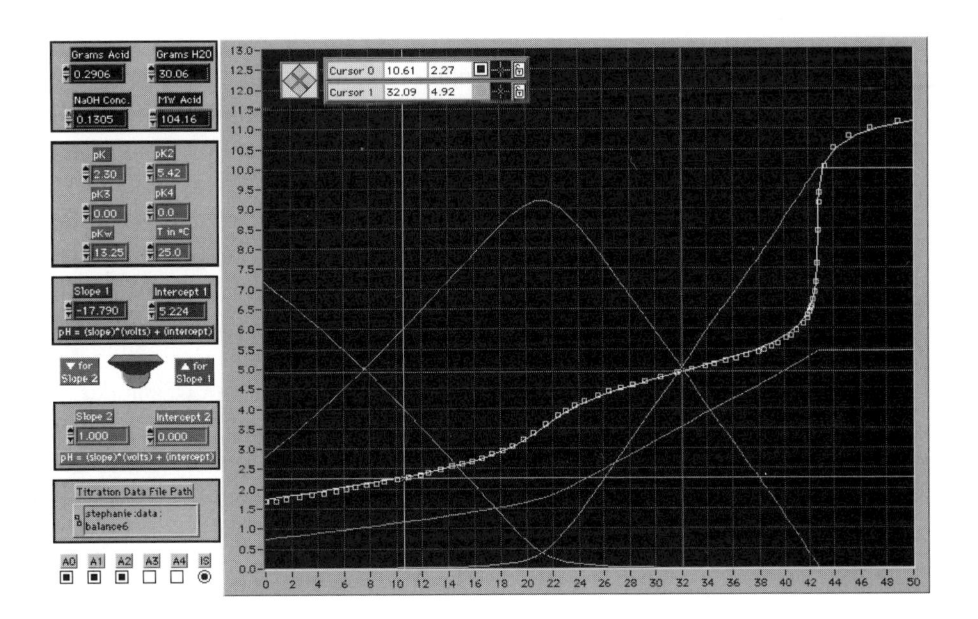

FIGURE 7.30
Students use LabVIEW to explore the effects of key experimental variables on using a weak acid titration for pharmaceutical quality control analysis.

Key to the success of the teaching method at St. Olaf is the software. The role-playing structure would be difficult to implement without LabVIEW as a main environment for the Software role. LabVIEW makes it possible for the students playing the Hardware and Software roles to cleverly interact in the construction of virtual instruments (VIs) to run the more traditional analytical instruments, such as a charge-coupled device parallel readout UV/VIS spectrophotometer. Students also can engage in more fundamental interfacing work by using LabVIEW to make their own potentiostats and other advanced interfaces using LabVIEW with National Instruments A/D boards, such as the NB-MIO-16.

In one common scenario, the Manager gathers and records data and plans with the Chemist on how to set up the best combinations of standard solutions and reagents to get the statistics needed for evaluating an analytical instrument. For the first time in many years, it is possible for a small group to learn analytical methods development by designing and doing rather than just button pushing. LabVIEW makes this practical in the short time frames that an undergraduate lab session imposes. Students analyze data on the laboratory floor as it is acquired, rather than after the lab in a dorm room.

Consider the following scenario: the Manager must decide what experimental variables need to be controlled if a weak acid titration is only to be used for an assay of future pharmaceutical materials. A LabVIEW titrator, shown in Figure 7.30, facilitates parameter evaluation by fitting experimental data to a theoretical model. Students can then explore the effects of suggested control variables, such as the ionic strength of the titration medium, by adjusting these parameters in the model, rather than in the lab. Huge time savings result, and creative decision making changes what used to be "just another boring titration" into an exciting group challenge.

LabVIEW has added a way of bringing young people into the decision-making loops that operate in the professional world. The use of LabVIEW at St. Olaf has made it possible for the faculty to show the students how a discipline that was once restricted to certain kinds of exactly manageable reaction chemistries has evolved into a multidisciplinary science that blends tools and techniques from physics, chemistry, engineering, and human resources. These team efforts result in problem-solving exercises under real-time deadlines.

For more information on this application of LabVIEW, read the *User Solutions* entitled "Learning Analytical Chemistry Using LabVIEW and the Macintosh," which can be found on the National Instruments website, or contact:

Professor John Walters
Department of Chemistry
St. Olaf College
1520 St. Olaf Avenue
Northfield, MN 55057-1098
e-mail: walters@stolaf.edu

7.7 SUMMARY

You can display plots of data using charts or graphs. Three types of charts were discussed—scope, strip, and sweep—and two types of graphs—waveform and XY. Both charts and graphs can display multiple plots at a time. Charts append new data to old data by plotting one point at time. Graphs display a full block of data. The waveform graph plots only single-valued functions with points that are evenly distributed. Conversely, the XY graph is a general-purpose Cartesian graph that lets you plot unevenly spaced, multivalued functions. You can customize the appearance of your charts and graphs using the legend and the palette.

The format and precision of the X and Y scale markers and zoom in and out are controlled with the palette even while the VI is executing.

KEY TERMS

Autoscaling: The ability of graphs and charts to adjust automatically to the range of plotted values.

Legend: An object owned by a chart or graph that displays the names and styles of the plots.

Plot: A graphical representation of data shown on a chart or graph.

Palette: An object owned by a chart or graph from which you can change the scaling and format options while the VI is running.

Scope chart: A chart modeled after the operation of an oscilloscope.

Strip chart: A chart modeled after a paper strip chart recorder, which scrolls as it plots data.

Sweep chart: Similar to a scope chart, except that a line sweeps across the display to separate new data from old data.

Waveform chart: Displays one point at a time on one or more plots.

Waveform graph: Plots single-valued functions or array(s) of values with points evenly distributed along the x-axis.

XY graph: A general-purpose, Cartesian graph used to plot multi-valued functions, with points evenly sampled or not.

EXERCISES

E7.1 Open Multiple Graphs Demo.vi in Learning\Chapter 7. Pop up on each graph and select **Plot Legend** from the **Visible Items** menu. Change the line style, point style, and color of each plot by popping up on each line in the legend and selecting **Line Style**, **Point Style**, and **Color**, respectively. Next, pop up on each graph and select **Cursor Legend** from the **Visible Items** menu. With the **Operating** tool, click on the cursor movement selector in each row of each cursor display to create a graph cursor. Graph cursors help identify the x and y values of data points on the graph. Move cursor 0 to the highest value on Plot 0, click on the lock icon and select **Lock to plot**. Repeat the same process with Plot 1 and cursor 1.

E7.2 Create a new VI where temperature data, created with the Digital Thermometer.vi in the Activity directory, is displayed on a graph. At the beginning of each execution of the VI, clear the graph.

E7.3 Create a VI that plots an ellipse

$$r^2 = \frac{A^2 B^2}{A^2 \sin^2 \phi + B^2 \cos^2 \phi}$$

where r, A, and B are input parameters and $0 \leq \phi \leq 2\pi$.

E7.4 Create a VI that graphs the function $\sin x$ where $x = 0 \ldots n\pi$ and the integral $y = \int_0^{n\pi} \sin x dx$. The value of n should be an input on the front panel.

PROBLEMS

P7.1 Create a new VI to plot a circle using an XY Graph.

P7.2 Create a new VI where temperature data, created with the Digital Thermometer.vi in the Activity directory, is displayed on a strip chart. Compute and display the running average of the temperature data.

P7.3 Complete the crossword puzzle.

Across

1. Plots single-valued functions or array(s) of values with points evenlydistributed along the x-axis.
3. An object owned by a chart or graph that displays the names and styles of the plots.
4. Similar to a scope chart, except that a line sweeps across the display to separate new data from old data.
5. A graphical representation of data shown on a chart or graph.
7. The ability of graphs and charts to adjust automatically to the range of plotted values.
8. A general-purpose, Cartesian graph used to plot multivalued functions, with points evenly sampled or not.

Down

1. Displays one point at a time on one or more plots.
2. An object owned by a chart or graph from which you can change the scaling and format options while the VI is running.
4. A chart modeled after the operation of an oscilloscope.
6. A chart modeled after a paper strip chart recorder, which scrolls as it plots data.

CHAPTER 8

Data Acquisition

The subject of this chapter is data acquisition (DAQ). Focusing on the use of the Easy I/O VIs, the basic notions associated with analog and digital I/O are presented. Some common terms and concepts associated with data acquisition are also discussed, including the components of a DAQ system, signal conditioning, and the types of signals encountered.

GOALS

1. Review some basic notions of signals and signal acquisition.
2. Introduce the organization of the DAQ VIs.
3. Understand the basics of analog and digital input and output using Easy I/O VIs.

8.1 COMPONENTS OF A DAQ SYSTEM

In this chapter, we introduce some of the basic notions associated with **data acquisition**. The focus is on the LabVIEW VIs that can be used in a DAQ system—analog and digital signal **input/output** VIs. The subject of data acquisition cannot be adequately covered in one chapter, although the fundamental ideas can be introduced and discussed in enough detail to generate enthusiasm for pursuing other sources of information.[1] The most effective way to learn about data acquisition is by doing it. Reading about it is not enough.

Your LabVIEW Student Edition CDRom contains a folder at the top level called Chapter 8. *This is where the latest DAQVIs are stored. Copy the files from that folder to the* Chapter 8 *folder in the* Learning *directory. If you do not have easy access to the CDRom, download the latest VIs from the website by using the* ***help*** *pull-down menu and selecting* ***Student Edition Web Resources****.*

At its most basic level, the task of a DAQ system is the measurement or generation of physical signals. Two options for constructing a DAQ system are illustrated in Figure 8.1. Some students think that having a plug-in DAQ board

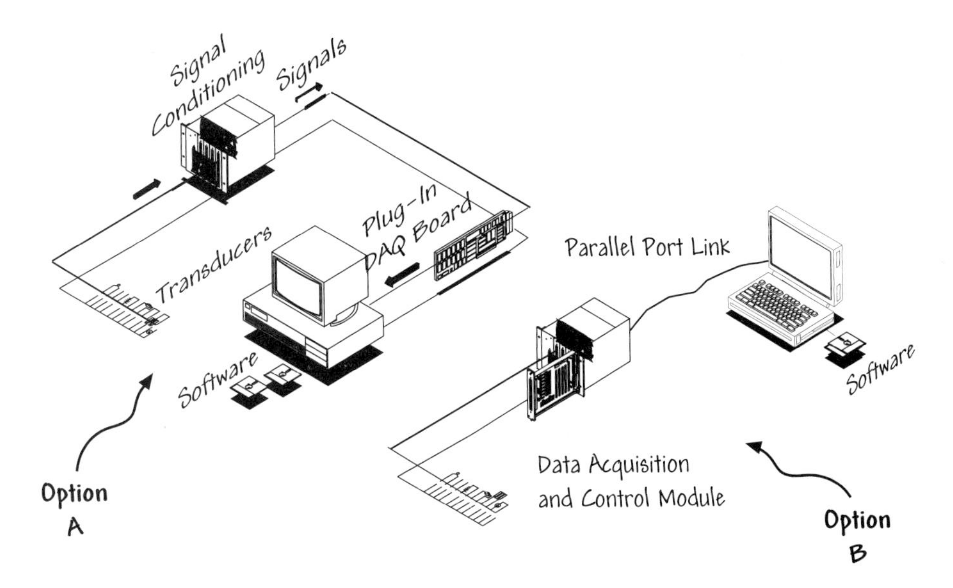

FIGURE 8.1
Two options for a DAQ system.

1. A good source of information for the beginner is the *Data Acquisition Basics Manual*, available from National Instruments, Inc. Part Number 320997C-01. For the advanced students, see Chapters 6 and 7 of *LabVIEW Graphical Programming, 2nd ed.*, by Gary W. Johnson (McGraw-Hill, New York, 1997).

properly installed in a personal computer is equivalent to having a complete DAQ system. This is not the case at all. In fact, the plug-in DAQ board is only one component of the system. A DAQ system generally has (in addition to the plug-in DAQ board) sensors and transducers, signal conditioning, and a suite of software for acquiring and manipulating the raw data, analyzing and displaying (and storing) the data.

In Option A (see Figure 8.1), the plug-in DAQ board resides in the computer. This computer can be a tower, desktop model, or a laptop with PCMCIA slots. In Option B, the DAQ board is external to the computer. With the latter approach you can build DAQ systems using computers without available plug-in slots (such as some laptops), and the computer and DAQ module communicate through various buses, such as the parallel port. Option B systems are usually more practical for remote data acquisition and control applications where you want to bring the DAQ system into the field.

Before a computer-based system can measure a physical signal, a sensor (or transducer) must convert the physical signal into an electrical signal (such as voltage or current). Generally, you cannot connect signals directly to a plug-in DAQ board in a computer, as you can with most stand-alone instruments. In many cases, the measured physical signal is very low-voltage and susceptible to noise. In these situations, the measured signal may need to be amplified and filtered before conversion to a digital format for use in the computer. A signal-conditioning accessory conditions measured signals before the plug-in DAQ board converts them to digital information. More will be said on signal conditioning later in the chapter. Software controls the DAQ system–acquiring the raw data, analyzing the data, and presenting the results.

8.2 TYPES OF SIGNALS

The concepts of signals and systems arise in a wide variety of fields, including science, engineering, and economics. In an effort to develop an analytic framework for studying certain natural phenomena, the notion of an input-output representation has arisen, and is illustrated in Figure 8.2. In the input-output

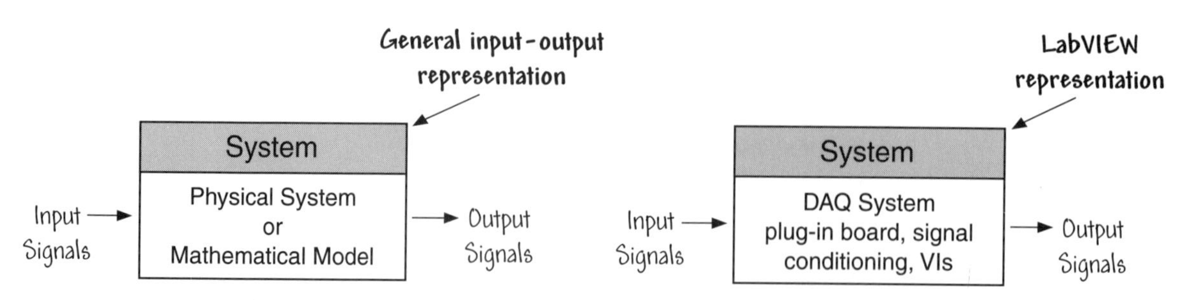

FIGURE 8.2
Modeling physical phenomena with input-output representations.

representation, the input signals are operated on by the system to produce the output signals. Signals are physical quantities that are functions of an independent variable (such as time) and contain information about a natural phenomena. For example, the input signals might be degraded and weak video signals received from a spacecraft approaching a distant asteroid. The system is the DAQ system, which includes the hardware and software to acquire, process and enhance the spacecraft signals (e.g., the spacecraft images of the asteroid can be enhanced to show features and colors more clearly), and the output signals would be the enhanced asteroid images. In general, physical signals are converted to electrical signals (such as voltage or current) before use by the signal-conditioning and DAQ hardware. This conversion is accomplished by some type of transducer. The list of common transducers includes video cameras, thermocouples, strain gauges, and thermistors. Once the physical signals are measured and converted to electrical signals, the information contained in them may be extracted and analyzed.

Five common classes of information that can be extracted from signals are illustrated in Figure 8.3. The signals evolve either continuously or only at discrete

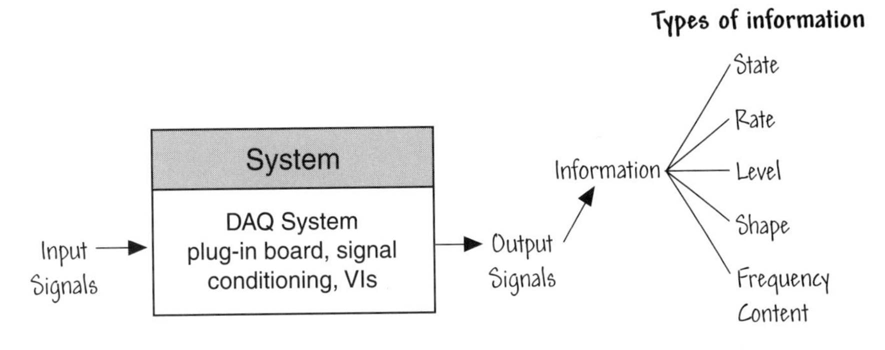

FIGURE 8.3
Measuring and analyzing signals provides information: state, rate, level, shape, and frequency content.

points in time and are known as continuous- or discrete-time signals, respectively. Room temperature is an example of a continuous-time signal (although we may discretize the temperature signal by recording the temperature only at specific sampling points). The end-of-day closing Dow-Jones stock index is an example of a discrete signal, taking on a new value once per working day.

For purposes of discussing data acquisition, we will use the following signal classifications:

- For digital signals, we have two types:
  - on-off
  - pulse train

- For analog signals, we have three types:
  - DC
  - time-domain
  - frequency-domain

The five signal types listed above are classified as analog or digital according to how they convey information. A digital signal has only two possible discrete levels: high level (on) or low level (off). An analog signal, on the other hand, contains information that varies continuously with respect to an independent variable (such as time).

A schematic with the various signal types is shown in Figure 8.4. Each signal type is unique in the information conveyed, and the five signal types closely parallel the five basic types of signal information: state, rate, level, shape, and frequency content.

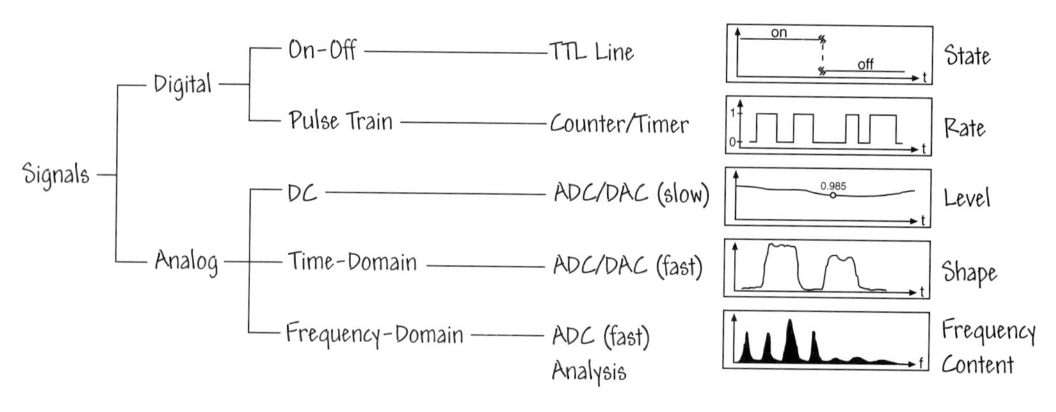

FIGURE 8.4
Signal types. (ADC → analog-to-digital converter, DAC → digital-to-analog converter, TTL → transistor-transistor logic)

8.2.1 Digital Signals

The two types of digital signals that we will consider here are the on-off switch and the pulse train signals. The on-off signal illustrated in Figure 8.5 conveys information concerning the immediate digital *state* of the signal. A simple digital state detector is used to measure this signal type. An example of a digital on-off signal is the output of a transistor-transistor logic (TTL) switch.

The second type of digital signal is the pulse train signal illustrated in Figure 8.6. This signal consists of a series of state transitions, and information is contained in the number of state transitions, the *rate* at which the transitions occur, and the time between one or more state transitions. The output signal of an optical encoder mounted on the shaft of a motor is an example of a digital pulse train signal.

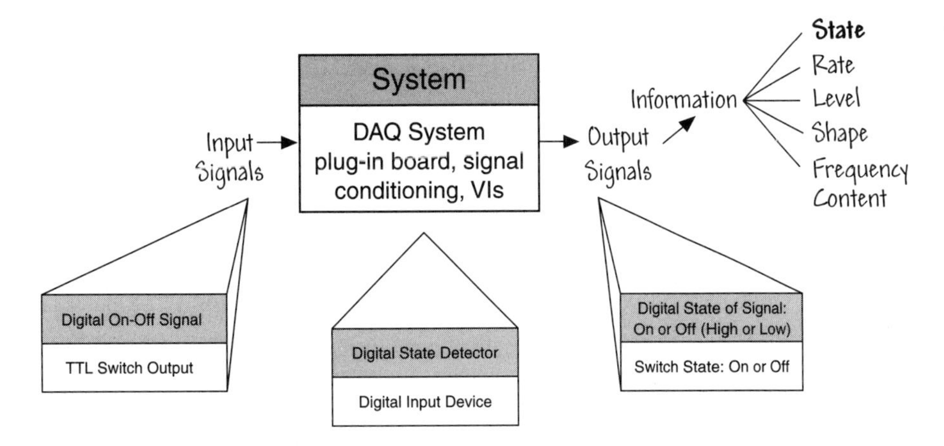

FIGURE 8.5
The on-off signal.

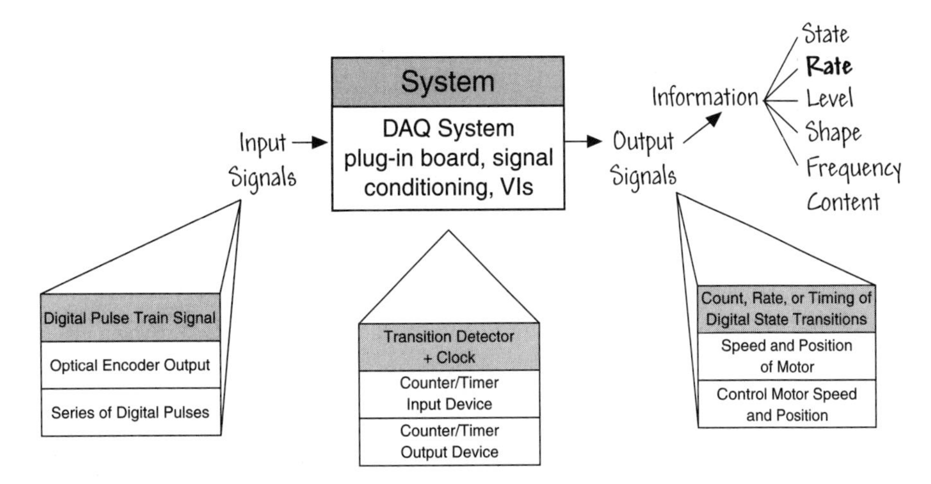

FIGURE 8.6
The pulse train signal.

8.2.2 Analog DC Signals

Analog DC signals are static (or slowly varying analog) signals that convey information in the *level* (or amplitude) of the signal at a given instant. The analog DC signal is depicted in Figure 8.7 in the context of the input-output representation. Since the analog DC signal is static or varies slowly, the accuracy of the measured level is of more concern than the time or rate at which you take the measurement. Common examples of DC signals include temperature, battery voltage, flow rate, pressure, strain-gauge output, and fluid level, as illustrated

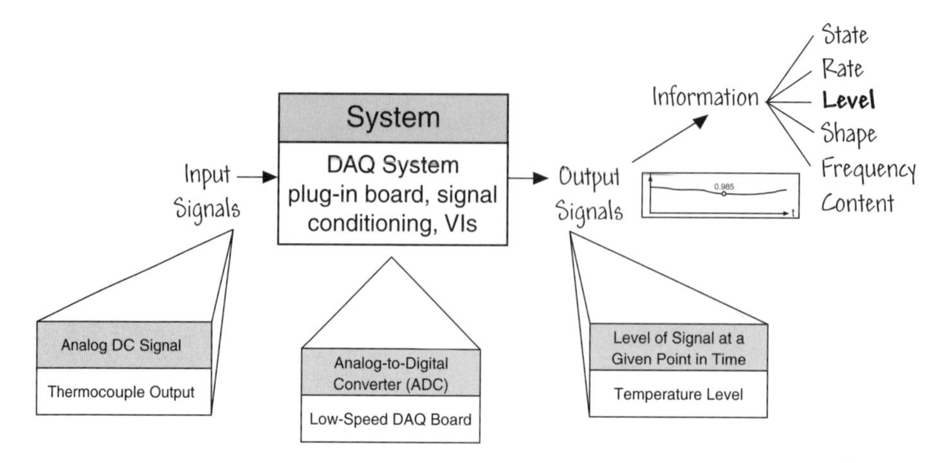

FIGURE 8.7
Analog DC signals are static or slowly varying analog signals.

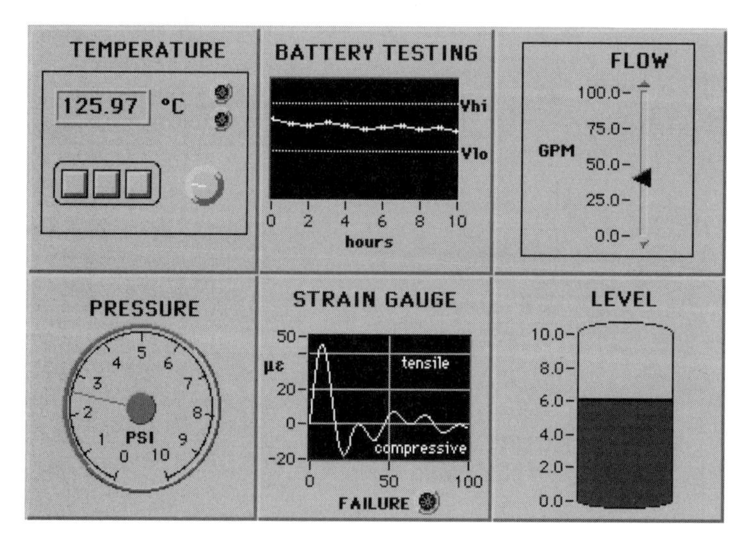

FIGURE 8.8
Common examples of DC signals include temperature, battery voltage, flow rate, pressure, strain-gauge output, and fluid level.

in Figure 8.8. The information contained in the signals is displayed on meters, gauges, strip charts, and numerical readouts.

The DAQ system will return a single value indicating the magnitude of the signal at the requested time. For the measurement to be an accurate representation of the signal, the DAQ system must possess adequate accuracy and resolution capability and adequate sampling rates (usually slow).

8.2.3 Analog Time-Domain Signals

Analog time-domain signals differ from other signals in that they convey useful information in signal level *and* the way this level varies with time. The information associated with a time-domain signal (also referred to as a waveform) includes such information as time to peak, peak magnitude, time to settle, slope, and shapes of peaks. An analog time-domain signal is illustrated in Figure 8.9.

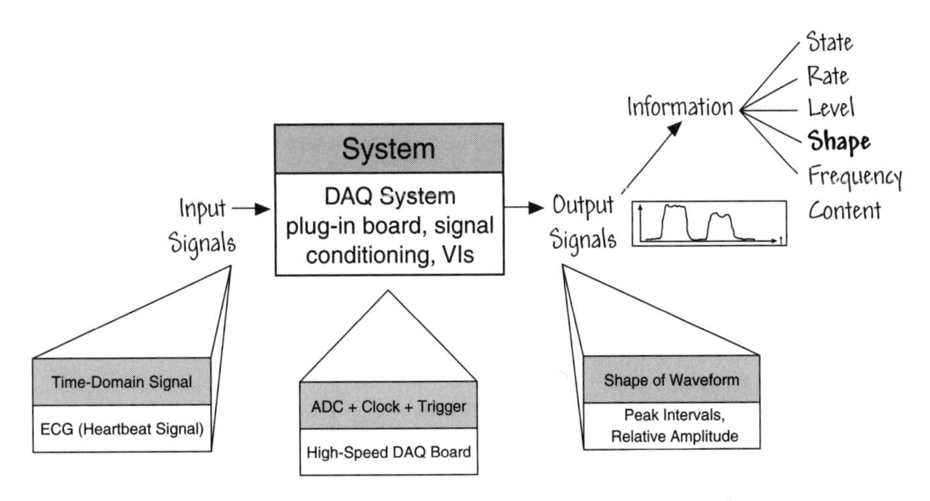

FIGURE 8.9
Analog time-domain signals convey information in the signal level and in the way this level varies with time.

You must take a precisely timed sequence of individual amplitude measurements to measure the shape of a time-domain signal. DAQ systems that are used to measure time-domain signals usually have an **analog-to-digital conversion** (ADC) component, a sample clock, and a trigger.

The physical signal must be sampled and measured at a rate that adequately represents the shape of the signal. Therefore, when acquiring analog time-domain signals, DAQ systems need a high-bandwidth ADC to sample the signal at high rates. The signal must also be sampled accurately without significant loss of precision.

Generally the signal measurements need to start at a specified time to guarantee that an interesting segment of the signal is acquired. The sample clock accurately times the occurrence of each analog-to-digital (A/D) conversion. In many situations, triggering is necessary to initiate the measurement process at a precise time. The trigger starts the measurement at the proper time based on some external condition specified by the user. There are an unlimited number of different time-domain signals, some of which are shown in Figure 8.10.

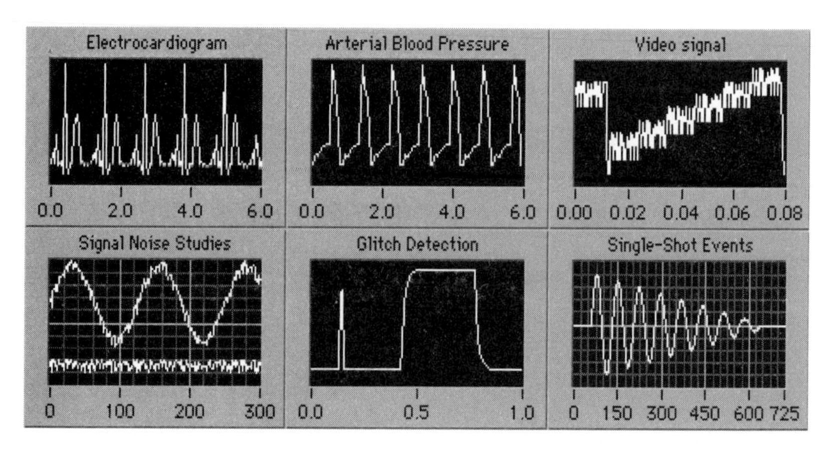

FIGURE 8.10
Six different time-domain signals.

8.2.4 Analog Frequency-Domain Signals

Analog frequency-domain signals are similar to time-domain signals in that they also convey information in the way the signals vary with time. However, the information extracted from a frequency-domain signal is based on the signal frequency content, as opposed to the shape of the signal. An analog frequency-domain signal is shown in Figure 8.11.

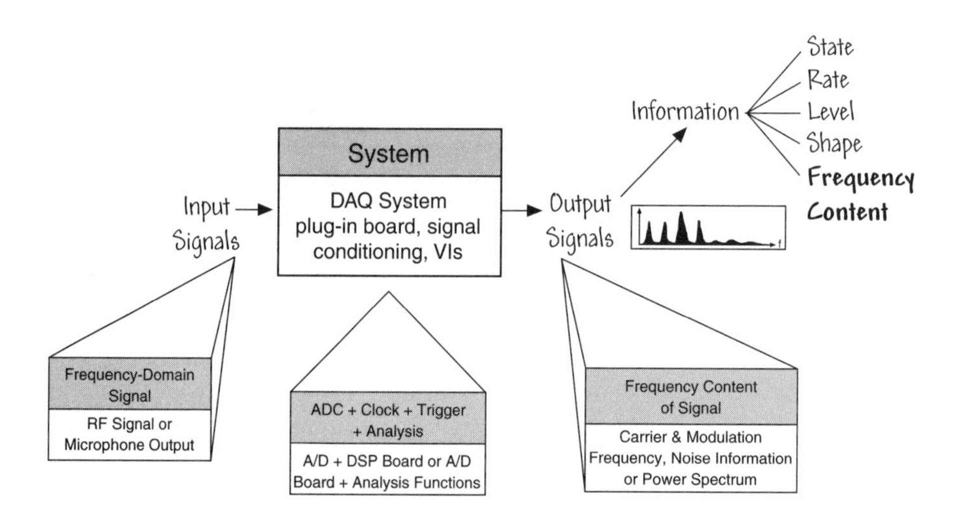

FIGURE 8.11
The information extracted from a frequency-domain signal is based on the signal frequency content.

As with DAQ systems that measure time-domain signals, a system used to measure a frequency-domain signal must also include an ADC, a sample clock, and a trigger to accurately capture the waveform. Additionally, the DAQ system must include the necessary analysis capabilities to extract frequency information from the signal. You can perform this type of digital signal processing (DSP) using application software or special DSP hardware designed to analyze the signal quickly and efficiently.

In short, a DAQ system that acquires analog frequency-domain signals should possess a high-bandwidth ADC capability to sample signals at high rates and an accurate clock to take measurements at precise intervals. Also, it is common to use triggers to initiate the measurement process precisely at prespecified times. Finally, a complete DAQ system should provide a library of analysis functions, including a function to convert time-domain information to frequency-domain information. Chapter 11 discusses some of the LabVIEW analysis functions, but time and space limitations make it impossible to cover all those available to VI developers.

Figure 8.12 shows some examples of frequency-domain signals. Each example in the figure includes a graph of the originally measured signal as it varies with respect to time, as well as a graph of the signal frequency spectrum. While you can analyze any signal in the frequency domain, certain signals and application areas lend themselves especially to this type of analysis. Among these areas are speech and acoustics analysis, geophysical signals, vibration, and studies of system transfer functions.

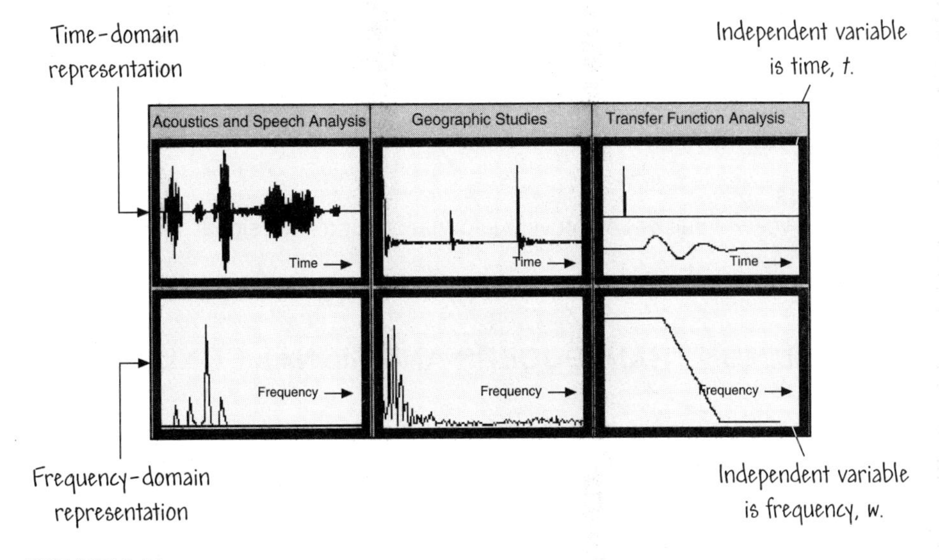

FIGURE 8.12
Three examples of frequency-domain signals: speech and acoustics analysis, geophysical signals, and transfer function frequency response.

8.2.5 One Signal—Five Measurement Perspectives

The five classifications of signals presented in the previous discussions are not mutually exclusive. A particular signal may convey more than one type of information, and you can classify a signal as more than one type and measure it in more than one way. In fact, you can use simpler measurement techniques with the digital on-off, pulse train, and DC signals, because they are just simpler cases of the analog time-domain signals.

The measurement technique you choose depends on the information you want to extract from the signal. In many cases you can measure the same signal with different types of systems, ranging from a simple digital input board to a sophisticated frequency analysis system. Figure 8.13 demonstrates how a series of voltage pulses can provide information for all five signal classes.

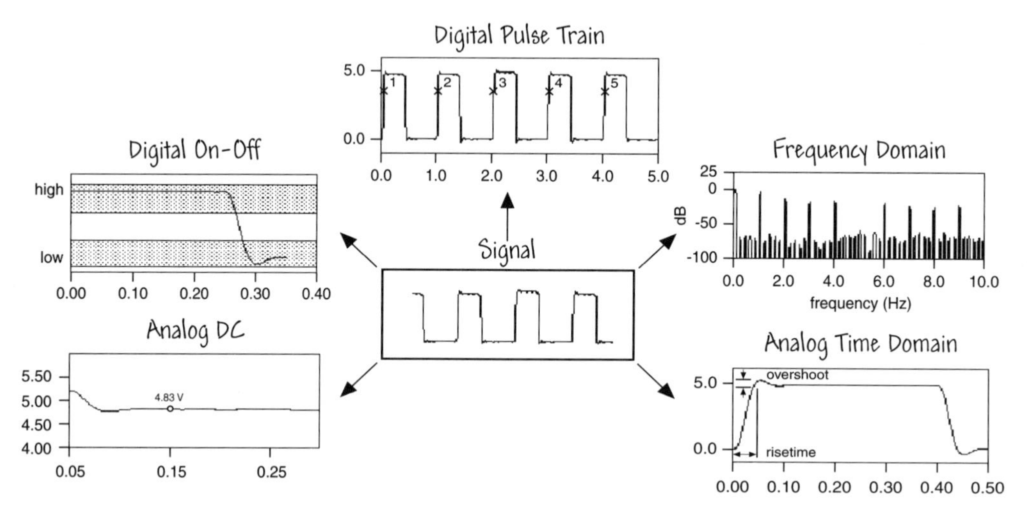

FIGURE 8.13
A series of voltage pulses can provide information for all five signal classes.

8.3 COMMON TRANSDUCERS AND SIGNAL CONDITIONING

When measuring a physical phenomena, a transducer must convert this phenomena (such as temperature or force) into a measurable electrical signal (such as voltage or current). Table 8.1 lists some common transducers used to convert physical phenomena into measurable quantities.

All transducers output an electrical signal that is often not well suited for a direct measurement by the DAQ system. For example, the output voltage of most thermocouples is very small and susceptible to noise, and often needs to be amplified and filtered before measuring. Figure 8.14 shows some common types of

TABLE 8.1 Phenomena and Transducers

Phenomenon	Transducer
Temperature	Thermocouples
	Resistance temperature detectors (RTDs)
	Thermistors
	Integrated circuit sensor
Light	Vacuum tube photosensors
	Photoconductive cells
Sound	Microphone
Force and pressure	Strain gauges
	Piezoelectric transducers
	Load cells
Position	Potentiometers
(displacement)	Linear voltage differential transformer (LVDT)
	Optical encoder
Fluid flow	Head meters
	Rotational flowmeters
	Ultrasonic flowmeters
pH	pH electrodes

transducer and signal pairs and the required signal conditioning. A highly expandable signal conditioning system—dubbed Signal Conditioning eXtensions for Instrumentation (**SCXI**) made by National Instruments, Inc.—conditions low-level signals in a noisy environment within an external chassis located near the sensors. The close proximity improves the signal-to-noise ratio of the signals reaching the DAQ boards. Of course, LabVIEW works with signal conditioning systems other than SCXI.

Some **signal conditioning** (such as linearization and scaling) can be performed in the software. LabVIEW provides several VIs in the **Data Acquisition** palette for such purposes. The remainder of this section is devoted to short descriptions of some of the basic ideas involved in signal conditioning. Further information on signal conditioning and SCXI hardware (i.e., setup procedures for SCXI hardware, hardware operating modes, and programming considerations for SCXI) can be found in other LabVIEW documentation (check the NI website).

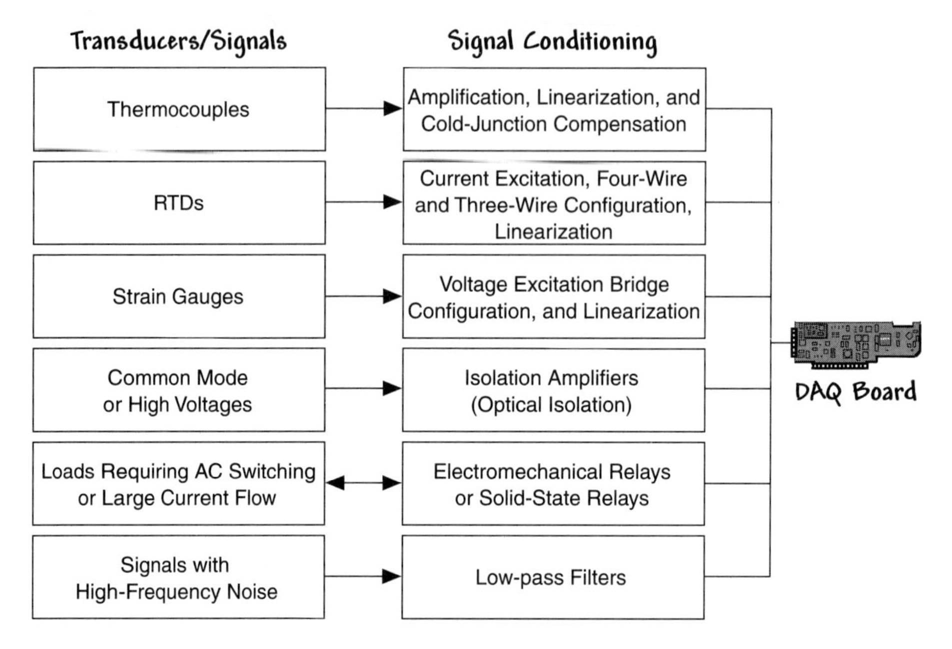

FIGURE 8.14
Common types of transducers/signals and the required signal conditioning.

Some common types of signal conditioning follow:

- Transducer excitation: Certain transducers (such as strain gauges) require external voltages or currents to excite their own circuitry in a process known as transducer excitation. The process is similar to a television needing power to receive and decode video and audio signals. The necessary excitation in a DAQ system can be provided by the plug-in DAQ boards and the signal-conditioning peripherals, and sometimes by external instruments.
- Linearization: Common transducers (such as strain gauges, thermistors, RTDs, and thermocouples) generate voltages that are nonlinear with respect to the phenomena they represent. The LabVIEW DAQ Utility VIs can perform software **linearization** to scale a transducer voltage to the correct units of strain or temperature.
- Isolation: Another common use for signal conditioning is to isolate the transducer signals from the computer. For example, when the signal being monitored contains large voltage spikes that could damage a computer or harm a person, you should not connect the signal directly to a DAQ board without some type of isolation. Figure 8.15 shows two common methods for isolating signals.
- Filtering: Another form of signal conditioning is the filtering of unwanted signals from the desired signal. A common filter reduces 60 Hz AC power-line noise present in many signals. Other well-known types of filters

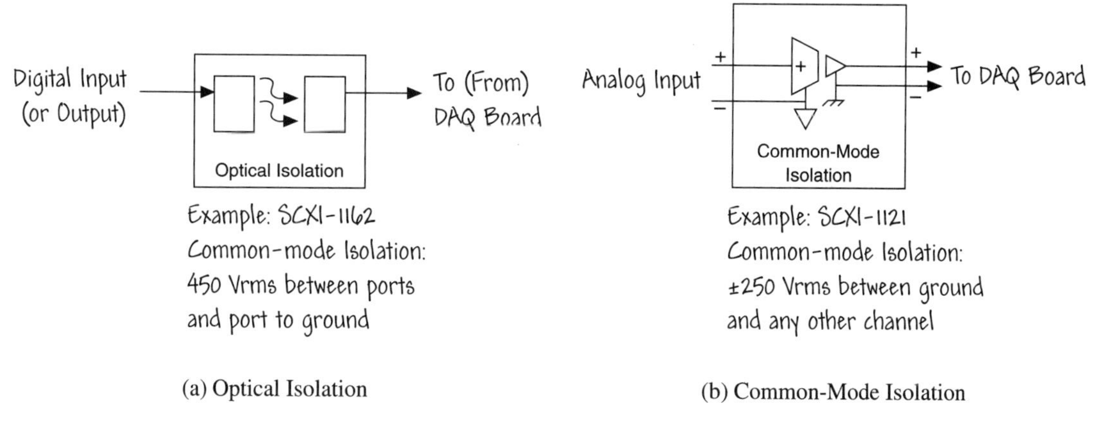

FIGURE 8.15
Two common methods for isolating signals.

include low-pass, high-pass, and notch filters. Some DAQ boards and signal-conditioning devices have built-in filters.

- Amplification: This is the most common type of signal conditioning. Amplification maximizes the use of the available voltage range to increase the accuracy of the digitized signal and to increase the signal-to-noise ratio (SNR). Low-level signals should be amplified at the DAQ board or at an external signal-conditioning peripheral positioned near the source of the signal, as shown in Figure 8.16. One reason to amplify low-level signals close to the signal source, instead of at the DAQ board, is to increase the signal-to-noise ratio. Consider the case where you amplify the signal only on the DAQ board. Then the DAQ board will also measure and digitize any noise that enters the lead wires along the path as the signal travels from the source to the DAQ board. On the other hand, the ratio of signal voltage to noise voltage that enters the lead wires is larger if you amplify the signal close to the signal source. Table 8.2 shows how the SNR changes with the location of amplification.

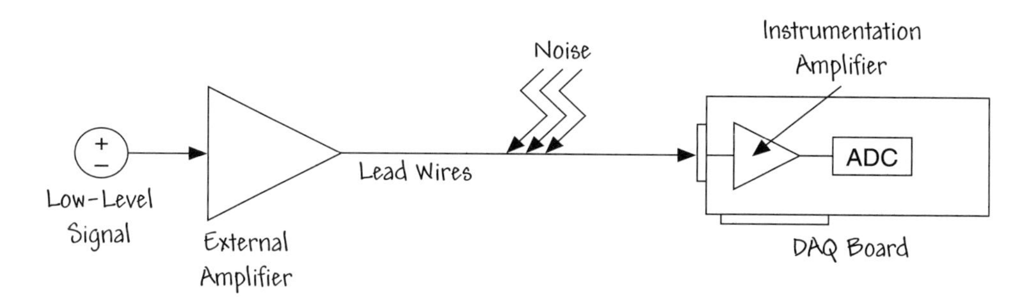

FIGURE 8.16
Low-level signals should be amplified at the DAQ board or at an external signal-conditioning peripheral positioned near the source of the signal.

TABLE 8.2 Effects of Amplification on Signal-to-Noise Ratio (S.C. → Signal-Conditioning Peripheral)

	Signal Voltage	S.C. Amplification	Noise in Lead Wires	DAQ Board Amplification	Digitized Voltage	SNR
Amplify only at DAQ board	0.01 V	None	0.001 V	×100	1.1 V	10
Amplify at S.C. and DAQ board	0.01 V	×10	0.001 V	×10	1.01 V	100
Amplify only at S.C.	0.01V	×100	0.001 V	None	1.001 V	1000

You can minimize the effects of external noise on the measured signal by using shielded or twisted-pair cables and by minimizing the cable length. Keeping cables away from AC power cables and computer monitors will also help minimize 50/60 Hz noise.

8.4 SIGNAL GROUNDING AND MEASUREMENTS

Up to this point, we have discussed three components of the DAQ system: signals, transducers, and signal conditioning. Now you might be tempted to think that all that remains is to wire the signal source to the DAQ board and begin acquiring data. However, a few important items must be considered:

- The nature of the signal source (grounded or floating)
- The grounding configuration of the amplifier on the signal conditioning hardware or DAQ board
- The cabling scheme to connect all the components together

A DAQ system is depicted in Figure 8.17 highlighting the signals and the cabling.

8.4.1 Signal Source Reference Configuration

Signal sources come in two forms: referenced and nonreferenced. Referenced sources are usually called **grounded signals**, and nonreferenced sources are called **floating signals**. A schematic of a grounded signal source is shown in Figure 8.18a.

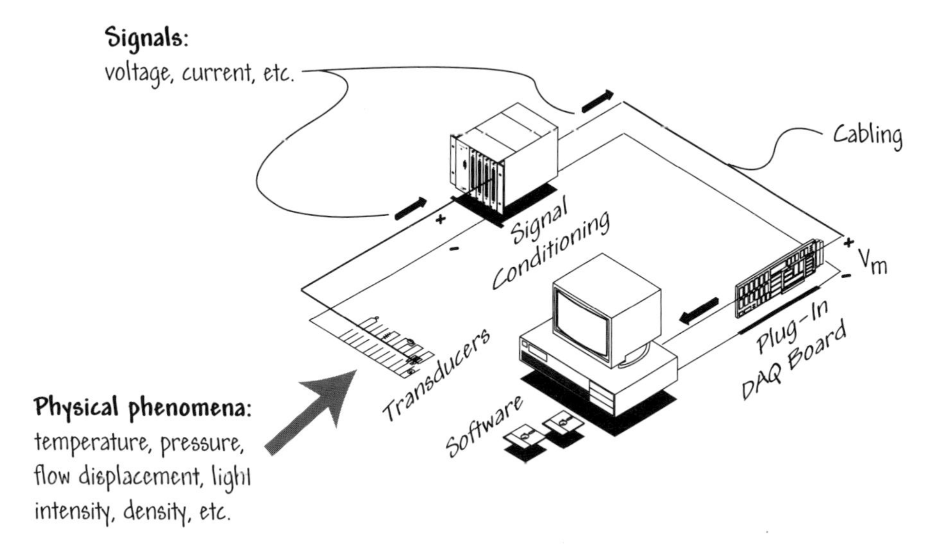

FIGURE 8.17
A DAQ system highlighting the signals and the cabling.

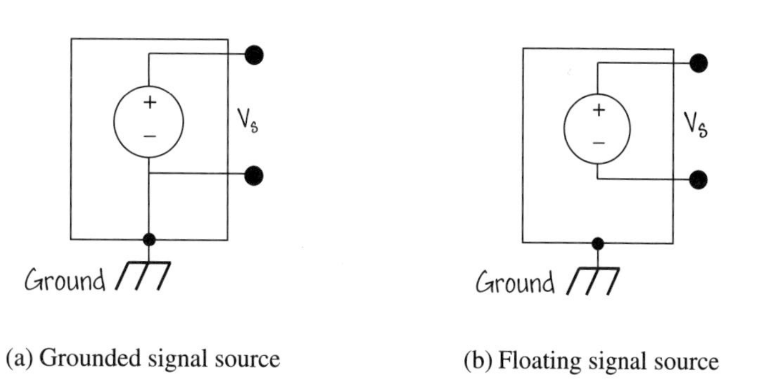

FIGURE 8.18
Grounded signal sources have voltage signals that are referenced to a system ground. Floating signal sources contain a signal that is not connected to an absolute reference.

Grounded signal sources have voltage signals that are referenced to a system ground, such as earth or a building ground. Devices that plug into a building ground through wall outlets, such as signal generators and power supplies, are the most common examples of grounded signal sources. Grounded signal sources share a common ground with the DAQ board.

Floating signal sources contain a signal (e.g., a voltage) that is not connected to an absolute reference, such as earth or a building ground. Some common examples of floating signals are batteries, battery-powered sources, thermocouples, transformers, isolation amplifiers, and any instrument that explicitly floats its output signal. As illustrated in Figure 8.18b, neither terminal of the floating source is connected to the electrical outlet ground.

8.4.2 Measurement System

A schematic of a measurement system is depicted in Figure 8.19. A measurement system can be placed in one of three categories:

- Differential
- Referenced single-ended (RSE)
- Nonreferenced single-ended (NRSE)

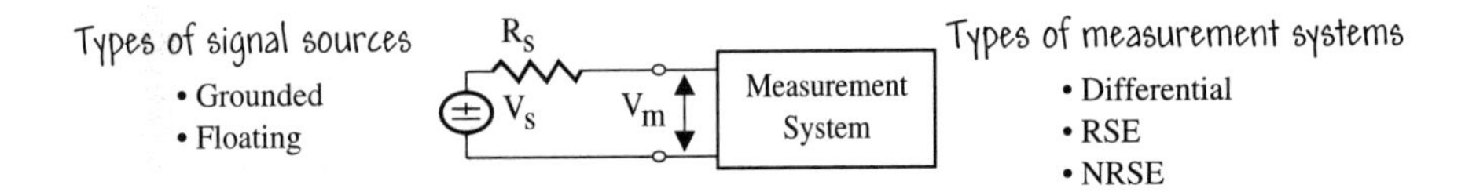

FIGURE 8.19
Types of signal sources and measurement systems.

In a **differential** measurement system, you do not need to connect either input to a fixed reference, such as earth or a building ground. DAQ devices with instrumentation amplifiers can be configured as differential measurement systems. Figure 8.20 depicts the 8-channel differential measurement system used in the MIO series devices. A **channel** is a pin or wire lead where analog or digital signals enter or leave a DAQ device. For this device, the analog input ground pin labeled AIGND is the measurement system ground. The analog multiplexers (labeled MUX in the figure) increase the number of available measurement channels while still using a single instrumentation amplifier.

An ideal differential measurement system, shown in Figure 8.21, reads only the potential difference between its two terminals—the (+) and (−) inputs. It completely rejects any voltage present at the instrumentation amplifier inputs with respect to the amplifier ground. In other words, an ideal differential measurement system completely rejects the common-mode voltage.

A **referenced single-ended** (RSE) measurement system measures a signal with respect to building ground and is sometimes called a grounded measurement system. Figure 8.22 depicts a 16-channel version of an RSE measurement system.

DAQ devices often use a **nonreferenced single-ended** (NRSE) measurement system, which is a variation of the RSE measurement system. In an NRSE measurement system, all measurements are made with respect to a common reference, because all of the input signals are already grounded. Figure 8.23 depicts an NRSE measurement system. AISENSE is the common reference for taking measurements and all signals in the system share this common reference. AIGND is the system ground.

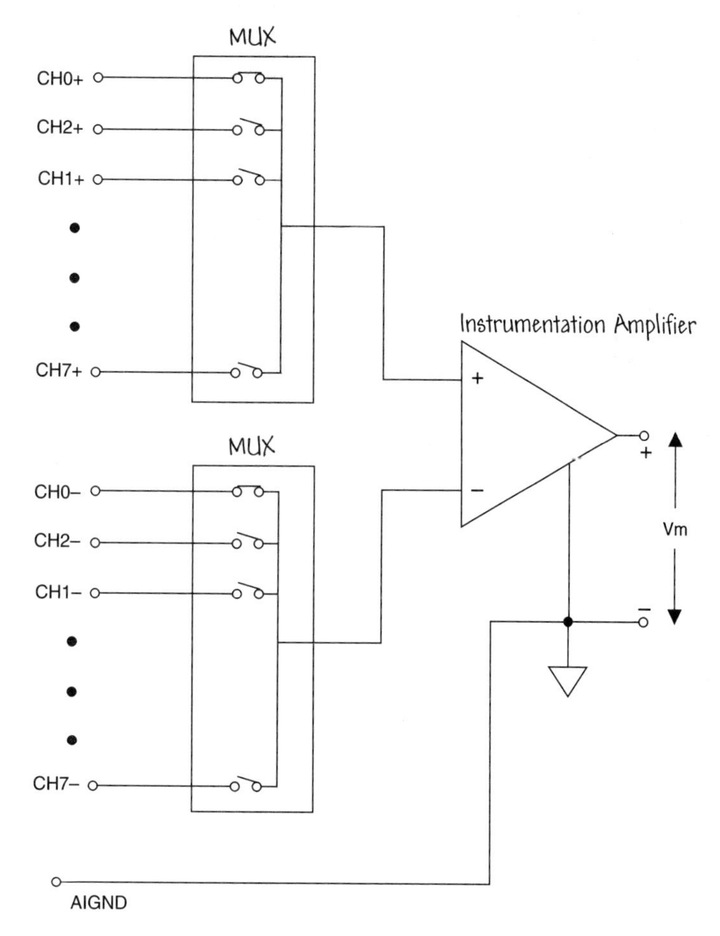

FIGURE 8.20
An 8-channel differential measurement system.

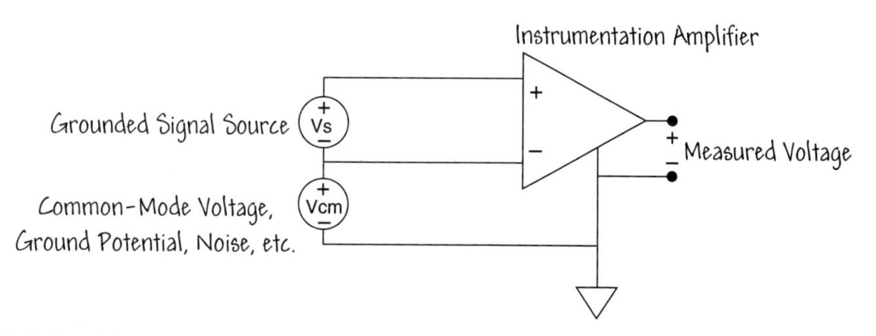

FIGURE 8.21
An ideal differential measurement system completely rejects common-mode voltage.

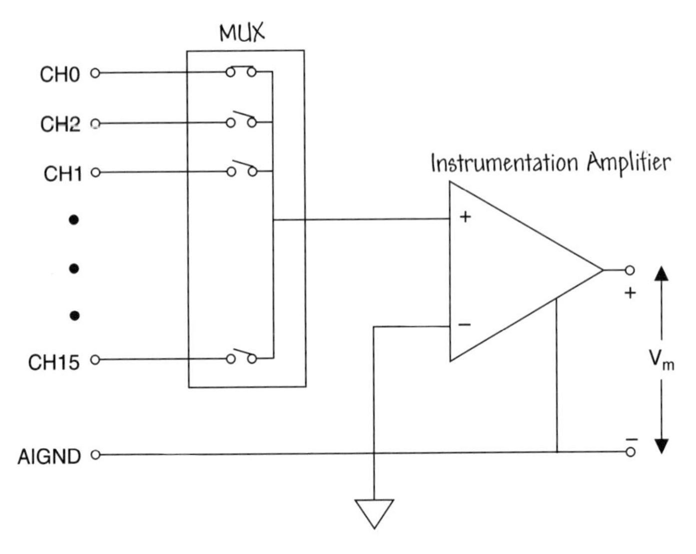

FIGURE 8.22
A 16-channel RSE measurement system.

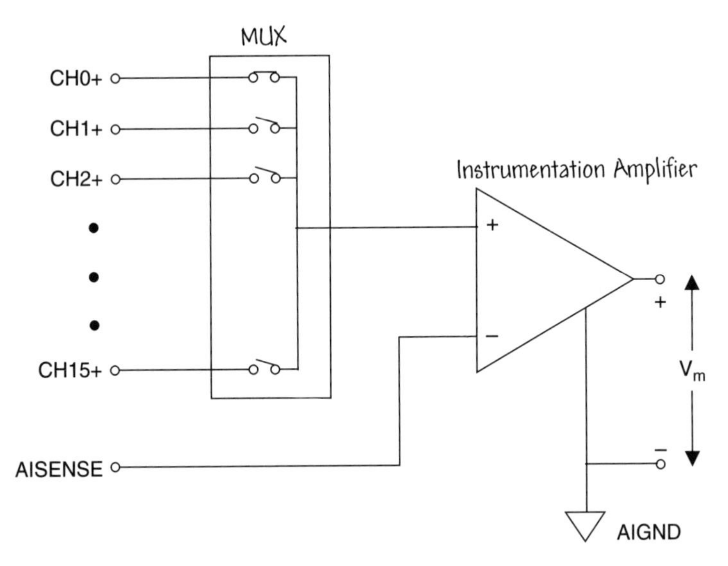

FIGURE 8.23
An NRSE measurement system.

A grounded signal source is best when using a differential or NRSE measurement system. An RSE measurement system can be used with a grounded signal source if the signal levels are high and the cabling has a low impedance. The measured signal voltage will be degraded with the RSE system, but the degradation is usually small relative to the signal voltage. Floating signal sources can be measured with differential, RSE, and NRSE measurement systems. In general, a differential measurement system is preferable because it rejects the ground loop-induced errors and reduces the noise picked up in the environment. On the other hand, the single-ended configuration allows for twice as many measurement channels and is acceptable when the magnitude of the induced errors is smaller than the required accuracy of the data. You can use single-ended measurement systems when all input signals meet the following criteria:

1. High-level signals (normally, greater than 1 V).
2. Short or properly shielded cabling traveling through a noise-free environment (normally less than 15 ft).
3. All signals can share a common reference signal at the source.

A summary of analog input connections is given in Figure 8.24.

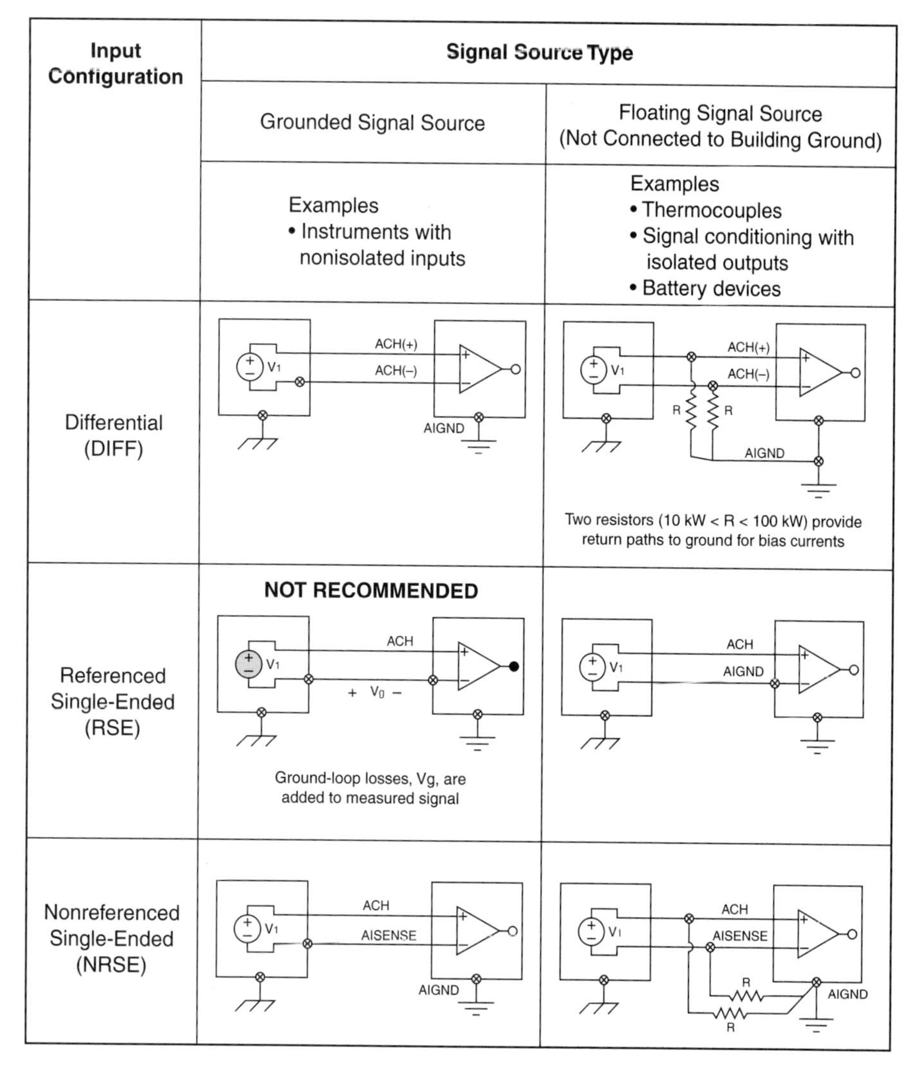

FIGURE 8.24
Summary of analog input connections.

8.5 ANALOG I/O CONSIDERATIONS

When preparing to configure a DAQ board you need to consider the quality of the analog-to-digital conversion. There are many questions that arise when making analog signal measurements with your DAQ system. For example, what are the signal magnitude limits? How fast does the signal vary with time? The latter question is important because the sample rate determines how often the A/D conversions take place. The four parameters of concern are

- Resolution
- Range
- Signal limit settings
- Sampling rate

Depending on the type of DAQ board you have, these four parameters will be set either in the hardware (using dip switches and jumpers on the board) or set using software (with the **Measurement and Automation Explorer** discussed later in this chapter).

The number of bits used to represent an analog signal determines the **ADC resolution** of the analog-to-digital conversion. You can compare the resolution on a DAQ device to the number of divisions on a ruler. For a fixed ruler length, the more divisions you have on the ruler, the more precise the measurements that you can make. A ruler marked off in millimeters can be read more accurately than a ruler marked off in centimeters. Similarly, the higher the ADC resolution, the higher the number of divisions of the ADC range and, therefore, the more accurately the analog signal can be represented. For example, a 3-bit ADC divides the signal range into 2^3 divisions, with each division represented by a binary or digital code between 000 and 111. The ADC then translates measurements of the analog signal to one of the digital divisions. Figure 8.25 shows a sine wave digital image obtained by a 3-bit ADC. Clearly, the digital signal does not represent the original signal adequately because there are too few divisions to represent the varying voltages of the analog signal. By increasing the resolution to 16 bits, however, the ADC's number of divisions increases from 2^3 to 2^{16}. The ADC can now provide an adequate representation of the analog signal.

The **range** refers to the minimum and maximum analog signal levels that the ADC can handle. You should attempt to match the range to that of the analog input signal to take best advantage of the available resolution. Fortunately, many DAQ devices feature selectable ranges. Consider the example shown in Figure 8.26a, where the 3-bit ADC has eight digital divisions in the range from 0 to 10 volts. If you select a range of -10.00 to $+10.00$ volts, as shown in Figure 8.26b, the same ADC now separates a 20-volt range into eight divisions. The

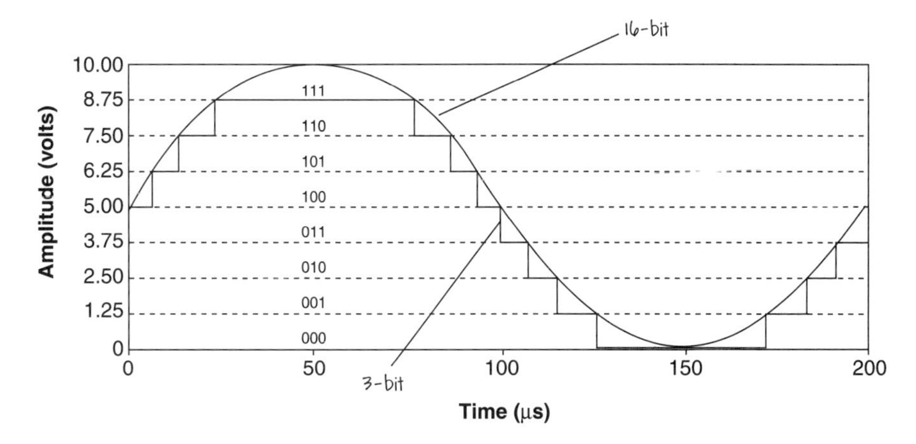

FIGURE 8.25
16-bit versus 3-bit resolution (5 kHz sine wave).

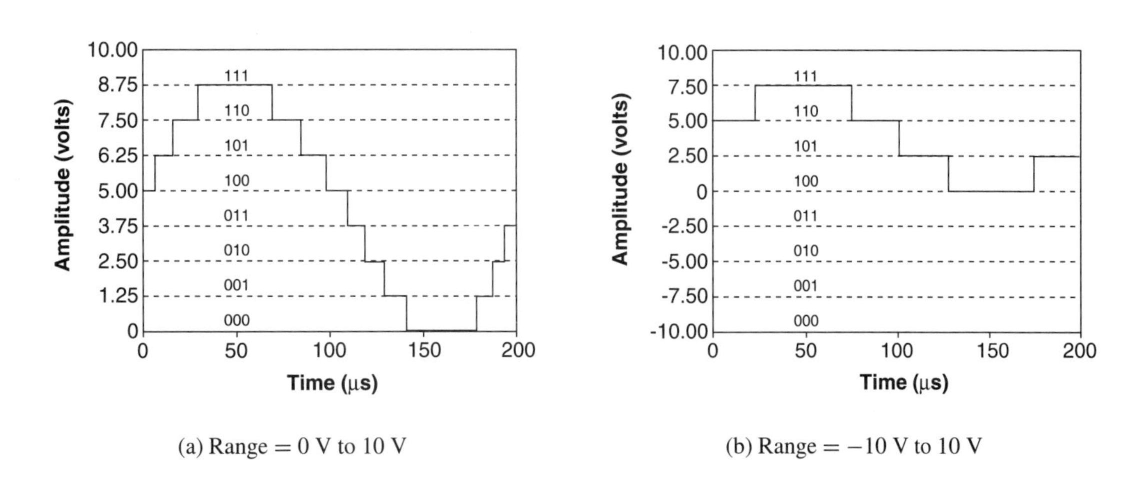

FIGURE 8.26
The 3-bit ADC in (a) has eight digital divisions in the range from 0 to 10 volts—if you select a range of −10.00 to 10.00 volts as in (b), the same ADC now separates a 20-volt range into eight divisions.

smallest detectable voltage increases from 1.25 to 2.50 volts, and you now have a much less accurate representation of the signal.

The **limit settings** are the maximum and minimum values of the signal you are measuring. The closer the limit setting is to the incoming analog signal maximum and minimum, the more digital divisions will be available to the ADC to represent the signal. Using a 3-bit ADC and a range setting of 0.00 to 10.00 volts, we see in Figure 8.27 the effects of a limit setting between 0 and 5 volts and 0 and 10 volts. With a limit setting of 0 to 10 volts, the ADC uses only four of the eight divisions in the conversion; with a limit setting of 0 to 5 volts, the ADC now has

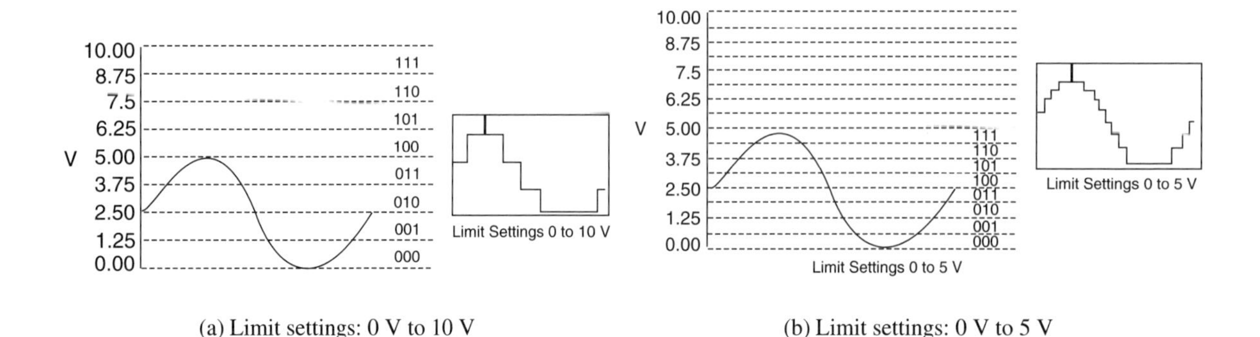

(a) Limit settings: 0 V to 10 V

(b) Limit settings: 0 V to 5 V

FIGURE 8.27
Precise limit setting allows the ADC to use more digital divisions to represent the signal.

access to all eight digital divisions. This makes the digital representation of the signal more accurate.

The resolution and range of a DAQ board and the limit settings determine the smallest detectable change in the input voltage. This change in voltage represents 1 **least significant bit** (LSB) of the digital value and is often called the **code width.** The smallest code width is calculated with the following formula:

$$V_{cw} = \frac{range}{2^{resolution}},$$

where the resolution is given in bits. For example, a 12-bit DAQ board with a 0 to 10-V range detects a 2.4-mV change. This is calculated as follows:

$$V_{cw} = \frac{range}{2^{resolution}} = \frac{10}{2^{12}} = 2.4 \text{ mV},$$

while the same board with a −10 to 10-V range detects only a change of 4.8 mV:

$$V_{cw} = \frac{range}{2^{resolution}} = \frac{20}{2^{12}} = 4.8 \text{ mV}.$$

A high-resolution ADC provides a smaller code width for a given range. For example, consider the two preceding examples, except that the resolution is now 16-bit. A 16-bit DAQ board with a 0 to 10-V range detects a 0.15-mV change:

$$V_{cw} = \frac{range}{2^{resolution}} = \frac{10}{2^{16}} = 0.15 \text{ mV},$$

while the same board with a −10 to 10-V range detects only a change of 0.3 mV:

$$V_{cw} = \frac{range}{2^{resolution}} = \frac{20}{2^{16}} = 0.3 \text{ mV}.$$

You also need to determine whether your signal is unipolar or bipolar, as this affects the code width. **Unipolar** signals range from 0 V to a positive value (e.g., 0 V to 10 V). **Bipolar** signals range from a negative value to a positive value (e.g., −5 V to 5 V). A smaller code width is obtained by specifying the range to be unipolar, if indeed the signal is unipolar; and conversely, specifying the range as bipolar if the signal is bipolar. If the maximum and minimum variation of the analog signal is smaller than the value of the range, you will need to set the limit settings to more accurately reflect the analog signal variation.

Some confusion may exist about selecting the limit settings rather than selecting the gain. *When you set the signal limit settings, you are effectively selecting the gain for that signal. Limit settings automatically magnify the magnitude of the signal to create more precise analog-to-digital conversions.*

The **sampling rate** is the rate at which the DAQ board samples an incoming analog signal. Figure 8.28 shows an adequately sampled signal as well as the effects of undersampling. The sampling rate determines how often an analog-to-digital (A/D) conversion takes place. Computing the proper sampling rate requires knowledge of the maximum frequency of the incoming signal and the accuracy required of the digital representation of the analog signal. It also requires some knowledge of the noise affecting the incoming signal and the capabilities of your hardware. A fast sampling rate acquires more points in a given time, allowing, in general, a better representation of the original signal than a slow sampling rate would allow. In fact, sampling too slowly may result in a misrepresentation of the incoming analog signal.

The effect of undersampling is that the signal appears as if it has a different frequency than it truly does. This misrepresentation is called an *alias*. To prevent

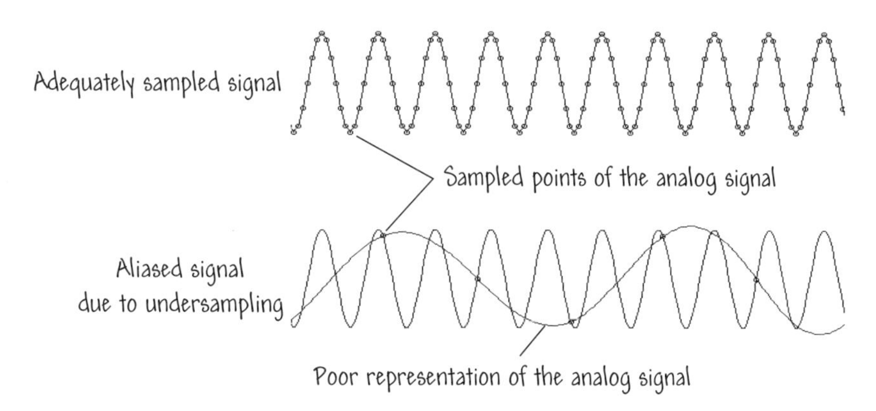

FIGURE 8.28
Sampling too slowly may result in a poor representation of the analog signal.

undersampling, you must sample, at a minimum, twice the rate of the maximum frequency component of the incoming signal. One way to deal with aliasing is to use low-pass filters that attenuate any frequency components in the incoming signal above the Nyquist frequency (defined to be one-half the sampling frequency). These filters are known as *antialiasing filters*.

The kind of calculations required to select the signal limits for a DAQ application are illustrated in this example. Keep in mind that the objective is to minimize the code width while making sure that the entire signal fits within the allowable device range.

Assume your transducer output is a sine wave with an amplitude of ±30 mV and an offset of 10 mV. Your board has limit settings of 0 to +10 V, ±10 V, and ±5 V. What limit setting would you select for maximum precision if you use a DAQ board with 12-bit resolution?

Table 8.3 shows how the code width of a hypothetical 12-bit DAQ device varies with range and limit settings. The values in Table 8.3 depend on the hardware, and you should consult your DAQ hardware documentation to determine the available device voltage range and limit settings for your device.

Step 1: Select a range.
Since the signal input is bipolar (the sine wave contains both positive and negative voltages), we must choose one of the bipolar ADC ranges. For our board, there are two, so to start we choose ±5 V, for a range of 10 V.

Step 2: Determine the code width.
Using the formula for determing the code width, we compute

$$V_{cw} = \frac{range}{2^{resolution}} = \frac{10}{2^{12}} = 2.4 \text{ mV}.$$

Step 3: Choose a limit setting.
Since our signal has a maximum magnitude of +30 mV, we must choose input settings that allow us to read ±30 mV. Referring to Table 8.3, we find that (for a range of −5 to 5 V) a limit setting of −50 mV to 50 mV will cover the signal variation of ±30 mV. Choosing this limit setting yields a code width (or precision) of 24.4 μV.

Step 4: Repeat steps 1 through 3 for range of ±10 V.
We must also try the other range of ±10 V to see if it yields a smaller code width. Repeating steps 1 through 3 gives us a limit setting of −0.1 V to 0.1 V and a code width of 48.8 μV. This does not improve our accuracy. Thus, we choose the settings as follows:

- Range = −5 to 5 V
- Limit setting = −50 mV to 50 mV

TABLE 8.3 Code Width for Various Ranges and Limit Settings

Voltage Range	Limit Settings	Code Width
0 to 10 V	0 to 10 V	2.44 mV
	0 to 5 V	1.22 mV
	0 to 2.5 V	610 μV
	0 to 1.25 V	305 μV
	0 to 1 V	244 μV
	0 to 0.1 V	24.4 μV
	0 to 20 mV	4.88 μV
−5 to 5 V	−5 to 5 V	2.44 mV
	−2.5 to 2.5 V	1.22 mV
	−1.25 to 1.25 V	610 μV
	−0.625 to 0.625 V	305 μV
	−0.5 to 0.5 V	244 μV
	−50 to 50 mV	24.4 μV
	−105 to 10 mV	4.88 μV
−10 to 10 V	−10 to 10 V	4.88 mV
	−5 to 5 V	2.44 mV
	−2.5 to 2.5 V	1.22 mV
	−1.25 to 1.25 V	610 μV
	−1 to 1 V	488 μV
	−0.1 to 0.1 V	48.8 μV
	−20 to 20 mV	9.76 μV

8.6 DAQ VI ORGANIZATION

The LabVIEW DAQ VIs are organized into palettes corresponding to the type of operation involved—analog input, analog output, or digital I/O, as illustrated in Figure 8.29. You access the palette by clicking on **Data Acquisition** in the **Functions** palette. The DAQ VIs are organized into six palettes:

- **Analog Input**
- **Analog Output**

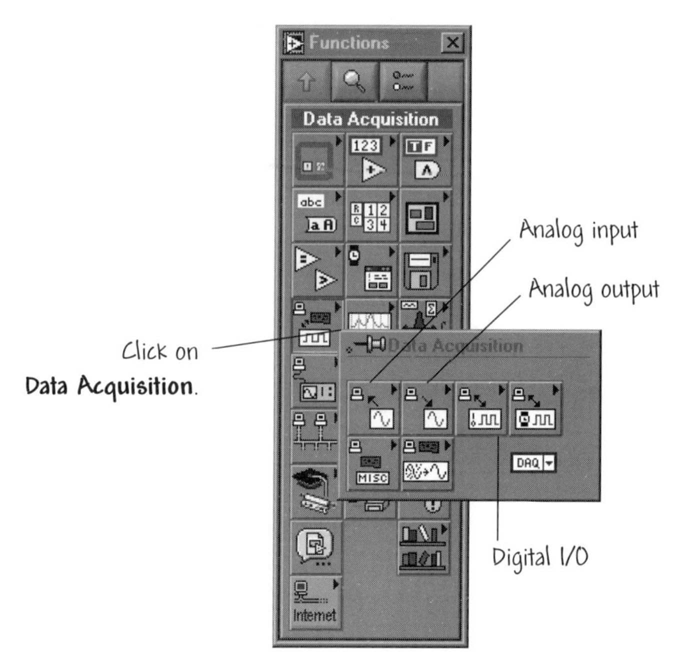

FIGURE 8.29
The **DAQ Acquisition** functions palette.

- **Digital I/O**
- **Counter**
- **Calibration and Configuration**
- **Signal Conditioning**

In this book, we will focus on the first three palettes: **Analog Input**, **Analog Output**, and **Digital I/O**.

Each palette contains VIs or palettes of VIs organized as Easy I/O, Intermediate, Utility, and Advanced VIs. The **Analog Input** palette shown in Figure 8.30 illustrates the organization. The top tier of VIs contains Easy I/O VIs, and the bottom tier contains Intermediate VIs. There are also two subpalettes: one for the **Analog Input Utility** VIs and one for the **Advanced Analog Input** VIs.

The Easy I/O VIs consist of high-level VIs that perform basic analog input, analog output, and digital I/O. They are ideal for getting started with DAQ in LabVIEW. The Easy I/O VIs include a simplified error-handling method so that when a DAQ error occurs in your VI, a dialog box shows error information. When the box appears, you will have the option to halt execution of the VI or to ignore the error.

Compared to the Easy I/O VIs, the Intermediate VIs have more hardware functionality and flexibility. They feature capabilities that the Easy I/O VIs lack, such as external timing and triggering. As you become more familiar with

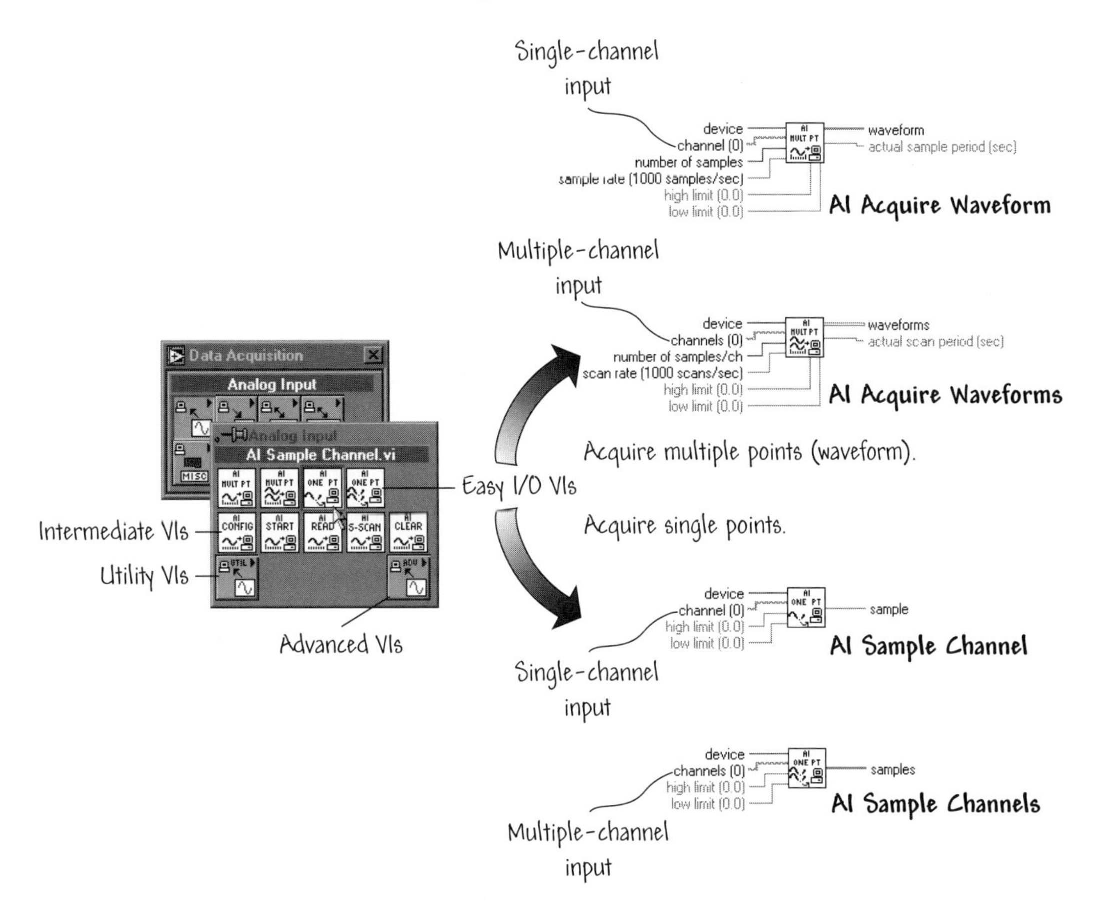

FIGURE 8.30
The **Analog Input** palette.

LabVIEW, you will discover that the Intermediate VIs are better suited for most of your applications. The Intermediate VIs feature more flexible error handling than the Easy I/O VIs in that you can pass error status information to other VIs and handle errors programmatically. The Utility VIs consist of convenient groupings of the Intermediate VIs. They will be helpful in situations where you need more control than the Easy I/O VIs provide.

The Advanced VIs are the lowest-level VIs and are required by few applications. In general, the Easy I/O and Intermediate VIs will suffice for most DAQ applications. However, with the Advanced VIs you have more control over the DAQ board operation, and you have access to more status information from the software driver.

The **Analog Output** palette is shown in Figure 8.31. The organization of the palette is similar to the **Analog Input** palette: the top tier of VIs contains Easy I/O VIs, and the bottom tier contains Intermediate VIs. There are also two subpalettes: one for the **Analog Output Utility** VIs and one for the **Advanced Analog Output** VIs.

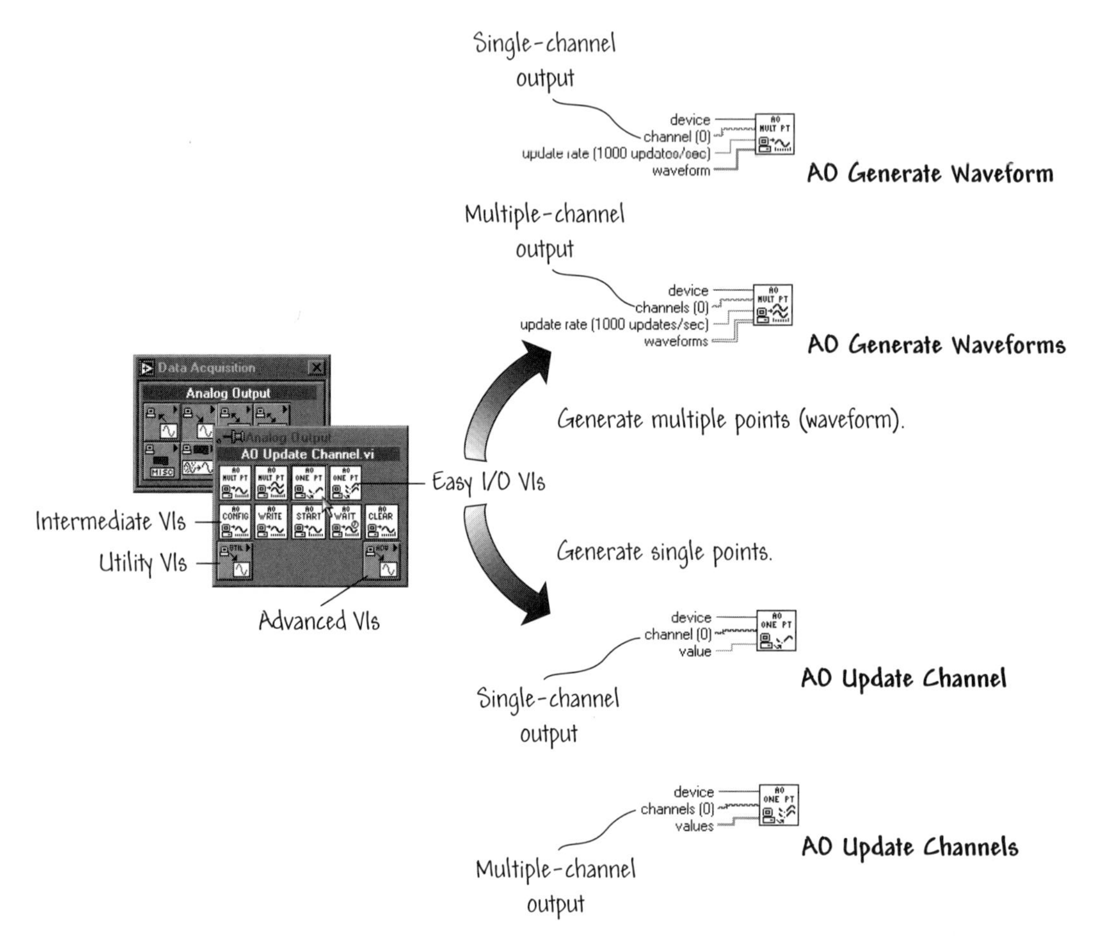

FIGURE 8.31
The **Analog Output** palette.

The **Digital I/O** palette is shown in Figure 8.32. The top tier of VIs contains Easy I/O VIs, and the bottom tiers contain Intermediate and Advanced VIs.

8.7 DAQ HARDWARE CONFIGURATION

LabVIEW provides utilities designed to help you define which signals are connected to which channels on your data acquisition board. In previous years, significant amounts of time were spent defining the signal types, connections, and transducer equations—and all this before beginning development of the actual DAQ system! For example, if you are using thermocouples, you must perform cold-junction compensation (CJC) calculations and apply appropriate scaling factors to convert raw measured voltages into actual temperature readings. This process is now one of entering the necessary information in dialog boxes to define

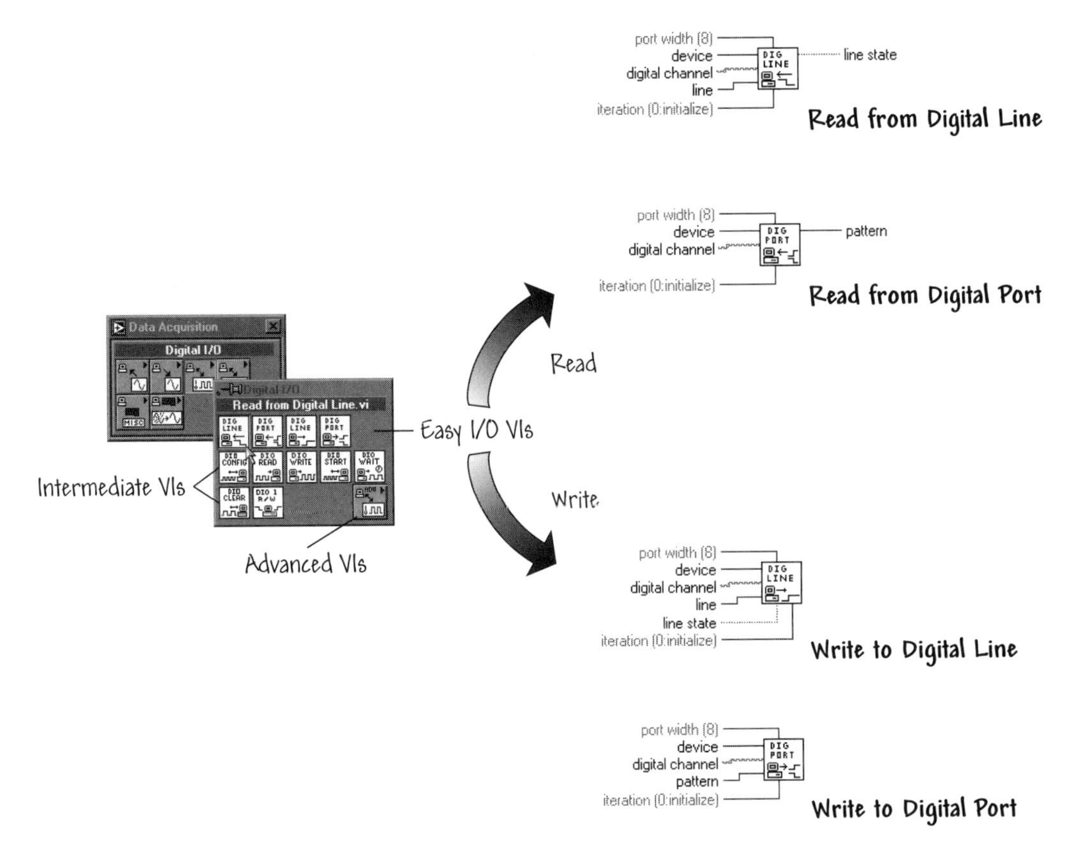

FIGURE 8.32
The **Digital I/O** palette.

an input signal, the type of transducer being used, any scaling factors required, CJC values, and the conversion factors. You can then reference the channel name (that you assign) for the input signal and the conversion from voltage to physical units is performed automatically (and transparently). You can save different configuration files for different settings or systems. Not only can you assign **channel names**, sensor types, engineering units, and scaling information to each channel using the LabVIEW utility, but you can also define the physical quantities you are measuring on each DAQ hardware channel. Once the software has been properly configured, the hardware will be configured correctly to make the measurement for each channel in terms of the physical quantity.

8.7.1 Windows

LabVIEW for Windows installs a configuration utility for establishing all board and channel configuration parameters. This utility is known as the **Measurement & Automation Explorer**, or MAX for short. After installing a DAQ board in

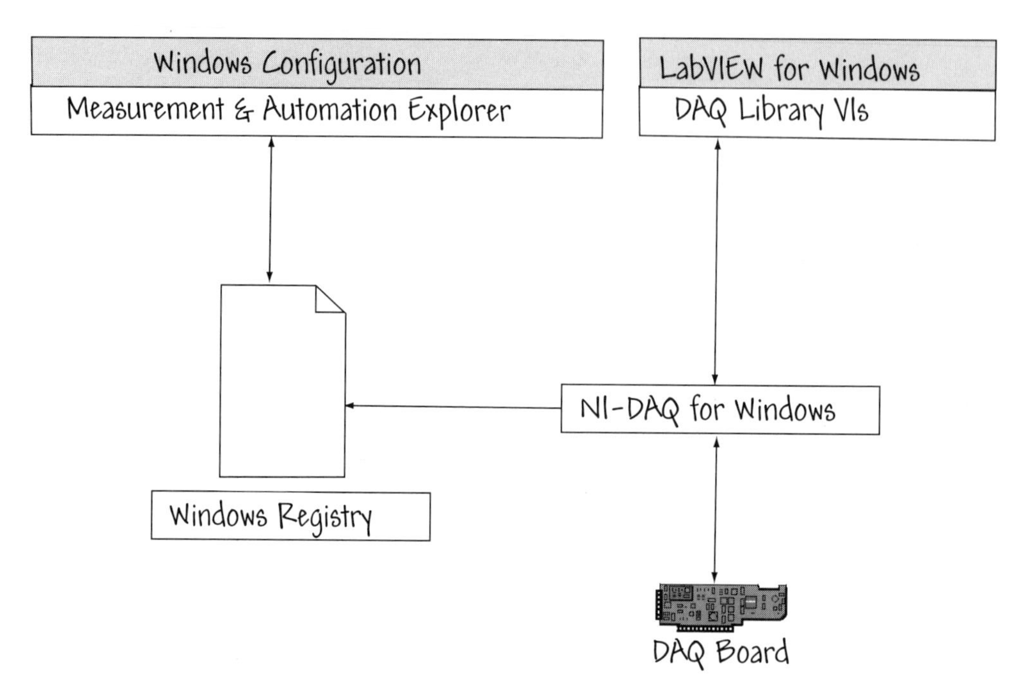

FIGURE 8.33
Windows configuration management.

your computer, you must run this configuration utility. The MAX utility reads the information the Device Manager records in the Windows registry and assigns a logical device number to each DAQ board. You use the device number to refer to the board in LabVIEW. Figure 8.33 shows the relationship between the DAQ board and MAX. The Windows Configuration Manager keeps track of all the hardware installed in your system, including National Instruments DAQ boards.

If you have a Plug & Play (PnP) board, the Windows Configuration Manager automatically detects and configures the board. If you have a non-PnP board (known as a Legacy device) you must configure the board manually using the ***Add New Hardware*** *option under the Control Panel.*

You can check the Windows Configuration by accessing the Device Manager (Start≫Settings≫Control Panel≫System≫Device Manager). You will find **Data Acquisition Devices**, which lists all DAQ boards installed in your computer as shown in Figure 8.34. Highlight a DAQ board and select **Properties** or double-click on the board, and you see a dialog window with tabbed pages. **General** displays overall information regarding the board. You use **Resources** to specify the system resources to the board such as interrupt levels, DMA, and base address for software configurable boards. **NI-DAQ Information** specifies

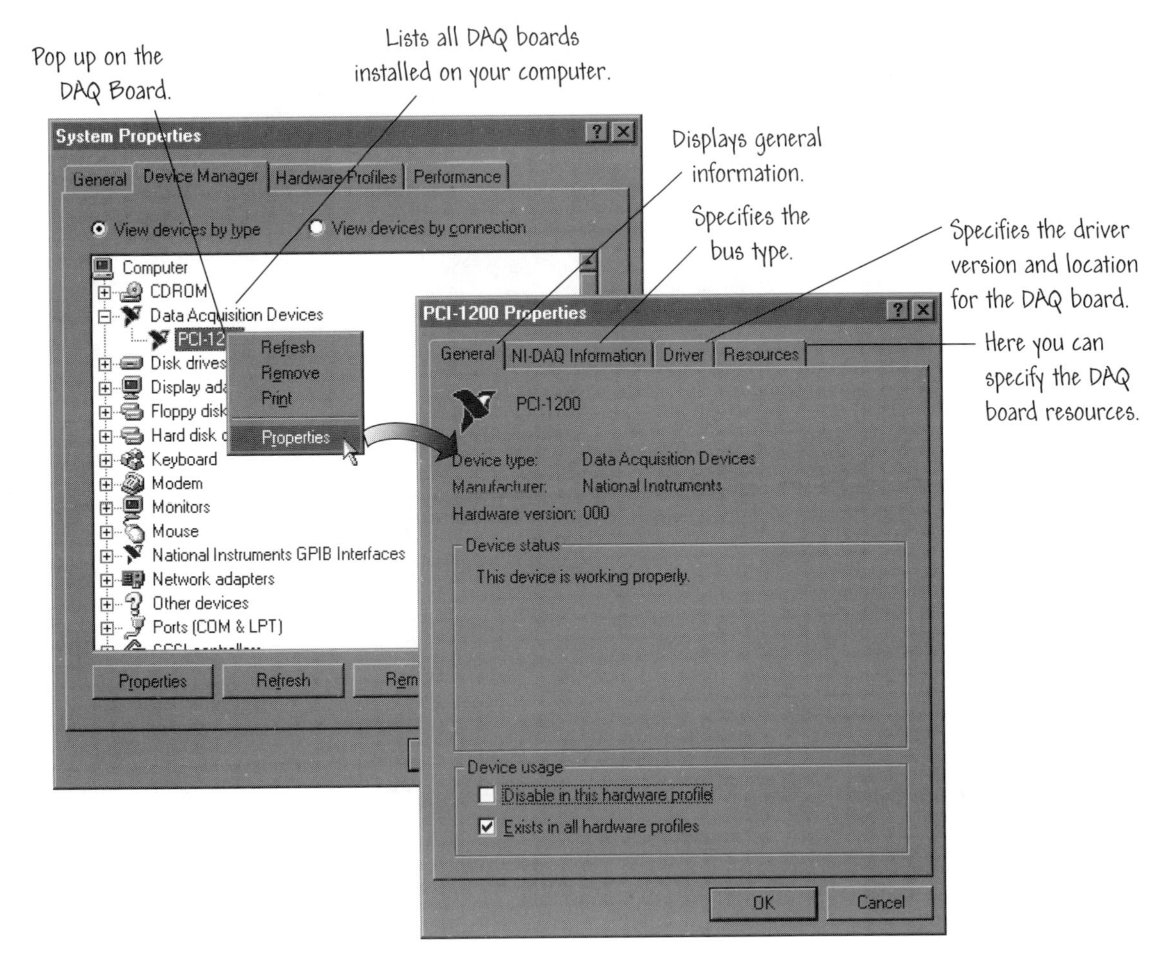

FIGURE 8.34
Checking the Windows Configuration by Accessing the Device Manager.

the bus type of your DAQ board. **Driver** specifies the driver version and location for the DAQ board.

LabVIEW for Windows DAQ VIs access the National Instruments standard NI-DAQ for Windows 32-bit dynamic link library (DLL). The LabVIEW setup program installs the NI-DAQ DLL in the WINDOWS\SYSTEM directory. NI-DAQ for Windows supports all National Instruments DAQ boards and SCXI.

The nidaq32.dll file, the high-level interface to your board, is loaded into the Windows\System directory. The nidaq32.dll file then interfaces with the Windows Registry to obtain the configuration parameters defined by the MAX. You access MAX either by double-clicking on its icon on the desktop or selecting **Measurement & Automation Explorer** from the **Tools** menu in LabVIEW, as illustrated in Figure 8.35.

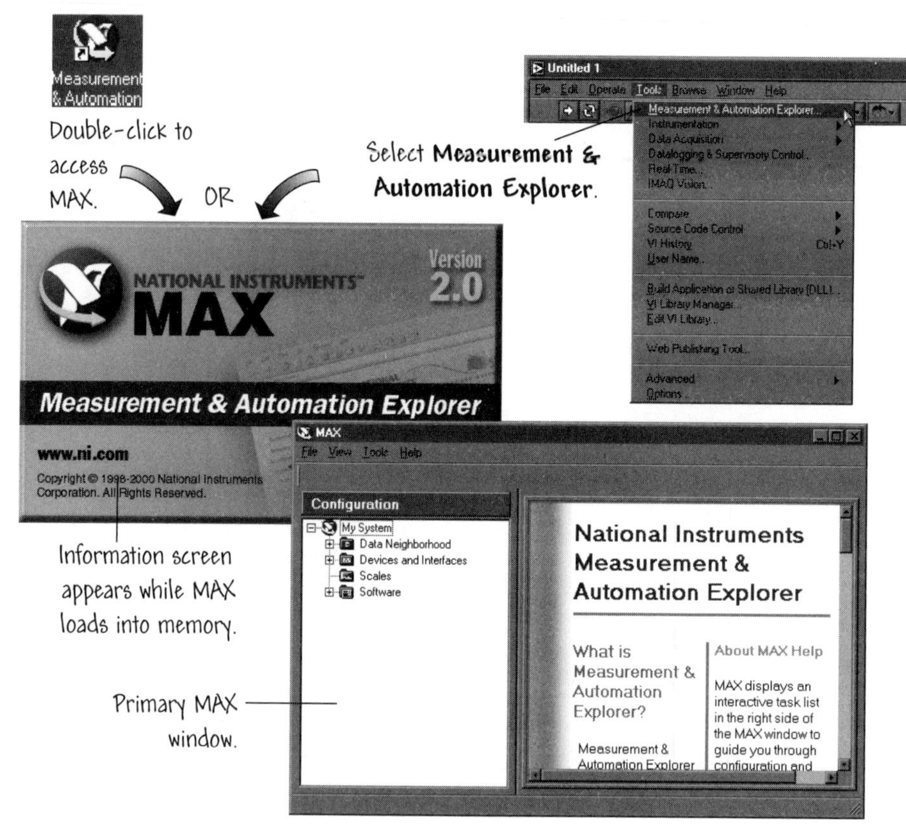

FIGURE 8.35
Accessing the Measurement & Automation Explorer (MAX).

8.7.2 Macintosh

The LabVIEW installation program installs the NI-DAQ for Macintosh software drivers necessary to communicate with National Instruments DAQ boards. You use the NI-DAQ Configuration utility to configure your DAQ board and accessories. Figure 8.36 shows the relationship between the DAQ boards and the NI-DAQ Configuration utility.

When you install NI-DAQ for Macintosh, install version 4.9 if you have an NB or a Lab Series board. Otherwise, install NI-DAQ version 6.6 or later for the PCI and DAQCard boards.

Using the MAX (Windows)

The objective of this exercise is to use the MAX to examine the configuration for the DAQ board in you computer and to configure one virtual channel.

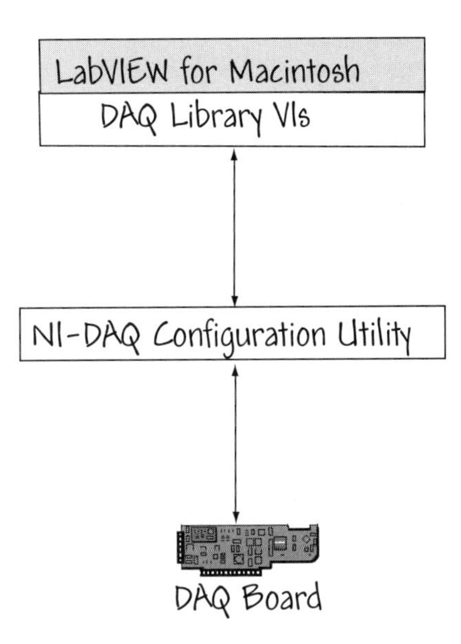

FIGURE 8.36
Macintosh configuration management.

Start MAX by double-clicking on the desktop icon or by selecting **Measurement & Automation Explorer** from the **Tools** menu in LabVIEW, as illustrated in Figure 8.35.

Depending on your system, MAX may be installed in a different location. On the Macintosh, the NI-DAQ configuration utility will be in a separate folder on your hard drive.

Open the Devices and Interfaces section as seen in Figure 8.37. The figure shows what MAX looks like for a PCI-1200 and a PCI-GPIB. The MAX window shows the National Instruments boards and software in your system. Note the Device number indicated in the parantheses after the DAQ board. The LabVIEW DAQ VIs use this Device number to determine which board performs DAQ operations. In Figure 8.37, we see that the DAQ board PCI-1200 is Device 1. You may have a different board installed and some of the options shown may be different. You can get more information about the board's configuration by examining its properties. With the DAQ board highlighted, select the **Properties** button just below the menu. A configuration window for the multiple input/output (MIO) board is shown in Figure 8.37.

The simulated DAQ is not configurable! To use the MAX you must have real hardware installed. If you are using simulated DAQ you will not see entries in the MAX for a simulated board or a data neighborhood. There are predefined

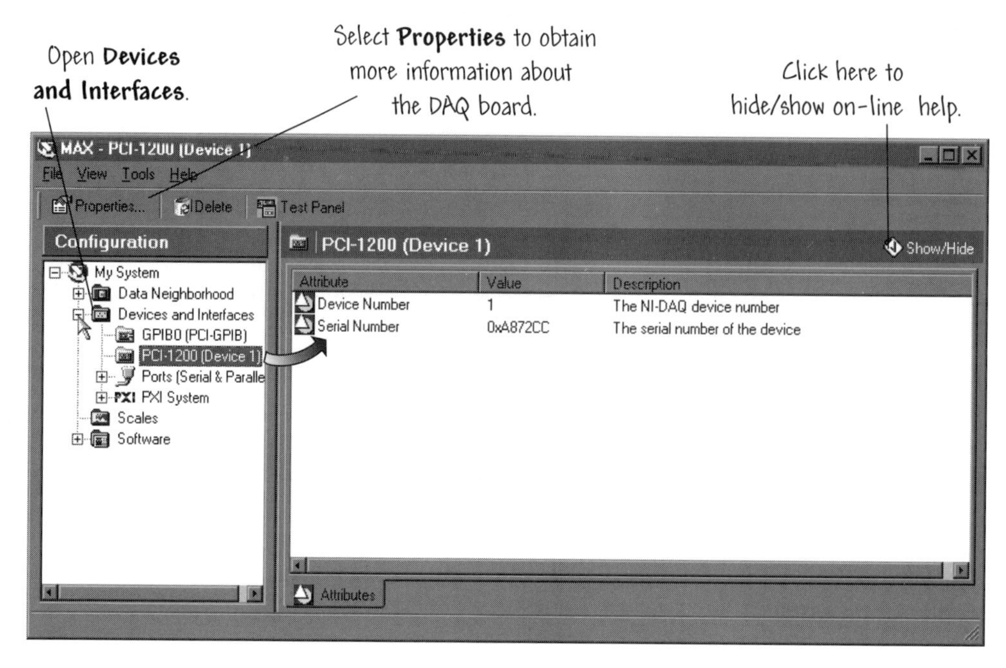

FIGURE 8.37
Checking the properties of the DAQ boards.

virtual channels which you can use in DAQ VIs, however these virtual channels cannot be modified.

*Press the **Show/Hide** button in the top right corner of the MAX window to hide the online help and show the DAQ board information.*

The configuration window depicted in Figure 8.38 contains several tabs. **System** reports the system resources assigned to the board through the Windows registry. The remaining tabs are used to configure the various analog input, output, and accessory parameters for the DAQ board. The **Test Resources** button tests the system resources assigned to the board according to the Windows Device Manager. Close the **Configuring Device 1:PCI-1200** window and refer the MAX window. Next, we want to configure a virtual channel. A virtual channel is a shortcut to configuring a channel which allows you to give descriptive names to physical channels. Right-click on the Data Neighborhood icon and choose **Create New...**. Select **Virtual Channel** and press the **Finish** button, as depicted in Figure 8.39. You will now configure a channel to take a reading from the temperature sensor (Analog Input Channel 0). After pressing the **Finish** button, a window for configuring a new channel will appear, and you can enter the virtual channel name as shown in Figure 8.40. Name the channel "Temperature" and add a short description as seen in Figure 8.41. Each new page of the setup is accessed by clicking on the **Next** button, and you can always go back to a previous page using the **Back** button. Once you have named the virtual channel, select **Next** and

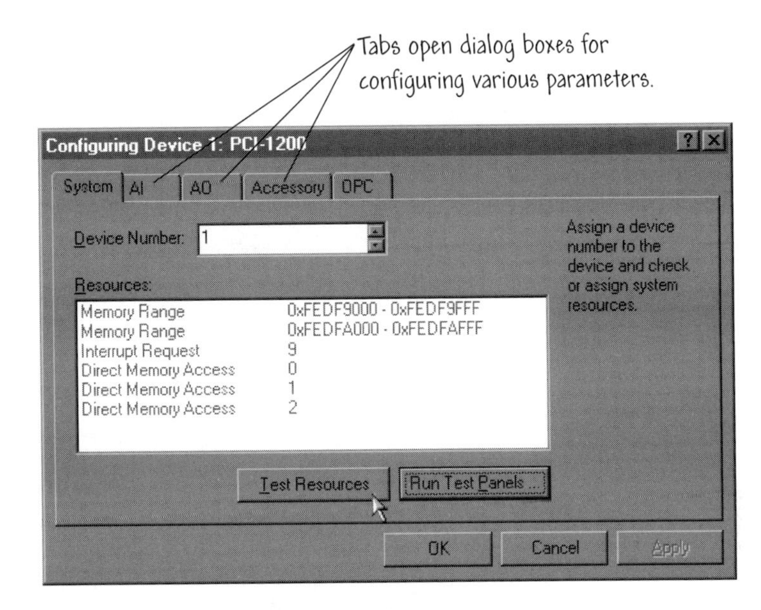

FIGURE 8.38
DAQ board properties.

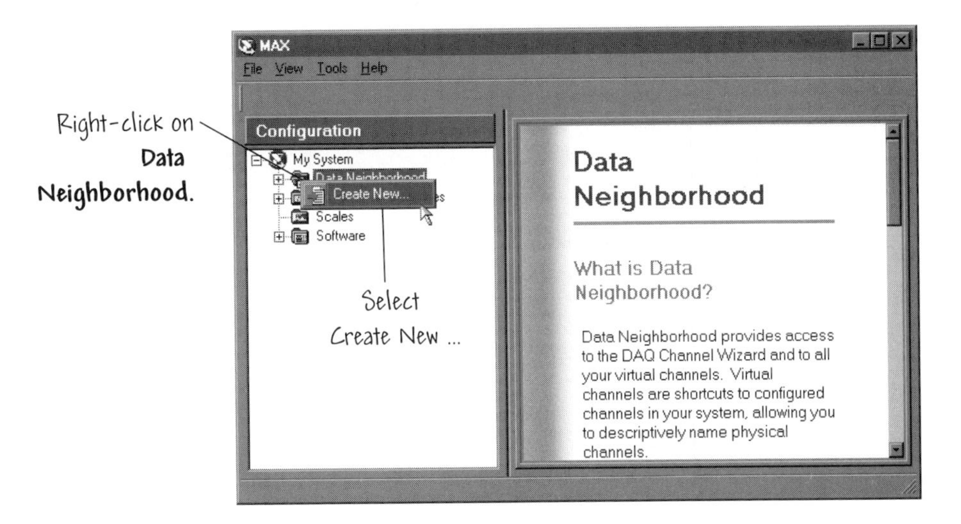

FIGURE 8.39
Configuring a new virtual channel.

move to the sensor type, as illustrated in Figure 8.42. You have many choices of sensor types, including voltage, current, resistance, frequency, thermocouple, RTD and accelerometer. In our example, we choose voltage. Once you have defined the physical quantity of the temperature sensor—click on the **Next** button. If your temperature sensor generates a voltage that is scaled to Deg C, you input the unit of measurement as Deg C, as shown in Figure 8.43. In this exercise use Deg C as the unit of measurement. The temperature range of 0 to 100° C may

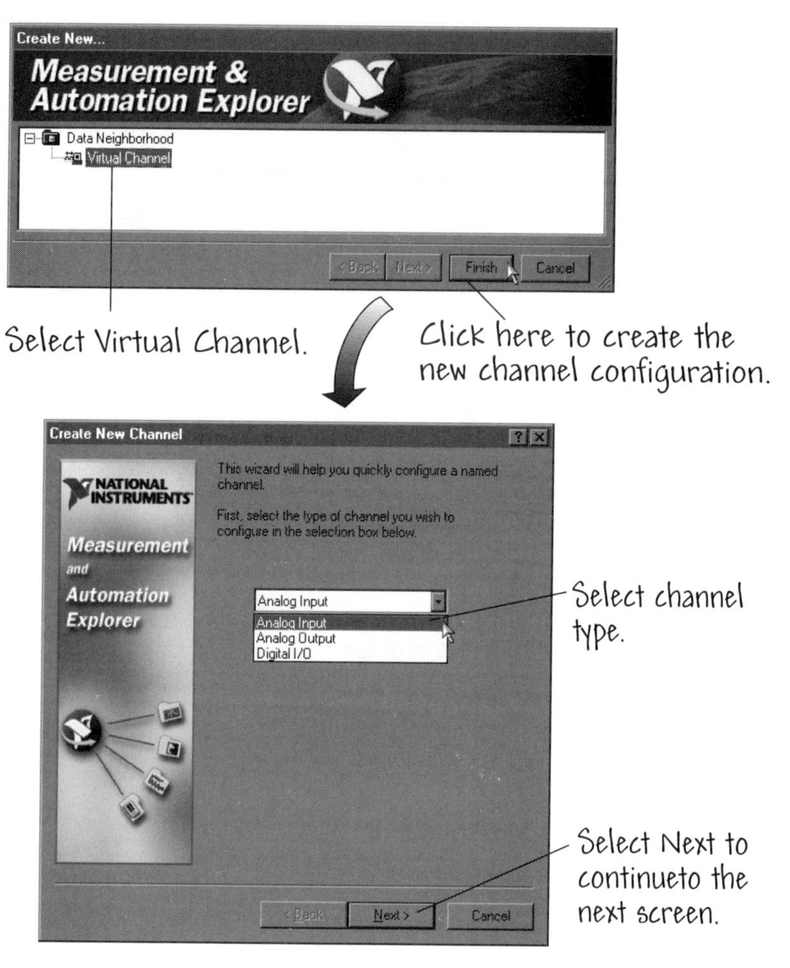

FIGURE 8.40
Creating a new virtual channel configuration.

be a wide range for measuring temperature. In general, you may need to expand (or reduce) the range depending on your application.

Click on the **Next** button to configure the scaling values for the temperature sensor. Select **Map Ranges** as the signal scale, as illustrated in Figure 8.44. This implies that you want to set the scaling range. Set the sensor range to 0 V to 1 V, and when finished with the input, click on **Next**.

You should now have arrived at the window that defines the hardware connection for the signal. The settings should match those shown in Figure 8.45, with the exception that the DAQ hardware should reflect the actual hardware that you have in your computer. For example, if you have a Lab-PC+ board, you should select that hardware in place of the PCI-1200 shown in Figure 8.45. You should select channel 0 in item 2 if the temperature sensor is connected to channel 0 of your DAQ board; otherwise select the channel number that you have wired to

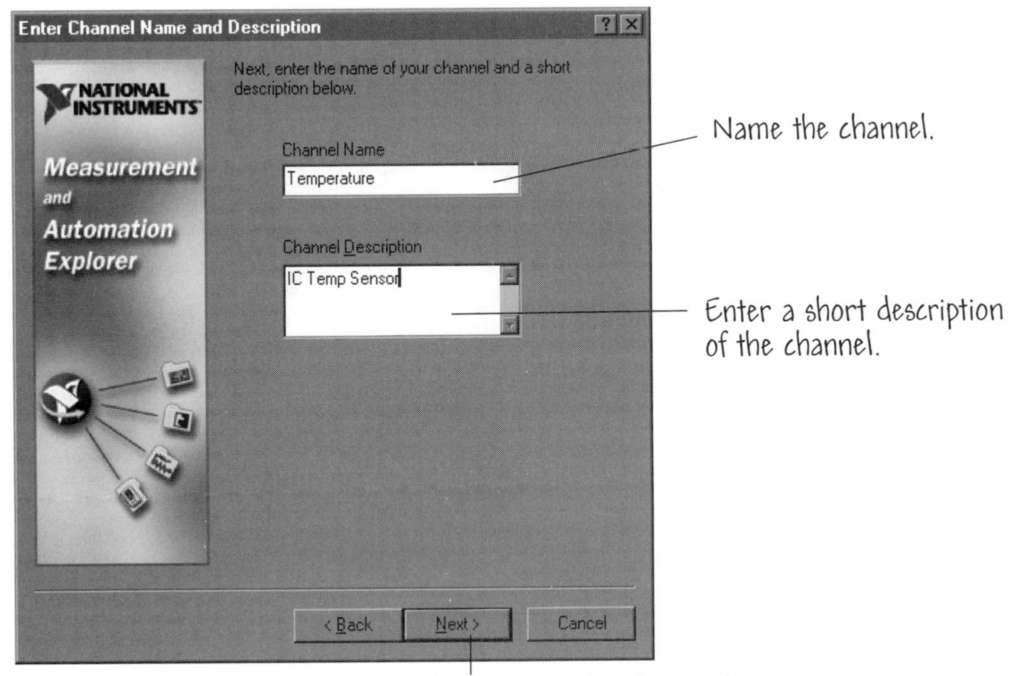

FIGURE 8.41
Naming the virtual channel and adding a short description.

your DAQ hardware. The analog input mode is also selected on this page. Refer to the previous discussion on differential, RSE, and NRSE modes for more information.

Click on the **Finish** button to view the channel you just defined. When you are finished defining the channel, MAX entries should appear as shown in Figure 8.46. In future sessions, you can edit the configurations through the MAX by highlighting the desired configuration and right-clicking on **Temperature**, as illustrated in Figure 8.47. Choose **Properties** to access the editing window. An editing window will appear and you can modify any of the configuration parameters. When you are finished configuring the virtual channel close the MAX window to exit.

◆

8.7.3 DAQ Channel Name Control

The DAQ Channel Name control is a LabVIEW data type used by the DAQ VIs for communicating to the DAQ boards. You can find the DAQ Channel Name control on the **I/O** subpalette of the **Controls** palette as shown in Figure 8.48.

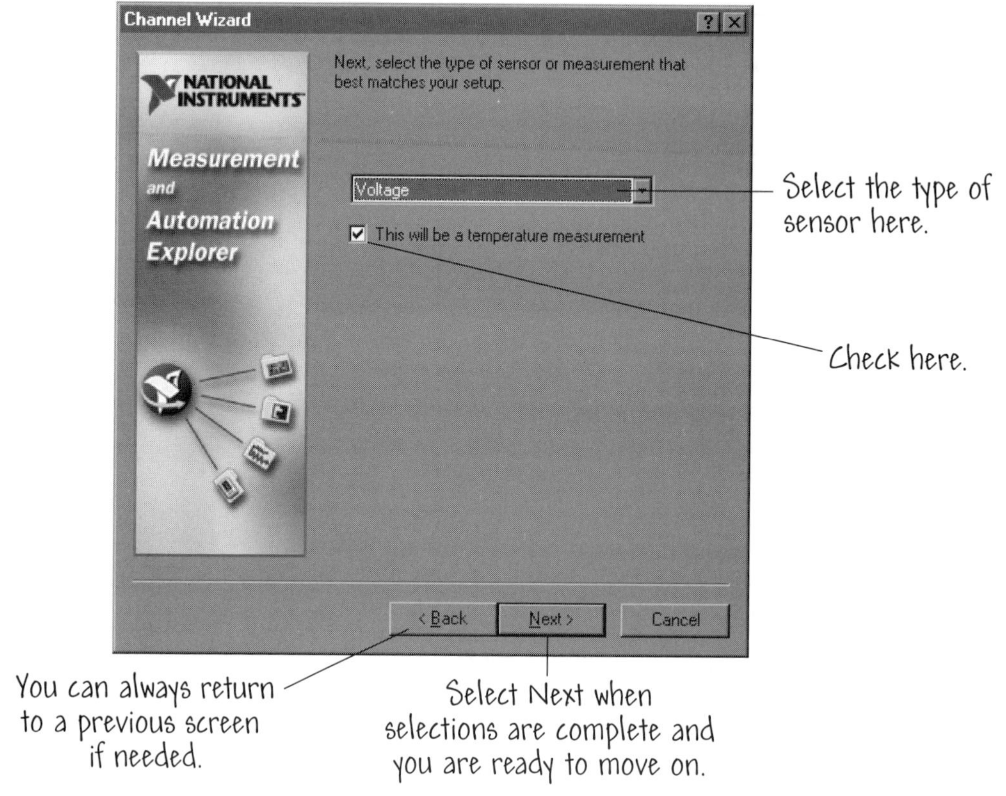

FIGURE 8.42
Selecting the sensor type.

The channel names can be entered into the control in two different ways. You can use the Operating tool to click on the DAQ Channel Name control and choose the channel name as defined in the MAX utility as shown in Figure 8.49. Another way to enter the DAQ channel name is to pop-up on the DAQ Channel Name control and select **Allow Undefined Names**. Then you can use the Labeling tool to enter the name or the channel number. The name will be one of the names of the virtual channels configured using the MAX, as discussed in the previous section.

8.7.4 The DAQ Wizards

The **DAQ Solution Wizard** helps you develop your LabVIEW applications by allowing you to choose among current data acquisition examples or to design a custom DAQ application. It works with analog input and output, digital I/O, and counter/timers. The DAQ Solution Wizard is an interactive utility that uses a series of windows that ask you about your applicaiton and creates an example

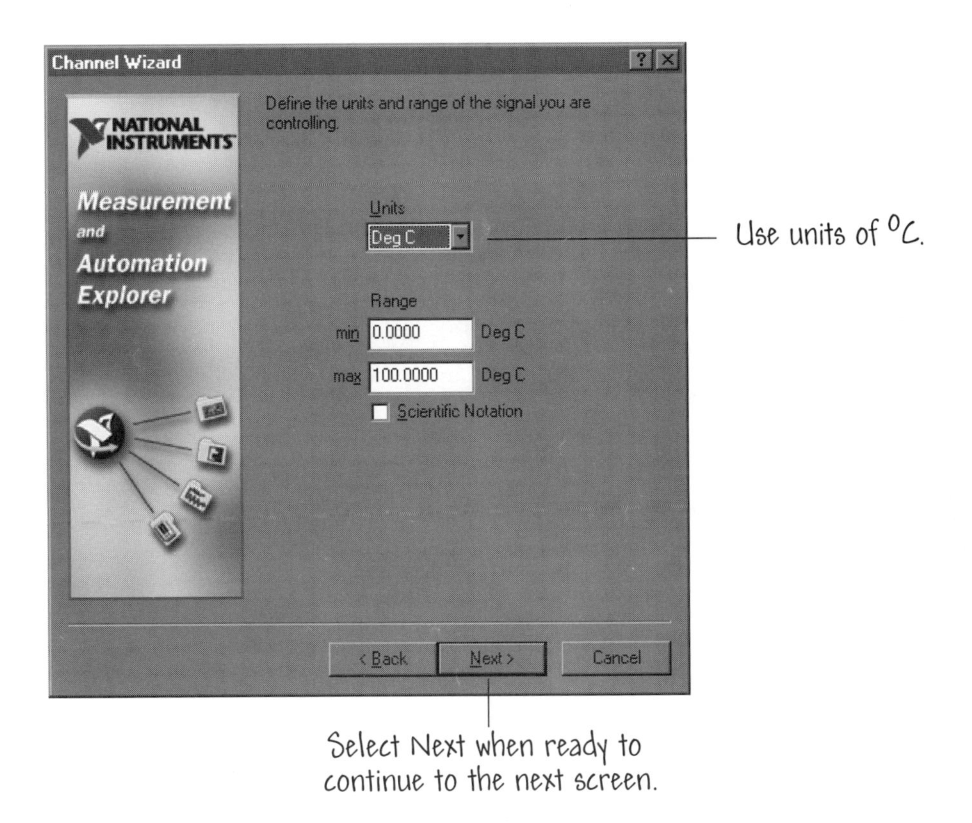

FIGURE 8.43
Defining the sensor units and scaling.

VI that you can save to a new location. The DAQ Solution Wizard is launched by selecting **DAQ Solution Wizard** from the Tools≫Data Acquisition menu in LabVIEW, as shown in Figure 8.50. You can see the virtual channels defined in the MAX by clicking on the **View Current Wizard Configuration** button. You can look at the channel definitions in more detail by clicking on the **Go to DAQ Channel Wizard** button to launch. The **DAQ Channel Wizard** is used to define which signals are connected to which channels on your DAQ board. The MAX utility opens when you press the **Go to DAQ Channel Wizard** button. You can then modify or add new virtual channels and scales for your data acquisition application. Subsequently, you can reference the channel names throughout the application, and all of the conversion processes are performed transparently.

In the open DAQ Solution Wizard window, make sure that the option **Use channel names specified in DAQ Channel Wizard** is selected. Selecting the **Next** button provides the option of either making a custom application of viewing the VIs in the Common Solutions Gallery. Select **Solutions Gallery** and press the **Next** button yielding the window shown in Figure 8.51.

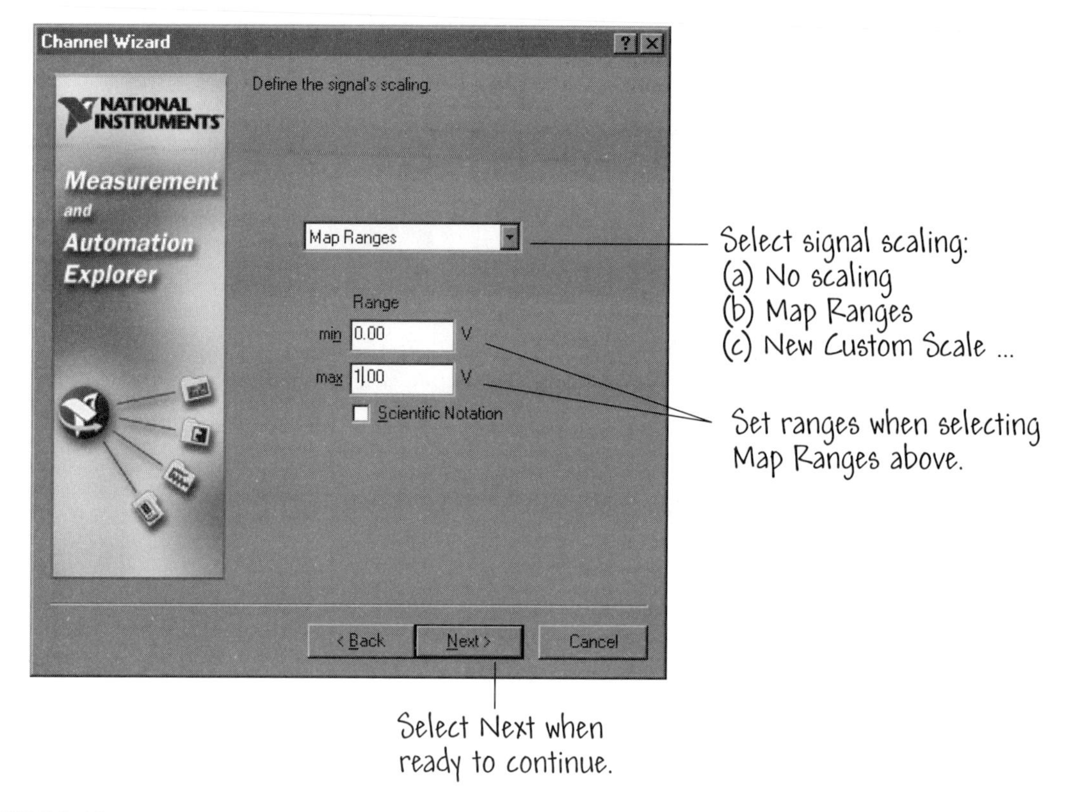

FIGURE 8.44
Selecting the signal scaling.

Select **Voltage & Current Measurement** from the Gallery Categories section and **Single-Point Voltage Measurement** from Common Solutions. When you press the **Next** button, you will see the available channels. Select the analog input channels by holding down <Shift> and clicking on the choice as illustrated in Figure 8.52. Clicking on the button will yield the VI shown in Figure 8.53.

Notice that the channels are already defined. Open and examine the block diagram. It uses the **AI One PT** VI to acquire the data. You can modify this VI just as you might any other VI. You will need to save the modified VI to hard disk if you want to utilize it in future sessions.

8.8 ANALOG INPUT

There are four classes of Analog Input VIs found in the **Analog Input** palette:

1. Easy Analog Input VIs
2. Intermediate Analog Input VIs

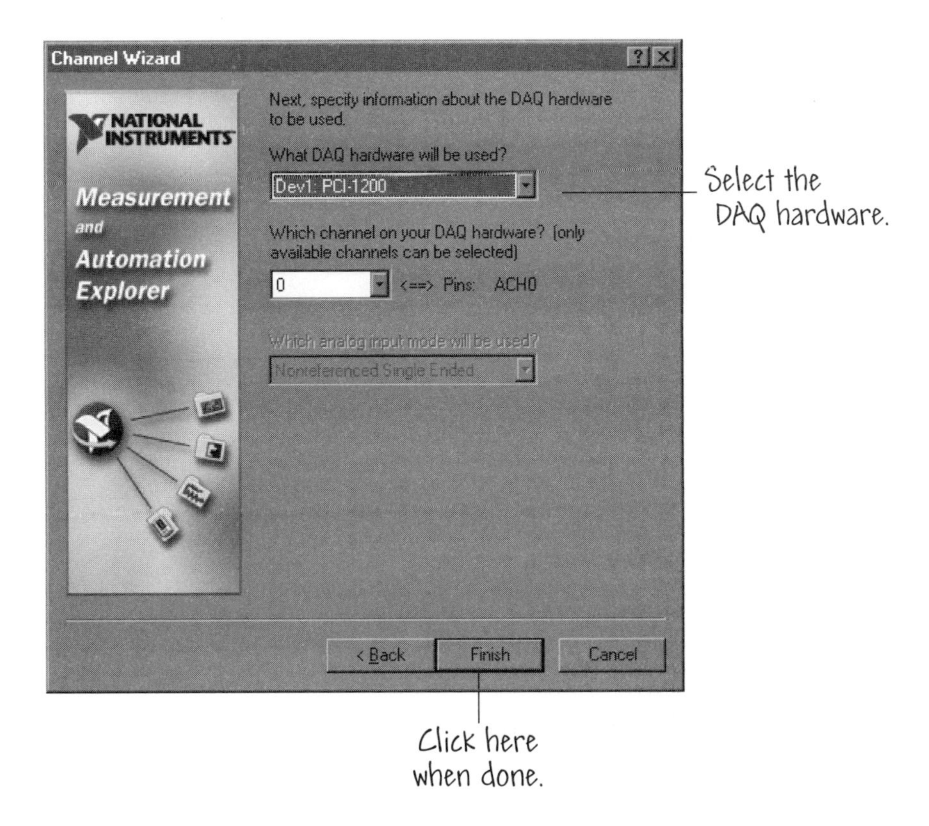

FIGURE 8.45
Selecting hardware.

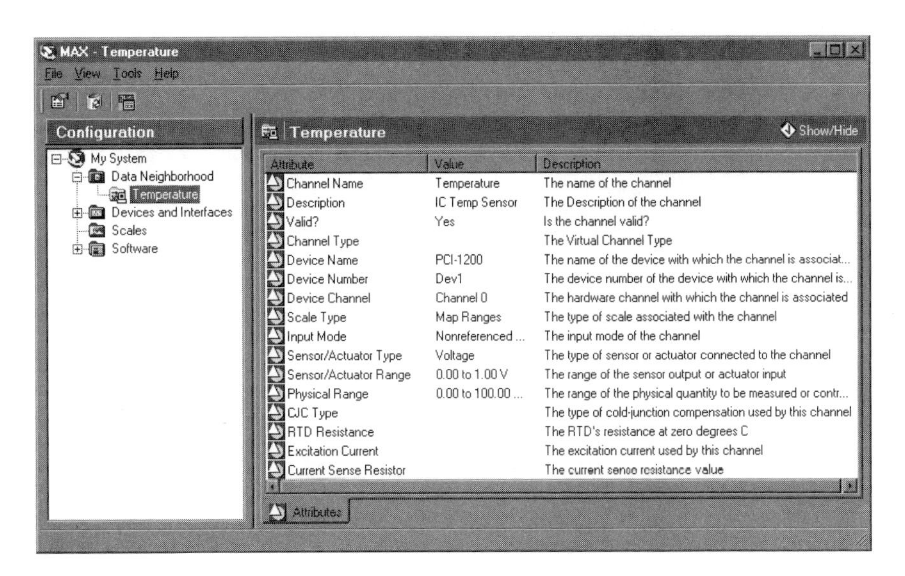

FIGURE 8.46
The virtual channel final configuration.

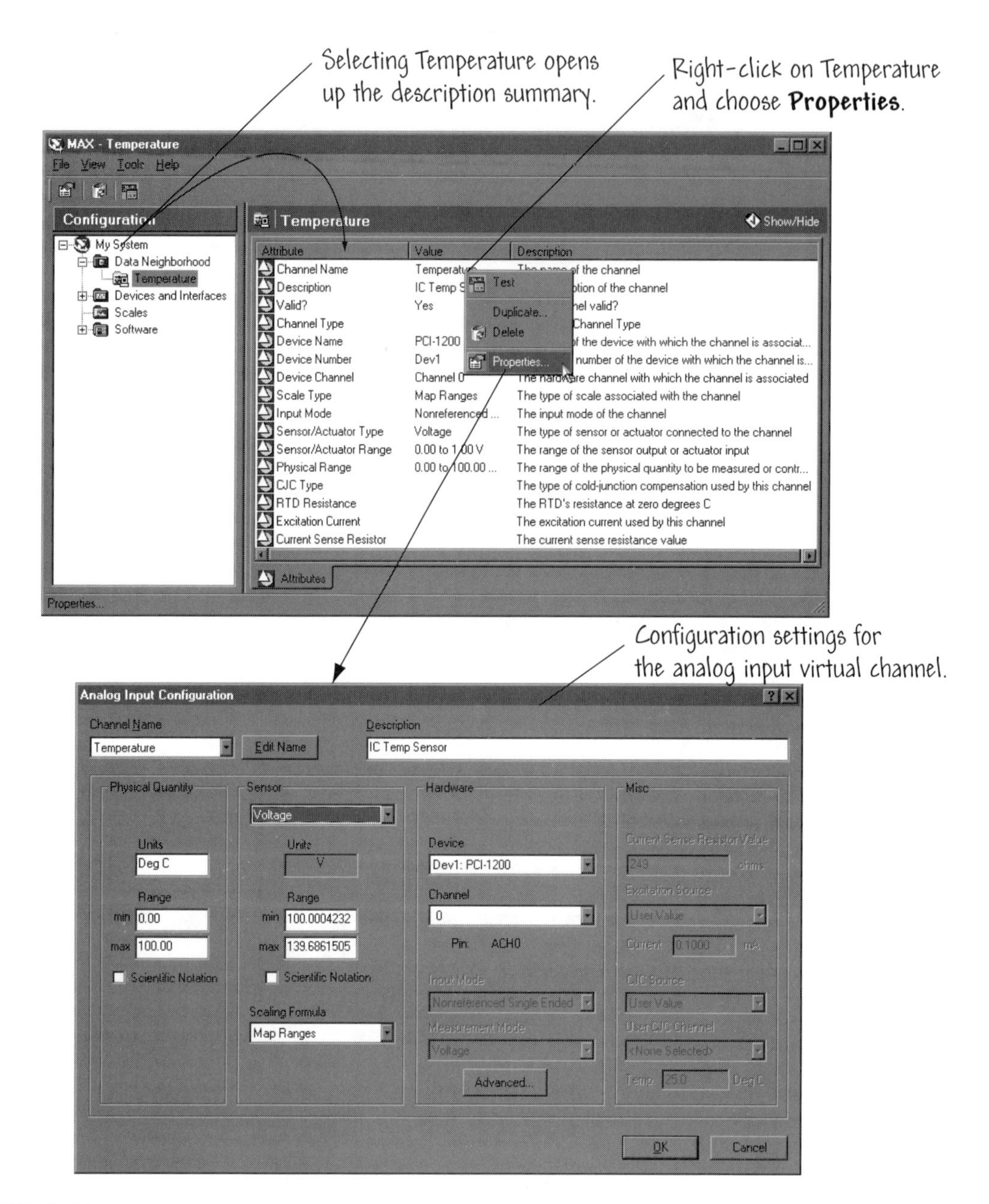

FIGURE 8.47
Editing the configuration.

3. Analog Input Utility VIs, and
4. Advanced Analog Input VIs

In *Learning with LabVIEW*, we consider only the use of the Easy Analog Input VIs, which perform simple analog input operations. You can run these VIs from the front panel or use them as subVIs in basic applications.

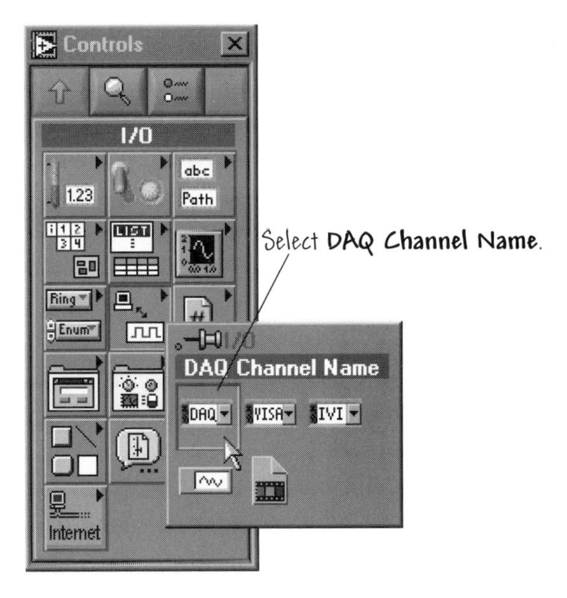

FIGURE 8.48
DAQ channel name control data type.

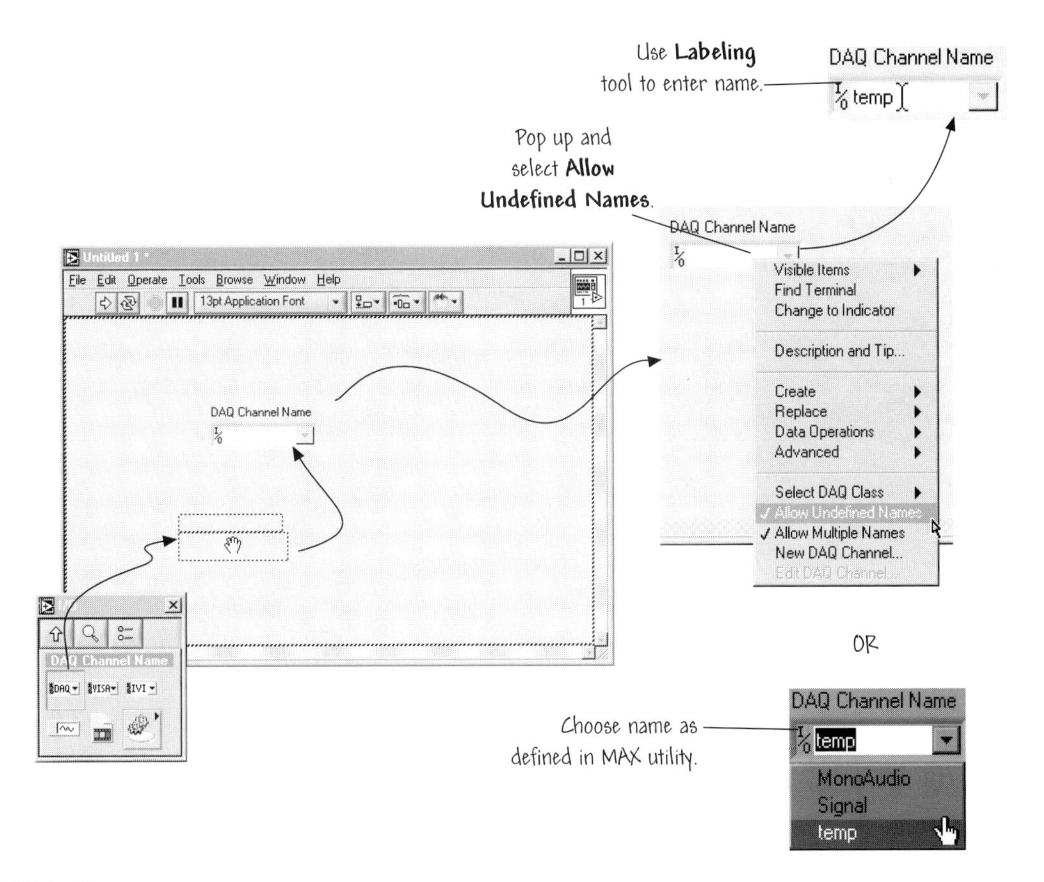

FIGURE 8.49
Entering DAQ channel names.

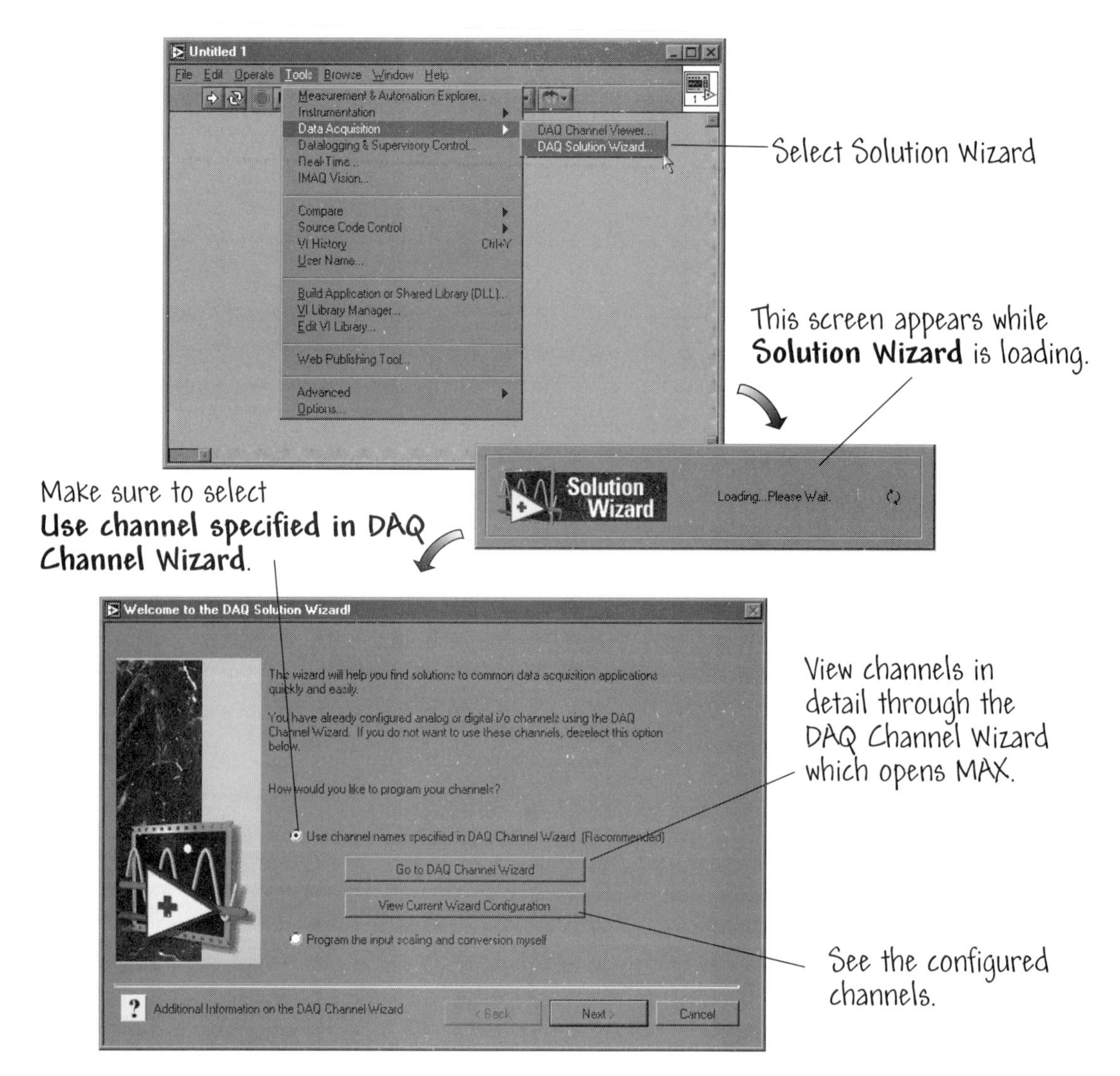

FIGURE 8.50
Accessing the DAQ Solution Wizard.

In this chapter, the exercises show real DAQ VIs with channels that have been configured using the Measurement and Automation Explorer. If you installed the simulation DAQ VIs, you need to wire values to the device and channel inputs, and you do not need to run the Measurement and Automation Explorer. For exercises that use one channel, enter 1 for device and 0 for channel. For exercises that use two channels, enter 1 for device and 0, 1 for channels. When you move your DAQ application to a machine with real VIs, you can continue to use numeric values for device and channel inputs, or you can run MAX and assign

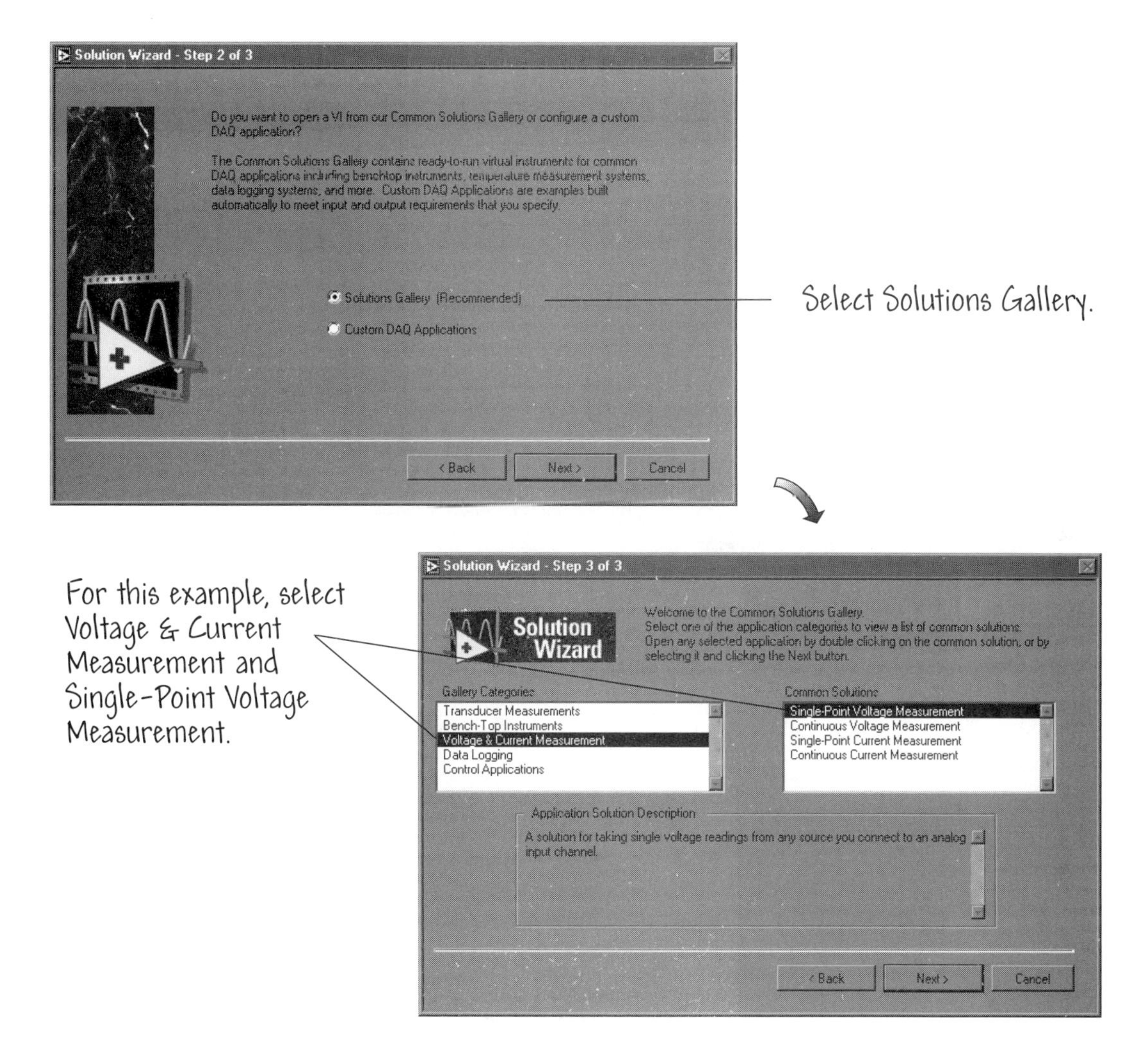

FIGURE 8.51
Accessing the Solutions Galley.

names to channels. When you use channel names configured with the MAX, you do not need a device value.

8.8.1 Single-Point Acquisition

A single-point analog input reads one value from the assigned input channel and immediately returns the value to the VI. This type of data acquisition is useful when you want to determine the magnitude of an acquired analog DC signal. An example of this would be if you want to periodically monitor the temperature in a room. You can connect the temperature transducer that produces a voltage (representing the temperature) to a single channel on your DAQ device and initiate

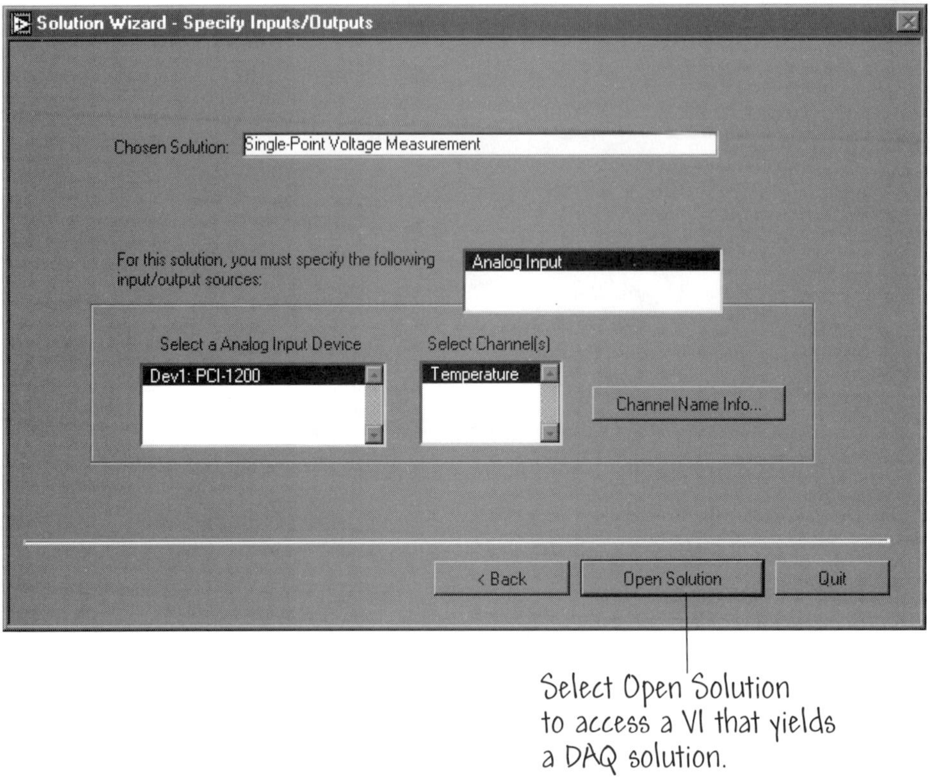

Solution Wizard - Specify Inputs/Outputs
Chosen Solution:
Single-Point Voltage Measurement
For this solution, you must specify the following input/output sources:
Analog Input
Select a Analog Input Device
Dev1: PCI-1200
Select Channel(s)
Temperature
Channel Name Info...
< Back
Open Solution
Quit
Select Open Solution to access a VI that yields a DAQ solution.

FIGURE 8.52
Selecting the inputs and outputs for the DAQ solution galley.

Single-Point Voltage Measurement *
File Edit Operate Tools Browse Window Help
13pt Application Font
MEASUREMENT DATA
0.00
0.00
0.00
0.00
Device
1
Channels
Temperature

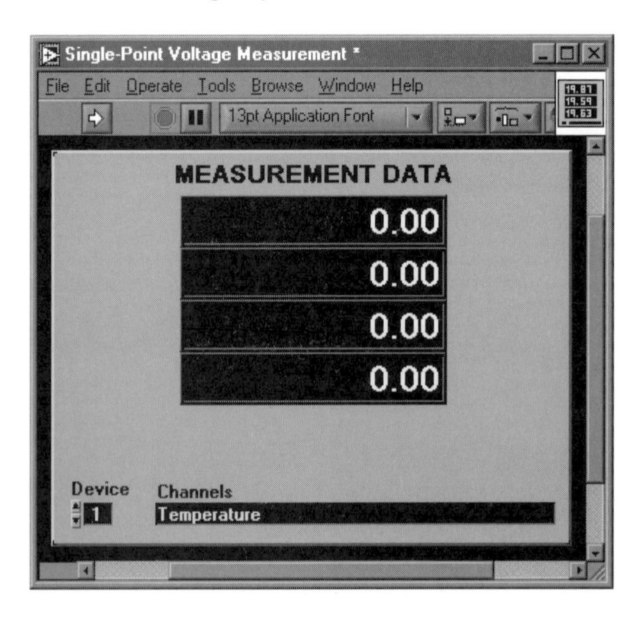

FIGURE 8.53
The DAQ Solution Galley VI.

(through the software) a single-point analog input data acquisition whenever you want to know the room temperature.

The **Analog Input** palette found on the **Data Acquisition** palette contains the VIs that perform the single-point acquisitions and other analog-to-digital (A/D) conversions (see Figure 8.30). The AI Sample Channel.vi is used for single-point acquisition; that is, it takes a single sample of the analog signal attached to a specified channel and returns the measured voltage. The two inputs for the VI are

- **Device**—the device number of the DAQ board. When you use a channel name configured with the MAX you do not need a device value. If you are using the simulation DAQ VIs, you must wire values to **device**.
- **Channel** specifies the analog input channel number or a channel name configured with the MAX. The default value is 0.

The other optional inputs are

- High limit and low limit—specify the limit settings of the input signal. The default values for high and low limit are +10 V and −10 V, respectively.

Do not wire values to these inputs if you are using the MAX, because this information has already been configured. Figure 8.54a illustrates how to wire AI Sample Channel.vi.

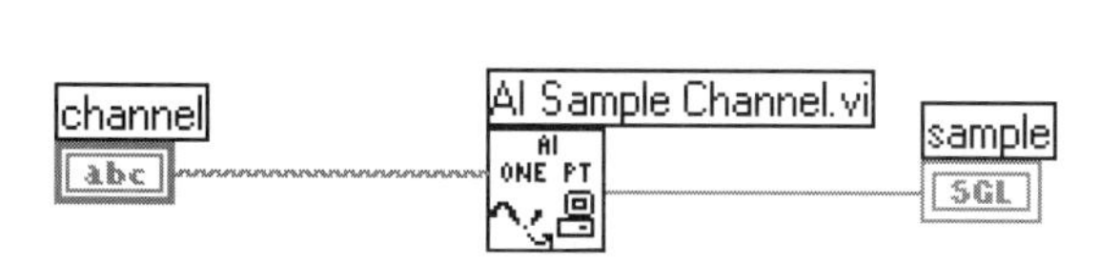

(a) Don't wire the input **device** when using real DAQ VIs and the DAQ Channel Wizard

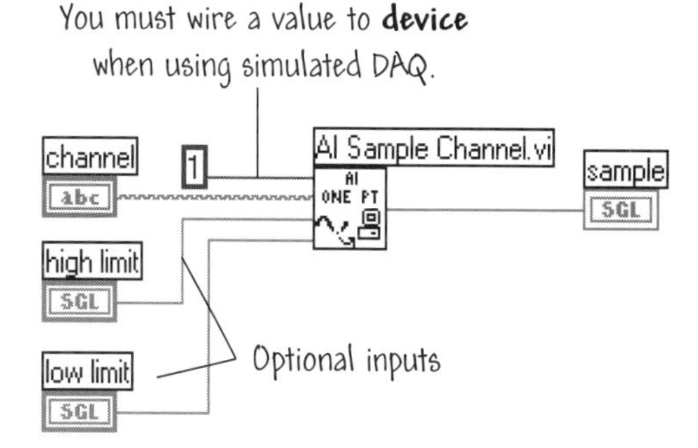

(b) Wire the input **device** when using simulated DAQ VIs

FIGURE 8.54
The AI Sample Channel.vi measures the signal attached to the channel you specify.

The output sample shown in Figure 8.54 is a scaled version of the input analog signal. If an error occurs during the operation of AI Sample Channel.vi, a dialog box displays the error code, and you have the option to abort the operation or to continue execution.

It is possible to acquire single-point data from several analog channels on a DAQ board. This is accomplished using AI Sample Channels.vi, which measures the signals attached to multiple channels and returns those measured values in an array. The VI inputs are the same as for AI Sample Channel.vi, except that in this case, the input channels is a string that specifies the list of analog input channels. The output of the VI is an array of measurements. The order of the measurements in the output array matches the order listed in the input channels string. For example, if channels is "1, 2, 4", samples[0] would be from CH 1, samples[1] from CH 2, and samples[2] from CH 4.

You do not need to enter the device or input limits if you set up your channels using the MAX. Instead, enter a channel name in the channel input, and the value returned is relative to the physical units you specified for that channel in the MAX.

Practice with Single-Point Acquisition

The objective of this example is to acquire an analog signal using a DAQ board. You will build a VI that measures the voltage of a temperature sensor. The temperature sensor outputs a voltage proportional to the temperature and is hard-wired to Channel 0 of the DAQ board.

Open a new VI and construct a front panel similar to the one depicted in Figure 8.55. The input variable **channel** is a DAQ Channel Name data type (as discussed in the previous section). Use the **Labeling** tool to input the text **Temperature**. Switch to the block diagram window and place the AI Sample Channel VI on the block diagram. Wire the block diagram using Figure 8.55 as a guide. In this exercise, the VI reads analog input Channel 0 and returns the measured voltage. The DAQ channels should have been configured using the MAX as discussed in the previous section (see Figure 8.46).

When the VI is ready to run, return to the front panel and select **Run**. The thermometer on the front panel will display the measured temperature in deg C. If an error occurs during the data acquisition process, the Easy I/O VIs automatically display a dialog box showing the error code and a description of the error. When you have finished experimenting with your VI, save it as AI Single Point.vi in the Users Stuff directory.

If you are having problems constructing the VI described above, you can find a working version of the VI shown in Figure 8.55 by opening the VI titled AI Single Point.vi located in the Chapter 8 folder in the Learning directory. An example of a VI that acquires single-point data from two analog channels is called AI

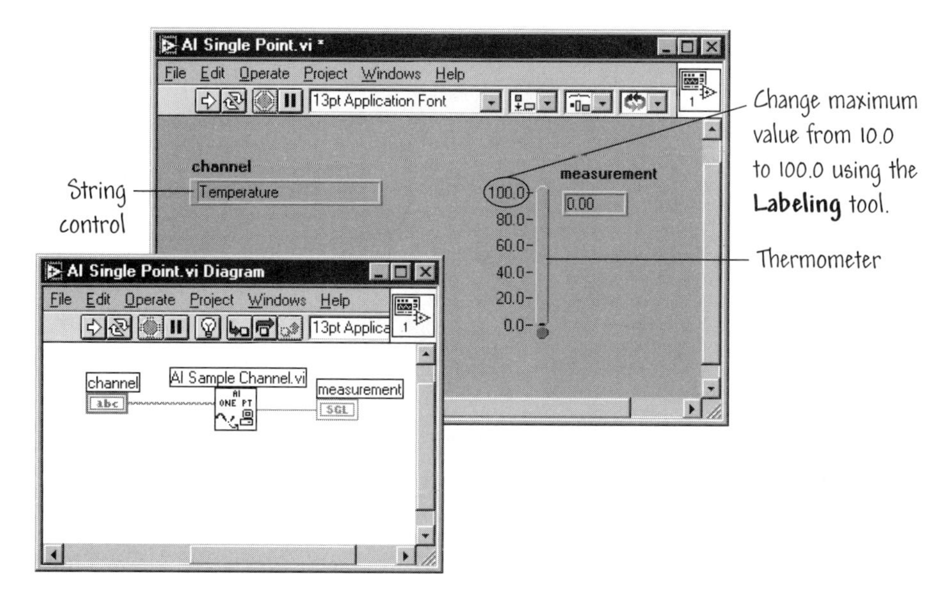

FIGURE 8.55
A VI designed to measure an analog temperature signal.

Single Pt Mult Ch.vi and can be found in the Chapter 8 folder in the Learning directory. ◆

8.8.2 Waveform Data Types

The waveform datatype is a LabVIEW datatype that combines the data read from the DAQ board with time information. The DAQ VIs return waveform data. You can place a waveform on the panel by selecting it from the **I/O** subpalette of the **Controls** palette as shown in Figure 8.56. When you wire the waveform output terminal of a DAQ VI directly to the waveform datatype, you will receive the starting time the data was acquired, the delta time for each data point, and an array of the data values. The waveform datatype can be wired directly to a waveform graph, and it will properly scale the X axis with the time data.

8.8.3 Waveform Acquisition

In some applications, acquiring one point at a time may not be fast enough. Also, with single-point acquisition it is difficult to obtain a constant sample interval between each point because the interval depends on a number of factors (e.g., loop execution speed and software overhead) that are not easily controlled. The AI Acquire Waveform.vi acquires a specified number of samples at a desired sample rate from a single input channel. The VI is shown in Figure 8.57.

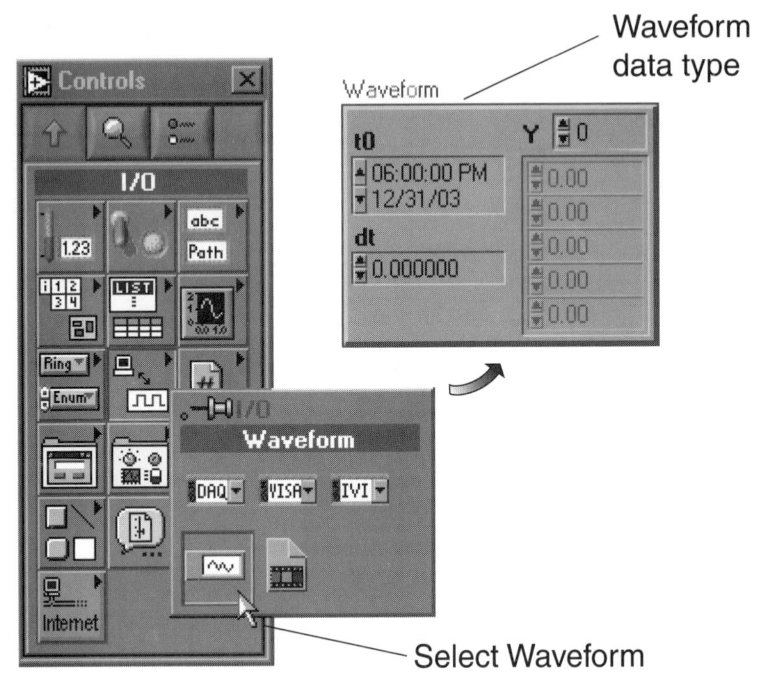

FIGURE 8.56
Waveform data type.

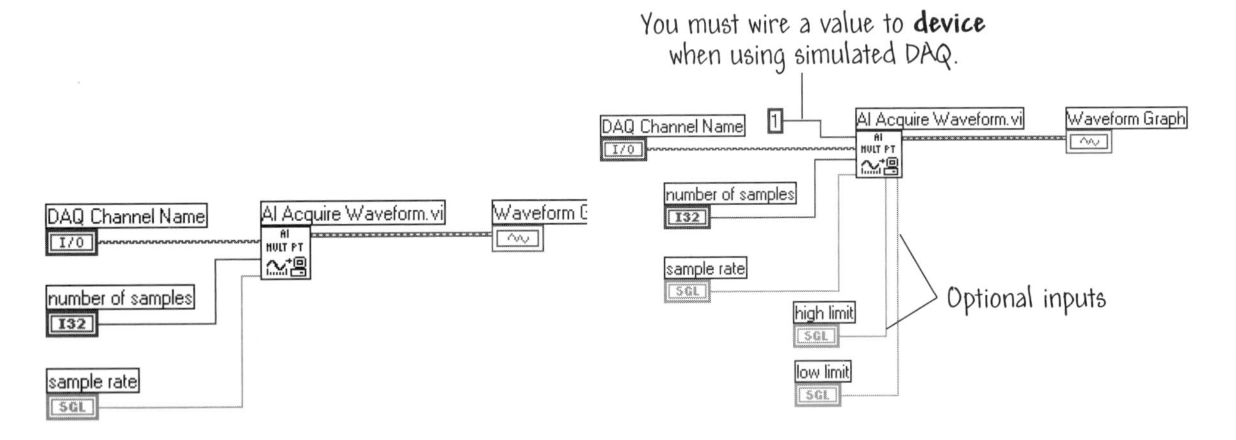

(a) Don't wire the input **device** when using real DAQ VIs and the DAQ Channel Wizard

(b) Wire the input **device** when using simulated DAQ VIs

FIGURE 8.57
The AI Acquire Waveform.vi acquires the specified number of samples at the specified sample rate from a single input channel and returns the acquired data.

Since this is an Easy I/O VI, it has the minimal number of inputs needed to acquire a waveform from a single channel. These nominal inputs are

- **Device**—the DAQ board device number. When you use channel names configured with the MAX, you do not need a device value. If you are using the simulation DAQ VIs, you must wire a value to **device**.
- **Channel** specifies the analog input channel number or a channel name configured with the MAX.
- **Number of samples**—the number of samples to acquire.
- **Sample rate**—the number of samples to acquire per second.

Figure 8.57 illustrates how to wire AI Acquire Waveform.vi. AI Acquire Waveform is a polymorphic VI that can be configured to output a **scaled array** or **waveform**. The outputs of the **AI Acquire Waveform.vi** are

- **Waveform**: contains the scaled analog input data, for both scaled array and waveform outputs.
- **Actual scan period**: the inverse of the actual sampling rate. Depending on the capabilities of your hardware, the actual sample rate may differ slightly from the requested sample rate. This applies to scaled array output only.

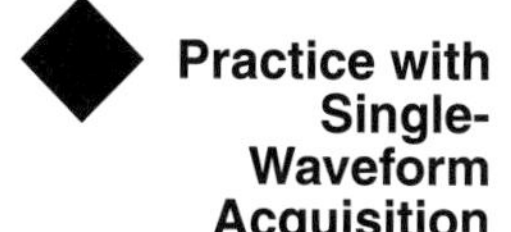

Practice with Single-Waveform Acquisition

The objective of this exercise is to acquire and display an analog waveform. The MAX will be used to configure the system foran input called **MonoAudio**. Your VI will use the DAQ VIs to acquire a signal and plot it on a graph. For this exercise, you will need to have an analog input signal available for assignment to channel 1.

Open a new VI and build the front panel shown in Figure 8.58. The input variable **channel** is labeled MonoAudio. The input **number of samples** control specifies the number of points to sample. The input **sample rate** specifies the sampling rate. When the VI is up and running, you can vary these input parameters and observe the effects on the waveform graph. After building the front panel, switch to the block diagram and wire the AI Acquire Waveform VI using Figure 8.58 as a guide.

Now you'll need to configure the DAQ system in preparation for acquiring the analog signal. Open the MAX, as shown in Figure 8.59. We will follow essentially the same steps through the MAX as we did previously in the example "Using the MAX." When the MAX opens up, it should resemble the one shown in Figure 8.46. The goal is to add to the channel configuration another analog input called **MonoAudio** assigned to channel 1. When the MAX screen appears, right-click on **Data Neighborhood** and select **Create New....** to initiate the process of

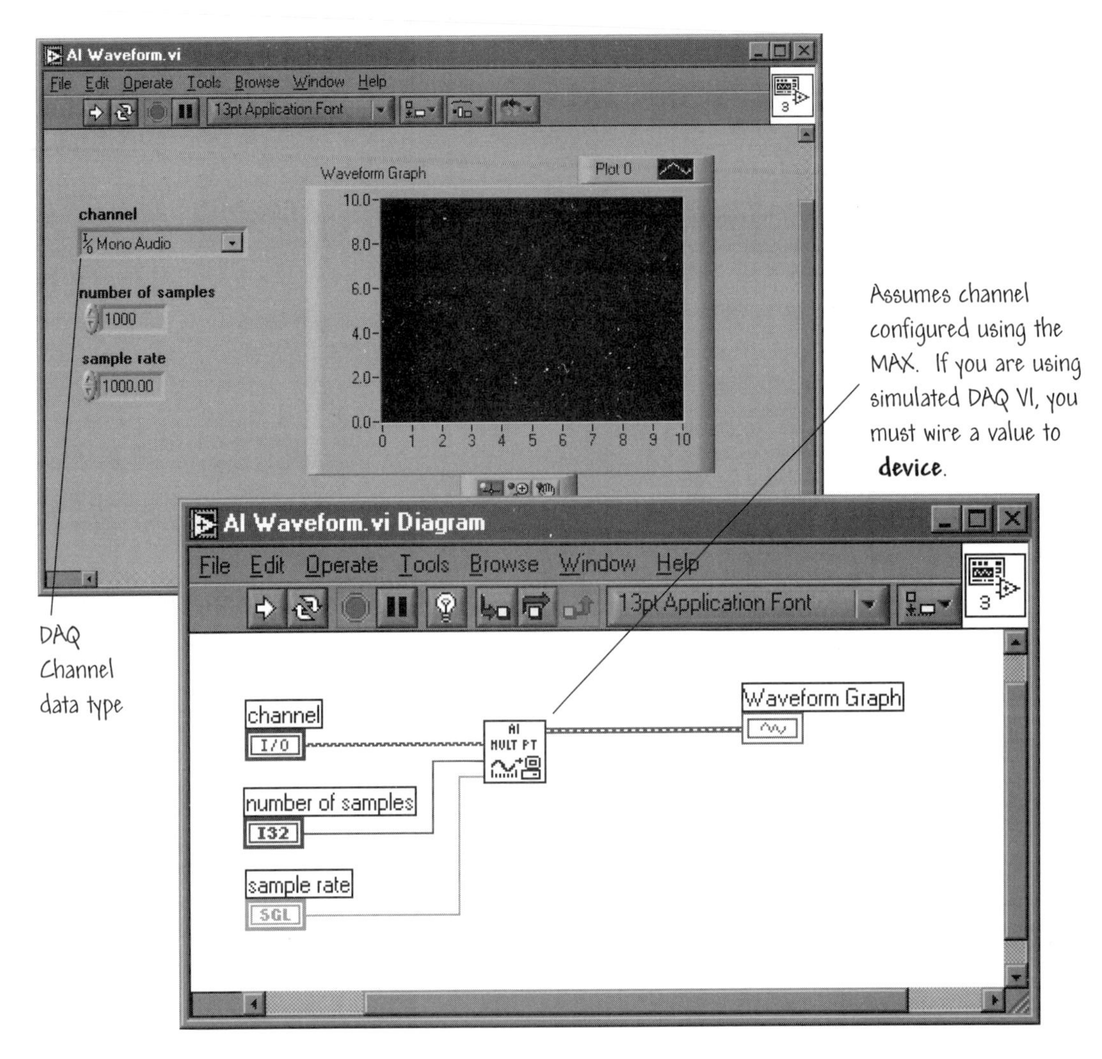

FIGURE 8.58
A VI designed to measure a waveform signal.

adding a new channel configuration, and remember that you are configuring an analog input.

In the screens that the MAX presents, fill in each item using Figure 8.60 as the guide. Some of the items that appear in MAX will already have default values that are acceptable. Otherwise, respond to the questions by entering the appropriate information. When you are finished entering the data, click on the **Finish** button located in the lower right corner of the fifth screen. The updated DAQ channel configuration should now include the **MonoAudio** on channel 1, as illustrated in Figure 8.61.

When you are finished, save the VI as AI Waveform.vi in Users Stuff in the Learning directory.

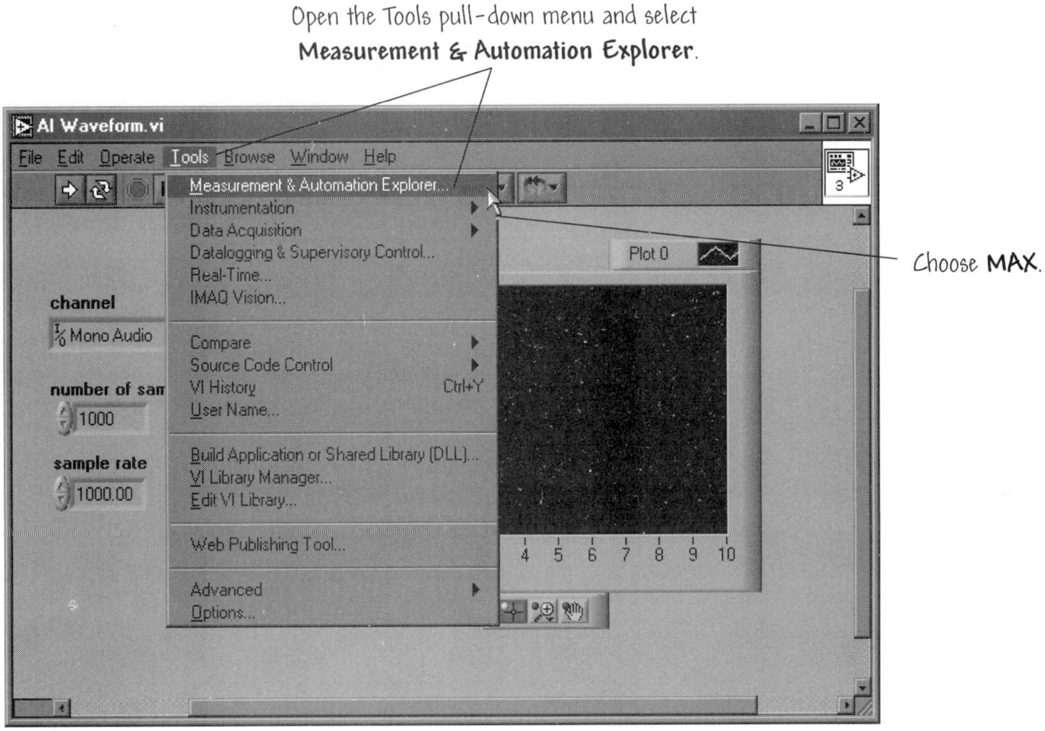

Open the Tools pull-down menu and select **Measurement & Automation Explorer**.

Choose **MAX**.

FIGURE 8.59
Opening MAX to configure the DAQ channel for acquiring waveform data.

Return to the front panel, enter the appropriate values for the controls, and run the VI. The graph plots the analog waveform. Try different values for the sampling rate and the number of samples.

If you are having problems with your VI, you can find a working version of the VI shown in Figure 8.58 in Chapter 8 *of the* Learning *directory. The VI is called* AI Waveform.vi*. An example of a VI that acquires multiple waveforms from two analog channels is called* AI Waveforms.vi *and can be found in the* Chapter 8 *folder in the* Learning *directory. If you use this VI, you will need to configure another analog input channel in the MAX.*

◆

8.9 ANALOG OUTPUT

There are two general methods for analog output—a single point at a time and many points at a time. Similiar to the situation with analog input, with analog output you can perform single updates on analog output channels (using no

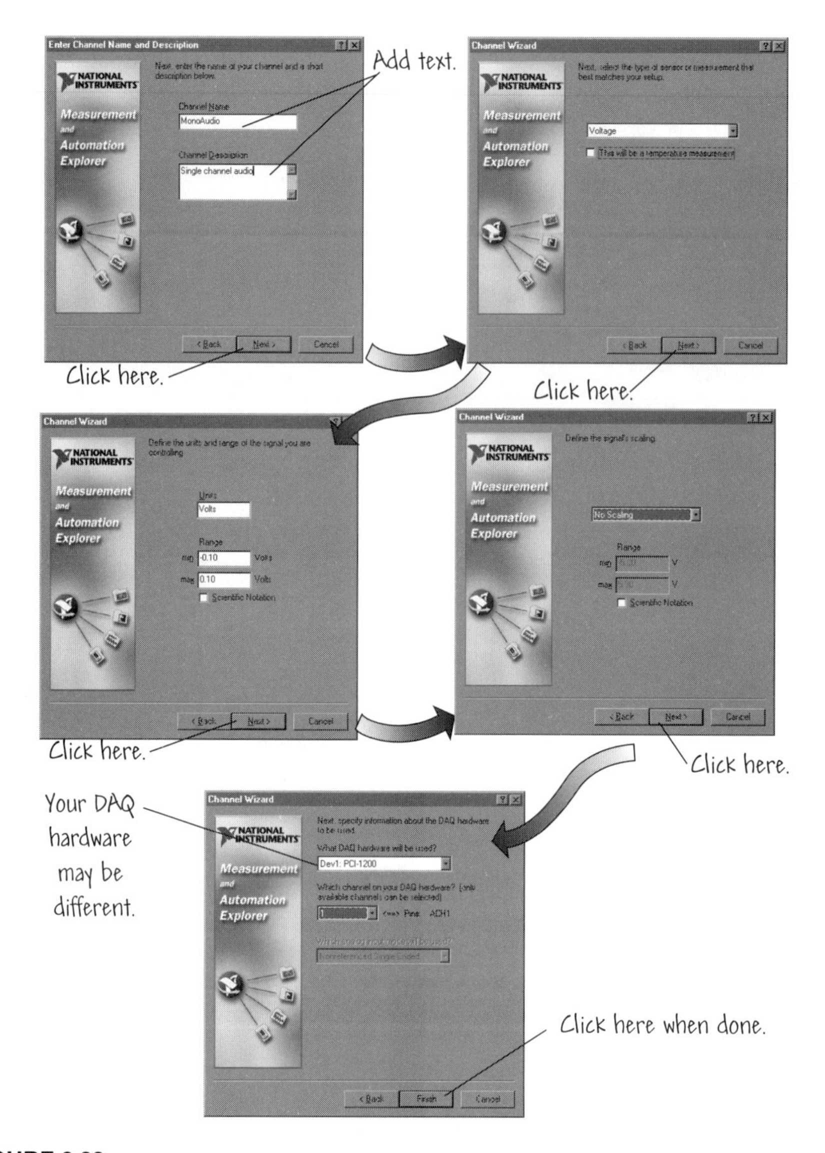

FIGURE 8.60
Configuring the MonoAudio analog channel using the MAX.

hardware clocks). You can update channels at a significantly faster and more precise rate using a buffer and an onboard counter to control how fast the updates occur.

In this chapter we discuss analog output and the role of VIs in the process. The VIs that control analog output operations are

- Easy Analog Output VIs: These VIs are appropriate for simple analog output applications, yet still possess built-in error-handling capability. The same limit settings are used for all output channels.

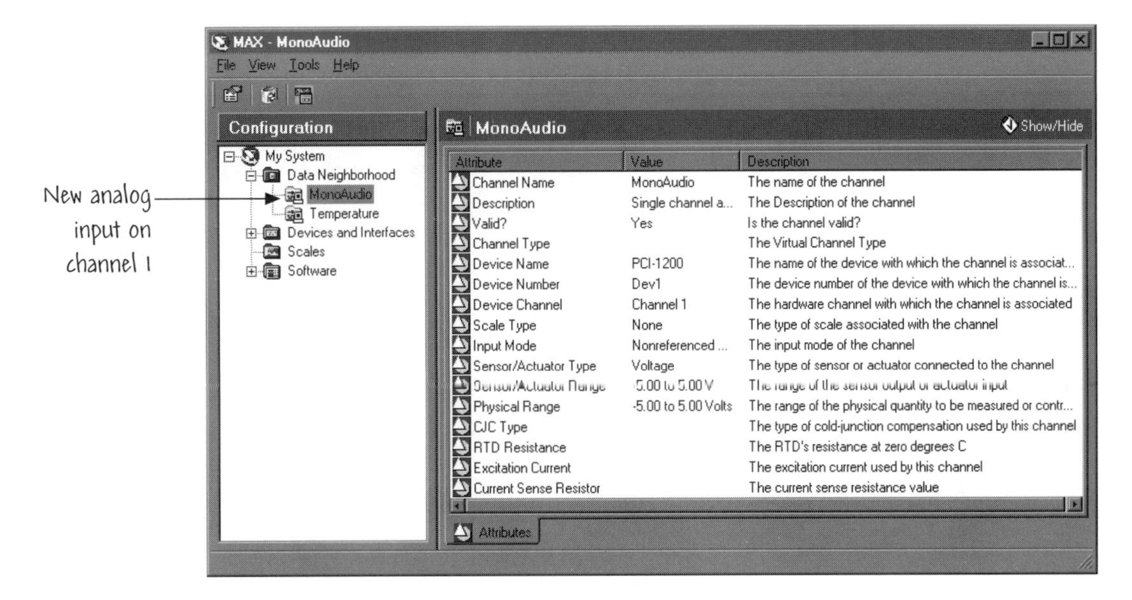

FIGURE 8.61
The MAX final configurations showing the **MonoAudio** input.

- Intermediate Analog Output VIs: These VIs are appropriate for most analog output applications that require special timing, user-defined error-handling routines, separate limit settings for each channel, or continuous data output.
- Analog Output Utility VIs: The Utility VIs are groupings of the intermediate VIs that perform single-point, waveform, and continuous analog output. These VIs combine the flexibility of user-defined error handling that come with using Intermediate VIs but with the convenience of calling just one VI per operation.
- Advanced Analog Output VIs: These are the building blocks for the other Analog Output VIs and are rarely needed since the Intermediate VIs provide most of the same functionality in a simpler format.

As is the case of analog input VIs, in *Learning with LabVIEW*, we consider only the use of the Easy Analog Output VIs that perform analog output operations. Much of the discussion that follows has the same flavor as the previous discussions on analog input—the process of configuring the channels and the functions of the associated VIs is similar. As before, the MAX is used to configure the channels, except that the configuration process begins by choosing **Analog Output**, as illustrated in Figure 8.62.

8.9.1 Single-Point Generation

The Analog Output library contains VIs that perform **digital-to-analog (D/A)** conversions. The AO Update Channel VI writes a specified value to an analog

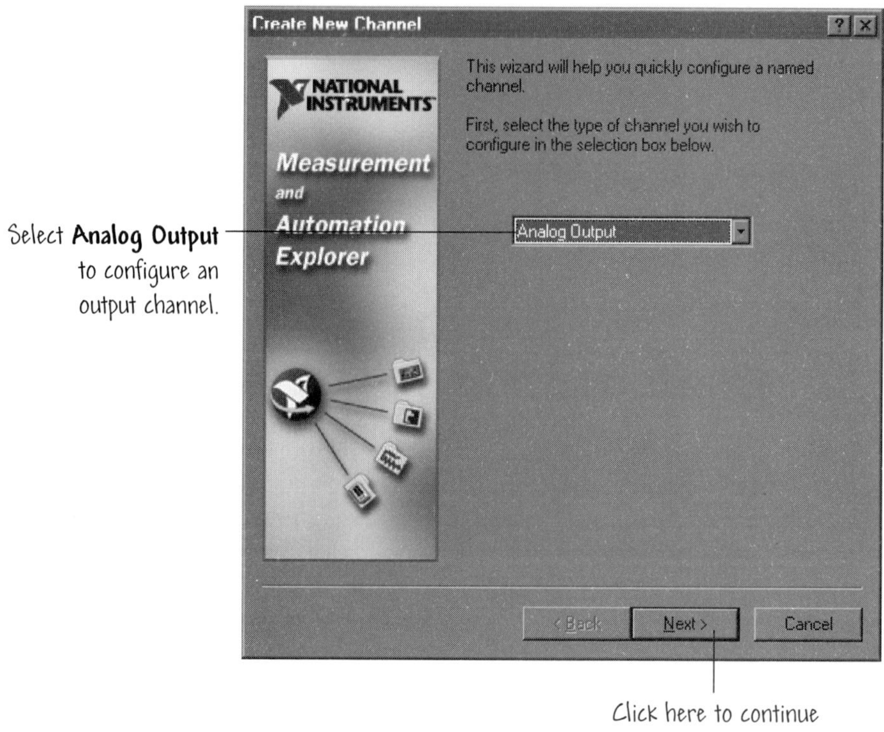

FIGURE 8.62
The MAX can configure the analog output channels—select **Analog Output** on the first MAX screen.

output channel. The VI has three inputs:

- **Device**: the device number of the DAQ board. You do not need a device value when you use channel names configured with the MAX. If you are using the simulation DAQ VIs, you must wire a value to **device**.
- **DAQ Channel Name**: specifies the analog output channel number.
- **Value**: the data to be output.

Figure 8.63 illustrates how to wire AO Update Channel.vi. If an error occurs during the operation of AO Update Channel, a dialog box displays the error code, and you have the option to abort the operation or continue execution.

Practice with Single-Point Generation

The objective of this exercise is to produce an analog output and to use the MAX to configure the system for a **Motor Speed** output.

Open a new VI and build the front panel shown in Figure 8.64. The input variable **channel** is a DAQ channel name control. The input **value** is determined by the user input on the front panel and is the data value that is output. Since Motor Speed is an analog output channel, after configuring the channel with the

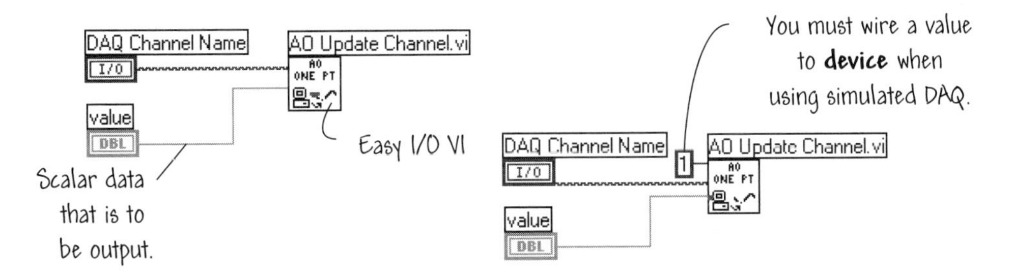

(a) Don't wire the input **device** when using real VIs and the DAQ Channel Wizard

(b) Wire the input **device** when using simulated DAQ VIs

FIGURE 8.63
The AO Update Channel.vi writes a single value to an analog output channel.

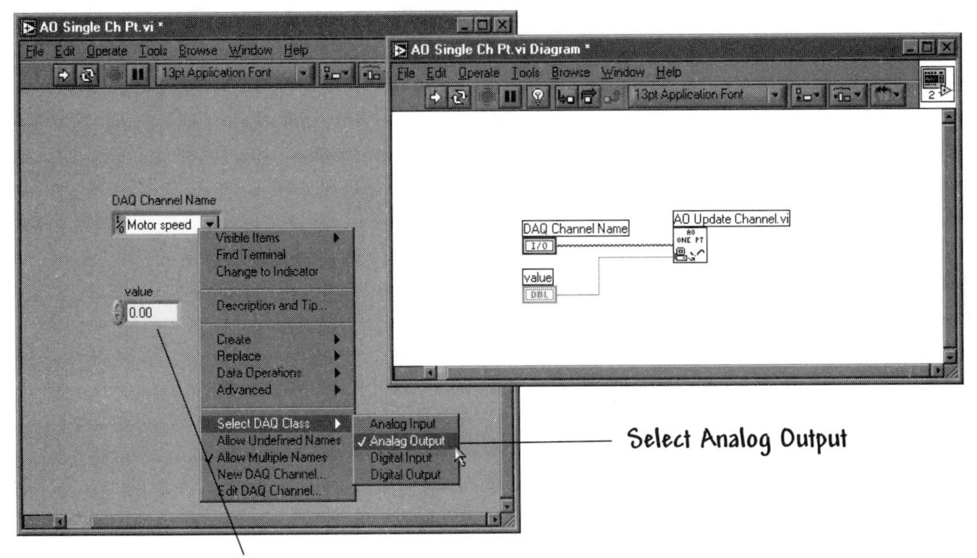

FIGURE 8.64
A VI designed to output a single point.

MAX, you will need to pop-up on the DAQ Channel Name control and choose **Select DAQ Class**. Then select **Analog Output** as shown in Figure 8.64. The result is that the analog output channel will be available for selection on the DAQ Channel Name menu. When the front panel is ready, switch to the block diagram and wire the VI AO Update Channel, using Figure 8.64 as a guide.

Open the MAX and follow essentially the same steps as with the analog input examples. The goal is to add to the channel configuration an analog output called

Motor Speed assigned to channel 0. Initiate the process of adding a new channel configuration, and remember that you are configuring an analog output channel. Right-click on **Data Neighborhood** in the MAX window and select **Create New...**. Then, in the next window select **Virtual Channel** and click on **Finish**.

At this point in the chapter, you should have configured two analog input channels **Temperature** and **MonoAudio**, and after this exercise is done, one analog output channel **Motor Speed**. In the screens that the MAX presents, fill in each item, using Figure 8.65 as the guide. Whe you are finished entering the

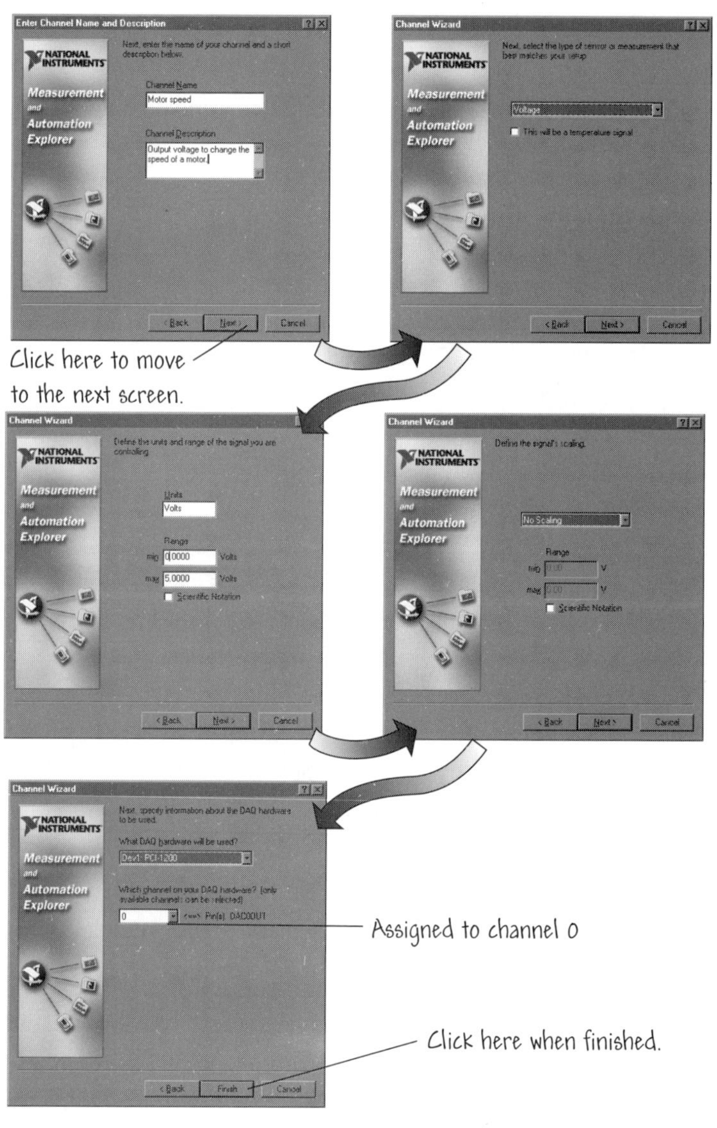

FIGURE 8.65
Configuring the analog output channel using the MAX.

data, click on **Finish**. You can access the **DAQ Channel Viewer** in Tools≫Data Acquisition≫DAQ Channel Viewer to see the current channel configuration, as shown in Figure 8.66.

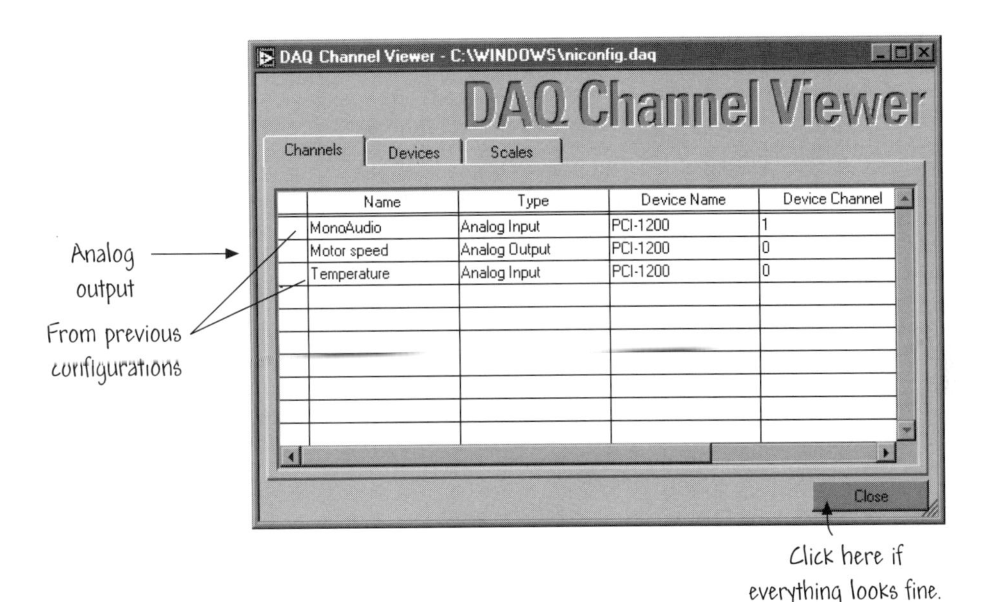

FIGURE 8.66
The MAX final configuration, including the **Motor Speed** output.

When you are finished constructing the VI, save it as AO Single Ch Pt.vi in the Users Stuff folder in the Learning directory. Modifying the VI to change from single-point output to a single channel to single-point outputs to multiple channels is an easy procedure. The AO Update Channels VI is used in place of the AO Update Channel VI, and of course, the number of assigned channels must be consistent with the desired number of output channels. Using the MAX makes the process of configuring the output channels a straightforward one.

If you are having problems with your VI, you can find a working version of the VI shown in Figure 8.64 in Chapter 8 of the Learning directory. The VI is called AO Single Ch Pt.vi. An example VI illustrating single-point outputs to multiple channels is AO Single Pt Mult Ch.vi located in the Chapter 8 folder. ◆

8.9.2 Waveform Generation

Sometimes when you generate output signals you are concerned with the rate of output. For example, you might want your DAQ system to act as a signal generator, and you want to control the rate of output and also the sample interval between points. Generating one point at a time may not be fast enough, and it is difficult to obtain a constant sample interval between each point because the interval depends on a number of factors over which you have limited control. With the AO Generate Waveform VI, you can generate multiple points at rates greater than the AO Update Channel VI can achieve, and you specify the sampling rates. The AO Generate Waveform VI has four inputs:

- **Device**: the device number of the DAQ board. When you use channel names configured with the MAX, you do not need a device value. If you are using the simulation DAQ VIs, you must wire a value to **device**.
- **Channel**: specifies the analog output channel number.
- **Update rate**: the number of voltage updates to generate per second.
- **Waveform**: a 1D array that contains data to be written to the analog output channel.

Figure 8.67 illustrates how to wire AO Generate Waveform.vi.

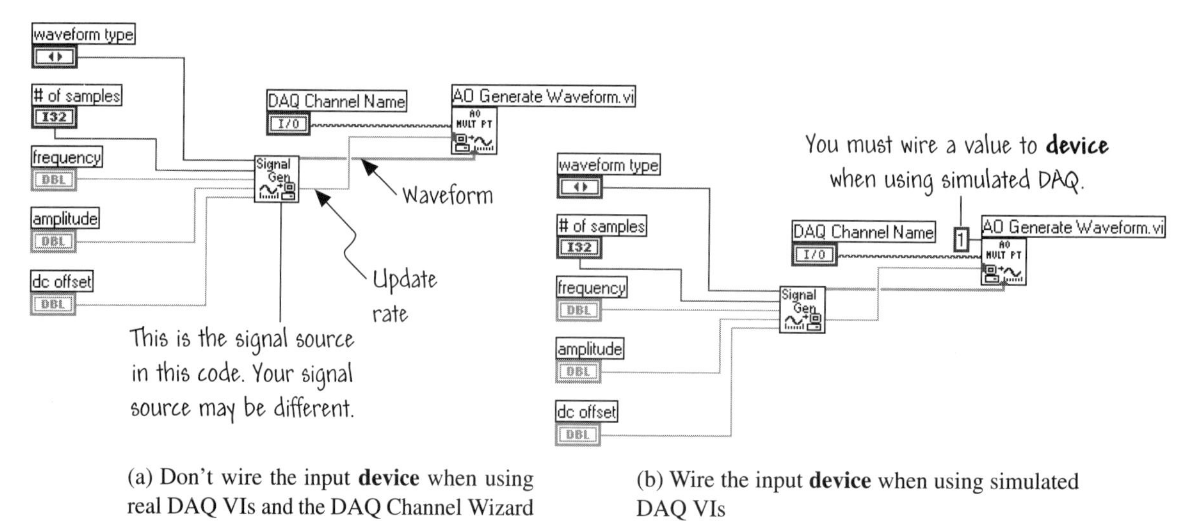

(a) Don't wire the input **device** when using real DAQ VIs and the DAQ Channel Wizard

(b) Wire the input **device** when using simulated DAQ VIs

FIGURE 8.67
The AO Generate Waveform.vi writes multiple values to an analog output channel.

Practice with Waveform Generation

The objective of this example is to produce an analog waveform output using AO Generate Waveform.vi. You will need to again use the MAX to configure the channels.

Open a new VI and build the front panel shown in Figure 8.68. The input variable DAQ Channel Name specifies the output analog channel configured using the MAX. When the front panel is ready, switch to the block diagram and wire the AO Generate Waveform VI, using Figure 8.68 as a guide.

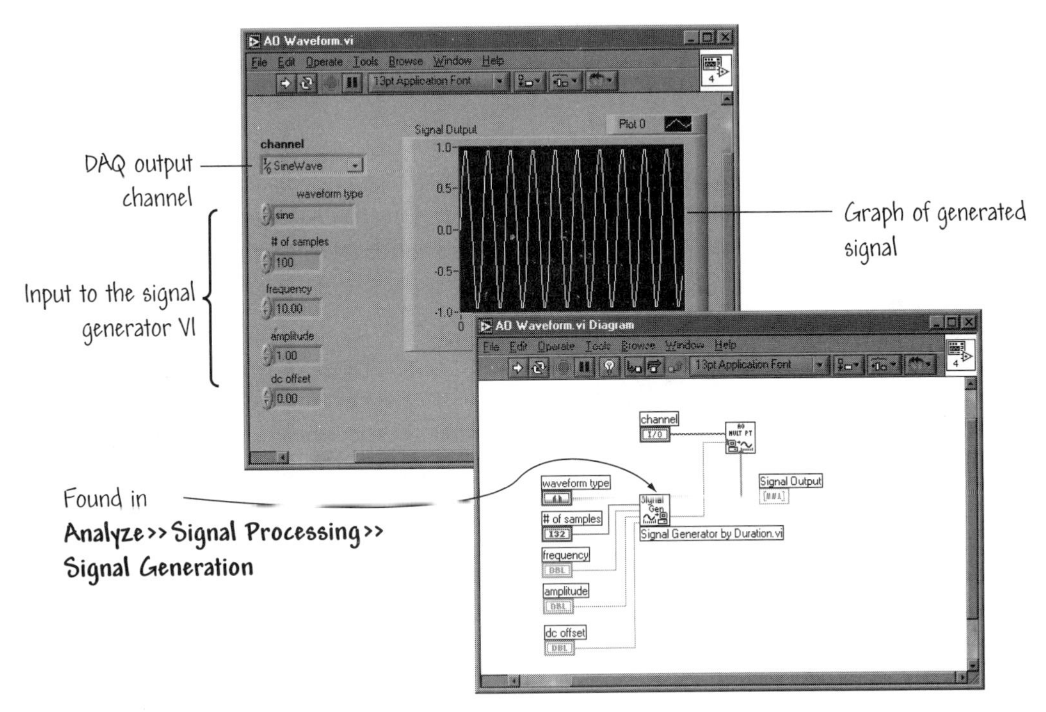

FIGURE 8.68
A VI designed to output multiple points—a single waveform.

To generate a waveform signal within the code, a VI called Signal Generation by Duration.vi is used. This VI can be found in the **Analyze** palette, as seen in Figure 8.68. Figure 8.69 illustrates the path to locating the signal-generation VI.

Open the MAX. The DAQ channel configuration needs to be modified to include the **Sine Wave** on channel 1, as illustrated in the DAQ Channel Viewer in Figure 8.70. To accomplish this task, you can follow the steps shown in Figure 8.71. Initiate the process of adding a new channel configuration, and remember to configure an analog output channel. Use Figure 8.71 as a guide to complete the channel configuration. When you are finished entering the data, click on the **Finish** button located in the lower right corner of the fifth screen.

When you are finished wiring the VI, save it as AO Waveform.vi in User's Stuff in the Learning directory.

If you are having problems with your VI, you can find a working version of the VI shown in Figure 8.68 in Chapter 8 of the Learning directory. The VI is called AO Waveform.vi.

A more advanced version of AO Waveform.vi would allow the DAQ system to output multiple-waveforms. A VI that performs multiple-waveform output is

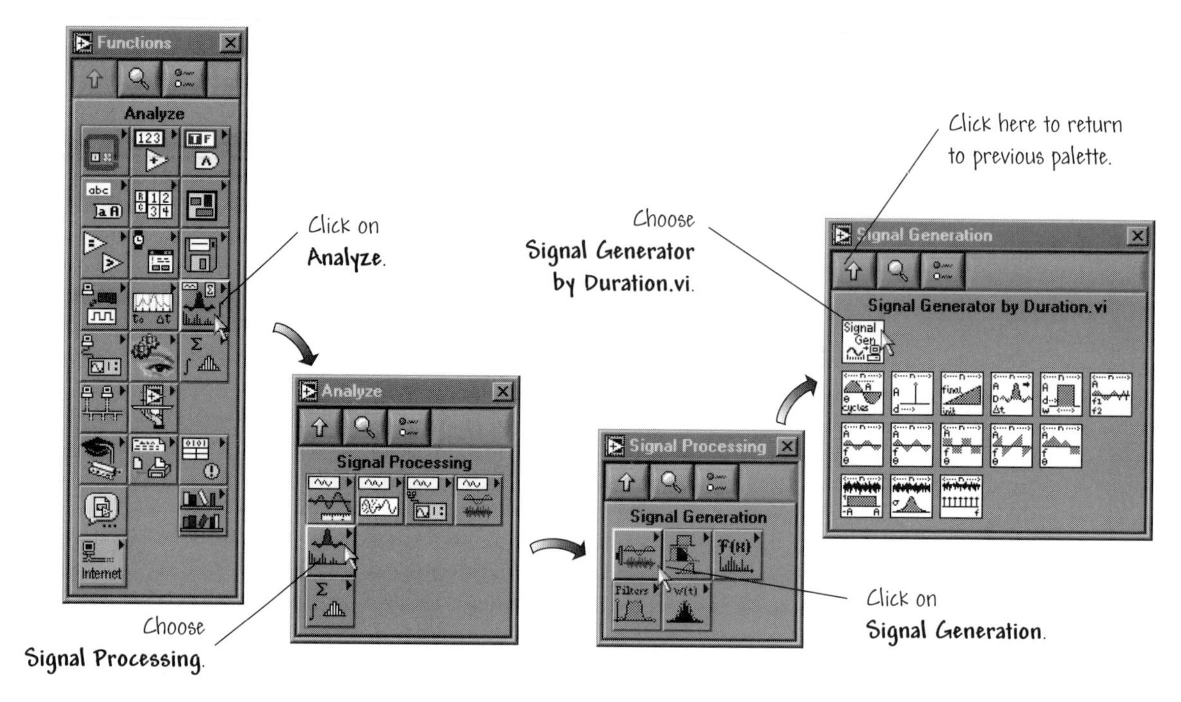

FIGURE 8.69
Using a VI to generate a sine wave signal (and other signals as well).

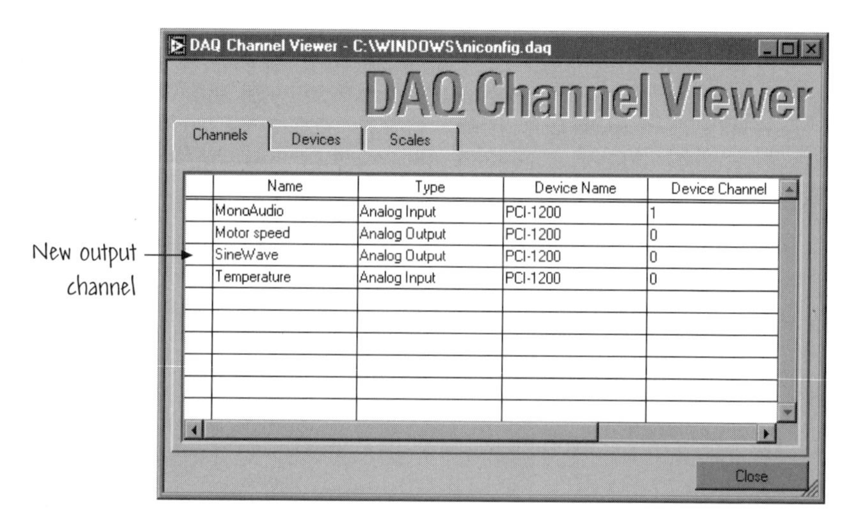

Name	Type	Device Name	Device Channel
MonoAudio	Analog Input	PCI-1200	1
Motor speed	Analog Output	PCI-1200	0
SineWave	Analog Output	PCI-1200	0
Temperature	Analog Input	PCI-1200	0

FIGURE 8.70
The **DAQ Channel Viewer** final configuration, including the **SineWave** output.

shown in Figure 8.72. In this situation, we use the signal-generation VI twice to generate multiple waveforms and the Build Array function to assemble the multi-dimensional array of input data. If you attempt to construct this VI or use the one provided in the Learning directory, you will need to configure the MAX for the

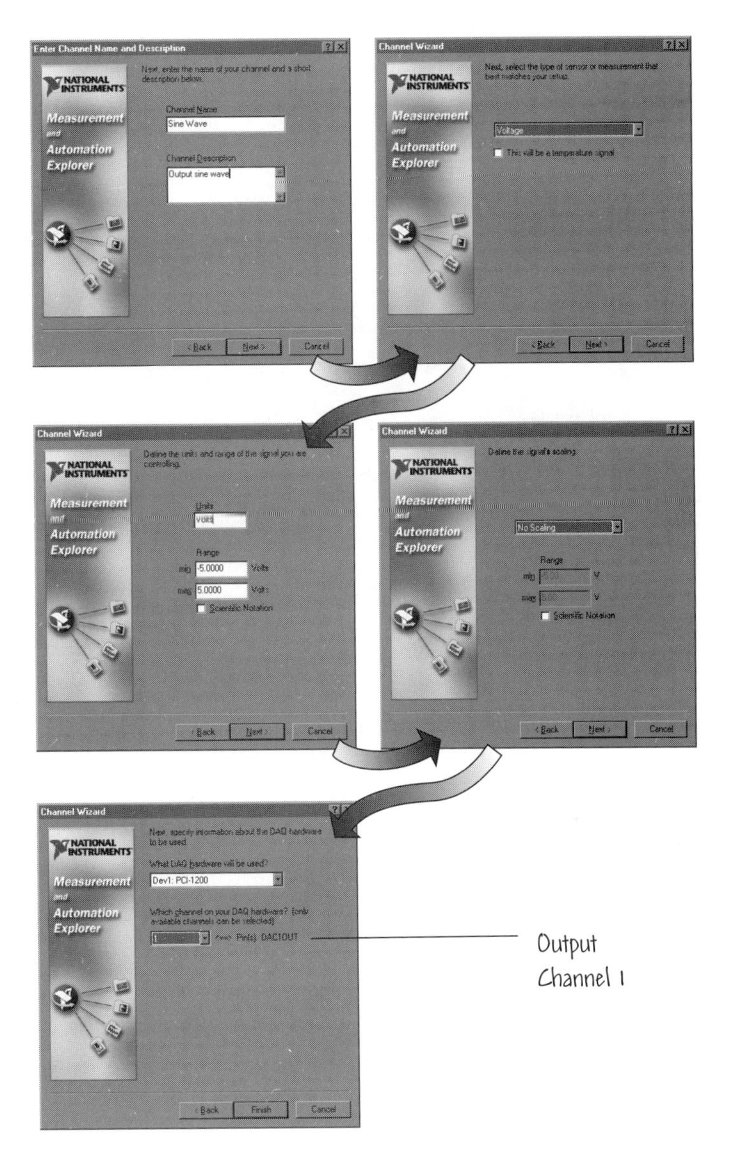

FIGURE 8.71
Configuring the analog output channel using the MAX.

additional output. A version of the VI shown in Figure 8.72 is called AO Waveforms.vi and is located in Chapter 8 of the Learning directory. ◆

8.10 DIGITAL I/O

Digital I/O components on DAQ devices consist of parts that generate or accept binary on/off signals that are often used to control processes, generate patterns for testing, and communicate with peripheral equipment. Digital lines are grouped

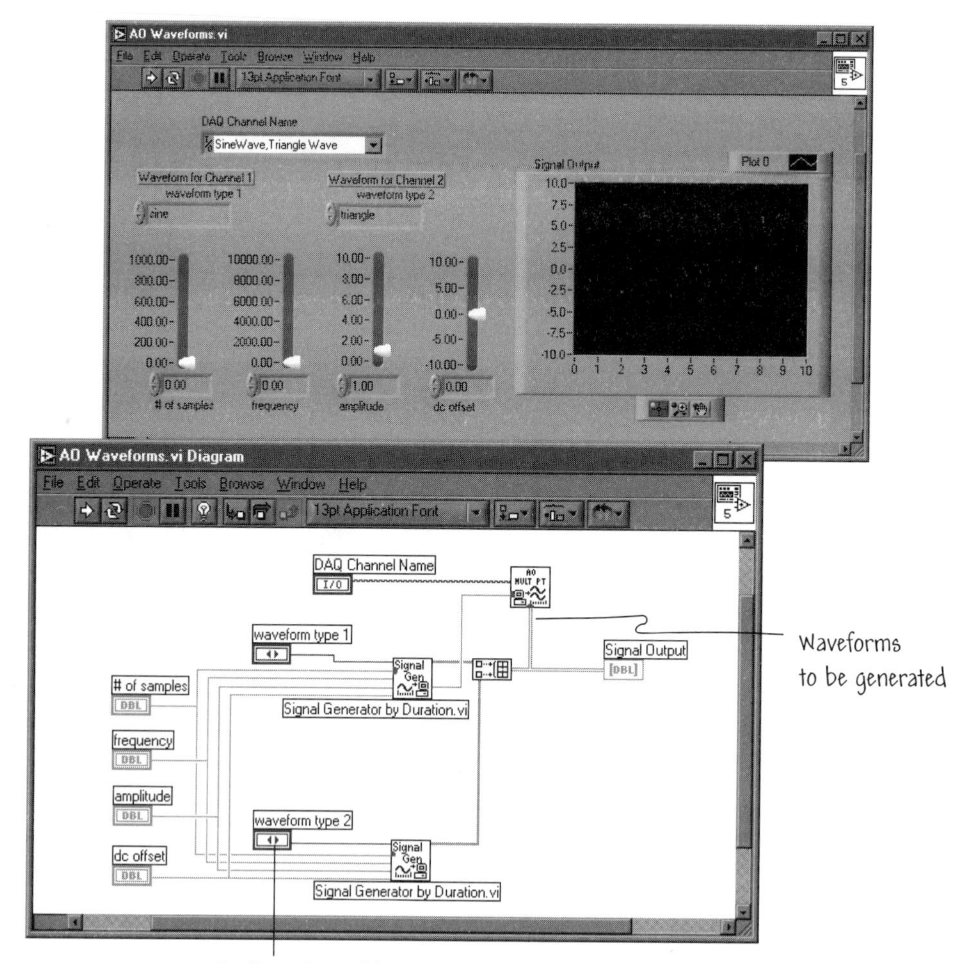

FIGURE 8.72
A VI designed to output multiple waveforms.

into ports generally consisting of four or eight lines per port, as depicted in Figure 8.73. Usually all lines within the same port must all be either input lines or output lines. Since a port contains multiple digital lines, by writing to or reading from a port you can set or retrieve simultaneously the states of multiple lines.

There are two types of digital DAQ acquisition: **immediate** (or nonlatched) and **handshaked** (or latched). With immediate digital I/O, the DAQ system updates the digital lines immediately. This is the only type of digital I/O considered in this chapter. Handshaked digital I/O occurs when a device accepts or transfers data after a digital pulse (that is, the handshake) has been received. There are two

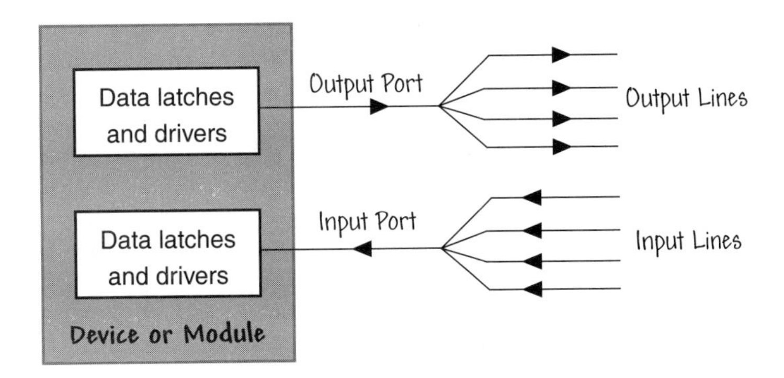

FIGURE 8.73
Digital I/O components on DAQ devices consist of hardware parts that generate or accept binary on/off signals.

types of handshaked digital I/O: **nonbuffered** and **buffered**. Not all devices and modules support latched (handshaked) digital I/O—refer to the documentation for your specific hardware for more information.

In this chapter we focus on the issues surrounding the transfer of digital data across a single port. The most common way to use digital lines is with immediate digital I/O, and all DAQ devices with digital components support this mode. When your VI calls a function in immediate digital I/O mode, the digital line or port states are immediately updated or the current digital value of an input line is returned. You can use the Easy I/O VIs for immediate digital I/O.

LabVIEW inputs or outputs only one value on each digital line in the immediate digital I/O mode. You can completely configure the port (and sometimes the line) direction in software, and you can switch directions repeatedly in a program. A typical example of when you might use immediate digital I/O is in controlling or monitoring relays. You can also use multiple ports or groups of ports to perform digital I/O functions; however, in order to group digital ports, you must use Intermediate or Advanced VIs.

Figure 8.74 shows the Easy I/O VIs and how their various inputs and outputs are wired. The four Easy I/O VIs can read data from or write data to a single digital line or to an entire port immediately.

*Unless you wire a positive value to the **iteration** input on the digital I/O VIs, all of these functions will automatically configure the DAQ board for the appropriate operation each time you call the VI. If you are using the Easy I/O VIs in a loop, consider wiring the **iteration** input to the loop iteratation terminal—this will avoid unneccesary port reconfiguration on every iteration.*

Practice with Digital I/O

The objective of this exercise is to practice with digital input and output using the four LabVIEW Easy I/O VIs. Before proceeding to constructing your VIs you

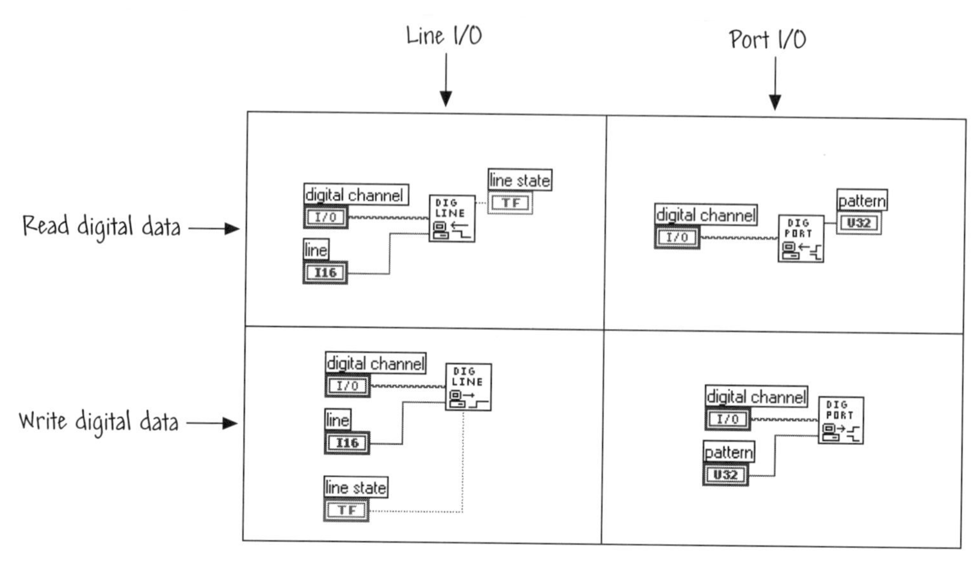

FIGURE 8.74
Wiring the digital Easy I/O VIs.

will need to use the MAX to configure the channels for digital I/O to both a line and to a port.

If you have configured channels using the MAX in the previous examples in this chapter, you are familiar with the process—the process for configuring channels for digital I/O is similar to the process for analog signals. Refer to the various sections in the earlier parts of this chapter on using the MAX if you need to refresh your memory. When you are ready to begin configuring the system, open the MAX, right-click on **Data Neighborhood** and select **Create New...**. As before, at the next screen select **Virtual Channel** and then press **Finish**. When the window **Create New Channel** appears, select **Digital I/O** from the ring. Click **Next** to continue the configuration process.

The digital channel name may refer to either a port or a line in a port. You do not need to specify device, line, or port width, since these inputs are not used by LabVIEW if a channel name is specified in the digital channel input.

Follow the steps shown in Figure 8.75 to configure the **Relay** channel as a **Write to Line** digital channel type. When that is complete, continue the configuration process by following the steps shown in Figure 8.76 to configure the **Relay Group** channel as **Write to Port** digital channel type.

In some applications, you may need to invert the digital values, that is, change a signal from on to off, and vice versa. This is a method of scaling the digital signals. With the MAX you can specify inversion on a per-line basis, so that the scaling of individual lines in a port can have different inversion.

When you are finished configuring the channels for digital output, exit the MAX and construct the two VIs:

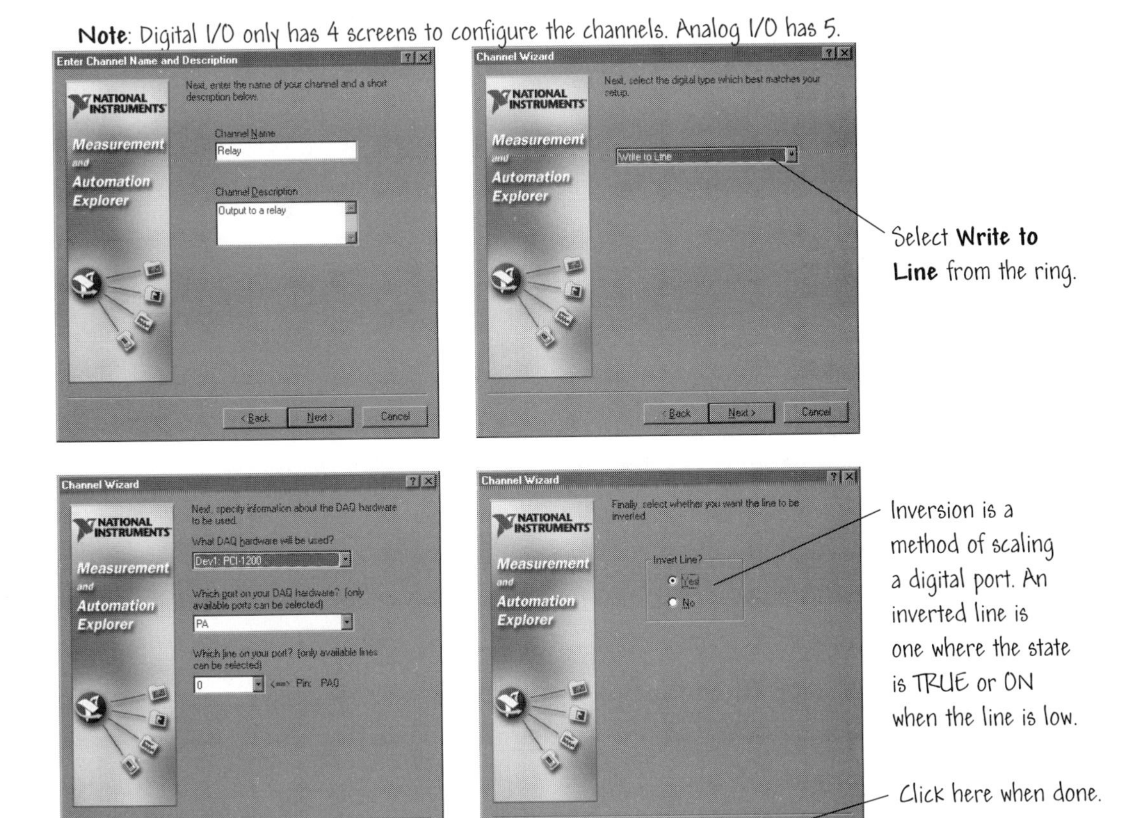

FIGURE 8.75
Configuring the DAQ system to write to a digital line.

- DIO Write 1 Line.vi: Sets a line on a port to a logical high or low state.
- DIO Write Port.vi: Outputs a digital pattern to the specified port.

One new object used in the VI is the Boolean Array to Number function shown on the block diagram in Figure 8.77. This function is located on the **Boolean** subpalette of the **Functions** palette. Use the online help to read about this function. These two VIs write to a line and port, respectively. When you are finished constructing the VIs, save them as DIO Write 1 Line.vi and DIO Write Port.vi in Users Stuff in the Learning directory.

If you are having problems with your VI, you can find working versions of the VIs shown in Figure 8.77 in Chapter 8 of the Learning directory. The VIs are called DIO Write 1 Line.vi and DIO Write Port.vi.

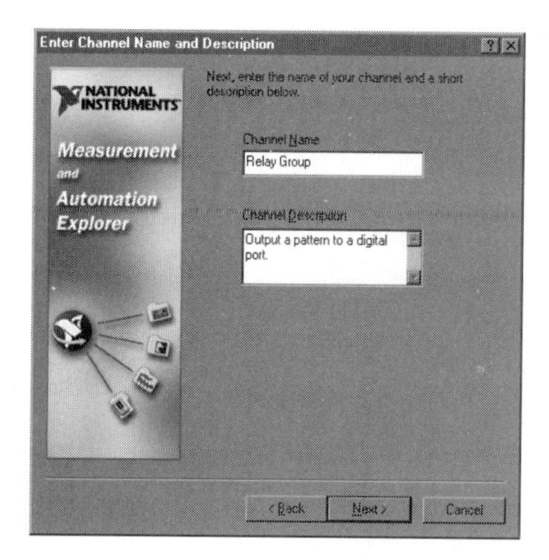

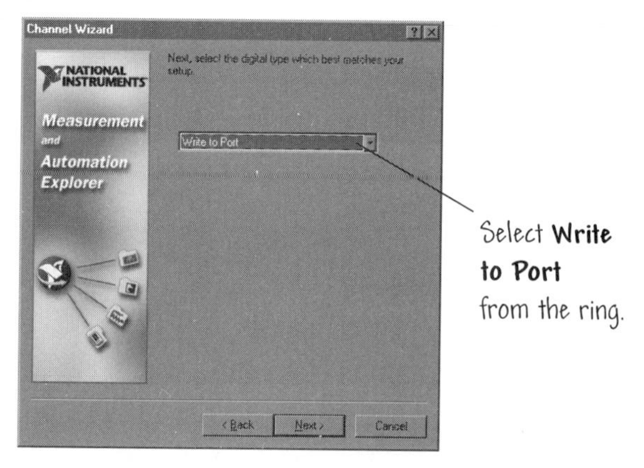

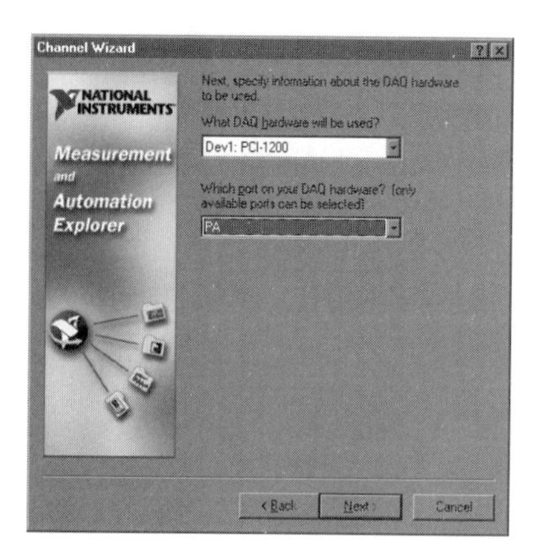

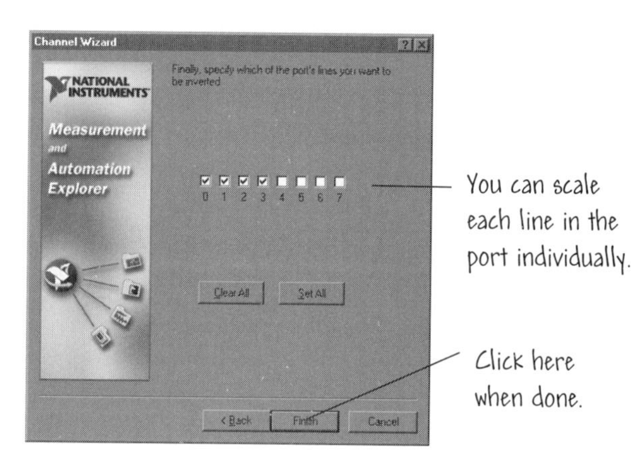

FIGURE 8.76
Configuring the DAQ system to write to a digital port.

A new object used in the VI is the Number to Boolean Array function shown in the block diagram in Figure 8.78. This function can be found on the **Boolean** subpalette of the **Functions** palette. Use the online help to read about this function.

If you are interested in practicing with digital read from a line or port, use the MAX to configure the system for digital read from a line and from a port. Then build the VIs shown in Figures 8.78.

- DIO Read 1 Line.vi: Read the state of a digital line.
- DIO Read Port.vi: Reads the state of all lines in a port.

If you are having problems with your VIs, you can find working versions of the VIs shown in Figure 8.78 in Chapter 8 *of the* Learning *directory. The VIs are called* DIO Read 1 Line.vi *and* DIO Read Port.vi. ◆

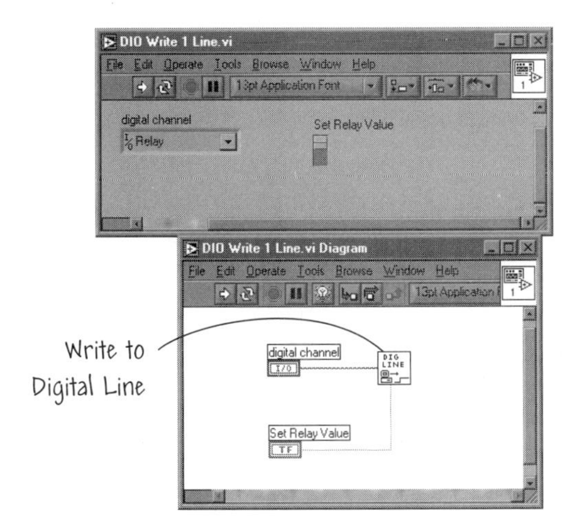

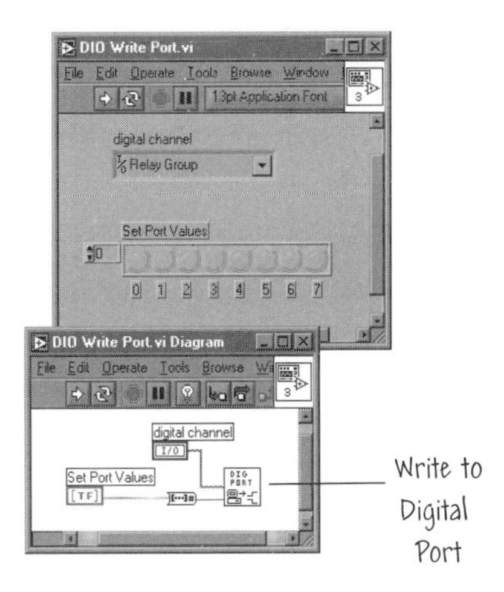

FIGURE 8.77
Two VIs to write to a line and port, respectively.

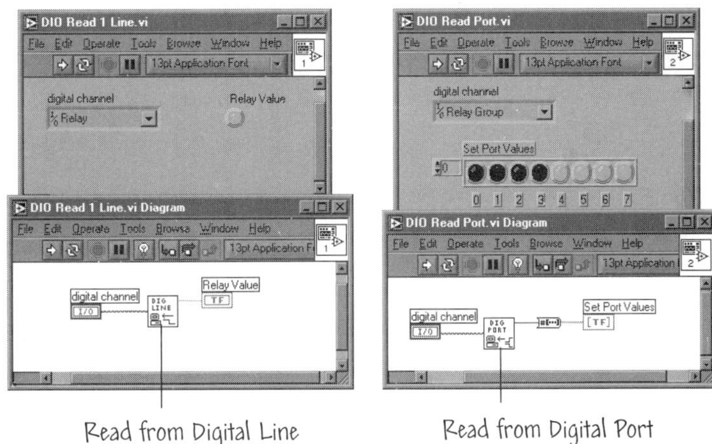

FIGURE 8.78
Two VIs to read from a line and port, respectively.

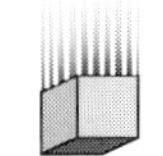

BUILDING BLOCK

8.11 BUILDING BLOCKS: DIGITAL ALARMS

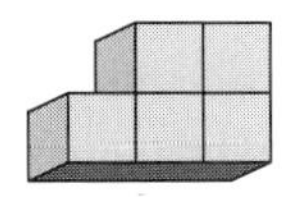

We depart now from the building block exercise based on the volume-measuring system used in Chapters 3–7 and turn to an exercise using data acquisition. The goal is to develop a VI to read from an analog input channel and to write to a digital channel. You should use the MAX to configure the channcls.

A real world example where you can use this VI is in measuring the temperature of a room. If the temperature rises above a certain value, you turn on the air conditioning.

If you have simulated VIs, you must wire a control or constant to the device input on the AI Sample Channels.vi and Write to Digital Port.vi.

You can use the front panel and block diagram of the VI illustrated in Figure 8.79 as a guide for your own work.

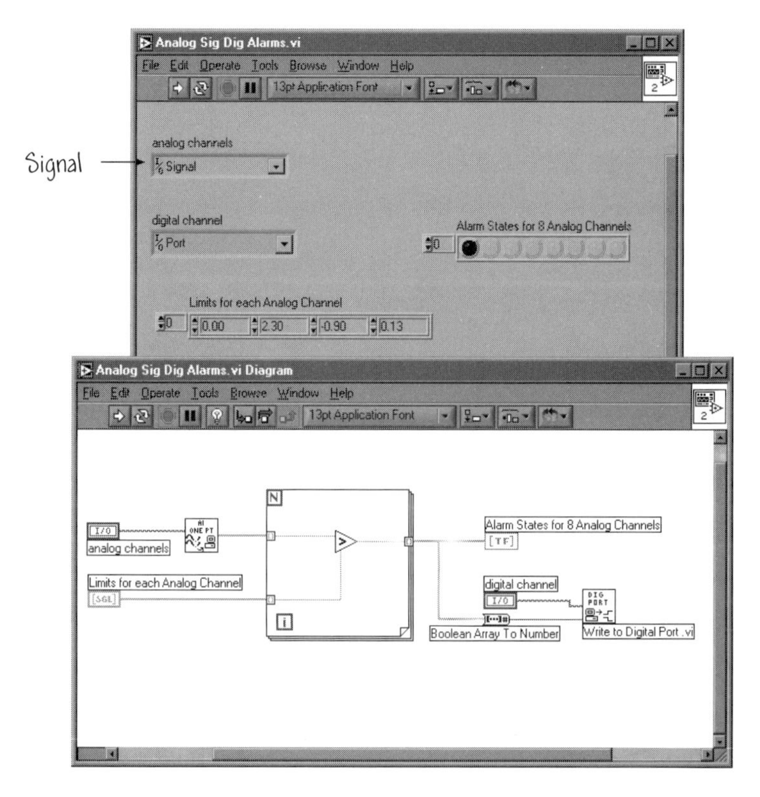

FIGURE 8.79
The front panel and block diagram for the Analog Sig Dig Alarms.vi.

The VI depicted in Figure 8.79 currently has only one analog channel for input and one digital channel for output. You can change that by listing the channel names separated by a comma in **analog channels**. You will have to configure the channels appropriately with the MAX. The analog input should be named **Signal** and the digital output named **Port**. The two inputs **analog channels** and **digital channel** are DAQ Channel name controls.

Before running the VI, you will need to set values for the limits for each analog channel. Run the VI to verify that you are acquiring data properly and writing out to the digital port. You might consider placing probes at strategic locations on the block diagram during the debugging phase. For example, placing a probe prior to the Boolean Array to Number function and one after the same function will help you to visualize the operation of the function. When you are finished with your VI, close it and save it as Analog Sig Dig Alarms.vi in the Users Stuff folder in the Learning directory.

8.12 RELAXING READING: USING DAQ IN STUDENT LABORATORIES

Electrical engineering students at the University of Pennsylvania use virtual instrumentation in a variety of laboratory classes. The Undergraduate Electrical Engineering Laboratory is the primary teaching laboratory and the place where most students are exposed to instrumentation for the first time. In a typical laboratory assignment the students design a virtual instrument that reads and writes analog data, plots the data on an XY graph and also writes the data to a spreadsheet file. An example of the VI used in the circuits and systems subject area is shown in Figure 8.80.

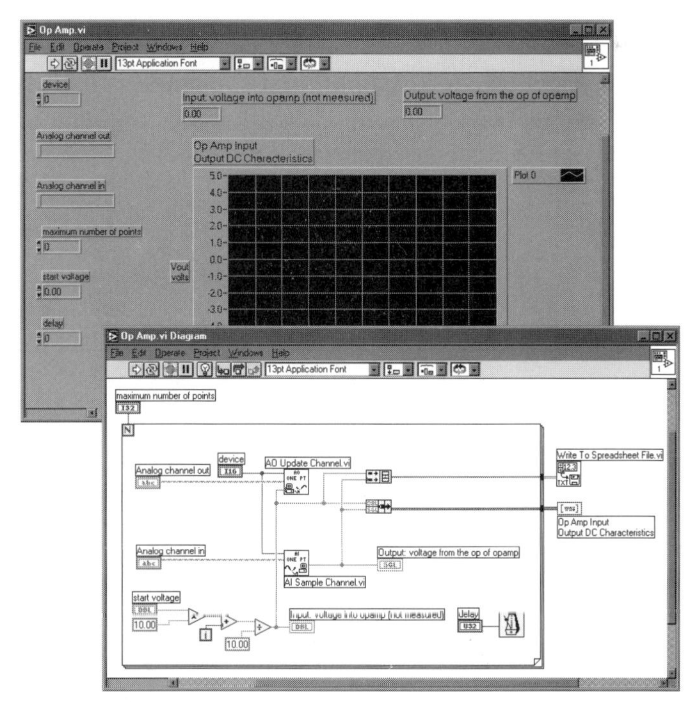

FIGURE 8.80
The front panel and block diagram for the Op Amp.vi

The VI shown in Figure 8.80 is located in the Chapter 8 *folder in the* Learning *directory. It is called* Op Amp.vi. *If you attempt to use the VI remember that you must configure the channels properly (using the Channel Wizard).*

The stated goals of incorporating LabVIEW in the laboratory at the University of Pennsylvania are to introduce students to

- Computer-based data processing concepts
- Laboratory automation concepts
- The idea that software is a powerful tool in instrumentation

The emphasis is on *building* VIs to satisfy the laboratory objectives, rather than using existing VIs.

Future plans include the introduction of the Internet so that students can explore remote data acquisition, and a possible next step would be to integrate *Hi Q* into the curriculum as a way to assist in report writing (more on *Hi Q* in Chapter 12). Some excellent resources for learning more about how virtual instrumentation can be used successfully in the classroom are the following websites:

http://www.ee.upenn.edu/rcal
and
http://www.ee.upenn.edu/rca/software/labview.html

The upenn websites have links to other interesting LabVIEW-oriented websites, and you might consider surfing around to see how other colleges and universities are utilizing virtual instruments.

For more information, contact:

Professor Siddharth Deliwala
Dept. of Electrical Engineering
Room 329, Moore School of Electrical Engineering
University of Pennsylvania
200 S 33rd Street
Philadelphia, PA 19104-6390
e-mail: deliwala@ee.upenn.edu

8.13 SUMMARY

DAQ systems consist of the following elements:

- Signals
- Transducers
- Signal-conditioning hardware

- DAQ board or module
- Application software

There are two types of signals:

- Analog—provides level, shape, or frequency content information
- Digital—provides state or rate information

With many transducers it is necessary to provide for signal conditioning. Some types of signal conditioning are

- Amplification
- Transducer excitation
- Linearization
- Isolation
- Filtering

There are two types of signal sources:

- Grounded sources—devices that plug into the building ground
- Floating sources—isolated from the building ground system

There are three types of measurement systems:

- Differential—use this type whenever possible!
- Referenced single-ended—use single-ended types if you require more channels
- Nonreferenced single-ended

Multifunction DAQ boards typically include:

- Analog-to-digital convertors (ADCs)
- Digital-to-analog convertors (DACs)
- Digital I/O ports
- Counter/timer circuits

When configuring the DAQ board, you should consider how the following parameters will affect the quality of the digitized signal:

- Resolution: Increasing resolution increases the precision of the ADC.
- Range: Decreasing range increases precision.
- Limit settings: Changing limit settings to reflect the signal range increases precision.

The Measurement & Automation Explorer is a utility that helps configure the channels on the DAQ board according to the sensors to which they are connected.

The LabVIEW DAQ VIs are organized into palettes corresponding to the type of operation involved—analog input, analog output, counter operations, or digital I/O. Under this palette, the DAQ VIs are organized into six palettes (we covered the first three topics in this chapter):

- **Analog Input**
- **Analog Output**
- **Digital I/O**
- **Counter**
- **Calibration and Configuration**
- **Signal Conditioning**

Each palette contains VIs or palettes of VIs organized as Easy I/O, Intermediate, Utility, and Advanced VIs. The Easy I/O VIs consist of high-level VIs that perform basic analog input, analog output, and digital I/O. We discussed the Easy I/O VIs only—they are ideal for simple analog and digital I/O or for getting started with DAQ in LabVIEW.

KEY TERMS

Analog-to-digital converter (ADC): An electronic device (often an integrated circuit) that converts an analog voltage to a digital number.

ADC resolution: The resolution of the ADC measured in bits. An ADC with 16 bits has a higher resolution (and thus a higher degree of accuracy) than a 12-bit ADC.

Bipolar: A signal range that includes both positive and negative values (e.g., −5 V to 5 V).

Channel: Pin or wire lead where analog or digital signals enter or leave a data acquisition device.

Channel name: A unique name given to a channel configuration in the Measurement & Automation Explorer.

Code width: The smallest detectable change in an input voltage of a DAQ device.

Digital-to-analog convertor (DAC): An electronic device (often an integrated circuit) that converts a digital number to an analog voltage or current.

DAQ Channel Wizard: Utility that guides you through naming and configuring your DAQ analog and digital channels.

DAQ Solution Wizard: Utility that generates solutions for DAQ applications.

Data acquisition (DAQ): Process of acquiring data from plug-in devices.

Differential measurement system: A method of configuring your device to read signals in which you do not connect inputs to a fixed reference (such as the earth or a building ground).

Floating signal sources: Signal sources with voltage signals that are not connected to an absolute reference or system ground. *Also* called nonreferenced signal sources.

Grounded signal sources: Signal sources with voltage signals that are referenced to system ground, such as the earth or building ground. *Also* called referenced signal sources.

Handshaked digital I/O: A type of digital I/O where a device accepts or transfers data after a signal pulse has been received. *Also* called latched digital I/O.

Immediate digital I/O: A type of digital I/O where the digital line or port is updated immediately or returns the digital value of an input line. *Also* called nonlatched digital I/O.

Input/output (I/O): The transfer of data to or from a computer system involving communication channels and DAQ interfaces.

Limit settings: The settings that you specify as the maximum and minimum voltages on analog input signals.

Linearization: A type of signal conditioning in which the voltage levels from transducers are linearized, so that the voltages can be scaled to measure physical phenomena.

LSB: Least significant bit.

Measurement & Automation Explorer: Provides access to all National Instruments DAQ, GPIB, IMAQ, IVI, Motion, VISA, and VXI devices.

Non-referenced single-ended (NRSE) measurement system: All measurements are made with respect to a common reference, but the voltage at this reference may vary with respect to the measurement system ground.

Range: The minimum and maximum analog signal levels that the analog-to-digital convertor can digitize.

Referenced single-ended (RSE) measurement system: All measurements are made with respect to a common reference or ground. *Also* called a grounded measurement system.

Sampling rate: The rate at which the DAQ board samples an incoming signal.

SCXI: Signal Conditioning eXtensions for Instrumentation. The National Instruments product line for conditional low-level signals within an external chassis near the sensors so that only high-level signals in a noisy environment are sent to the DAQ board.

Signal conditioning: The manipulation of signals to prepare them for digitizing.

Unipolar: A signal range that is always positive (e.g., 0 V to 10 V).

Update rate: The rate at which voltage values are generated per second.

EXERCISES

E8.1 Assume you are sampling a transducer that varies between 80 mV and 120 mV. Your board has a voltage range of 0 to +10 V, ±10 V, and ±5 V. What limit setting would you select for maximum precision if you use a DAQ board with 12-bit resolution? Use the limit setting values in Table 8.3 for your calculations.

E8.2 From the LabVIEW startup screen, click on the **Search Examples** button. Click on the following links: I/O Interfaces≫Data Acquisition (DAQ)≫Software-

Created Instruments≫One Channel Simple Oscilloscope. This VI is a good example of how virtual instruments emulate actual instruments. Run this VI.

E8.3 If you have **Search Examples** open to Software-Created Instruments, click on the Benchtop Function Generator link. To open the **Search Examples**, refer to the previous exercise. This VI is another good example of how virtual instruments emulate actual instruments. Run this VI.

PROBLEMS

P8.1 You are a production engineer at a plant and you want to continuously measure the flow of fluid through pipes. The sensor that measures flow produces an analog signal that you can acquire. Create a VI that continuously acquires analog data until you press a stop button on the front panel. As you acquire data, display it on a chart. Hint: You will need to use the intermediate analog input VIs.

P8.2 You are acquiring analog data that measures the volume of a tank. You only want to take continual measurements when the machinery is on. Turning on the machinery can act as a trigger, specifically a digital trigger, to start measuring data. Write a VI that acquires analog data only after a digital trigger has occurred.

P8.3 Complete the crossword puzzle.

Across

7. The settings specifed as max and min voltages on analog input signals.
9. The min and max analog signal levels that the a-to-d convertor can digitize.
10. All measurements are made with respect to a common reference or ground.
13. The manipulation of signals to prepare them for digitizing.
14. A unique name given to a channel configuration in the MAX.
15. The smallest detectable change in an input voltage of a DAQ device.

Down

1. A signal range that is always positive.
2. All measurements are made with respect to a common reference, but the voltage may vary with respect to the ground.
3. Process of acquiring data from plug-in devices.
4. A type of signal conditioning in which the voltage levels from transducers are linearized.
5. A signal range that includes both positive and negative values.
6. Least significant bit.
8. The rate at which the DAQ board samples an incoming signal.
11. Pin or wire lead where signals enter or leave a DAQ device.
12. The rate at which voltage are generated.

CHAPTER 9

Strings and File I/O

This chapter introduces strings and file input/output (I/O). You have used strings in a limited fashion throughout the book, but here we discuss them more formally. In instrument control applications, numeric data is commonly passed as character strings, and LabVIEW has many built-in string functions that allow you to manipulate the string data. Writing and reading data from files also utilizes strings. We will discuss how to use high-level file I/O VIs to save data to and retrieve data from a disk file. High-level VIs perform the three basic functions associated with file I/O: opening or creating a file, writing or reading data from the file, and closing the file.

GOALS

1. Practice creating string controls and indicators.
2. Be able to convert a number to a string, and vice versa.
3. Learn to use the file I/O VIs to write and read data to a disk file.
4. Understand how to write data to a file in a format compatible with many common spreadsheet applications.

9.1 STRINGS

A **string** is a sequence of characters—these can be displayable or nondisplayable characters. Strings are often used in **ASCII** text messages. In LabVIEW, strings are also used in instrument control when numeric data is passed as character strings and subsequently converted back to numbers. Another situation requiring the use of strings is in storing numeric data, where numbers are first converted to strings before writing them to a file on disk. In this chapter, we will discuss strings and their various uses in file input/output (I/O).

String controls and indicators are in the **String & Path** subpalette of the **Controls** palette, as illustrated in Figure 9.1. As discussed in previous chapters (see

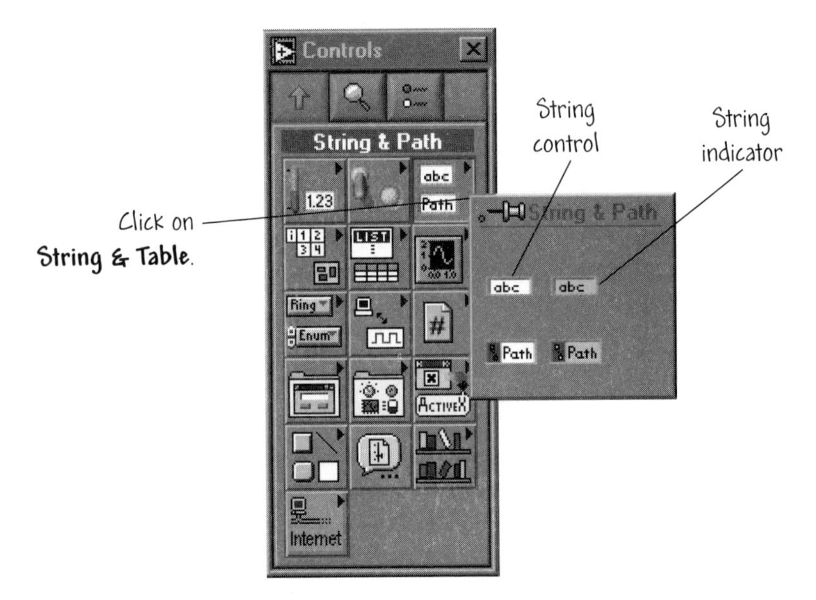

FIGURE 9.1
The **String & Path** subpalette.

Chapter 3 for example), you enter and change text inside a string control using the **Operating** tool or the **Labeling** tool. If there is not enough room to fit your text in the default size of the string control, you can enlarge the string controls and indicators by dragging a corner with the **Positioning** tool. If front panel space is limited, you can use a scrollbar to minimize the space that a front panel string control or indicator occupies, as illustrated in Figure 9.2. The Visible Items≫Scrollbar option is located on the string short cut menu. If there is not enough room to place the scrollbar within the string control or indicator, the scrollbar option will be dimmed. This indicates that you must increase the vertical size if you want a scrollbar.

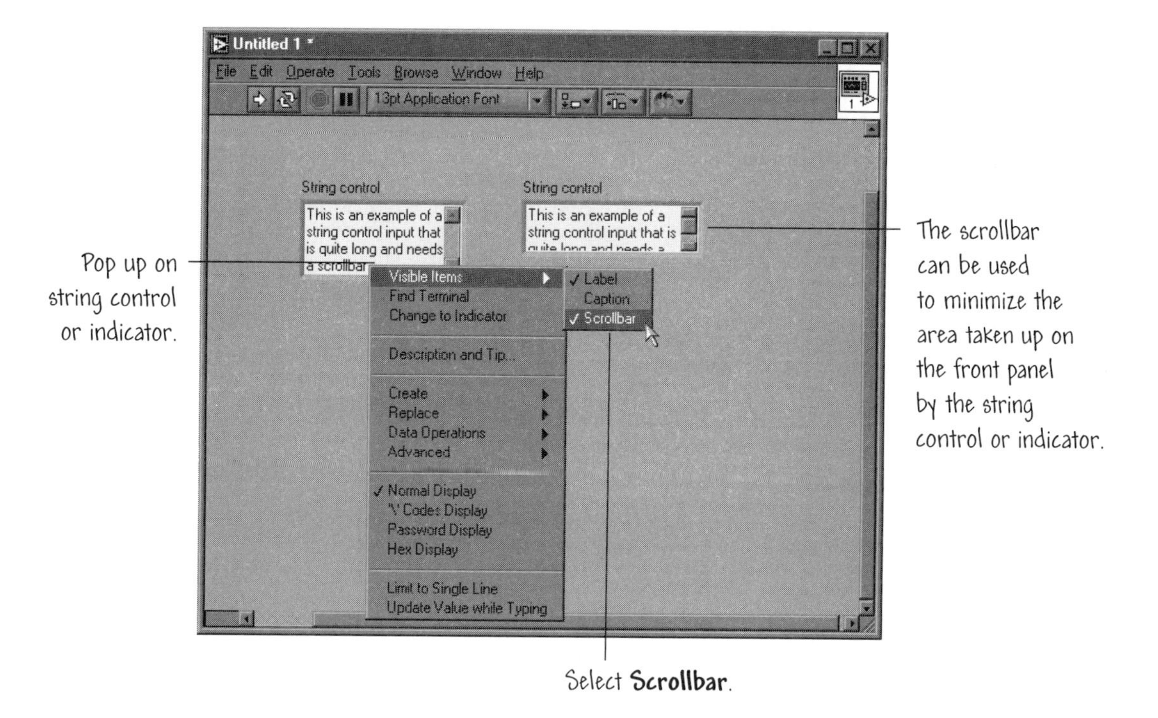

FIGURE 9.2
The scrollbar can be used to minimize the size of string controls and indicators.

You can configure string controls and indicators for different types of display, such as passwords, \ codes, and hex. String controls and indicators can display and accept characters that are usually **nondisplayable**—backspaces, carriage returns, tabs, and so on. Choose '/' **Codes Display** from the string's short cut menu to display these characters. In the '/' **Codes Display** mode, nondisplayable characters appear as a backslash followed by the appropriate code. A complete list of codes appears in Table 9.1.

To enter a nondisplayable character into a string control, click the appropriate key, like space or tab, or type the backslash character \, followed by the code for the character. As shown in Figure 9.3a, after you type in the string and click the **Enter** button, any nondisplayable characters appear in backslash code format.

The characters in string controls and indicators are represented internally in ASCII format. You can view the ASCII codes in hex by choosing **Hex Display** from the string's short cut menu, as shown in Figure 9.3b. You can also choose a password display by enabling the **Password Display** option from the string short cut menu, as shown in Figure 9.3c. With this option selected, only asterisks appear in the string's front panel display, although on the block diagram the string data reflects the input string. This allows you to set up a security system requiring a password key before proper operation of the VI.

TABLE 9.1 A list of backslash codes

Code	G Interpretation
\00 – \FF	Hex value of an 8-bit character; must be uppercase.
\b	Backspace (ASCII BS, equivalent to \08).
\f	Form feed (ASCII FF, equivalent to \0C).
\n	Linefeed (ASCII LF, equivalent to \0A).
\r	Carriage return (ASCII CR, equivalent to \0D).
\t	Tab (ASCII HT, equivalent to \09).
\s	Space (equivalent to \20).
\\	Backslash (ASCII \, equivalent to \5C).

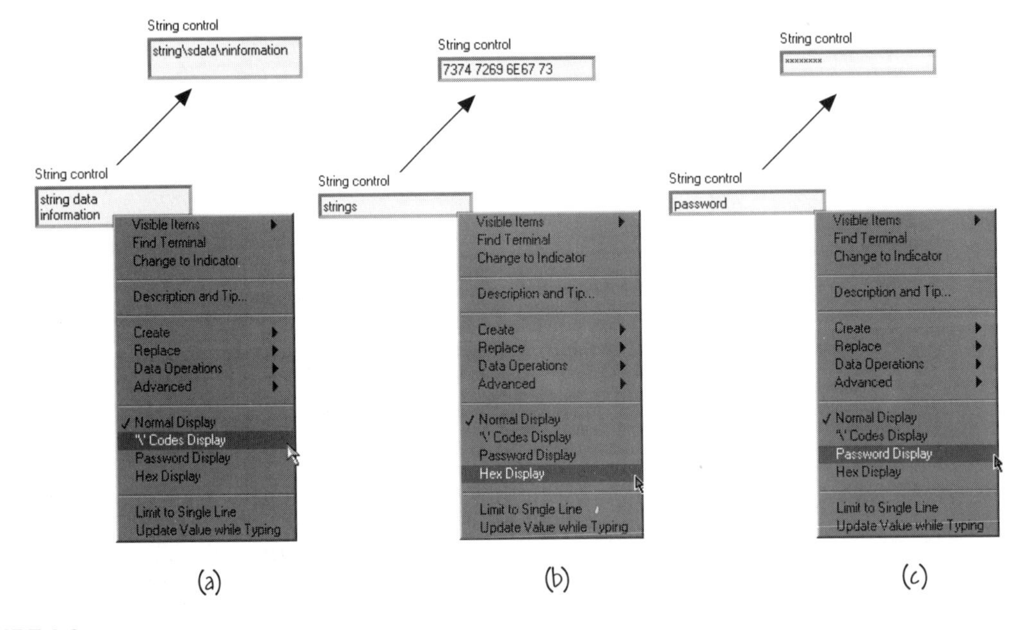

FIGURE 9.3
(a) Displaying characters that are usually nondisplayable. (b) Choosing **Hex Display** to view the ASCII codes in hex. (c) Choosing the **Password Display** option from the pop-up menu.

Practice with Manipulating Strings

In this exercise, you will practice with different ways to manipulate strings using three VIs:

- **Format Into String**: Concatenates and formats numbers and strings into a single output string.

- **Scan from String**: Scans a string and converts valid numeric characters (0 to 9, +, −, e, E, and period) to numbers.
- **Match Pattern**: Searches for an expression in a string, beginning at a specified offset, and if it finds a match, splits the string into three substrings—the substring before the matched substring, the matched substring itself, and the substring that follows the matched substring.

The VI that you will create is intended to interact with a digital multimeter (DMM). Open a new VI and construct a front panel using Figure 9.4 as a guide. Place two string controls, one numerical control, and one string indicator on the front panel. Switch to the block diagram. Select **Format Into String** in the **String** subpalette, as shown in Figure 9.4. In this exercise, this function converts the number you specify to a string and concatenates the inputs to form a command

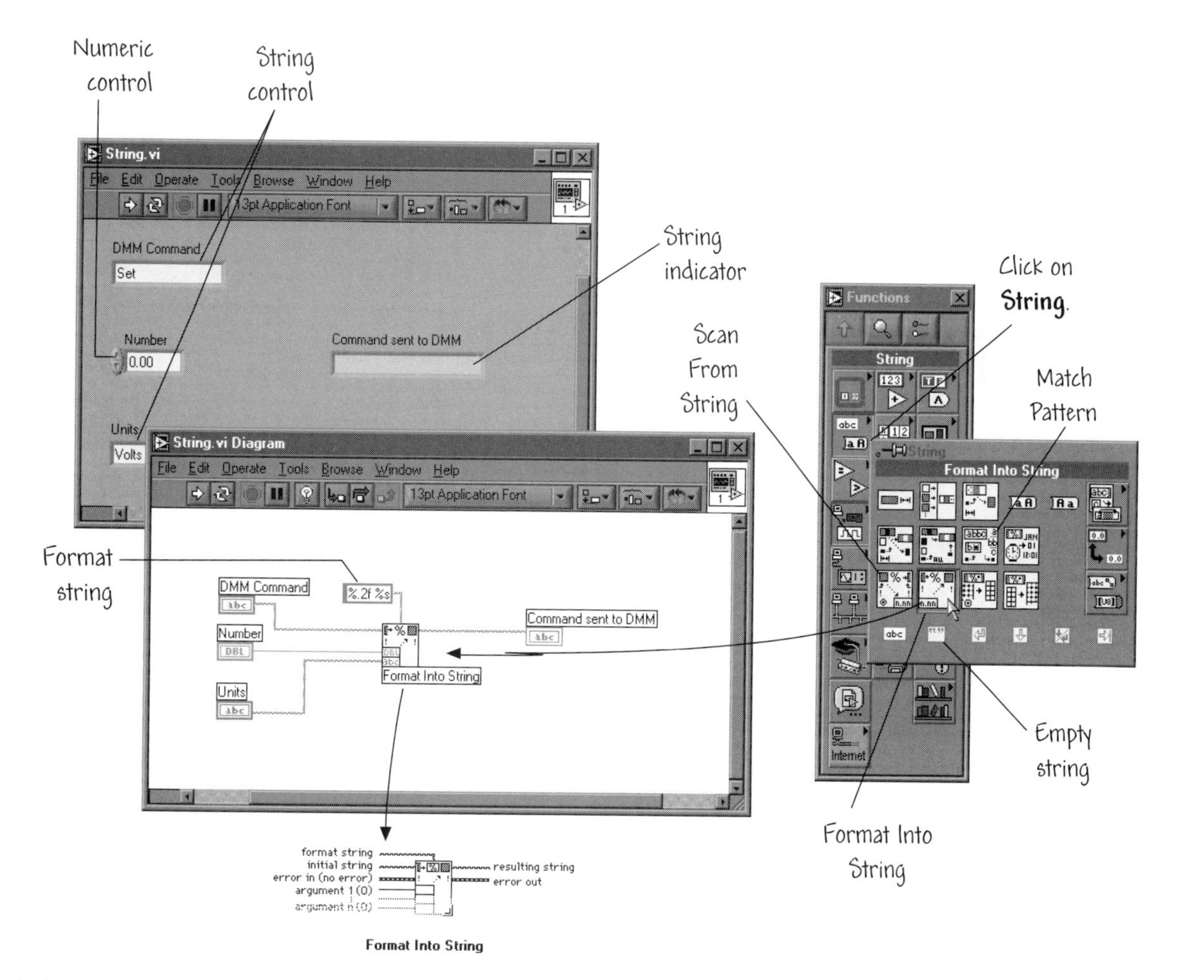

FIGURE 9.4
Constructing a VI using the Format Into String function to build a command to send to the DMM.

for the DMM. Using the **Positioning** tool, enlarge **Format Into String** so that two inputs appear in the lower left corner. Wire the numeric control **Number** to the first input argument (as shown in Figure 9.4). Wire the string control **Units** to the second input argument; the input type for Format Into String will automatically change from **DBL** to **abc** upon wiring. Wire the string control **DMM Command** to the **initial string** input at the top left. Finally, wire the output **resulting string** to the string indicator **Command sent to DMM**.

You can create strings according to a format specified using format strings. With format strings you can specify the format of arguments—the field width, base (hex, octal, and so on), and any text that separates the arguments. The format string can be seen in Figure 9.4 (wired at the top of the Format Into String function). A dialog box can be used to obtain the desired format string. On the block diagram, pop up on the Format Into String function and select **Edit Format String**, as illustrated in Figure 9.5. You can also double click on the node to access the dialog box. Notice that the **Current Format Sequence** contains the argument types in the order that you wired them—**Format fractional number** and **Format string**. The **Format fractional number** corresponds to the

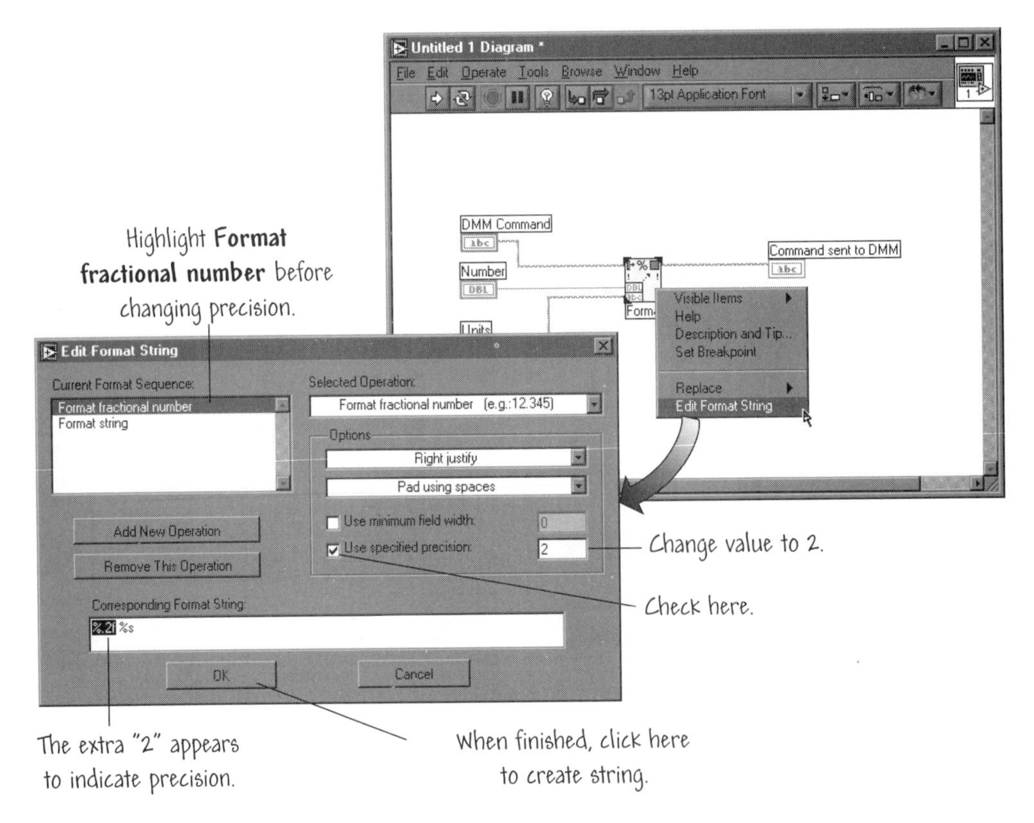

FIGURE 9.5
Setting the desired format string using the dialog box.

input value from the **Number** control and **Format string** corresponds to the input value from the **Units** control.

The only change we want to make is to set the precision of the numeric to 2. To do this, highlight **Format fractional number** in the **Current Format Sequence** list box, click in the **Use Specified Precision** checkbox, and type in the number 2 in the box. When you are finished, press <Enter> (**Windows**) or <return> (**Macintosh**) and you should see the **Corresponding Format String** (near the bottom of the dialog box). Press the **Create String** button to automatically insert the correct format string information and wire the format string to the function, as shown in Figure 9.4.

Return to the front panel and type text inside the two string controls and a number inside the digital control, and run the VI. You now have a VI that can concatenate strings and numbers to form a command that can be sent to an external instrument, such as the digital multimeter. Save the VI as String.vi in the Users Stuff directory.

To continue the exercise, we want to add the capability to scan a string and convert any valid numeric characters (0 to 9, +, −, e, E, and period) to numbers. Use Figure 9.6 as a guide and add a string control and numeric indicator to the front panel. Then switch to the block diagram and wire the Scan from String

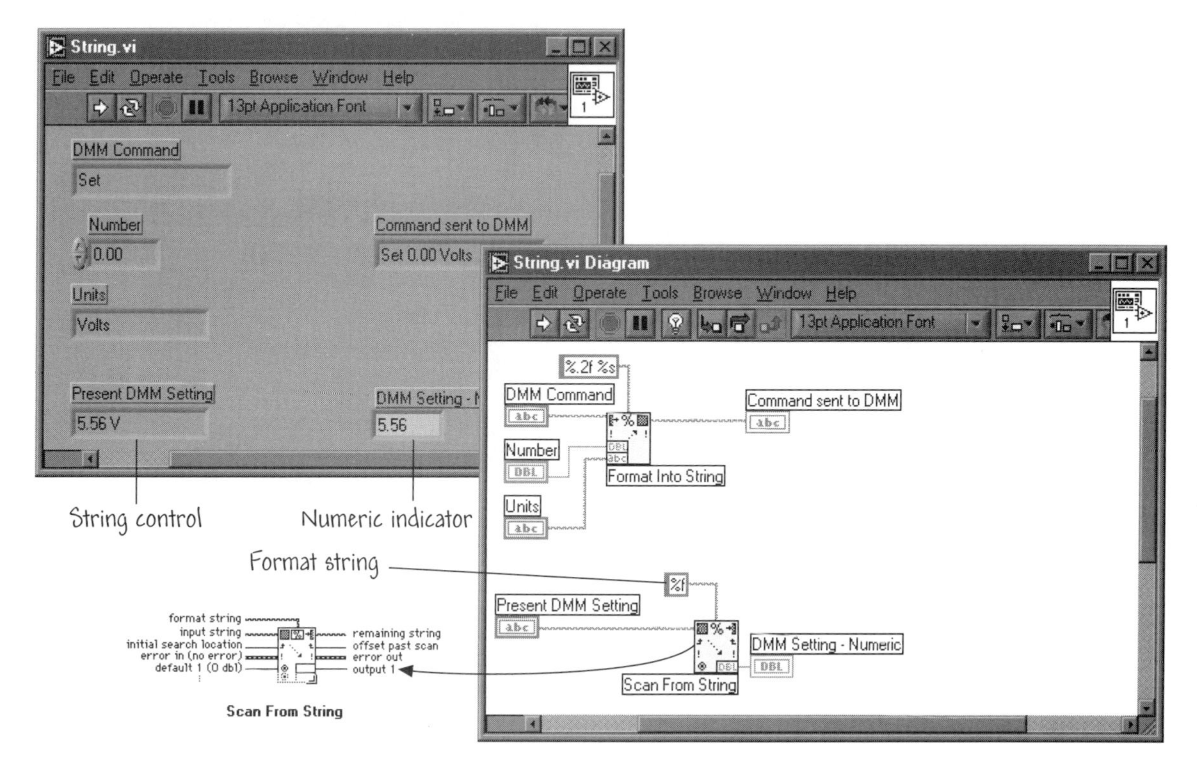

FIGURE 9.6
Constructing a VI using the Scan from String function to extract the present DMM setting from a string.

function, as shown in Figure 9.6. The function itself is located on the **String** palette, as illustrated in Figure 9.4. After the wiring is complete, double click on the function (or pop up on the function and choose **Edit Format String**) to open the dialog box. The current scan sequence should indicate **Scan Number** and the corresponding scan string should show %f. This is fine, so just select **OK** and make sure that the **format string** input (at the top of the **Scan from String** function) is properly wired (see Figure 9.6). To test the VI, type in **5.56 V** in the Present DMM Setting string control and run the VI. You should find that the VI extracts the numeric value 5.56 and displays that value in the numeric indicator DMM Setting - Numeric. Try other voltage values in the input.

Finish with this VI, using Figure 9.7 as a guide. The last addition to the VI uses the Match Pattern function in conjunction with the Scan From String function to detect and extract a series of DMM data points. The Match Pattern function detects the matched substring (in this case, a comma) and outputs the substring before the comma to the **Scan From String** function which scans the substring and outputs a detected number. The For Loop iterates until the end of the string is detected—in other words, an empty string is found. You can find the empty string constant on the **Strings** palette, as illustrated in Figure 9.4.

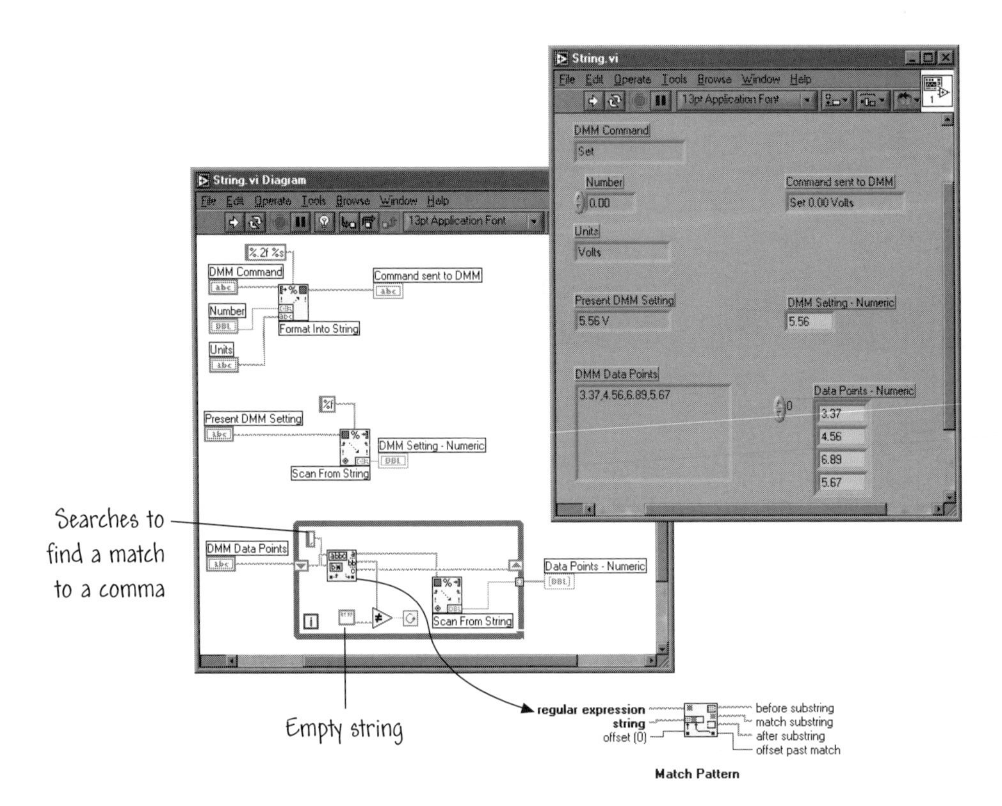

FIGURE 9.7
Constructing a VI using the Match Pattern function to detect and extract a series of DMM data points.

When your VI has been wired properly, enter a few numbers in the string control **DMM Data Points**, such as 3.37,4.56,6.89,5.67. Run the VI and verify that the output array **Data Points - Numeric** contains the four numbers.

A working version of String.vi can be found in Learning\Chapter 9. Refer to the working version if your VI is not working as you think it should. ◆

9.2 FILE I/O

File input and output (I/O) operations are used to store and retrieve data from files on disk. These operations generally involve three basic steps:

- Open an existing file or create a new file.
- Write to or read from a file.
- Close a file.

LabVIEW provides file I/O functions that comprise a powerful and flexible set of tools for working with files. Figure 9.8 shows the **File I/O** palette. You can

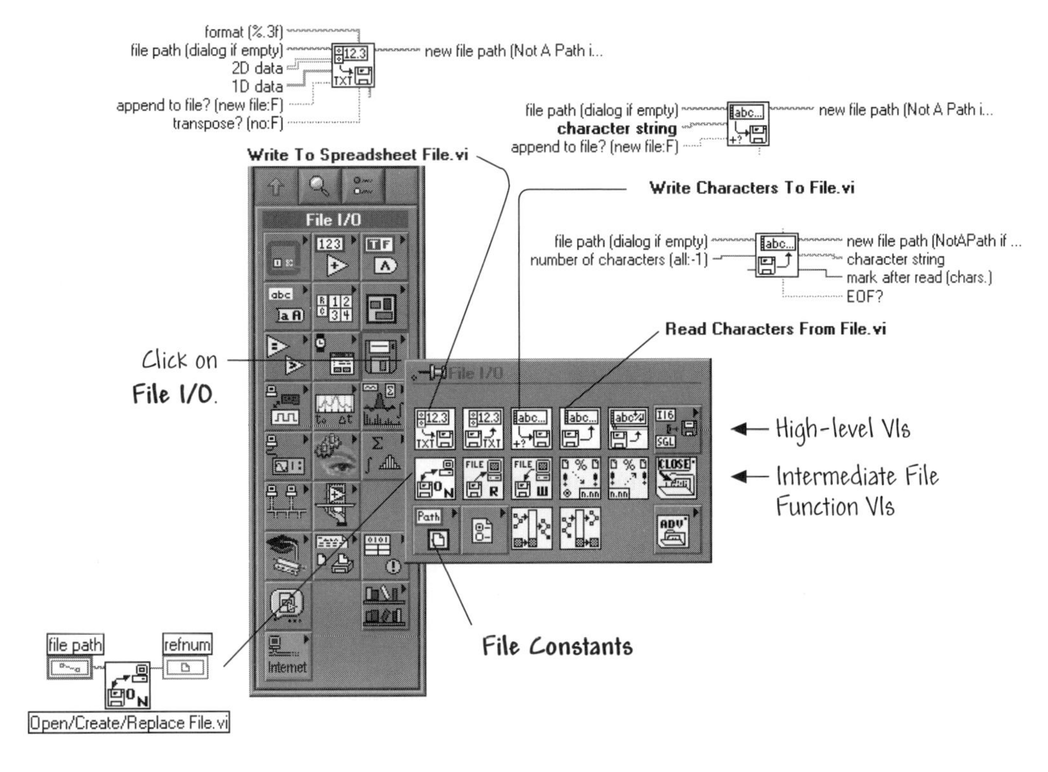

FIGURE 9.8
The **File I/O** palette.

use the file VIs to write or read the different types of data:

- Strings to text files
- One-dimensional (1D) or two-dimensional (2D) arrays of single-precision numbers to spreadsheet text files.
- 1D or 2D arrays of numbers to binary files.

The file I/O VI library of functions comprises high-level and intermediate File Function VIs. The high-level file VIs call the intermediate File Function VIs to perform complete, easy-to-use file operations. The high-level VIs open or create a file, write or read to it, and close it. If an error occurs, these VIs display a dialog box that describes the problem and gives you the option to halt execution or to continue. The high-level file VIs are located on the top row of the palette (see Figure 9.8) and consist of the following VIs:

- **Write Characters To File**: Writes a character string to a new byte stream file or appends the string to an existing file.
- **Write To Spreadsheet File**: Converts a 2D or 1D array of single-precision (SGL) numbers to a text string and writes the string to a new byte stream file or appends the string to an existing file.
- **Read Characters From File**: Reads a specified number of characters from a byte stream file, beginning at a specified character offset.
- **Read From Spreadsheet File**: Reads a specified number of lines or rows from a numeric text file, beginning at a specified character offset, and converts the data to a 2D, single-precision array of numbers.
- **Read Lines From File**: Reads a specified number of lines from a text file, beginning at a specified character offset.
- **Binary File** VIs: A subpalette that contains VIs that read and write 16-bit (word) integers or single-precision floating-point numbers to binary files.

The intermediate File Function VIs perform one file operation at a time. These VIs perform error detection in addition to their other functions. The most commonly used intermediate File Function VIs are located on the second row of the palette. The remaining functions are located in the **Advanced** subpalette. In addition to reading and writing data, the file I/O functions can move and rename files and directories, return volume information about a file or directory, and return the end-of-file location.

For example, to create a new file or replace an existing file, you can use the intermediate File Function VIs to Open/Create/Replace File, as shown in Figure 9.8. With file I/O, you need to define a path to the file. This can be accomplished by popping up on the left side of the function and selecting **Create Control** to create an input name **file path**. When the file is opened, a reference number is created, and you can read or write to the file. Remember to close an open file before your application completes execution. Many of LabVIEW's high-level I/O

VIs automatically open or create new files, read or write data to those files, and close the files upon completion. In the next three subsections we will use three VIs that take care of these three basic operations. In the last subsection, we will discuss briefly an application that uses different VIs to perform the basic file I/O operations individually.

You can store or retrieve data from files in three different formats.

1. **Text (ASCII) Byte Stream**—You should store data in ASCII format when you want to access it from another software package, such as a word processing or spreadsheet program. To store data in this manner, you must convert all data to ASCII strings.
2. **Binary Byte Stream**—These files are the most compact and fastest method of storing data. You must convert the data to binary string format, and you must know exactly what data types you are using to save and retrieve the data to and from files.
3. **Datalog files**—These files are in binary format that only G can access. Datalog files are similar to database files because you can store several different data types into one (log) record of a file. You must use the intermediate File Function VIs when accessing these files.

When dealing with files, you will frequently see the terms **end-of-file**, **refnum**, **not-a-path**, and **not-a-refnum**. The end-of-file (EOF) is the character offset of the end of the file relative to the beginning of the file. Refnum is an identifier that G associates with a file when opened. Not-a-path and not-a-refnum are predefined values that indicates that a path is invalid and that a refnum associated with an open file is invalid, respectively.

9.2.1 Writing Data to a File

In this section we will discuss writing data to a file using the Write Characters To File VI. This VI writes a character string to a new byte stream file or appends the string to an existing file, depending on the VI input parameters. The Write Characters To File VI opens or creates a file, writes the data, and then closes the file.

In this exercise, the objective is to create a VI to append temperature data to a file in ASCII format. This VI uses a For Loop to generate temperature values and store them in a file. During each iteration, the VI converts the temperature data to a string, adds a comma as a delimiting character, and then appends the string to a file.

Open a new front panel and place the objects as shown in Figure 9.9. The front panel contains a digital control and a waveform chart. The chart displays the temperature data. Pop up on the waveform chart and select Visible Items≫Digital Display and de-select Visible Items≫Plot Legend and Visible Items≫

Graph Palette. Also make sure that either the waveform chart has y-axis autoscale enabled or that the maximum value of the y-axis is set to at least 90 degrees. The Number of Points control specifies how many temperature values to acquire and write to file. Pop up on the Number of Points and choose Representation≫I32.

Switch to the block diagram and wire the code as shown in Figure 9.9. Add a For Loop and make it large enough to encompass the various components. Place the Digital Thermometer.vi on the block diagram—remember that this VI is located in the Activity directory, which you can select through Functions ≫ Select a VI..., and returns a simulated temperature measurement from a temperature sensor.

Add a shift register to the loop by popping up on the loop border. This shift register contains the path name to the file. It is initially set to **Empty Path**, which can be found in Functions≫File I/O≫File Constants, as shown in Figure 9.9. The **Empty Path** function initializes the shift register so that the first time you try to write a value to file, the path is empty. When the Write Characters to File VI encounters an empty path for the path name of the file, it will display a dialog box from which you can select a file name. Once that file name has been selected via the dialog box, the shift register will pass the file name on to subsequent iterations of the For Loop.

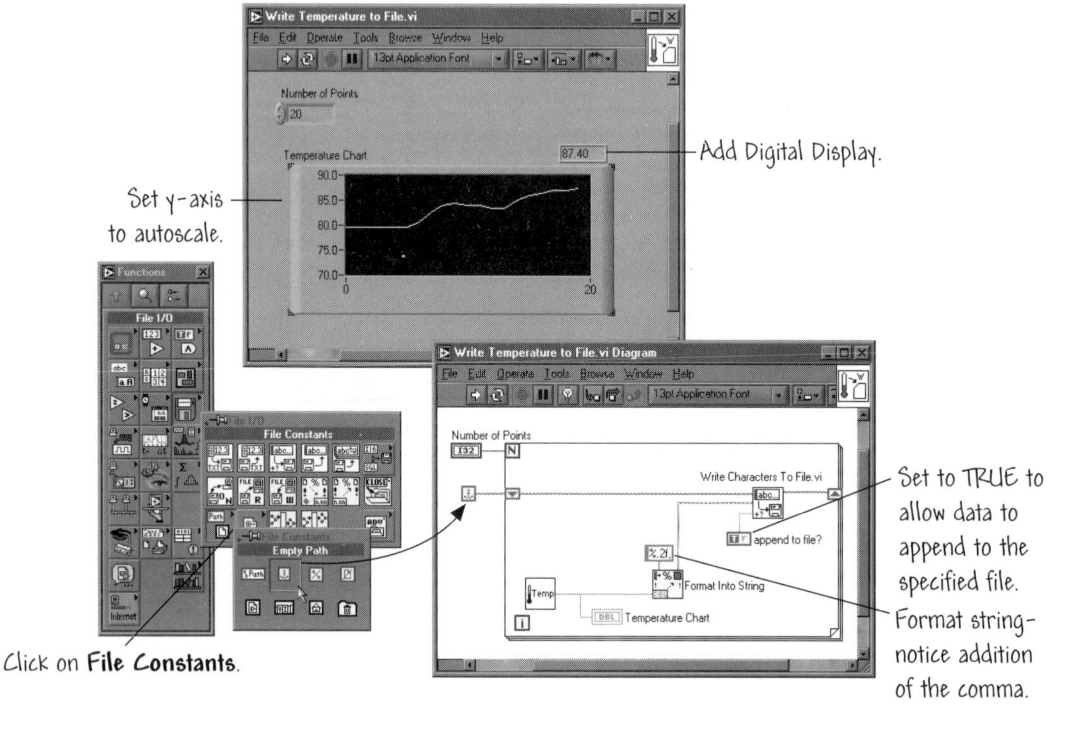

FIGURE 9.9
Using the Write Characters to File VI to write temperature data to a file.

The Format Into String function converts the temperature measurement (a number) to a string and concatenates the comma that follows it. We need to modify the format string to change the precision to 2 digits after the decimal point and to add the comma after each data point. The format string can be seen in Figure 9.9 (wired at the top of the Format Into String function). On the block diagram, pop up on the Format Into String function and select **Edit Format String** or double click on the node to access the dialog box. As shown in Figure 9.10a, the **Current Format Sequence** contains the **Format fractional number**. The first thing to do is to set the precision to 2, as illustrated in Figure 9.10b. Then select **Add New Operation** and in the **Selected Operation** pull-down menu, choose **Output exact string**. Make sure that the exact string that you input is

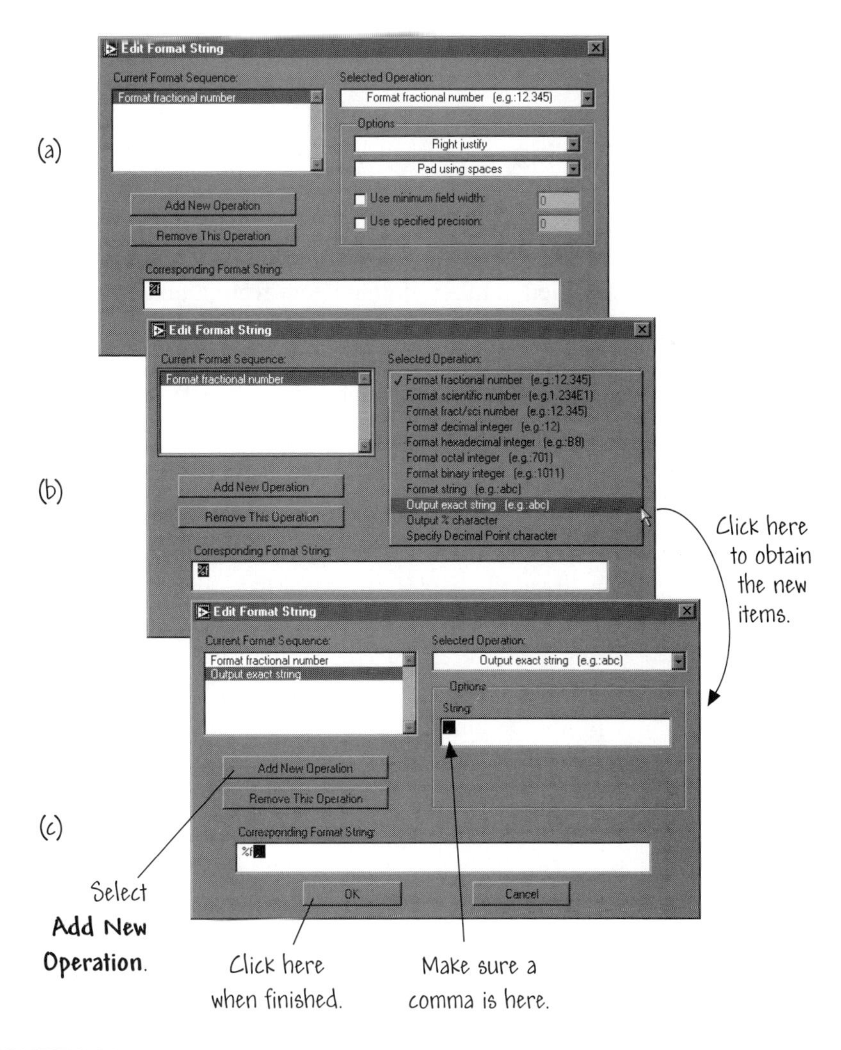

FIGURE 9.10
Setting the desired format string using the dialog box.

a comma (as shown in Figure 9.10c). When finished, select **Create String** and verify that the string format is "%.2f,"—as desired.

Finish wiring the objects, return to the front panel, and run the VI with the number of points set to 20. A file dialog box prompts you for a file name—select a file name such as Test. When you enter a file name, the VI starts writing the temperature values to that file as each point is generated. Save the VI as Write Temperature to File.vi in the Users Stuff folder. Use any word processing software, such as Notepad for Windows or Teach Text for Macintosh, to open that data file and view the contents. You should get a file containing twenty data values (with a precision of two places after the decimal point) separated by commas.

A working version of Write Temperature to File.vi *can be found in the* Chapter 9 *folder in the* Learning *directory.* ◆

9.2.2 Reading Data from a File

As a natural follow-on to writing data to a file, in this section we discuss reading data from a file using the Read Characters From File VI. This function reads a specified number of characters from a text byte stream file beginning at a specified character offset. The VI opens the file before reading from it and closes it afterwards. In the following example, you will get a chance to construct a VI to read the temperature data that you wrote to a file in the previous exercise.

Read Data from a File

The goal here is to construct a VI that reads the temperature data file you wrote in the previous example and displays the data on a waveform graph. You must read the data in the same data format in which you saved it. The data was originally saved in ASCII format using string data types, so it must read in as string data.

Open a new front panel and build the front panel shown in Figure 9.11. You will need to place a string indicator and a waveform graph to display the temperature data that is read.

Build the block diagram as shown in Figure 9.11. The Read Characters From File VI reads the data from the file and outputs the information in a string. If no path name is specified, a file dialog prompts you to enter a file name.

The Extract Numbers VI is located in the Examples\General\Strings.llb. This VI takes an ASCII string containing numbers separated by commas, line feeds, or other non-numeric characters and converts them to an array of numbers. It uses the Spreadsheet String To Array function to convert a spreadsheet

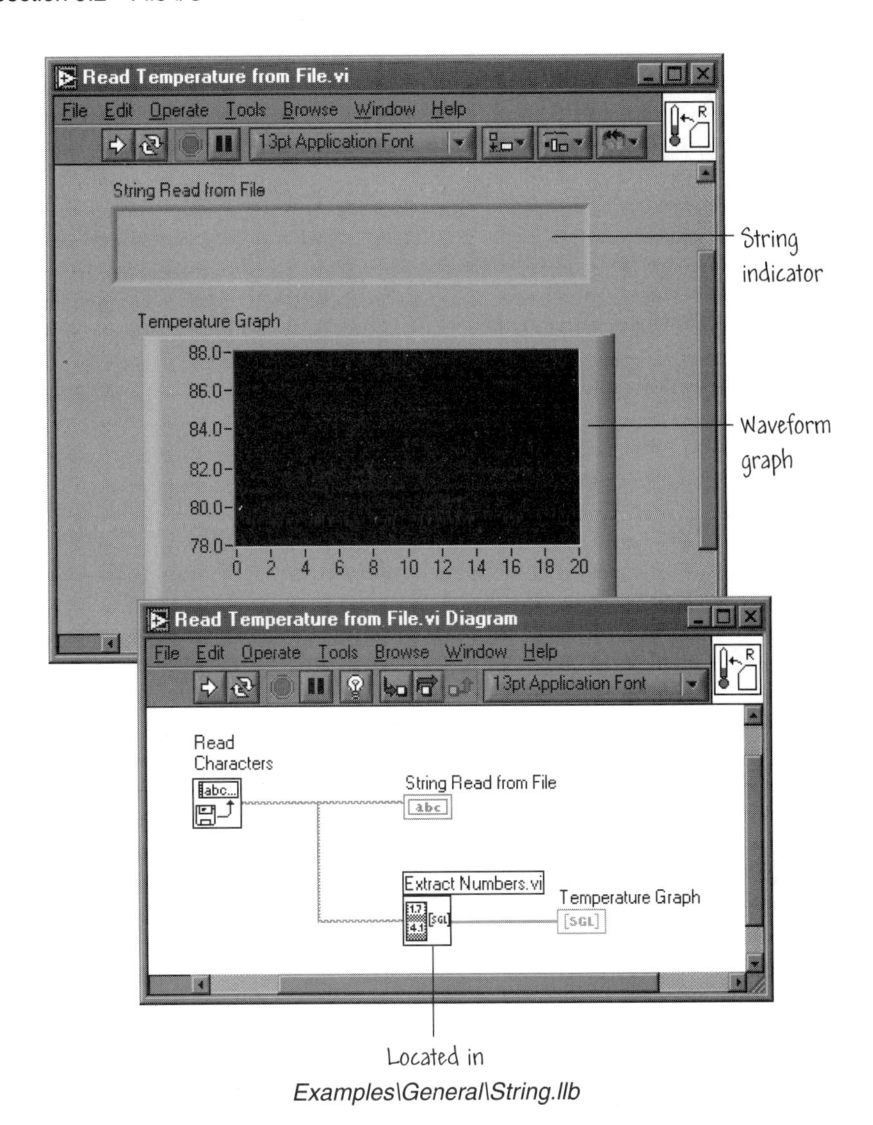

FIGURE 9.11
Using the Read Characters From File VI to read temperature data from a file.

string (that is, delimiter-separated columns with end-of-line characters between rows) into an array of numbers (by default) or strings. You can access the Extract Numbers VI on Functions≫Select a VI....

Complete the wiring of the block diagram, return to the front panel, and run the VI. When prompted for a file name, select the data file name that contains the temperature data from the previous exercise. You should see the same temperature data values displayed in the graph as you saw in the previous exercise. Save the VI as Read Temperature from File.vi in the Users Stuff and close the VI.

A working version of Read Temperature from File.vi *can be found in the* Chapter 9 *folder in the* Learning *directory.* ◆

9.2.3 Manipulating Spreadsheet Files

In many instances it is useful to be able to open saved data in a spreadsheet. In most spreadsheets, tabs separate columns and EOL (end-of-line) characters separate rows, as shown in Figure 9.12a. Opening the file using a spreadsheet program yields the table shown in Figure 9.12b.

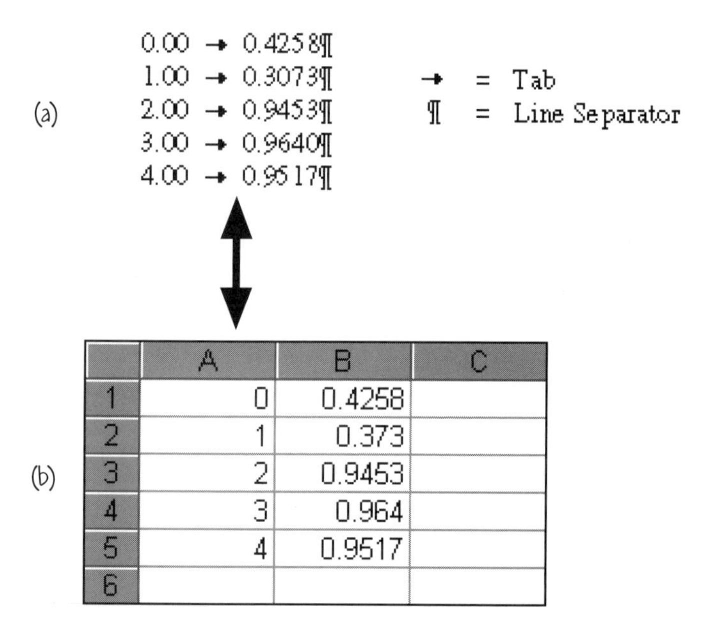

FIGURE 9.12
A common spreadsheet data format.

We will concentrate in this section on writing data to a spreadsheet format, with the idea that you will be accessing that data using a spreadsheet application outside of LabVIEW. However, if you want to read spreadsheet data from within LabVIEW, it is possible to read data in text format from a spreadsheet using the Read From Spreadsheet File VI. This VI reads a specified number of lines or rows from a spreadsheet file, beginning at a specified character offset, and converts the data to a 2D, single-precision array of numbers. This is a high-level VI; hence it opens the file beforehand and closes it afterwards.

Write to a Spreadsheet File

The objective of this exercise is to construct a VI that will generate and save data to a new file in ASCII format. This data can then be accessed by a spreadsheet application.

Open a new VI and construct a front panel, as shown in Figure 9.13. This VI generates two data arrays and plots them on a graph. You modify this VI to write the two arrays to a file where each column contains a data array. The front panel contains only one waveform graph.

Open the block diagram and modify the VI by adding the block diagram functions shown in Figure 9.13. The Write to Spreadsheet File VI (for location of this VI see Figure 9.8) converts the 2D or 1D array of single-precision numbers to a text string and writes the string to a new text byte stream file or appends the string to an existing file. If you have not specified a path name, then a file dialog box appears and prompts you for a file name. You can write either a 1D or 2D array to file—in this exercise we have a 2D array of data, so the 1D input is not wired. With this VI, you can use a spreadsheet delimiter such as tabs or commas in your data. The format string default is %.3f, which creates a string long enough to contain the number with three digits to the right of the decimal point.

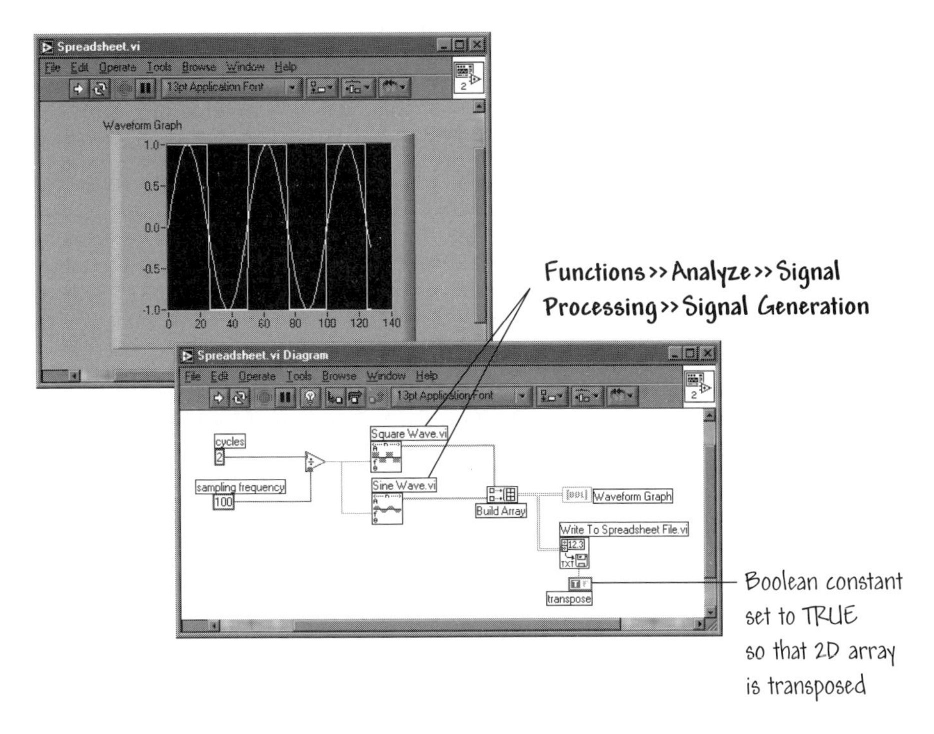

FIGURE 9.13
Using the Write To Spreadsheet File VI to write data in ASCII format to a file.

This VI is a high-level VI that opens or creates the file before writing to it and closes it afterwards. You can use Write to Spreadsheet File VI to create a text file readable by most spreadsheet applications.

The Boolean constant (see Figure 9.13) connected to the Write To Spreadsheet File VI controls whether or not the 2D array is transposed before writing it to file. To change the value to TRUE, click on the constant with the **Operating** tool. Generally each row of a spreadsheet file contains a data array; thus in this case you want the data transposed because you want data arrays in each column.

After finishing up the wiring on the block diagram, return to the front panel and run the VI. After the data arrays have been generated, a file dialog box prompts you for the file name of the new file you are creating. Type in a file name and click on **OK**.

Do not attempt to write data in VI libraries. Doing so may result in overwriting your library and losing your previous work.

Save the VI in Users Stuff and name it Spread-sheet.vi. You now can use a spreadsheet application or a text editor to open and view the file you just created.

In this example, the data was not converted or written to file until the entire data arrays had been collected. If you are acquiring large buffers of data or would like to write the data values to disk as they are being generated, then you must use a different File I/O VI.

A working version of Spreadsheet.vi *can be found in the* Chapter 9 *folder in the* Learning *directory.*

◆

9.2.4 Writing Waveform Data to File

You can use the built-in **Waveform File I/O** VIs located in the Waveform subpalette of the **Function** palette to write waveform data to a file, as shown in Figure 9.14. The **Write Waveforms to File** and **Read Waveforms from File** VIs write data in a datalog file. You use the **Export Waveforms to Spreadsheet File** VI to write the waveform data to a spreadsheet format in an operation similar to the high-level **Write to Spreadsheet File** VI. The VI opens a data file specified by the **file path** input or opens a dialog box if the path is empty. You wire the **waveform** directly to the input, and this VI converts the data to spreadsheet format using a **Tab** delimiter as the default. You can select to append to an existing file or write to a new file. You can also add a header to the file or write multiple time columns to the file. The file is closed after the data is written to the file.

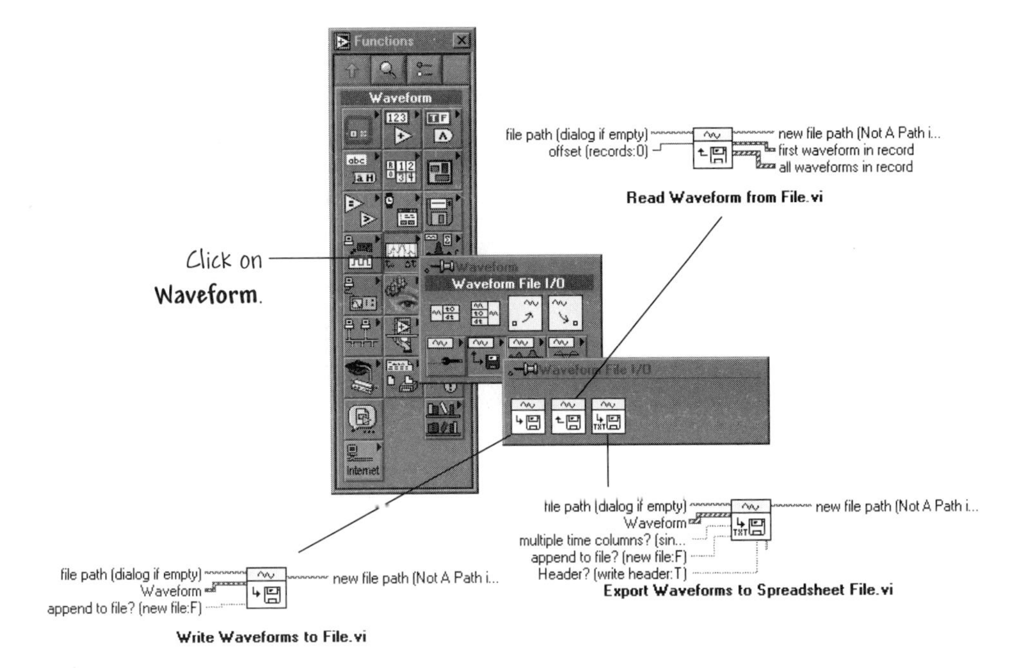

FIGURE 9.14
Writing and reading waveform data.

The goal here is to construct a VI that writes data to a spreadsheet. To begin, open a new front panel and build the front panel shown in Figure 9.15. Place a channel name control, waveform graph, waveform data type indicator, and two numeric controls on the front panel.

Continue by building the block diagram as shown in Figure 9.15. Export waveform to Spreadsheet File.vi on the **Functions** palette in Waveform≫Waveform File I/O. This VI writes the waveform data to a spreadsheet format. Since the **file path** input is unspecified, a dialog box will open and prompt you to enter a filename. Choose a name of your choice and save it in the Users Stuff directory. You will also need to select a channel to read the data. In the previous chapter you used MAX to configure several input channels—choose one of these.

Finish wiring the objects, return to the front panel, and run the VI with the number of samples set to 10 and the samples/sec set to 2. When the file dialog box prompts you for a file name—select a file name such as Test. When you enter a file name, the VI starts writing the data to that file as each point is generated. Save the VI as Write Waveform Data To Sprd.vi in the Users Stuff folder. Use any spreadsheet software, such as Excel, to open the data file and view the contents. You should get a file containing twenty data values.

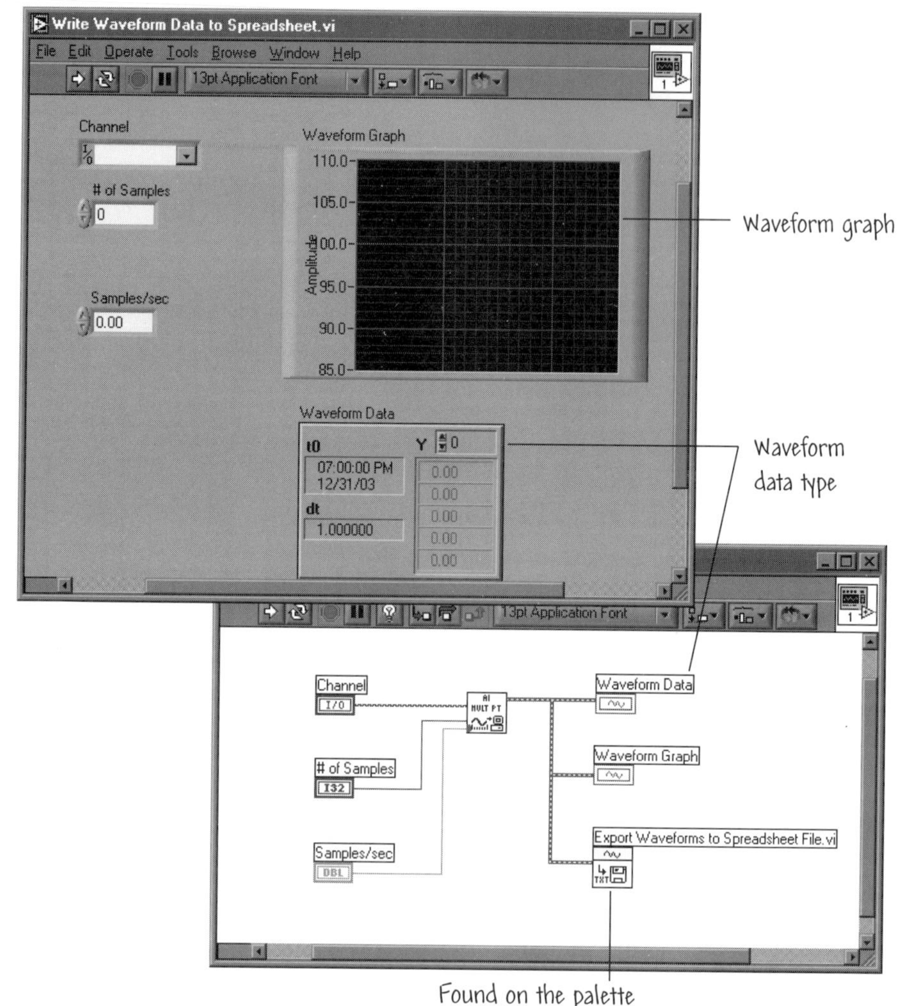

FIGURE 9.15
Writing data to a file in a spreadsheet format.

A working version of Write Waveform Data To Sprd.vi *can be found in the Chapter 9 folder in the* Learning *directory.*

◆

9.2.5 Advanced File I/O VIs

Sometimes the high-level file I/O functions do not provide the necessary level of functionality. For additional file I/O functions, consider the Intermediate File Function VIs located along the second row of the **File I/O** palette, as illustrated in Figure 9.8.

An example of using the intermediate File Function VIs to write data to a file is shown in Figure 9.16, which shows the Write To Text File VI. This VI writes an ASCII text file that contains data values with time-stamps. Open the VI located in the **Chapter 9** folder in the **Learning** directory and study the block diagram. This VI performs the three file operations—open, write, and close—using individual VIs. It also includes error checking and handling.

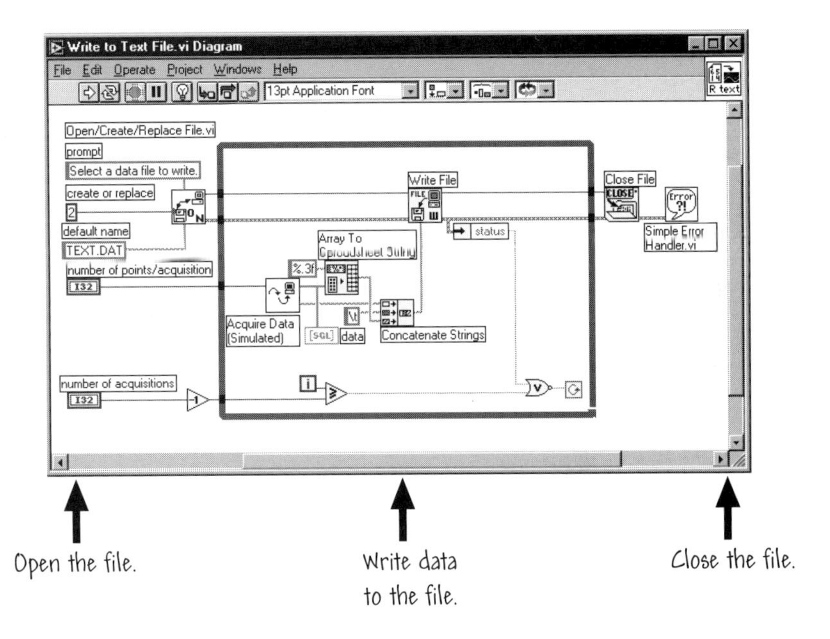

FIGURE 9.16
Advanced file figure.

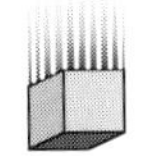

BUILDING BLOCK

9.3 BUILDING BLOCKS: MEASURING VOLUME

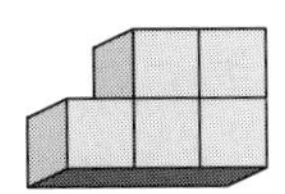

In this exercise you will create a VI that writes data to a file. Open the **Volume Chart.vi** and modify the block diagram using Figure 9.17 as a guide. The **Volume Chart.vi** should have been saved in **Users Stuff** folder in the **Learning** directory as part of the Chapter 7 "Building Blocks" exercise. If you did not complete the exercise in Chapter 7 you can build the VI from scratch, but refer to the "Building Blocks" section in that chapter for a view of the front panel.

The block diagram uses two string functions: Array To Spreadsheet String and Concatenate Strings. To get the time and date, use the Get Date/Time String function located on the **Time & Dialog** subpalette on the **Functions** palette. This function is used to time tag the data written to a file with the current time.

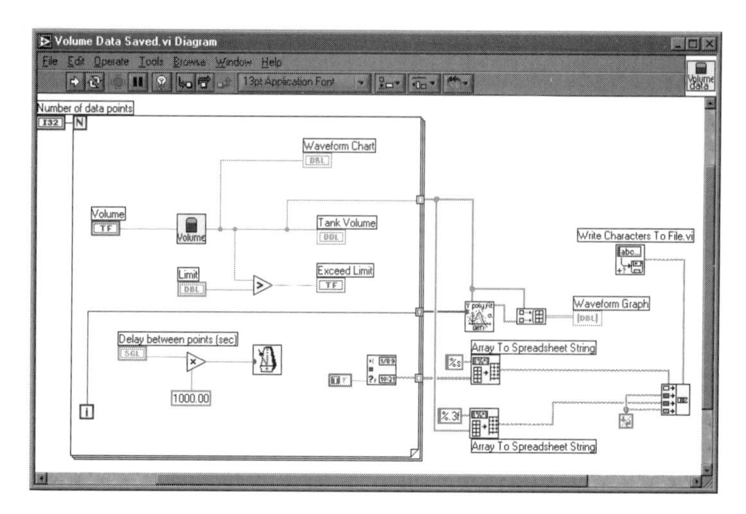

FIGURE 9.17
The Volume Data Saved.vi block diagram.

When the VI is complete, click on **Run** and watch the VI execute. You will be prompted for the name of the file to which the volume data is to be written. Type in the name of your choice and save the data file in the Users Stuff folder. When the VI is finished executing, you can view the data file in (**Windows**) Notepad or (**Macintosh**) SimpleText. Close the VI when you are done experimenting and save it as Volume Data Saved.vi in the Users Stuff folder.

9.4 RELAXING READING: INTEGRATED LABORATORY INSTRUCTION

The faculty at Bethel College adopted several basic software packages—including LabVIEW—for use across departments. One goal of the professors is to bring students to the point of being able to use LabVIEW competently in independent research projects. The idea is that exposure to common software tools will foster cross-disciplinary collaboration. LabVIEW is utilized as a uniform interface to laboratory instruments in the Departments of Physics, Chemistry, Biology, and Psychology. A few of the interesting laboratory experiments developed for the students include physiological data acquisition, electroencephalographic recording, and controlling operant chambers in animal-conditioning experiments.

One of the principal applications of LabVIEW is physiological data acquisition, particularly in a neurobiology course. The DAQ system is designed to measure contractions of isolated smooth or skeletal muscle preparations. The contractions are measured through a force transducer and can be filtered to remove unwanted frequency components. With this system, students can readily quantify effects of various hormones, neurotransmitters, and pharmacological agents. Students can write data to files for further analysis after the experiment and they

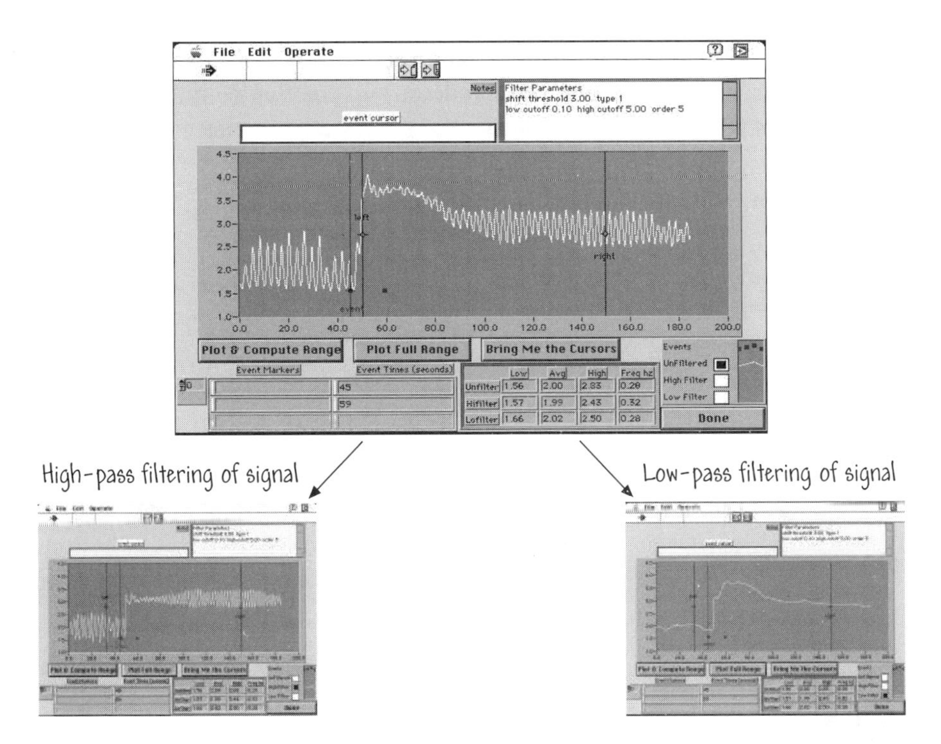

FIGURE 9.18
The front panel of a physiological data analysis subVI displays the effect on isolated duodenum of introducing acetylcholine into the medium at time unit 50.

can transfer graphical records such as those illustrated in Figure 9.18 from LabVIEW into a word processor, thereby facilitating laboratory report preparation. The front panel of the data analysis VI is shown in Figure 9.18. The strip chart shows the effect on duodenal contractions of introducing acetylcholine into the medium (at about time unit 50). The same data following either high-pass or low-pass filtering is also shown in Figure 9.18.

Another interesting application of LabVIEW at Bethel College uses the same hardware described above for electroencephalographic recording. Visual stimuli are presented to a human subject on one computer, and a digital pulse is sent out via an NB-MIO-16 board to trigger recording for 1–2 seconds from scalp electrodes on another computer running LabVIEW. A signal-averaging VI produces the composite result of numerous stimulus presentations—in psychophysiological parlance, an "event-related potential."

A third example of how LabVIEW is being used in laboratory instruction is controlling operant chambers in animal-conditioning experiments. A VI records incoming digital pulses generated by an animal pressing a lever. Results are displayed graphically in a manner that mimics standard strip chart output, called "cumulative records" in the instrumental conditioning literature. Digital pulses are also sent out to activate a food-pellet dispenser. All input and output signals are handled through a NB-MIO-16 board.

For more information on this application of LabVIEW, read the *User Solutions* entitled "Integrated Laboratory Instruction with LabVIEW," which can be found on the National Instruments website, or contact:

Dwight Krehbiel, Ph.D.
Department of Psychology
Bethel College
North Newton, KS 67117
e-mail: krehbiel@bethelks.edu

9.5 SUMMARY

File input and output operations store and retrieve information from a disk file. LabVIEW supplies you with simple functions that take care of almost all aspects of file I/O. You can store or retrieve data from files in ASCII byte stream, binary byte stream, or datalog files. This allows you to interact with word processing programs, spreadsheet programs, or with other VIs in LabVIEW.

KEY TERMS

ASCII: American Standard Code for Information Interchange.

Binary byte stream files: Files that store data as a sequence of bytes.

Datalog files: Files that store data as a sequence of records of a single, arbitrary data type that you specify when you create the file.

EOF: End-of-file. Character offset of the end of a file relative to the beginning of the file (that is, the EOF is the size of the file).

Hex: Hexadecimal. A base-16 number system.

Nondisplayable characters: ASCII characters that cannot be displayed, such as null space, backspace, and tab.

Not-a-path: A predefined value for the path control that means that a path is invalid.

Not-a-refnum: A predefined value that means the refnum associated with an open file is invalid.

Refnum: An identifier that G associates with a file when you open it. You use the file refnum to indicate that you want a function or VI to perform an operation on the open file.

String: A sequence of displayable or nondisplayable ASCII characters.

Text (ASCII) byte stream files: Files that store data as a sequence of ASCII characters.

EXERCISES

E9.1 Open the Learning\Chapter 8\AI Waveform.vi. Select Write to Spreadsheet File.vi from the Functions≫File I/O palette. Wire the array of information from AI Acquire Waveform.vi to the 1D data input of Write to Spreadsheet File.vi. Run the VI. Enter the file name you want the waveform data to be saved to. Open this file in (**Windows**) Notepad or (**Macintosh**) SimpleText to view the data. Save this VI as AI Waveform&SpreadFile.vi in Learning\Users Stuff.

E9.2 Open the Learning\Chapter 8\AI Waveform.vi. Select Write to SGL File.vi from Functions≫File I/O≫Binary File VIs. Wire the array of information from AI Acquire Waveform.vi to the 1D array input of Write to SGL File.vi. Run the VI. Enter the file name you want the waveform data to be saved to. Save this VI as AI Waveform&BinaryFile.vi in Learning\Users Stuff directory. Select Help≫Examples...≫Fundamentals. Click File I/O≫Binary File Examples≫Read Binary File links. Run this VI to read the file you just create with AI Waveform&Binary File.vi.

PROBLEMS

P9.1 Complete the crossword puzzle.

Across

4. Files that store data as a sequence of records of a single, arbitrary data type that you specify when you create the file.
6. A predefined value for the path control that means that a path is invalid.
7. Files that store data as a sequence of ASCII characters.
9. A base-16 number system.
10. An identifier that G associates with a file when you open it.

Down

1. Files that store data as a sequence of bytes.
2. American Standard Code for Information Interchange.
3. ECharacter offset of the end of a file relative to the beginning of the file.
5. A predefined value that means the refnum associated with an open file is invalid.
8. A sequence of displayable or nondisplayable ASCII characters.

P9.2 Open the Learning\Chapter 8\AI Single Point.vi. Change this VI so you take 50 temperature measurements. Also, bundle the time and date with each measurement. Hint: To get the time and date, use the Get Date/Time String function. Write the 50 measurements with time and date to a datalog file. To find out more about datalog files, refer to the online help and examples.

P9.3 Create a VI that reads the datalog file created in the previous exercise and displays that information on the front panel.

CHAPTER 10

Instrument Control

This chapter introduces the concept of communicating with and controlling external instruments. In this chapter we focus on the GPIB (General Purpose Interface Bus) and RS-232 (a serial interface bus standard). A brief introduction to the main components of an instrument control system are described. The application of the **Measurement & Automation Explorer (MAX)** in detecting instruments and installing instrument drivers is discussed. The notion of an instrument driver is an important topic woven throughout the chapter.

GOALS

1. Learn about GPIB and serial instrument control.
2. Understand how to interact with your instruments using the MAX.
3. Gain some experience with instrument drivers (for the HP34401A multimeter).

10.1 COMPONENTS OF AN INSTRUMENT CONTROL SYSTEM

LabVIEW communicates with and controls external instruments (such as oscilloscopes and digital multimeters) using GPIB (General Purpose Interface Bus), RS-232 (a serial interface bus standard), VXI (VME eXtensions for Instrumentation), and other hardware standards. In this chapter we focus on the GPIB and serial communication.

10.1.1 What Is GPIB?

Hewlett Packard developed the General Purpose Interface Bus (or **GPIB**) standard in the late 1960s to interconnect and control its line of programmable instruments. The interface bus was originally called HP-IB. In this context, a **bus** is the means by which computers and instruments transfer data. National Instruments made GPIB available to users of non-Hewlett-Packard equipment.

At the time it was developed, GPIB provided a much-needed specification and protocol to govern this communication. Figure 10.1 shows a typical GPIB system. While the GPIB is one way to bring data into a computer, it is fundamentally different from data acquisition with boards that plug into the computer. Using a special protocol, GPIB brings data that has been acquired by another computer or instrument into the computer using a "handshake," while data acquisition involves connecting a signal directly to a DAQ board in the computer.

The original purpose of GPIB was to provide computer control of test and measurement instruments. The GPIB was soon applied to intercomputer communication and control of scanners, film recorders, and other peripherals because of its 1 Mbyte/sec maximum data transfer rates. The Institute of Electrical and Electronic Engineers (**IEEE**) standardized the GPIB in 1975, and it became accepted

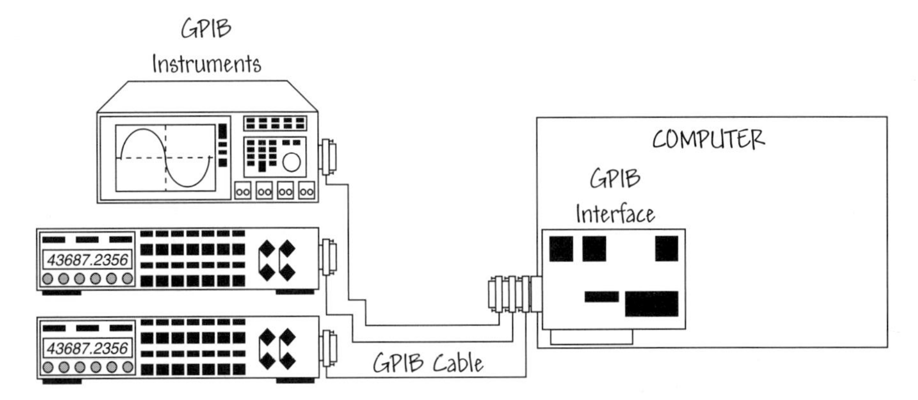

FIGURE 10.1
A typical GPIB system.

as IEEE Standard 488-1975. The standard has since evolved into ANSI/IEEE Standard 488.2-1987. The GPIB functions for LabVIEW follow the IEEE 488.2 specification. The terms GPIB, HP-IB, and IEEE 488 are synonymous.

10.1.2 GPIB Messages

The GPIB carries two types of messages:

- **Device-dependent messages** contain device-specific information such as programming instructions, measurement results, machine status, and data files. These are often called **data messages**.
- **Interface messages** manage the bus itself, and perform such tasks as initializing the bus, addressing and unaddressing devices, and setting device modes for remote or local programming. These are often called **command messages**.

Physically, the GPIB is a digital, 24-conductor, parallel bus. It comprises 16 signal lines and 8 ground-return lines. The GPIB connector is depicted in Figure 10.2. The 16 signal lines are divided into three groups:

- **Eight data lines**: The eight data lines (denoted DIO1 through DIO8) carry both data and command messages. GPIB uses an eight-bit parallel, asynchronous data transfer scheme where whole bytes are sequentially handshaked across the bus at a speed determined by the slowest participant in the transfer. Because the GPIB sends data in bytes (1 byte = 8 bits), the messages transferred are frequently encoded as ASCII character strings. All commands and most data use the 7-bit ASCII or International Standards Organization (ISO) code set, in which case the eighth bit, DIO8, is unused or is used for parity.
- **Three handshake lines**: The three handshake lines asynchronously control the transfer of message bytes among devices. These lines guarantee that message bytes on the data lines are sent and received without transmission error.
- **Five interface management lines**: The five interface management lines manage the flow of information across the interface from device to computer.

*The next generation in high-speed GPIB has been proposed by National Instruments, and is known as **HS488**. Speed increases up to 8 Mbyte/sec have been achieved by removing the propagation delays associated with the three handshake lines.*

10.1.3 GPIB Devices and Configurations

You can have many GPIB devices (that is, many instruments and computers) connected to the same GPIB. Typical GPIB linear and star configurations are

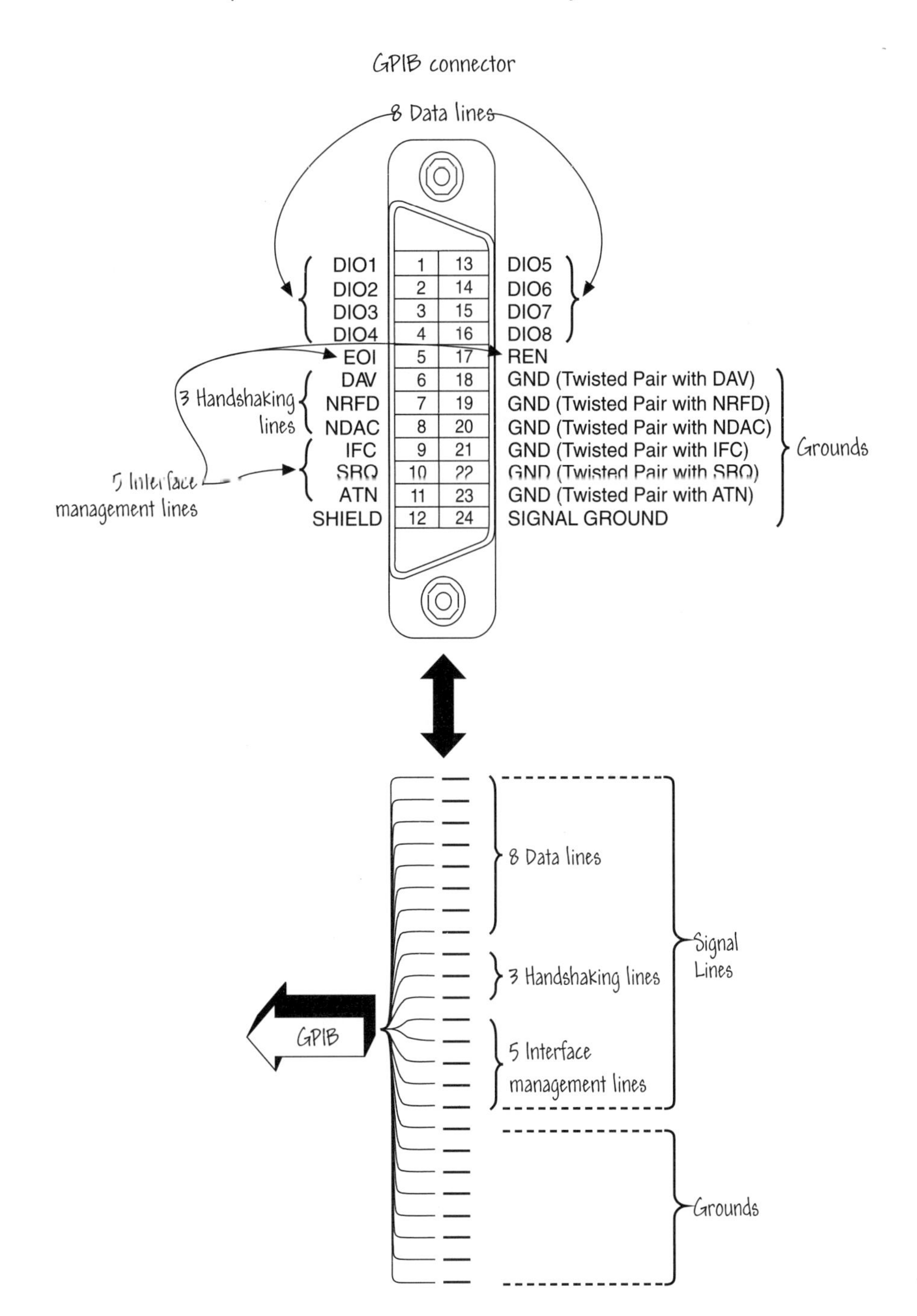

FIGURE 10.2
The GPIB connector.

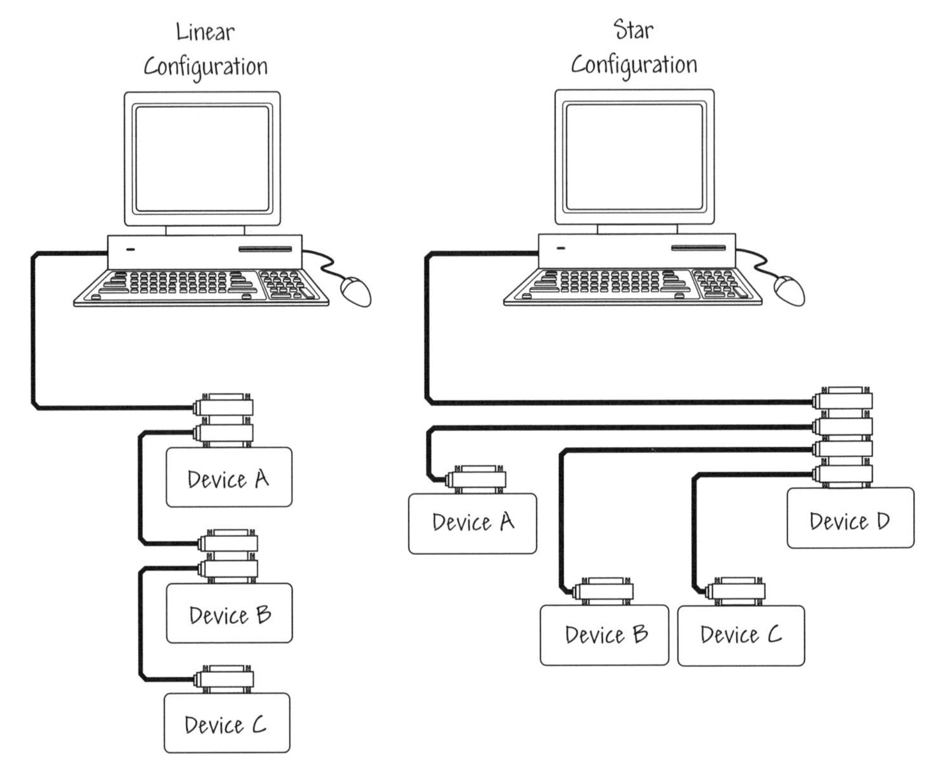

FIGURE 10.3
Typical GPIB configurations.

illustrated in Figure 10.3. You can even have more than one GPIB board in your computer. A typical multiboard GPIB configuration is illustrated in Figure 10.4. GPIB devices are grouped in three categories:

- **Talkers**: send data messages to one or more listeners.
- **Listeners**: receive data messages from the talker.
- **Controllers**: manage the flow of information on the GPIB by sending commands to all devices.

GPIB devices can fall into multiple categories. For example, a digital voltmeter can be both a talker and a listener.

The GPIB has one controller (usually a computer) that controls the bus. The role of the GPIB controller is similar to the role of a central processing unit in a computer. A good analogy is the switching center of a city telephone system. The communications network (the GPIB) is managed by the switching center (the controller). When a party (a GPIB device) wants to make a call (that is, to send a data message), the switching center (the controller) connects the caller (the talker) to the receiver (the listener). To transfer instrument commands and

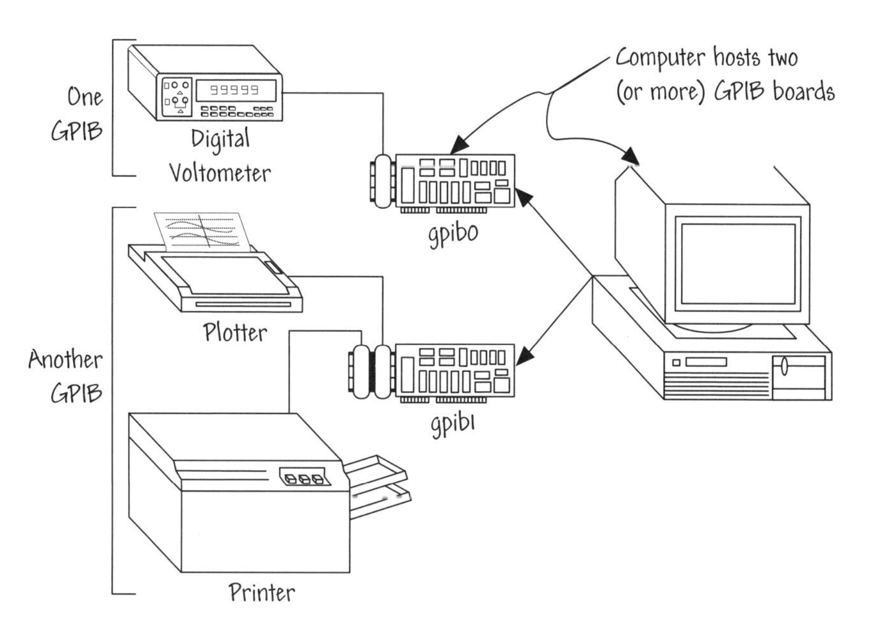

FIGURE 10.4
A typical multiboard GPIB configuration.

data on the bus, the controller addresses one talker and one or more listeners. The controller must address the talker and a listener before the talker can send its message to the listener. The data strings are then handshaked across the bus from the talker to the listener(s). After the talker transmits the message, the controller may unaddress both devices. LabVIEW provides VIs that automatically handle these GPIB functions.

Some bus configurations do not require a controller. For example, one device may always be a talker (called a talk-only device) and there may be one or more listen-only devices. A controller is necessary when you must change the active or addressed talker or listener. A computer usually handles the controller function. With the GPIB board and its software, the personal computer plays all three roles: Controller, Talker, and Listener.

There can be multiple controllers on the GPIB but only one controller at a time is active. The active controller is called the **controller-in-charge** (CIC). Active control can be passed from the current CIC to an idle controller. Only the system controller (usually the GPIB board) can make itself the CIC.

With GPIB, the physical distance separating the GPIB devices matters. To achieve high-rate data transfers, we must accept a number of physical restrictions. Typical numbers are as follows:

- A maximum separation of 4 meters between any two devices and an average separation of 2 meters over the entire bus. For high-speed applications, you should have at least one device per meter of cable.

- A maximum total cable length of 20 meters. A maximum of 15 meters is desirable for high-speed applications.
- A maximum of 15 devices connected to each bus, with at least two-thirds of the devices powered on. For high-speed applications, all devices should be powered on.

Bus extenders and expanders can be used to increase the maximum length of the bus and the number of devices that can be connected to the bus. You can also communicate with GPIB instruments through a TCP/IP network. For more information, refer to the National Instruments website.

10.1.4 Serial Port Communication

Serial communication is another popular means of transmitting data between a computer and another computer, or between a computer and a peripheral device. However, unlike GPIB, a serial port can communicate with only one device, which can be limiting for some applications. Serial port communication is also very slow.

Most computers and many instruments have built-in serial port(s). Since serial communication uses the built-in serial port in your computer, you can send and receive data without buying any special hardware. Serial communication uses a transmitter to send data—one bit at a time—over a single communication line to a receiver. This method works well when sending or receiving data over long distances and when data transfer rates are low. It is slower and less reliable than GPIB, but you do not need a special board in your computer, and your instrument does not need to conform to the GPIB standard. Figure 10.5 shows a typical serial communication system.

There are different serial port communication standards. Developed by the Electronic Industries Association (EIA) to specify the serial interface between

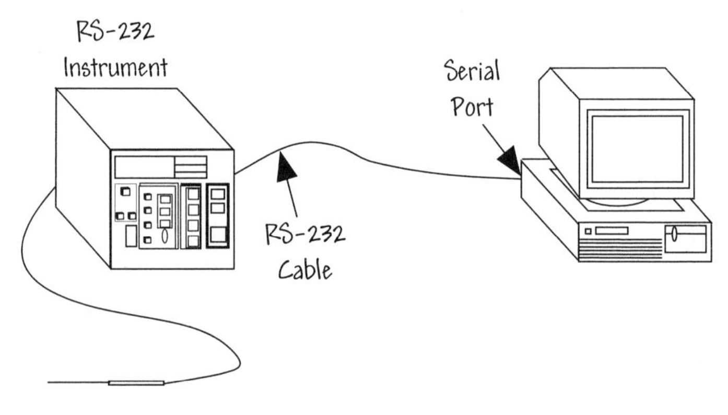

FIGURE 10.5
A typical serial communication system.

equipment known as Data Terminal Equipment (DTE), such as modems and plotters, and Data Communications Equipment (DCE), such as computers and terminals, the RS-232 standard includes signal and voltage characteristics, connector characteristics, individual signal functions, and recipes for terminal-to-modem connections. The most common revision to this standard is the RS-232C used in connections between computers, printers and modems.

The RS-232 serial port connectors come in two varieties: the 25-pin connector and the 9-pin connector. Both connectors are depicted in Figure 10.6. The 9-pin connector has two data lines (denoted TxD and RxD in the figure) and five handshake lines (denoted RTS, CTS, DSR, DCD, and DTR in the figure). This compact connector is occasionally found on smaller RS-232 laboratory equipment and has enough pins for the "core" set used for most RS-232 interfaces. The 25-pin connector is the "standard" RS-232 connector with enough pins to cover all the signals specified in the RS-232 standard. Only the "core" set of pins is labeled in Figure 10.6.

Serial communication requires that you specify four parameters: the baud rate of the transmission, the number of data bits encoding a character, the sense of the optional parity bit, and the number of stop bits. Each transmitted character is packaged in a character frame that consists of a single start bit followed by the data bits, the optional parity bit, and the stop bit or bits. The baud rate is a measure of how fast data is moving between instruments that use serial communication.

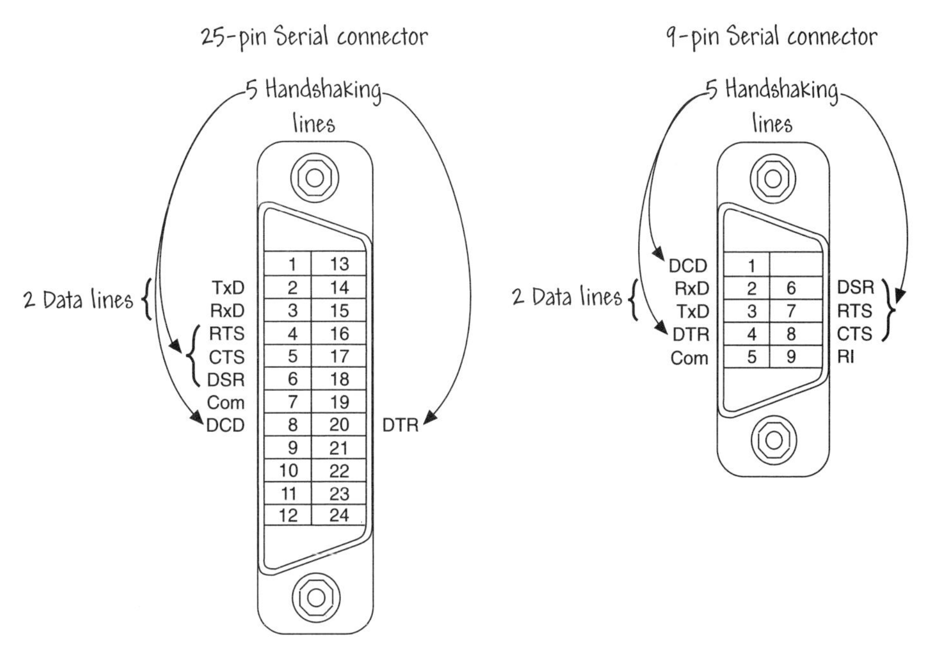

FIGURE 10.6
The RS-232 serial port connectors.

10.2 DETECTING AND CONFIGURING INSTRUMENTS

With the **Measurement & Automation Explorer (MAX)** you can automatically detect connected instruments, install the required instrument drivers, and manage the instrument drivers already installed. The software architecture for GPIB instrument control using LabVIEW is similar to the architecture for DAQ (see Chapter 8.1). Figure 10.7 shows the software architecture on the Windows platforms. Instrument drivers are LabVIEW applications written to control a specific instrument. More on instrument drivers in the next section. You can use the instrument drivers or parts of the instrument drivers to develop your own application quickly. You can also set naming aliases for your instruments for easier instrument access. If you are using a Macintosh, the NI-488.2 Configuration Utility (available at Start≫Settings≫Control Panel configures the parameters for the GPIB devices installed in you Macintosh computer. The Macintosh OS automatically recognizes GPIB devices. You can view or modify the default configuration settings using the NI-488.2 configuration utility.

In this example, you will use the MAX to configure and test the GPIB interface. MAX interacts with the various diagnostic and configuration tools installed

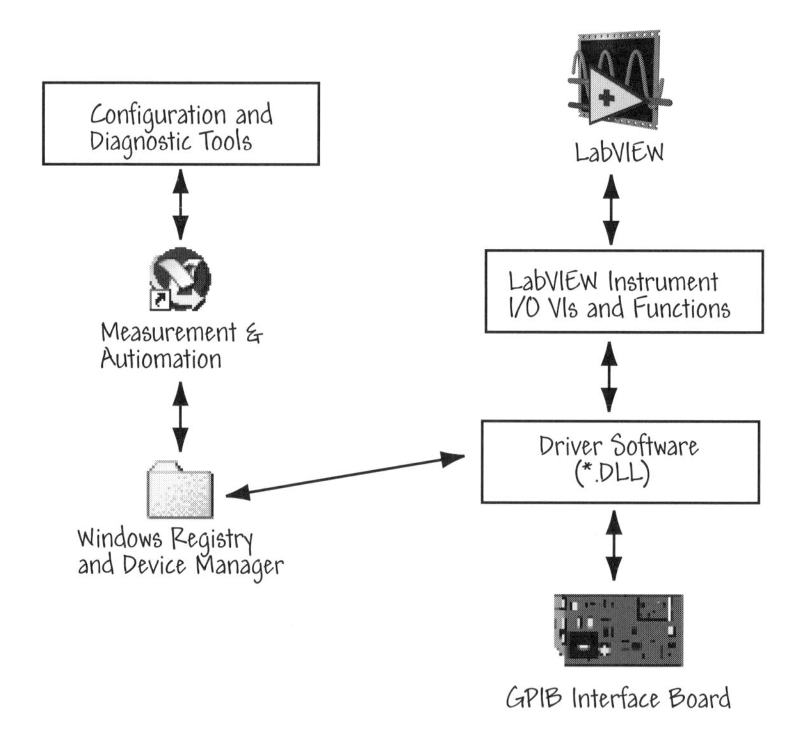

FIGURE 10.7
Software architecture for Windows platforms.

with the driver and also with the Windows Registry and Device Manager (see Figure 10.7). The driver-level software is in the form of a dynamically linked library (DLL) and contains all the functions that directly communicate with the GPIB board. The LabVIEW Instrument I/O VIs and functions directly call the driver software.

You configure the objects listed in the MAX by right-clicking on the item and making a selection from the shortcut menu. Figure 10.8 shows the GPIB interface board in the MAX utility and the results of pressing the **Scan for Instruments** button at the top of the window.

The configuration utilities and hierarchy described in this example are specific to Windows platforms. If you are using a Macintosh or other operating system, refer to the manuals that came with you GPIB interface board for the appropriate information for configuring and testing that board.

As discussed in Chapter 8, the MAX is a configuration utility for your software and hardware. MAX executes system diagnostics, adds new channels, interfaces, and virtual channels, and views devices and instruments connected to your system. Open MAX by double-clicking on its icon on the desktop or by selecting **Measurement & Automation Explorer** from the **Tools** menu.

The four possible selections in MAX are:

- **Data Neighborhood**—Use this selection to create virtual channels, aliases, and tags to your channels or measurements configured in Devices and Interfaces as you did in the DAQ Chapter.
- **Devices and Interfaces**—Use this selection to configure resources and other physical properties of your devices and interfaces. Using this selection, you can view attributes of one or more multiple devices, such as serial numbers.
- **Scales**—Use this selection to set up simple operations to perform on your data, such as scaling.
- **Software**—Use this selection to determine which drivers and application software are installed and their version numbers.

If you do not have an instrument attached to your computer, the MAX will obviously not find any instruments. In this example, the attached instrument is a National Instruments GPIB Device Simulator, denoted by **Instrument0** in MAX as seen in Figure 10.8.

Click **Communicate with Instrument** to communicate with the instrument, as shown in Figure 10.9. By default, the string *IDN? appears in the command to send to this GPIB device. The string *IDN? is an IEEE 488.2 standard identification request to the instrument. Click **Query** to send the string. In this example the string is sent to the NI GPIB Device Simulator. The instrument replies with the string "National Instruments GPIB Device Simulator Rev A.2." The next time

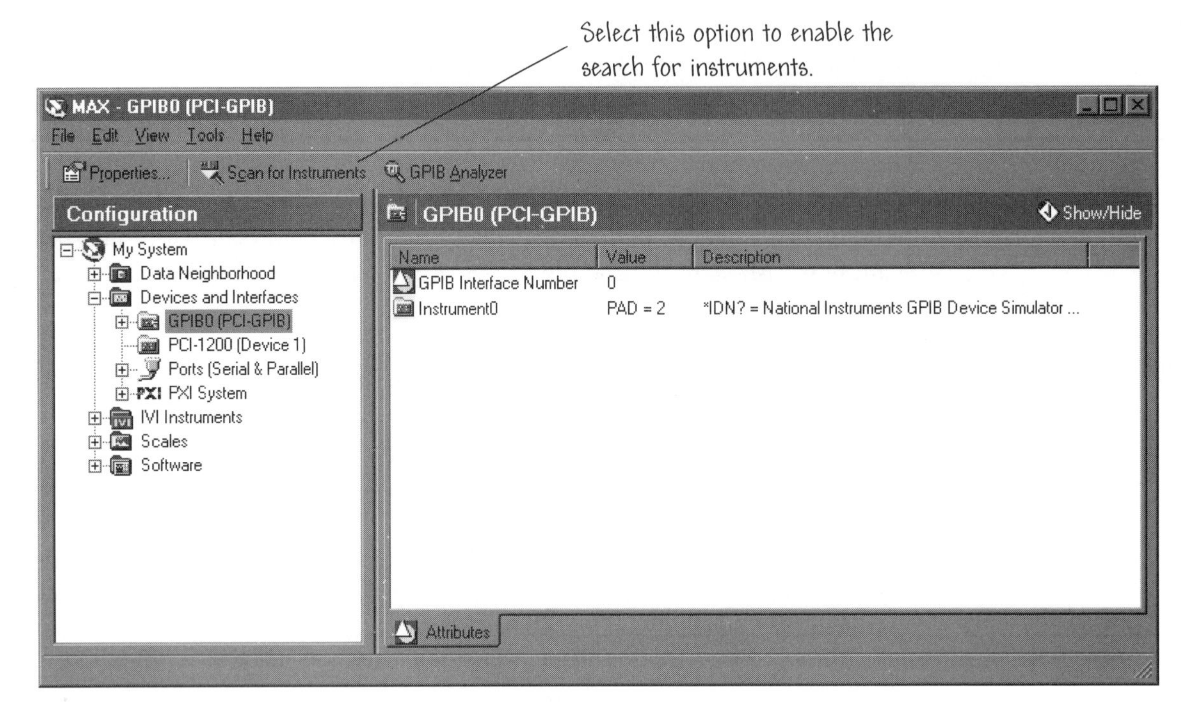

FIGURE 10.8
Using MAX to search for instruments.

you launch the MAX, it will automatically contain your stored instrument configuration.

You can also examine the properties of your GPIB board. On the MAX screen, highlight the GPIB hardware and then choose **Properties**, as illustrated in Figure 10.10. A window will open which provides all the information about your GPIB board.

10.3 INSTRUMENT DRIVERS

An **instrument driver** is a piece of software (that is, a VI) that controls a particular instrument. Instrument drivers eliminate the need to learn the complex, low-level programming commands for each instrument. LabVIEW is ideally suited for creating instrument drivers since the VI front panel can simulate the operation of an instrument's front panel. The block diagram sends the necessary commands to the instrument to perform the various operations specified on the front panel. When using an instrument driver, you do not need to remember the commands necessary to control the instrument—this is specified via input on the front panel.

LabVIEW provides many VIs that can be used in the development of an instrument driver for your hardware. These VIs can be grouped into the following

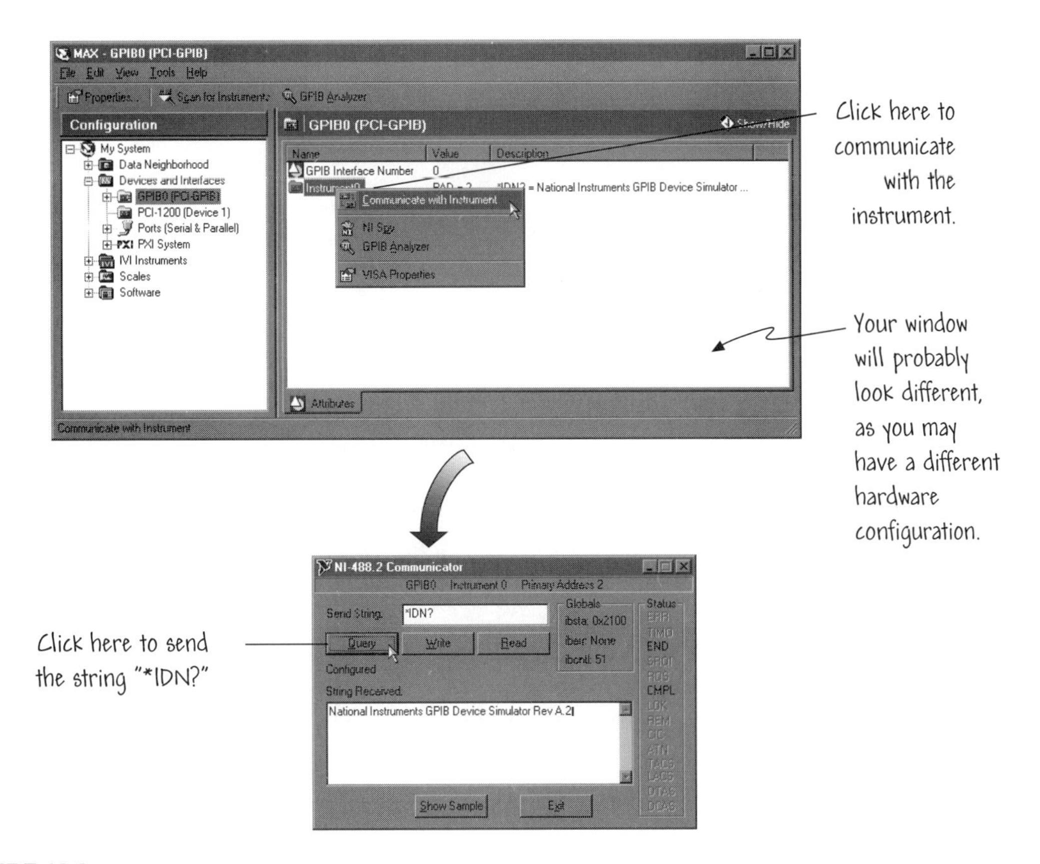

FIGURE 10.9
A National Instruments GPIB device simulator is detected or GPIB address 4.

categories:

- Standard VISA I/O functions
- Traditional GPIB functions and added capability via the GPIB 488.2
- Serial port communication functions

VISA stands for Virtual Instrument Software Architecture. In essence, VISA is a VI library for controlling GPIB, serial, or VXI instruments and making the appropriate calls depending on the type of instrument. VISA by itself does not provide instrumentation programming capability—it is a high-level application programming interface (API) that calls lower-level code to control the hardware. Each VISA instrument driver VI corresponds to a programmatic operation, such as configuring, reading from, writing to, and triggering an instrument. The GPIB and serial port functions provide similar capabilities.

Two questions come to mind:

1. When should you attempt to develop your own instrument driver from "scratch" using the VISA, GPIB, or serial port functions?

And if you must develop your own instrument driver,

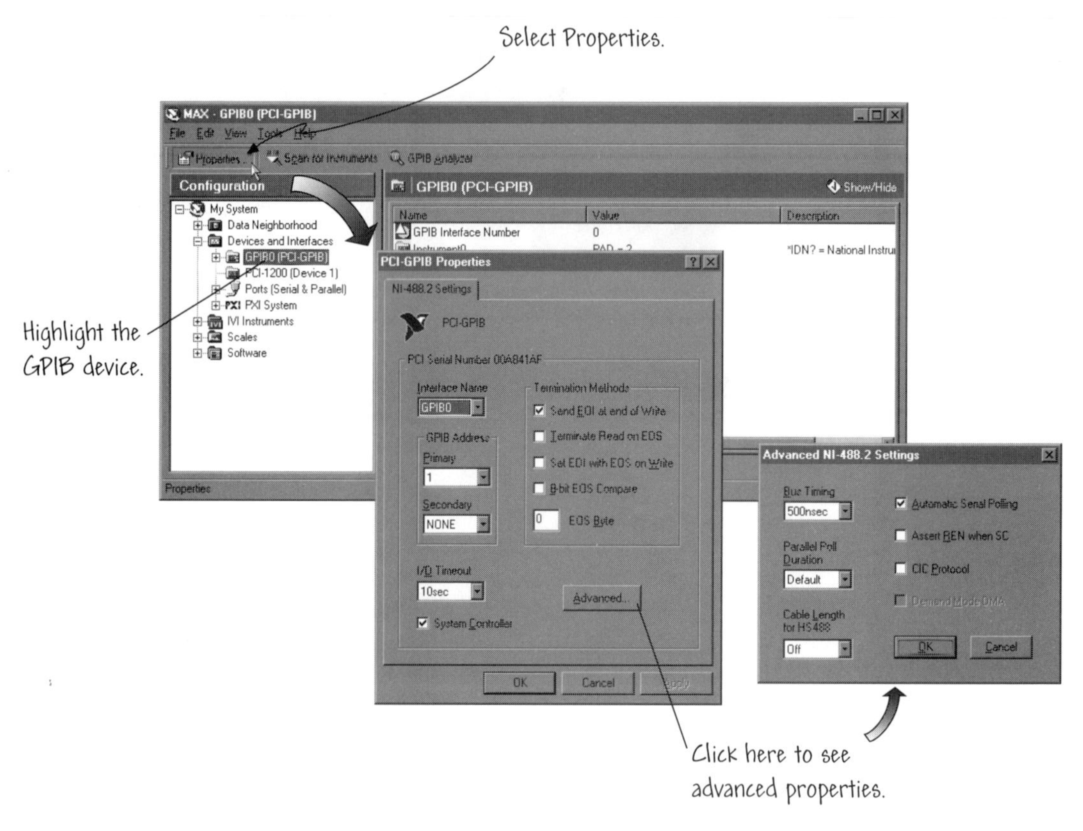

FIGURE 10.10
Examining the properties of the GPIB devices.

2. When should you use the VISA functions, and when should you use the GPIB functions?

The answer to the first question is simple—students new to LabVIEW should not attempt to develop their own instrument drivers! They should use the instrument drivers developed by National Instruments, rather than attempt to develop drivers from "scratch." Instrument drivers can be downloaded from the National Instruments website using the Instrument Driver Network. To access the driver network, connect to

http://www.ni.com/idnet

directly or use Tools≫Instrumentation≫Instrument Driver Network... to automatically connect to the network as illustrated in Figure 10.11. ◆

The LabVIEW instrument driver library contains instrument drivers for a variety of programmable instruments that use the GPIB, serial, or VXI interfaces. You can use an instrument driver from the library to control your instrument, or

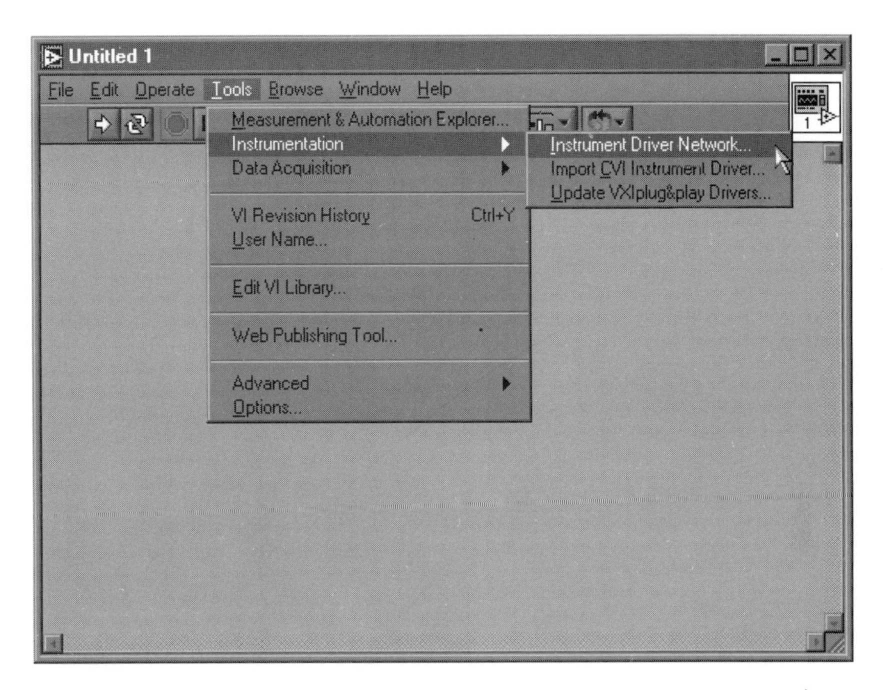

FIGURE 10.11
Link automatically to the National Instruments Instrument Driver Network.

you can customize the instrument drivers, since they are distributed with their block diagram source code. Once you have properly installed the appropriate instrument drivers using the MAX, you can access the instrument drivers on the **Functions** palette, as shown in Figure 10.12.

If you have a web browser installed, you can link automatically to the National Instruments Instrument Driver Network, as shown in Figure 10.11. The Instrument Driver Network provides the complete library of available instrument drivers for LabVIEW. In this web page, you can search for your instrument among over 600 drivers available with free source code. You then can download the instrument driver you need and install it into the instr.lib *folder in the LabVIEW root directory.*

If you decide to build your own instrument driver, VISA is the standard API throughout the instrumentation industry. In addition, the VISA functions can control a suite of instruments of different types, including GPIB, serial, or VXI. In other words, VISA provides interface independence. Students that need to program instruments for different interfaces only need to learn one API.

The HP34401A Instrument Driver

In this example we will take a look at the HP34401A instrument driver. The HP 34401 A digital multimeter instrument driver is installed automatically as

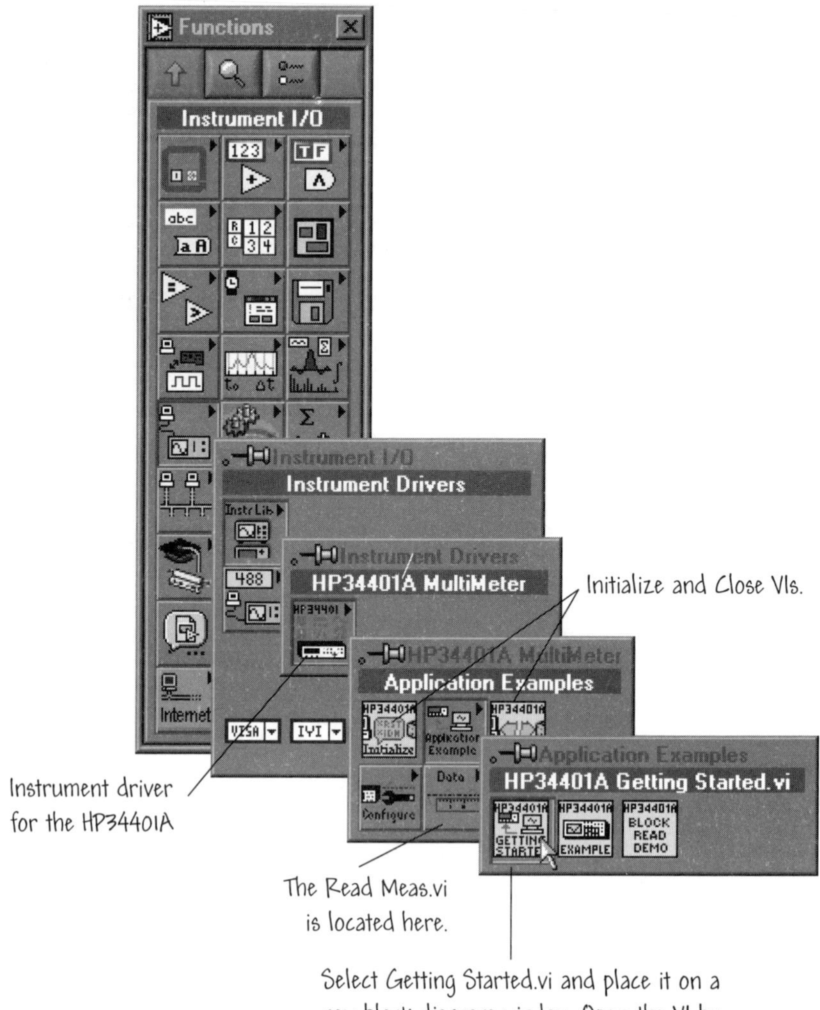

FIGURE 10.12
Locating the installed instrument drivers on the **Functions** palette.

part of the LabVIEW Student Edition. Begin by first opening a new VI. Using Figure 10.12 as a guide to locating the HP34401A Getting Started VI, drop the VI on the block diagram. Then double-click on the VI, which should open up the front panel. The Getting Started VIs are used to verify communication with your instrument and test a typical programmatic instrument operation. The HP34401A Getting Started VI front panel is shown in Figure 10.13.

With the exception of the address field, the defaults for most controls on the front panel will be sufficient for your first run. You will need to set the address

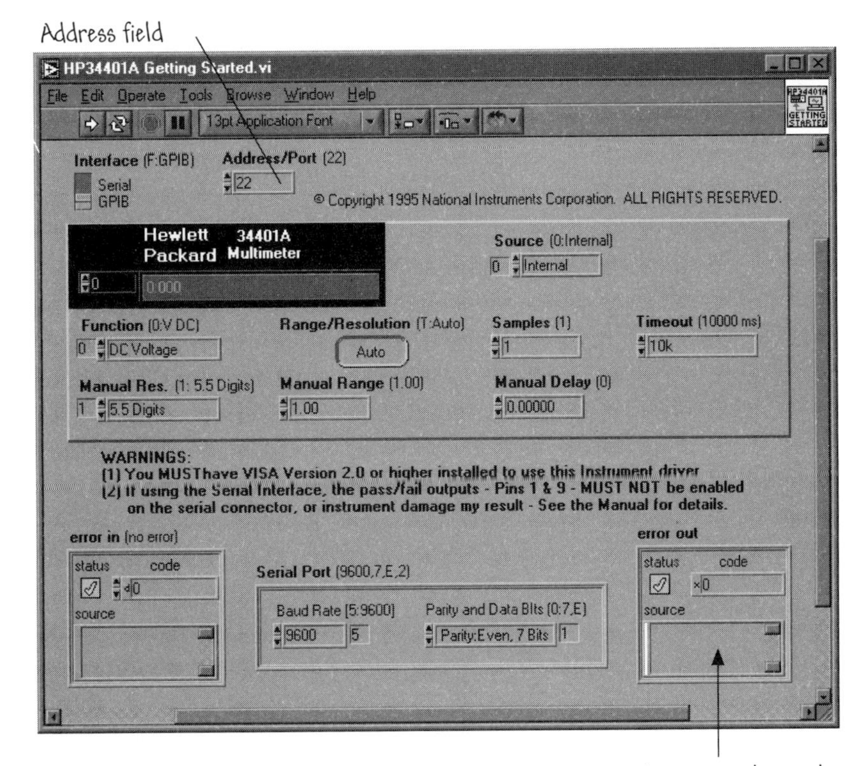

FIGURE 10.13
The HP34401a Getting Started VI front panel.

appropriately. If you do not know the address of your instrument, refer to the MAX for help. After running the VI, check to see that reasonable data was returned and an error was not reported in the error cluster (bottom right corner of the front panel).

If your Getting Started VI doesn't work, you need to check that

- NI-VISA is installed. If you did not choose this as an option during your LabVIEW installation, you will need to install it before rerunning your Getting Started VI.
- The instrument address is correct.
- The selected instrument driver supports the exact instrument model you are using.

Once you have verified basic communication with your instrument using the Getting Started VI, you may want to customize instrument control for your particular needs. If your application needs are similar to the Getting Started VI, the simplest means of creating a customized VI is to save a copy of the Getting Started VI by selecting **Save As...** from the **File** menu. You can change the default values on the front panel by selecting **Make Current Values Default** from the **Operate** menu.

The block diagram of an instrument driver generally consists of three VIs: the Initialize VI, an application function VI, and the Close VI. A simple instrument driver can be assembled from the VIs provided by LabVIEW for the HP34401A. Observing Figure 10.12, we see that the three VIs we need are HP34401A Initialize.vi, HP34401A Read Meas.vi, and HP34401A Close.vi. A straightforward instrument driver comprised of the three basic VIs is shown in Figure 10.14. Build the block diagram shown in Figure 10.14, if you have the HP34401A or do not have an instrument. If you are using another instrument driver, build a VI similar to Figure 10.14 using the Initialize, Read Meas, and Close VIs for that instrument. ◆

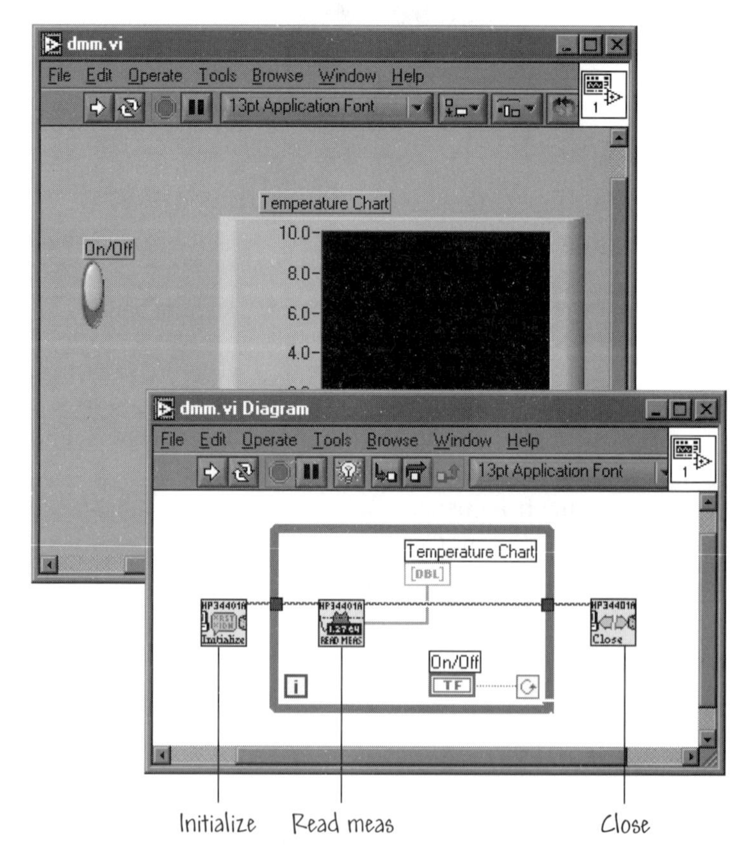

FIGURE 10.14
A simple HP34401A instrument driver.

10.3.1 Developing Your Own Instrument Driver

Make sure to check thoroughly for existing instrument drivers for your instrument before starting to build your own from scratch. You should check first at the National Instruments website as discussed in the previous sections. If that search fails, you should then search the website of the manufacturer of your equipment—they may have developed the instrument driver already. During your website searches, be on the lookout for instrument drivers that support a similar instrument, since instruments from the same model series often have similar command sets. If you find any such drivers, download them and assess the similarity of their command sets to that of your instrument—they may work directly or with minor modifications. For instruments from the same model series, you might need to contact the manufacturer and ask for details on the differences between the command sets.

If an instrument driver for your particular instrument does not exist, you can do one of several things (in rank order):

1. Use a driver for a similar instrument. Often similar instruments from the same manufacturer have similar if not identical command sets. The degree to which an instrument driver will need to be changed will depend on how similar the instruments and their command sets are. If the command sets are very different, you may be better off starting from scratch.
2. Create a simple instrument driver using the guidelines in the *Developing a Simple On-line help Driver* topic found in the LabVIEW **Online Reference**.
3. Develop a complete, fully functional instrument driver. To develop a National Instruments quality driver, you can download **Application Note AN006** *Developing a LabVIEW Instrument Driver* from the National Instruments website at

   http://www.ni.com/appnotes.nsf/

   This application note will help you to develop a complete instrument driver. To aid in the development of your instrument driver, National Instruments has created standards for instrument driver structure, device management, instrument I/O, and error reporting. The **Application Note AN006** describes these standards, as well as the purpose of a LabVIEW instrument driver, its components, and the integration of these components. In addition, this application note suggests a process for developing useful instrument drivers.

As to the question of whether to use VISA, GPIB, or serial functions, the recommendation is to rely on VISA. There are four major reasons for using VISA over GPIB. VISA

- is the industry standard,
- provides interface independence,

- provides platform independence, and
- will be easily adaptable in future instrumentation control applications.

VISA uses the same operations to communicate with instruments regardless of the interface type. VISA is designed so that programs written using VISA function calls are easily portable from one interface type to another like GPIB, serial or VXI. To ensure platform independence, VISA strictly defines its own data types. The VISA function calls and their associated parameters are uniform across all platforms, so that software can be ported to other platforms and then recompiled. A LabVIEW program using VISA can be ported to any platform supporting LabVIEW. A final advantage of using VISA is that it is an object-oriented API that will easily adapt to new instrumentation interfaces as they are developed in the future, making application migration to the new interfaces easy.

If you must develop your own instrument driver, you should use VISA functions rather than GPIB because of the versatility of VISA.

10.4 FUTURE OF INSTRUMENT DRIVERS AND INSTRUMENT CONTROL

Current instrument drivers use VISA, which is a common software interface for controlling GPIB, serial, or VXI instruments. At the present time and into the future, there will be one instrument driver for all oscilloscopes, for instance, no matter the manufacturer, model, or hardware interface (such as GPIB). These new instrument drivers are called Interchangeable Virtual Instruments (IVI) drivers and are supported by the IVI Foundation. The IVI Foundation is comprised of end-user test engineers and system integrators with many years of experience building GPIB and VXI-based test systems. By defining a standard instrument driver model that enables engineers to swap instruments without requiring software changes, the IVI Foundation members believe that significant savings in time and money will result because of

- Software that does not change when instruments become obsolete.
- A single software application that can be used on a system with different instrument hardware, which maximizes existing resources.
- Portable code that can be developed in test labs and hosted on different instrument in the production environment.

The following instrument types or classes have been defined by the IVI Foundation: oscilloscope, DMM, arbitrary waveform generator, switch, and power supply. More instrument types will be defined in the future. For more information on the IVI drivers and the IVI Foundation, refer to the **Application Note AN 121** "Using IVI Drivers to Build Hardware-Independent Test Systems with Lab-

VIEW and LabWindows/CVI" on the National Instruments website, www.ni.com/appnotes.nsf. You may also want to visit the IVI Foundation website at

http://www.ivifoundation.org/

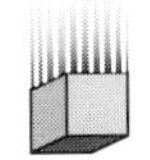

BUILDING BLOCK

10.5 BUILDING BLOCKS: DEMO SCOPE

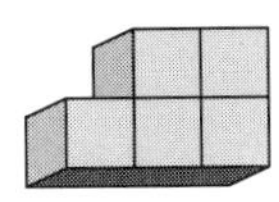

If you do not have instrument I/O hardware installed, you can use the Demo Scope VI as a substitute to learn about instrument control. The Demo Scope VI is the demonstration equivalent of a Getting Started VI for an actual instrument driver.

Open the Demo Scope VI instrument driver located in the library

vi.lib\tutorial.llb

Click on **Run** to begin the process of acquiring data from one or two channels on your oscilloscope, as shown in Figure 10.15. You can play around with changing the time and volts per division settings. When you are finished acquiring data, click the **Stop[F4]** button to stop the VI, as shown in Figure 10.15.

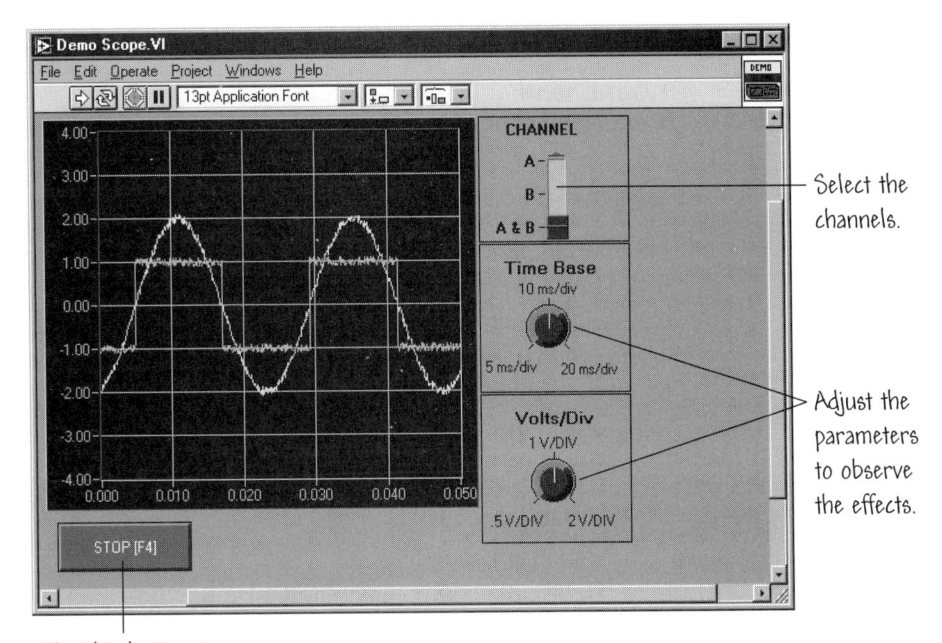

FIGURE 10.15
The Demo Scope VI front panel.

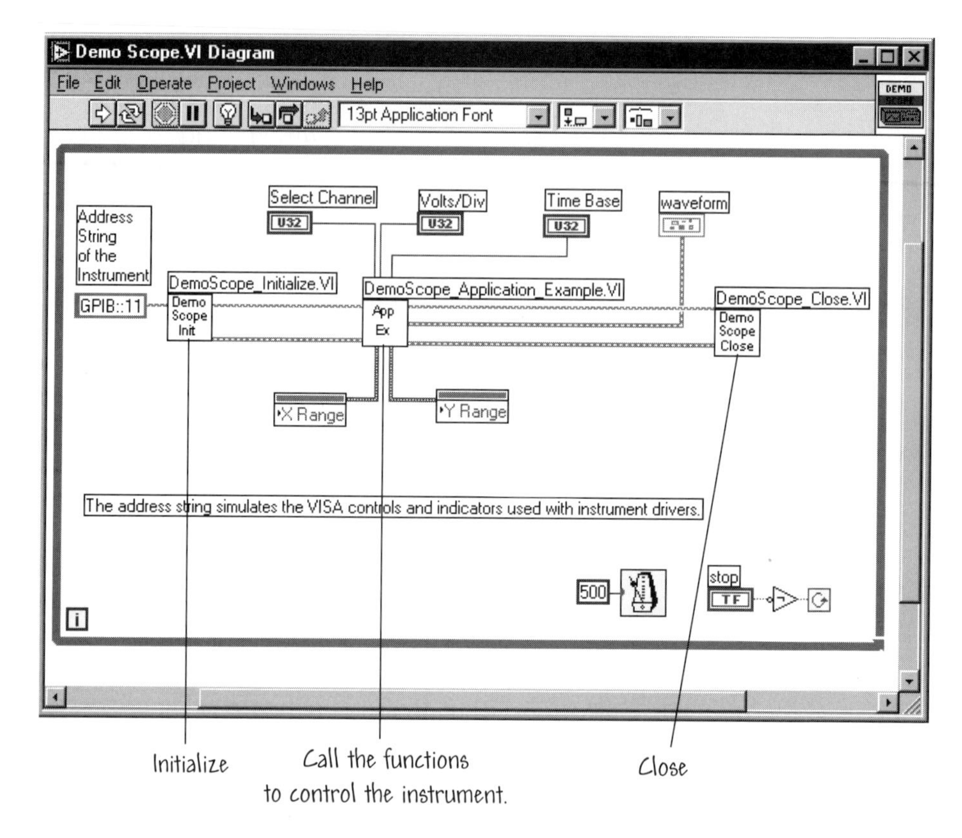

FIGURE 10.16
The Demo Scope VI block diagram.

Switch to the block diagram window, shown in Figure 10.16. Notice that the first function is the initialization, followed by the commands to send to the instrument in the DemoScope_Application_Example.vi. The DemoScope_Close.vi then closes communication with the instrument. LabVIEW instrument drivers follow this model—initializing the instrument, then calling the functions to control the instrument, and finally closing the instrument communication.

10.6 RELAXING READING: MONITORING INTERACTIONS BETWEEN MOLECULES IN THIN FILMS

Films of organic compounds one molecule thick are frequently used in chemical sensors and optical devices and as models for biological membranes. At the School of Chemistry and Biochemistry at the Georgia Institute of Technology, LabVIEW was utilized to construct an instrument that simultaneously assembles two different monomolecular films while monitoring the light-absorbing characteristics of each. The researchers assembled a hybrid instrument from commercial components, relying on LabVIEW to unite these components into an effective single instrument by providing the means for integrating different instrument components and interfacing protocols, as well as for performing unattended instrument operation. The process of monitoring interactions between molecules in thin films is controlled using an RS-232 serial port and a GPIB interface. The optical detector used for monitoring the light-absorbing characteristics of the films is controlled with the GPIB interface.

The researchers at Georgia Tech used their new "composite" instrument to monitor the assembly and light-absorbing characteristics of two different monomolecular films simultaneously. Film A contained a single type of organic dye molecule (3,3-dioctadecyloxacarbocyanine perchlorate). Film B contained a 50/50 mixture of the dye molecule and a phospholipid (dipalmitoylphosphatidyl choline). Dye molecules of this type are important tools in biological investigations because they can accumulate in cell membranes, where they undergo light-absorption and emission changes based on membrane electrical potential.

The absorption spectra shown in Figure 10.17 were collected every 3 mN/m of surface pressure change for Film A. According to the researchers, analysis of

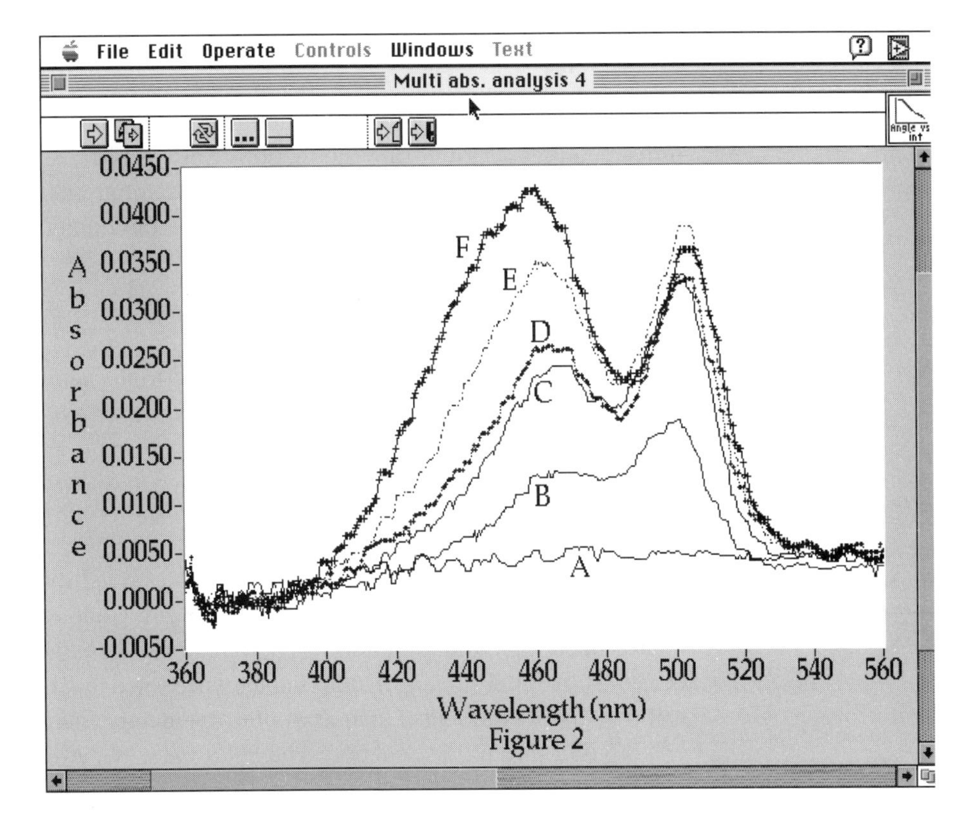

FIGURE 10.17
The light-absorbing characteristics of a pure monolayer of 3,3'-dioctadecyloxacarbocyanine perchlorate are displayed in a multiplot graph. This multiplot was created as the isotherm for the monolayer film was constructed. (A) 0.0 mN/m; (B) 3.0 mN/m; (C) 9.0 mN/m; (D) 15.0 mN/m; (E) 21.0 mN/m; (F) 27.0 mN/m.

the light-absorbing characteristics of these two films reveals two distinct absorption features, that is, two peaks at 500 nm and 490 nm, respectively. The absorption maximum at 500 nm results from monomer (or single-dye molecule) absorption. The absorption maximum at 469 nm results from dimer absorption (which is the very close association of two monomer dye molecules). A plot similar to the one shown in Figure 10.17 was obtained for Film B. Comparing data from the two films, the researchers observed that the ratio of dimers to monomers is different in each film. Film A (with a single component) had a larger proportion of dimers, especially at higher surface pressures, than Film B (the mixed component film). This information provides a means for understanding the characteristics of the film as well as guidance for future film construction parameters.

For further reading on the technical aspects of the project:

1. A. Ulman, *An Introduction to Ultrathin Organic Films from Langmuir-Blodgett to Self-Assembly*, Academic Press, Boston, 1991.
2. G. Roberts, *Langmuir-Blodgett Films*, Plenum Press, New York, 1990.
3. J. L. Moore, B. A. DeBry, and K. D. Hughes, "Routine Absorbance Spectroscopy at the Air/Water Interface," *Applied Spectroscopy* 49, no. 3 (1995): 386–391.

For more information on this application of LabVIEW, read the *User Solutions* entitled "Monitoring Interactions Between Molecules in Thin Films" by Ken Hughes and Jeff Moore of the Georgia Institute of Technology, which can be found on the National Instruments website, or contact:

Kenneth D. Hughes
Associate Professor of Chemistry
Kennesaw State University
Marietta, GA 30144
e-mail: kenneth.hughes@chemistry.gatech.edu

10.7 SUMMARY

As we already know from the chapter on data acquisition, LabVIEW can communicate with external devices. As discussed in Chapter 8, you can use DAQ boards in conjunction with LabVIEW to read and generate analog input, analog output, and digital signals. In this chapter we learned that LabVIEW can also control external instruments (such as digital voltmeters and oscilloscopes) over the GPIB bus or through a serial port. The Instrument Wizard is used to detect instruments and to install instrument drivers. An instrument driver is a VI that controls a particular instrument. Students should use the instrument drivers developed by National Instruments rather than attempt to develop drivers from "scratch."

Instrument drivers can be downloaded from the National Instruments website using the Instrument Driver Network.

KEY TERMS

Bus: The means by which computers and instruments transfer data.

Controller: A GPIB device that manages the flow of information on the GPIB by sending commands to all devices. A computer usually handles the controller function.

Controller-in-charge: The active controller in a GPIB system.

Device-dependent messages: Messages that contain device-specific information such as programming instructions, measurement results, machine status, and data files. These are often called **data messages**.

GPIB: General Purpose Interface Bus is the common name for the communications interface system defined in ANSI/IEEE Standard 488-1975 and 488.2-1987.

IEEE: Institute for Electrical and Electronic Engineers.

Instrument driver: A set of LabVIEW VIs that communicate with an instrument using standard VISA I/O functions.

Interface messages: Messages that manage the bus itself, and perform such tasks as initializing the bus, addressing and unaddressing devices, and setting device modes for remote or local programming. These are often called **command messages**.

Listener: A GPIB device that receives data messages from the talker.

MAX: A utility that provides information about connected instruments, and installs and manages instrument drivers.

Serial communication: A popular means of transmitting data between a computer and a peripheral device by sending data one bit at a time over a single communication line.

Talker: A GPIB device that sends data messages to one or more listeners.

VISA: Virtual Instrument Software Architecture. VISA is a VI library for controlling GPIB, serial, or VXI instruments.

EXERCISES

E10.1 A more sophisticated oscilloscope than the one discussed in "Building Blocks" is called Two Channel Oscilloscope.vi and is located in Examples\Apps\demos.llb. Open and run the oscilliscope example. You can select between displaying channels A, B, or both A and B concurrently. Once the channel(s) is selected, you can set the time base, volts.div, trigger source, slope, and level. This VI is intended to demonstrate the flexibility of LabVIEW for instrument control.

E10.2 From the LabVIEW startup screen, select a **New VI**. Select Tools≫Instrumentation≫Instrument Driver Network.... Once you have connected to the website, navigate around until you find and click on **Download Driver**. Locate an instrument you are using or have used and download the driver. Run the executable to install the VIs in the proper location. Close LabVIEW, then open LabVIEW. Select **New VI** and select Functions≫Instrument Drivers and the subpalette of the driver you just downloaded. You will notice that the VI organization is very similar to that of the HP34401A instrument driver discussed in this chapter.

PROBLEMS

P10.1 Open Frequency Response.vi located in Examples\Apps\freqresp.llb. This VI simulates an application that uses GPIB instruments to perform a frequency response test. What brand of digital multimeter is being simulated in this example? Switch to the block diagram and verify that For Loops, Formula Nodes, graphs, and arrays are some of the LabVIEW objects used in the code. Run the VI in continuous run mode and observe the effects of varying the amplitude and frequency of the signal.

P10.2 Complete the crossword puzzle.

Across

1. A popular means of transmitting data between a computer and a peripheral device by sending data one bit at a time over a single communication line.
5. A GPIB device that manages the flow of information on the GPIB by sending commands to all devices.
7. A GPIB device that receives data messages from the talker.
8. A GPIB device that sends data messages to one or more listeners.
9. A set of LabVIEW VIs that communicate with an instrument using standard VISA I/O functions.
11. The common name for the communications interface system defined in ANSI/IEEE Standard 488-1975 and 488.2-1987.
12. The means by which computers and instruments transfer data.

Down

2. Messages that manage the bus itself, and perform such tasks as initializing the bus and addressing devices.
3. The active controller in a GPIB system.
4. Institute for Electrical and Electronic Engineers.
6. A VI library for controlling GPIB, serial, or VXI instruments.
10. A utility that provides information about connected instruments, and installs and manages instrument drivers.

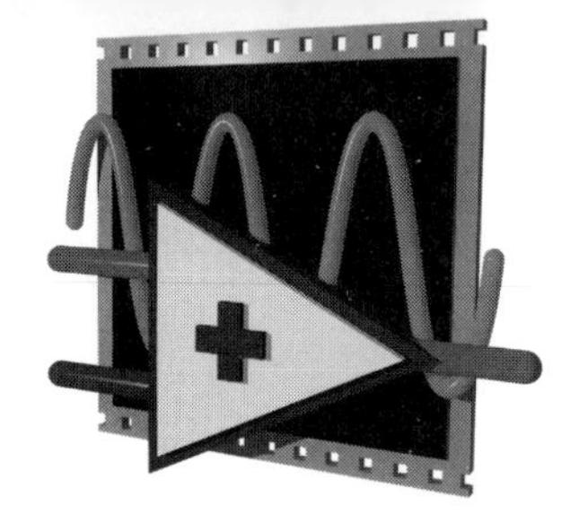

CHAPTER 11

Analysis

LabVIEW is an excellent environment for analysis of signals and systems. The G programming language is ideally suited to developing programs for solving linear algebraic systems of equations, curve fitting, integrating ordinary differential equations, computing function zeroes, computing derivatives of functions, integrating functions, generating and analyzing signals, computing discrete Fourier transforms, and filtering signals. This chapter presents an overview of these topics.

GOALS

1. Introduce some of LabVIEW's capabilities for analyzing signals and systems.
2. Study some of the mathematical analysis VIs made available by LabVIEW.

11.1 LINEAR ALGEBRA

LabVIEW provides a number of important VIs dedicated to solving systems of linear algebraic equations. These types of systems arise frequently in engineering and scientific applications. An entire branch of mathematics is dedicated to the study of *linear algebra*. We will only be able to touch on a few important subjects within the linear algebra arena in the space available to us here.

11.1.1 Review of Matrices

The basic element used in the computations is the so-called *matrix*. A matrix is represented by an $m \times n$ array of numbers:

$$\mathbf{A} = \begin{bmatrix} a_{0,0} & a_{0,1} & \cdots & a_{0,n-1} \\ a_{1,0} & a_{1,1} & \cdots & a_{1,n-1} \\ \cdots & \cdots & \cdots & \cdots \\ a_{m-1,0} & a_{0,1} & \cdots & a_{m-1,n-1} \end{bmatrix},$$

where n is the number of columns, and m is the number of rows. When $m \neq n$, the matrix is called a *rectangular* matrix; conversly when $m = n$, the matrix is called a *square* matrix. An $m \times 1$ matrix is called a *column vector*, and a $1 \times n$ matrix is called a *row vector*. Other special forms of the matrix are the *diagonal* matrix, the *zero* matrix, and the *identity* matrix. Examples of these three types of matrices are

$$\mathbf{A} = \begin{bmatrix} 2 & 0 & 0 \\ 0 & 4 & 0 \\ 0 & 0 & -6 \end{bmatrix}, \quad \mathbf{0} = \begin{bmatrix} 0 & 0 & 0 \\ 0 & 0 & 0 \\ 0 & 0 & 0 \end{bmatrix}, \quad \mathbf{I} = \begin{bmatrix} 1 & 0 & 0 \\ 0 & 1 & 0 \\ 0 & 0 & 1 \end{bmatrix},$$

respectively. The elements of a matrix, denoted by $a_{i,j}$, can be real or complex numbers.

Addition of two matrices is performed element by element. For example,

$$\begin{bmatrix} 2 & 3 & 1 \\ -7 & 4 & 5 \\ 2 & 0 & -6 \end{bmatrix} + \begin{bmatrix} 2 & -7 & 2 \\ 2 & 4 & 1 \\ 1 & 5 & -1 \end{bmatrix} = \begin{bmatrix} 4 & -4 & 3 \\ -5 & 8 & 6 \\ 3 & 5 & -7 \end{bmatrix}.$$

You can easily check that $\mathbf{A}+\mathbf{B}=\mathbf{B}+\mathbf{A}$ and that $(\mathbf{A}+\mathbf{B})+\mathbf{C}=\mathbf{A}+(\mathbf{B}+\mathbf{C})$. Therefore, matrix addition is commutative and associative.

If we multiply a matrix by a scalar, α, the result is obtained by multiplying each element of the matrix by the scalar, yielding

$$\alpha\mathbf{A}=\begin{bmatrix} \alpha a_{0,0} & \alpha a_{0,1} & \cdots & \alpha a_{0,n-1} \\ \alpha a_{1,0} & \alpha a_{1,1} & \cdots & \alpha a_{1,n-1} \\ \cdots & \cdots & \cdots & \cdots \\ \alpha a_{m-1,0} & \alpha a_{0,1} & \cdots & \alpha a_{m-1,n-1} \end{bmatrix}.$$

Multiplication of matrices $\mathbf{AB}$ requires the two matrices to be of compatible dimensions—the number of columns of $\mathbf{A}$ must be equal to the number of rows of $\mathbf{B}$. Thus if $\mathbf{A}$ is an $m \times n$ matrix, and $\mathbf{B}$ is an $n \times p$ matrix, the product $\mathbf{C}=\mathbf{AB}$ is an $m \times p$ matrix. The element $c_{i,j}$ of the matrix $\mathbf{C}$ is given by

$$c_{i,j}=a_{i,1}b_{1,j}+a_{i,2}b_{2,j}+\cdots+a_{i,m}b_{m,j}\ .$$

In general, matrix multiplication is not commutative; that is,

$$\mathbf{AB}\neq\mathbf{BA}\ .$$

The *transpose* of a real matrix (that is, a matrix comprised of only real numbers) is formed by interchanging the rows and columns. For example, if

$$\mathbf{A}=\begin{bmatrix} 2 & 3 & 1 \\ -7 & 4 & 5 \end{bmatrix}, \quad \text{then} \quad \mathbf{A}^T=\begin{bmatrix} 2 & -7 \\ 3 & 4 \\ 1 & 5 \end{bmatrix},$$

where the matrix $\mathbf{A}^T$ is the transpose of $\mathbf{A}$. A real matrix is called a *symmetric* matrix if $\mathbf{A}^T=\mathbf{A}$. If the elements of the matrix $\mathbf{C}$ are complex numbers, then we extend the notion of a transpose to *complex conjugate transpose*. This means that we transpose the matrix and then replace each element with its own complex conjugate. We denote the complex conjugate transpose as $\mathbf{C}^H$. A complex matrix is called a *Hermitian* matrix if $\mathbf{C}^H=\mathbf{C}$.

For square matrices, we can define the operations **trace**, **determinant**, and **inversion**. The trace of an $n \times n$ square matrix $\mathbf{A}$ is the sum of the diagonal elements,

$$\text{tr}\,\mathbf{A}=a_{1,1}+a_{2,2}+\cdots+a_{n,n}\ .$$

The determinant of a 2×2 matrix is given by

$$\det\mathbf{A}=\left\|\begin{bmatrix} a_{1,1} & a_{1,2} \\ a_{2,1} & a_{2,2} \end{bmatrix}\right\|=a_{1,1}a_{2,2}-a_{1,2}a_{2,1}\ .$$

If the determinant is identically equal to zero, then we say that the matrix is *singular*. In general, the determinant can be computed as a function of the *minors* and *cofactors* of the matrix. The minor of an element $a_{i,j}$ is the determinant of an $n-1 \times n-1$ matrix formed by removing the ith row and the jth column of the original matrix $\mathbf{A}$. For example, if

$$\det \mathbf{A} = \left\| \begin{bmatrix} a_{1,1} & a_{1,2} & a_{1,3} \\ a_{2,1} & a_{2,2} & a_{2,3} \\ a_{3,1} & a_{3,2} & a_{3,3} \end{bmatrix} \right\| ,$$

then the minor of the $a_{2,3}$ element is

$$M_{2,3} = \left\| \begin{bmatrix} a_{1,1} & a_{1,2} \\ a_{3,1} & a_{3,2} \end{bmatrix} \right\| = a_{1,1}a_{3,2} - a_{1,2}a_{3,1} .$$

The cofactor of $a_{i,j}$ is defined as

$$\alpha_{i,j} = \text{cofactor } a_{i,j} = (-1)^{i+j}\mathbf{M}_{i,j} .$$

In general, we can compute the determinant of an $n \times n$ square matrix $\mathbf{A}$ as

$$\det \mathbf{A} = \sum_{j=1}^{n} a_{i,j}\alpha_{i,j}$$

for any row i. Similarly, we can compute the determinant as

$$\det \mathbf{A} = \sum_{i=1}^{n} a_{i,j}\alpha_{i,j}$$

for any column j.

The *adjoint matrix* of an $n \times n$ square matrix $\mathbf{A}$ is formed by transposing the matrix and replacing each element $a_{i,j}$ with the cofactor $\alpha_{i,j}$. Therefore, we have

$$\text{adjoint } \mathbf{A} = \begin{bmatrix} \alpha_{1,1} & \alpha_{1,2} & \alpha_{1,3} \\ \alpha_{2,1} & \alpha_{2,2} & \alpha_{2,3} \\ \alpha_{3,1} & \alpha_{3,2} & \alpha_{3,3} \end{bmatrix}^T = \begin{bmatrix} \alpha_{1,1} & \alpha_{2,1} & \alpha_{3,1} \\ \alpha_{1,2} & \alpha_{2,2} & \alpha_{3,2} \\ \alpha_{1,3} & \alpha_{2,3} & \alpha_{3,3} \end{bmatrix} .$$

The matrix inverse is denoted by $\mathbf{A}^{-1}$ and can be computed as

$$\mathbf{A}^{-1} = \frac{\text{adjoint } \mathbf{A}}{\det \mathbf{A}} .$$

The matrix inverse must satisfy the relationship

$$\mathbf{A}^{-1}\mathbf{A} = \mathbf{A}\mathbf{A}^{-1} = \mathbf{I} .$$

The matrix inverse does not exist (that is, it is singular) when det $\mathbf{A} = 0$. If the matrix $\mathbf{A}$ is singular, then there exists a nonzero vector $\mathbf{v}$ such that $\mathbf{A}\mathbf{v} = \mathbf{0}$.

11.1.2 System of Algebraic Equations

Suppose that we want to solve the following system of algebraic equations:

$$\begin{aligned} 4x_1 + 6x_2 + x_3 &= 4 \\ x_1 + 2x_2 + 3x_3 &= 0 \\ 5x_2 - x_3 &= 1 \end{aligned}$$

The unknowns are the variables x_1, x_2, and x_3. We can identify the two column vectors $\mathbf{x}$ and $\mathbf{b}$ as

$$\mathbf{x} = \begin{bmatrix} x_1 \\ x_2 \\ x_3 \end{bmatrix} \quad \text{and} \quad \mathbf{b} = \begin{bmatrix} 4 \\ 0 \\ 1 \end{bmatrix} .$$

Then we can write the system of algebraic equations as

$$\mathbf{A}\mathbf{x} = \mathbf{b} ,$$

where

$$\mathbf{A} = \begin{bmatrix} 4 & 6 & 1 \\ 1 & 2 & 3 \\ 0 & 5 & -1 \end{bmatrix} .$$

Thus, we have rewritten the problem in a compact matrix notation. Can we solve for the vector $\mathbf{x}$? In this case, the matrix $\mathbf{A}$ is invertible (i.e., the inverse exists), so the solution is readily obtained as

$$\mathbf{x} = \mathbf{A}^{-1}\mathbf{b} = \begin{bmatrix} 0.9123 \\ 0.1228 \\ -0.3860 \end{bmatrix} .$$

If the matrix $\mathbf{A}$ is singular, then the number of solutions depends on the vector $\mathbf{b}$. If a solution does exist in the singular case, it is not unique!

For any given $n \times n$ square matrix $\mathbf{A}$, we would like to know if there exists a scalar λ and a corresponding vector $\mathbf{v} \neq \mathbf{0}$ such that

$$\lambda \mathbf{v} = \mathbf{A}\mathbf{v} .$$

This scalar λ is called the *eigenvalue*, and the corresponding vector $\mathbf{v}$ is called the *eigenvector*. Rearranging yields

$$\lambda \mathbf{v} - \mathbf{A}\mathbf{v} = (\lambda \mathbf{I} - \mathbf{A})\, \mathbf{v} = \mathbf{0} .$$

Therefore, a solution exists (for $\mathbf{v} \neq \mathbf{0}$) if and only if

$$\det\ (\lambda \mathbf{I} - \mathbf{A}) = 0 .$$

If $\mathbf{A}$ is an $n \times n$ matrix, then $\det\ (\lambda \mathbf{I} - \mathbf{A}) = 0$ is an nth-order polynomial (known as the *characteristic equation*) whose solutions are called the *characteristic roots* or eigenvalues. The eigenvalues of a square matrix are not necessarily unique and may be complex numbers (even for a real-valued matrix!). Given the eigenvalues of a matrix, we can compute the trace and determinant as

$$\mathrm{tr}\ \mathbf{A} = \sum_{i=1}^{n} \lambda_i$$

$$\det\ \mathbf{A} = \prod_{i=1}^{n} \lambda_i$$

We see that if any eigenvalue is zero, then the determinant is zero, and thus it follows that the matrix is singular.

An interesting fact regarding eigenvalues is that if a matrix is real and symmetric, then its eigenvalues will be real. If we compute the "square" of a real matrix $\mathbf{A}$ according to $\mathbf{B} = \mathbf{A}\mathbf{A}^T$, then $\mathbf{B}$ is real and symmetric. In fact, the matrix $\mathbf{B}$ is nonnegative, in the sense that, for any nonzero column vector $\mathbf{v}$ it follows that the scalar value $\mathbf{v}^T\mathbf{B}\mathbf{v} \geq 0$. Now if we compute the eigenvalues of $\mathbf{B}$, we find that they are all nonnegative and real. Taking the square root of each eigenvalue yields quantities known as *singular values*, and denoted here by β_i for $n = 1, 2, \ldots, n$. An $n \times n$ matrix has n nonnegative singular values. Two important singular values are the maximum and minimum values, $\beta_{\max}$ and $\beta_{\min}$, respectively. Singular values are important in computational linear algebra because they tell us something about how close a matrix is to being singular. We know that if the determinant of a matrix is zero, then it is a singular matrix. But what about if you compute the determinant numerically with the computer and determine that the determinant is 10^{-10}? Is this close enough to zero to say that the matrix is singular? The answer to this is provided by the

condition number of a matrix. It can be defined in different ways, but it is commonly defined as

$$\text{cond } \mathbf{A} = \frac{\beta_{\max}}{\beta_{\min}} .$$

The condition number can vary from 0 to ∞. A matrix with a condition number near 1 is closer to being nonsingular than a matrix with a very large condition number. The condition number is useful in assessing the accuracy of solutions to systems of linear algebraic equations.

As a final practical note, it is not generally a good idea to explicitly compute the matrix inverse when solving systems of linear algebraic equations, because inaccuracies are associated with the numerical computations—especially when the condition number is high. The preferred solution technique involves using *matrix decompositions*. Popular techniques include

- Singular Value Decomposition (SVD)
- Cholesky decomposition (or QR)

The idea is to decompose the matrix into component matrices that have "nice" numerical properties.

For example, suppose we decompose the matrix

$$\mathbf{A} = \mathbf{QR} ,$$

where $\mathbf{Q}$ is an *orthogonal* matrix (that is, $\mathbf{Q}^T\mathbf{Q} = \mathbf{I}$), and $\mathbf{R}$ is upper triangular (all the elements below the diagonal are zero). Then,

$$\mathbf{Ax} = \mathbf{QRx} = \mathbf{b} .$$

Multiplying both sides by $\mathbf{Q}^T$ yields

$$\mathbf{Rx} = \mathbf{Q}^T\mathbf{b} .$$

Then you use the fact that $\mathbf{R}$ is upper triangular to *back-substitute* and solve for $\mathbf{x}$ without ever computing a matrix inverse.

11.1.3 Linear System VIs

The *Student Edition of LabVIEW* comes with a complete set of VIs that can be used to perform all the matrix computations discussed previously (and much more!). You can find the VIs on the palette shown in Figure 11.1.

The Linear Algebra Calculator.vi is a useful VI that can be used to perform a variety of linear algebra operations on a matrix $\mathbf{A}$ or to solve systems of linear algebraic equations. This VI can take data as input on the front panel or read the data from a file (that is, from a spreadsheet file). The next example gives you the opportunity to experiment with linear algebra computations.

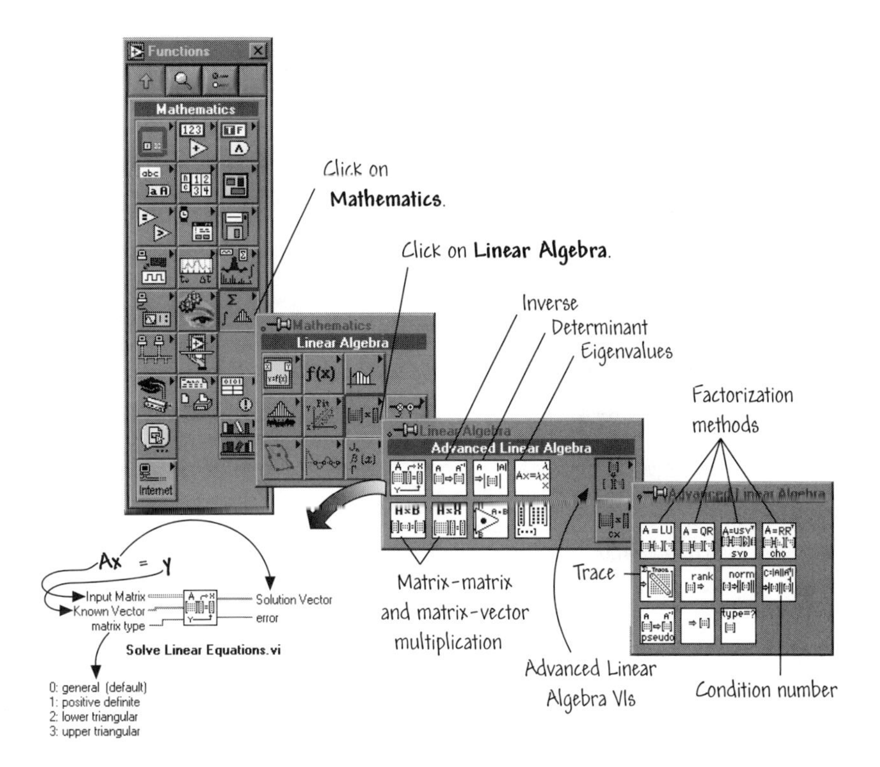

FIGURE 11.1
The linear algebra VIs.

Linear Algebra Calculator

Open Linear Algebra Calculator.vi located in Examples\Analysis\linaxmpl.llb. The front panel and block diagram are shown in Figure 11.2.

By default, the VI is set up to solve the system of linear algebraic equations

$$\mathbf{Ax} = \mathbf{b} \, ,$$

where

$$\mathbf{A} = \begin{bmatrix} 4 & 2 & -1 \\ 1 & 4 & 1 \\ 0.1 & 1 & 2 \end{bmatrix} , \quad \text{and} \quad \mathbf{b} = \begin{bmatrix} 2 \\ 12 \\ 10 \end{bmatrix} .$$

Choose **Solve Linear Equations** in the lower left side of the front panel and run the VI. Verify that you obtain the solution

$$\mathbf{x} = \begin{bmatrix} 0.59 \\ 1.84 \\ 4.05 \end{bmatrix} .$$

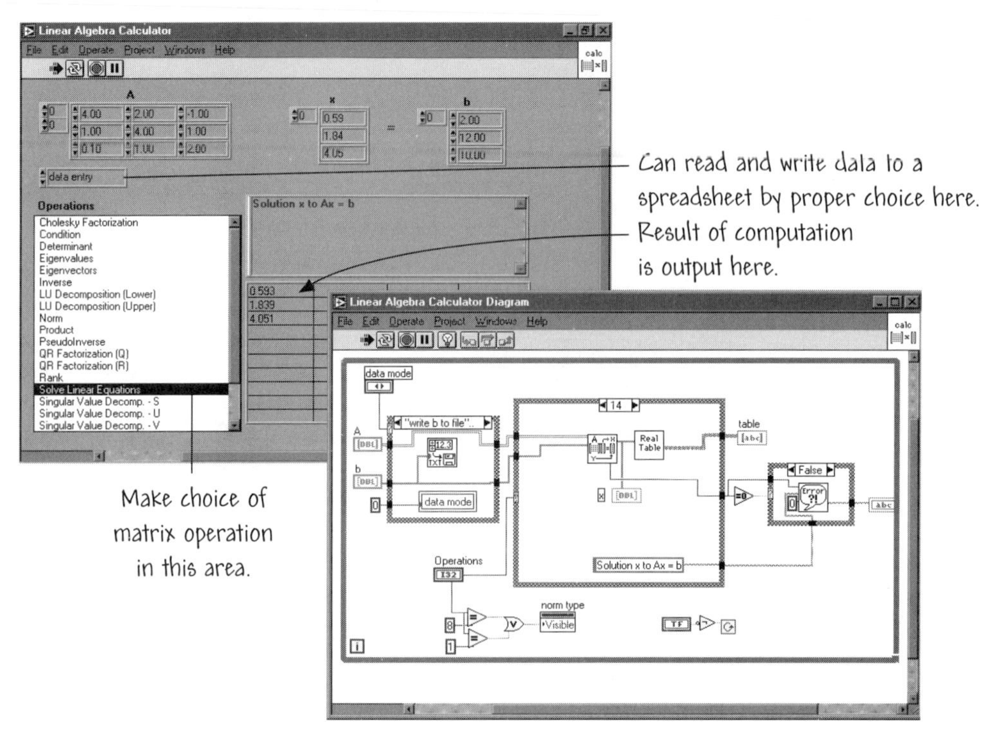

FIGURE 11.2
The linear algebra calculator.

Use the calculator to compute the determinant of the matrix **A**. You should get the result det $\mathbf{A} = 23.6$. Compute the condition number, the inverse, and the trace of the matrix **A** (condition number $= 4.418$ and trace $= 10$). The VI runs until the stop button is pressed.

Modify the input matrix to be

$$\mathbf{A} = \begin{bmatrix} 4 & 2 & -1 \\ 1 & 4 & 1 \\ 2 & 8 & 2 \end{bmatrix}.$$

Compute the condition number. You should find that the condition number is very, very large! Now compute the matrix inverse (by selecting **Inverse** from the menu). What happens? An error message appears that lets you know that the system of equations cannot be solved because the input matrix is singular. Vary the (3,3) element and observe the effect on the condition number. When the condition number reduces to less than 10, compute the matrix inverse again. Does it work in this case? It should!

Set the (3,3) term of the matrix to 2.01:

$$\mathbf{A} = \begin{bmatrix} 4 & 2 & -1 \\ 1 & 4 & 1 \\ 2 & 8 & 2.01 \end{bmatrix} .$$

Verify that the determinant is 0.14. Now compute the condition number—it should be quite high (above 2,500). What does this result lead you to conclude about the advisability of solving the system of linear algebraic equations by inverting the **A** matrix? Basically, when the condition number is high, it is not advisable to solve the system of equations by matrix inversion.

◆

11.2 CURVE FITTING

Curve fitting is a common technique used in science, engineering, business, medicine, and other fields in the analysis of data. The technique involves extracting a set of curve parameters (or coefficients) from the data set to obtain a functional description of the data set. Using curve fitting, digital data can be represented by a continuous model. For example, you may want to fit the data with a straight line model. The curve-fitting procedure would provide values for the linear curve fit in terms of slope and axis offset.

11.2.1 Curve Fits Based on Least Squares Methods

The main algorithm used in the curve-fitting process is known as the least squares method. Define the error as

$$e(\mathbf{a}) = [f(x, \mathbf{a}) - y(x)]^2 ,$$

where $e(\mathbf{a})$ is a measure of the difference between the actual data and the curve fit, $y(x)$ is the observed data set, $f(x, \mathbf{a})$ is the functional description of the data set (this is the curve-fitting function), and $\mathbf{a}$ is the set of curve coefficients that best describes the curve. For example, let $\mathbf{a} = (a_0, a_1)$. Then the functional description of a line is

$$f(x, \mathbf{a}) = a_0 + a_1 x .$$

The least squares algorithm finds $\mathbf{a}$ by solving the Jacobian system

$$\frac{\partial e(\mathbf{a})}{\partial \mathbf{a}} = 0 .$$

The curve-fitting VIs solve the Jacobian system automatically and return the set of coefficients that best describes the input data set. The automatic nature of this process provides the opportunity to concentrate on the results of the curve fitting rather than dealing with mechanics of obtaining the curve-fit parameters.

When we curve-fit data, we generally have available two input sequences, Y and X. The sequence X is usually the independent variable (e.g., time) and the sequence Y is the actual data. A point in the data set is represented by (x_i, y_i), where x_i is the ith element of the sequence X, and y_i is the ith element of the sequence Y. Since we are dealing with samples at discrete points, the VIs calculate the mean-square error (MSE), which is a relative measure of the residuals between the expected curve values and the actual observed values, using the formula

$$\text{MSE} = \frac{1}{n}\sum_{i=0}^{n-1}(f_i - y_i)^2 ,$$

where f_i is the sequence of fitted values, y_i is the sequence of observed values, and n is the number of input data points.

LabVIEW offers a number of curve-fitting algorithms, including:

- Linear Fit:

$$y_i = a_0 + a_1 x_i$$

- Exponential Fit:

$$y_i = a_0 e^{a_1 x_i}$$

- General Polynomial Fit:

$$y_i = a_0 + a_1 x_i + a_2 x_2 + \cdots$$

- General Linear Fit:

$$y_i = a_0 + a_1 f_1(x_i) + a_2 f_2(x_i) + \cdots$$

  where y_i is a linear combination of the parameters $a_0, a_1, \ldots$. This type of curve fit provides user-selectable algorithms (including SVD, Householder, Givens, LU, and Cholesky) to help achieve the desired precision and accuracy.

- Nonlinear Levenberg-Marquardt Fit:

$$y_i = f(x_i, a_0, a_1, a_2, \ldots)$$

  where $a_0, a_1, a_2, \ldots$ are the parameters. This method does not require y to have a linear relationship with $a_0, a_1, a_2, \ldots$. Although it can be used for linear curve fitting, the Levenberg-Marquardt algorithm is generally used for nonlinear curve fits.

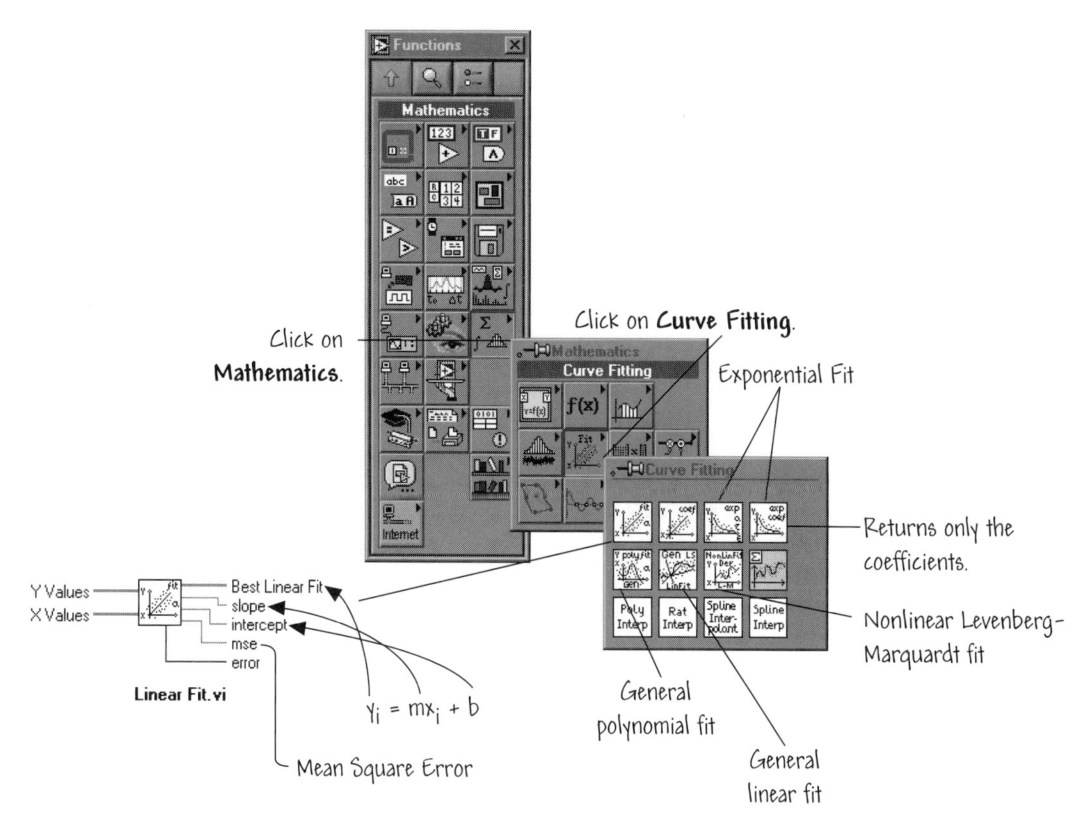

FIGURE 11.3
The curve-fitting VIs.

The curve-fitting VIs can be found on the **Curve Fit** palette, as illustrated in Figure 11.3.

Practicing with Curve Fitting

Open Regressions Demo.vi located in Examples\Analysis\regressn.llb. The front panel is shown in Figure 11.4. The VI generates noisy data samples that are approximately linear, exponential, or polynomial and then uses the appropriate curve-fitting VIs to determine the best parameters to fit the given data. You can control the noise level with the knob on the front panel. You can also select an algorithm and the number of samples to fit. For the polynomial curve fits, the order of the polynomial can be varied via front panel input.

Select *Linear* in the **Algorithm Selector** control and set the **Noise Level** to around 0.05. Run the VI and make a note of the computed error displayed in the **mse** indicator. Increase the noise level to 0.1 and again make a note of the computed error. Continue this process for noise levels of 0.15, 0.2, and 0.25. Did you detect any trends? You should have seen the MSE increase as the noise level increased.

Select Polynomial in the **Algorithm Selector** control and set the **Noise Level** to around 0.05. Set the **Order** control to 2. Run the VI and make a note

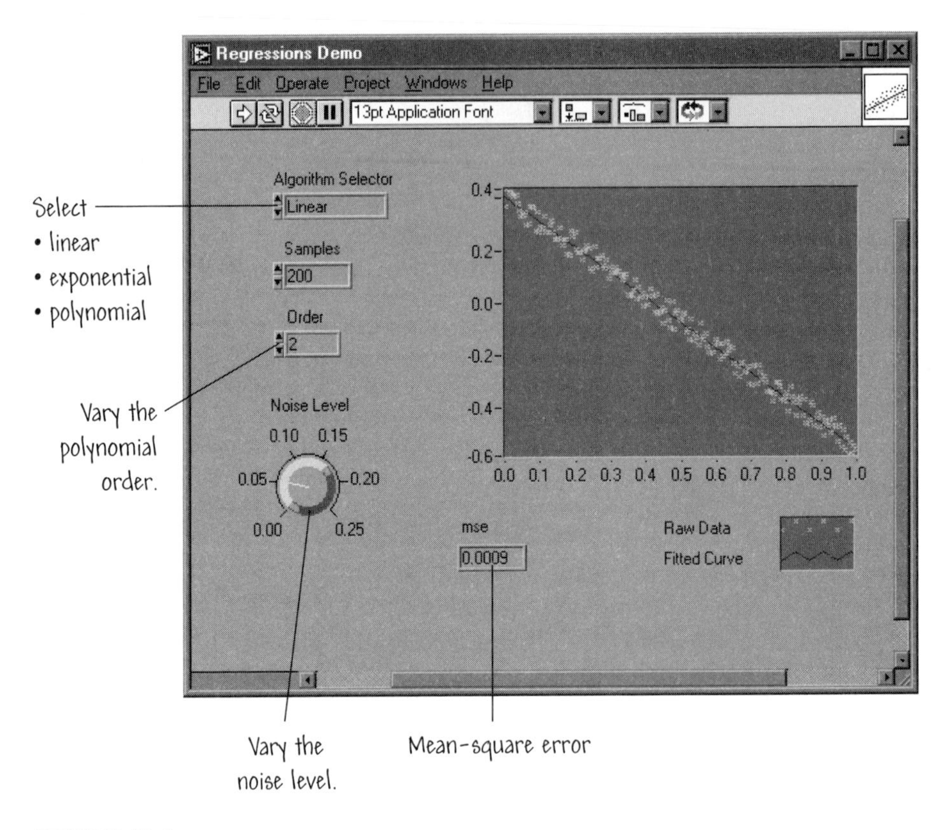

FIGURE 11.4
A demo to investigate curve fitting with linear, exponential, and polynomial fits.

of the computed error. Follow the same procedure as above. Did you obtain the same trends as the noise increased? Reduce the noise level to 0.15. This time run a computer experiment with the noise level fixed but increase the polynomial order from 2 to 6. Did you detect any trends as the polynomial order increased? In this case, as you increase the polynomial order, the computed error fluctuates, but remains basically on the same order of magnitude.

With the **Algorithm Selector** control set to Polynomial, run the VI with the polynomial order set to 0. Then change the polynomial order to 1 and run the VI. With polynomial order equal to 0 and 1, the fitted curve is a horizontal line and a straight line with a (generally) nonzero slope, respectively. Experiment with the Regressions Demo VI and see if you can discover new trends. Consider comparing the linear fit with the exponential fit.

◆

In Examples\Analysis\regressn.llb you will find other VIs that can be used to experiment with curve fits. The general linear fit and nonlinear Levenberg-Marquardt fit demonstration VIs are called General LS Fit Example.vi and Nonlinear Lev-Mar Exponential Fit.vi, respectively.

11.2.2 Fitting a Curve to Data with Normal Distributions

Real-world data is very often normally (or Gaussian) distributed. The mathematical description of a normal distribution is

$$f(x) = \frac{1}{\sigma\sqrt{2\pi}} \exp\left[-\frac{1}{2}\left(\frac{x-m}{\sigma}\right)^2 \right],$$

where m is the mean and σ is the standard deviation. Figure 11.5 shows the normal distribution with $m = 0$ and $\sigma = 0.5$, 1.0, and 2.0. As seen in the figure, the normal distribution is bell-shaped and symmetric about the mean, m. The area under the bell shaped curve is unity! The two parameters that completely describe the normally distributed data are the mean, m, and the standard deviation, σ. The peak of the bell shaped curve occurs at m. The smaller the value of σ, the higher the peak at the mean and the narrower the curve.

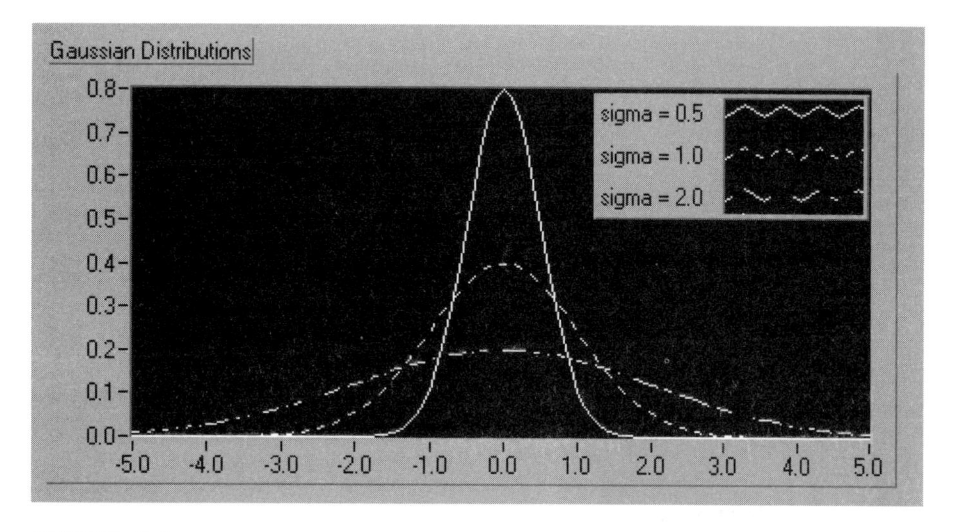

FIGURE 11.5
The normal distribution shape for three values of σ: 0.5, 1.0, and 2.0.

This normal distribution is illustrated in the Figure 11.6. The standard deviation is an important parameter that defines the "bounds" within which a certain percentage of the data values are expected to occur. For example:

- About two-thirds of the values will lie between $m - \sigma$ and $m + \sigma$.
- About 95% of the values will lie between $m - 2\sigma$ and $m + 2\sigma$.
- About 99% of the values will lie between $m - 3\sigma$ and $m + 3\sigma$.

Therefore, an interpretation of these values is that the probability that a normally distributed random value lies outside $\pm 2\sigma$ is approximately 0.05 (or 5%).

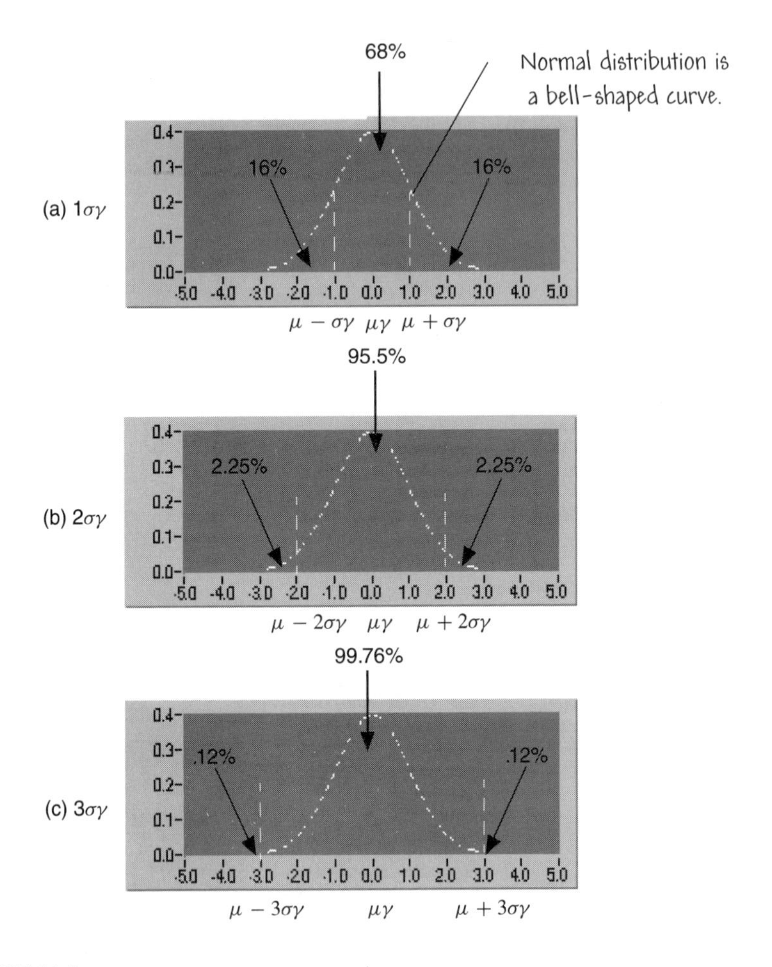

FIGURE 11.6
The normal distribution for 1σ, 2σ, and 3σ.

Normal Distributions

In the folder Chapter 11 in the Learning directory you will find a VI called Normal (Gaussian) Fit.vi. This VI generates a random data set and then plots the distribution. The front panel is shown in Figure 11.7. If you want to experiment with normally distributed data, open and run the VI. You can vary the number of samples and the standard deviation (that is, the σ). If you set $\sigma = 0.5, 1.0,$ and 2.0, you will be able to duplicate the graphs shown in Figure 11.5.

Run the VI and vary the number of samples from 10 to 10,000. You should notice that as the number of samples increases, the shape of the data distribution becomes more and more bell-shaped. Does this exercise relate in any way to your experience with grade distributions in class? Would you expect to have a bell-shaped grade distribution in a class of 5 students? 100 students? 500 students? (Answers: no, maybe, yes)

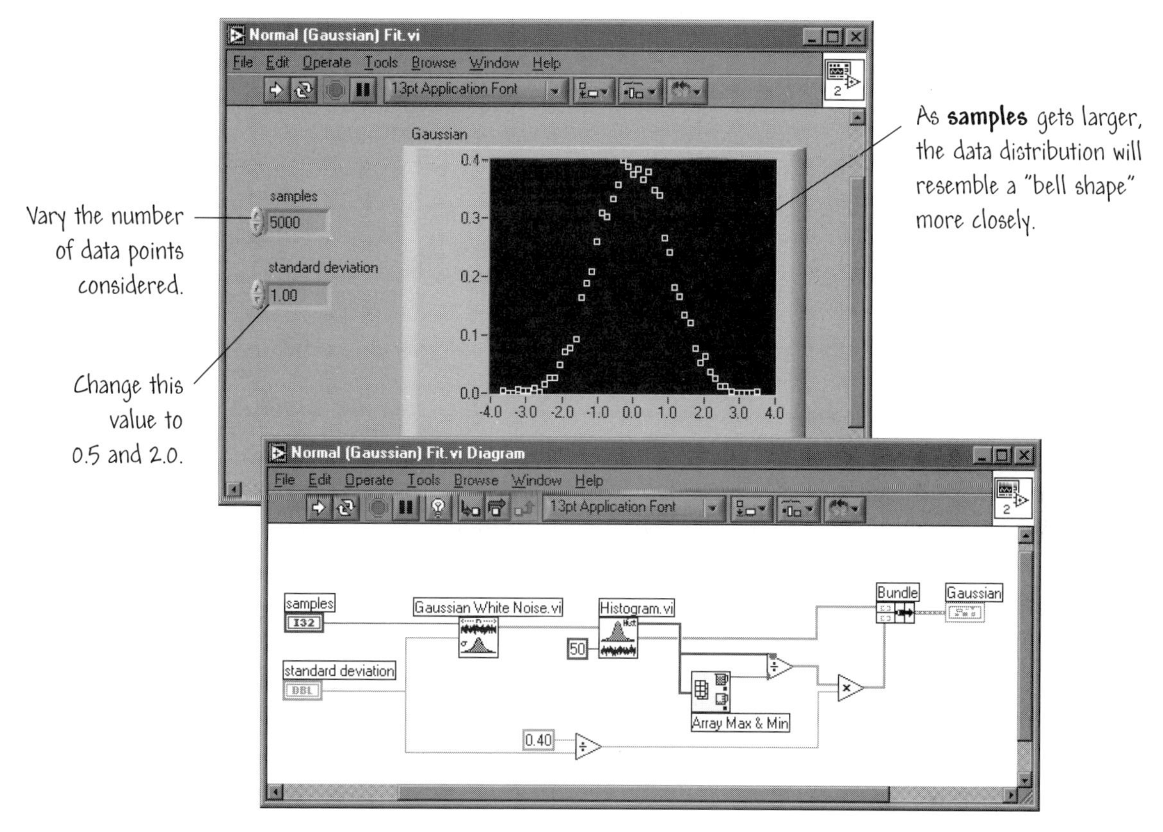

FIGURE 11.7
Normally distributed data.

Using Normal (Gaussian) Fit.vi as a starting point, construct a VI to curve-fit the normal distribution and to compute the mean and sigma from the curve-fit parameters.

◆

11.3 DISPLAYING FORMULAS ON THE FRONT PANEL

In Chapter 5 we discussed Formula Nodes. Recall that a Formula Node is a resizable box placed on the block diagram (similar to the Sequence Structure, Case Structure, For Loop, and While Loop) containing one or more formula statements. The formula statements use a syntax similar to most text-based programming languages for arithmetic expressions. In this section we will learn about placing formulas on the front panel (rather than on the block diagram, as with Formula Nodes) and then gaining access to a family of analysis VIs for optimization, integration, graphing, and many more. The formulas retain the familiar text-based programming language syntax.

There are a few differences between using parser VIs to place formulas on the front panel and Formula Nodes. The Formula Nodes have access to the binary functions—max, min, mod, rem—and the parser VIs do not have this access. However, the parser VIs can access more complex functions—Gamma, spike, Legendre elliptic integral 1st kind, square, and more—and have access to most of the same standard functions, such as $\sin x$, e^x, and $\tanh x$, as do the Formula Nodes. The Formula Nodes have access to a variety of logical, conditional, inequality, and equality operators, such as $<$, $>$, $==$, $\geq$, $\leq$, while the parser VIs do not. The parser VIs and the Formula Nodes play different roles in LabVIEW. We can learn about the parser VIs by working an example.

Practice with Parser VIs

Consider the VI shown in Figure 11.8, which shows how you can enter a formula on the front panel, evaluate it, and graph the results. Open a new VI and construct a VI using Figure 11.8 as a model. The parser VI at the center of this VI

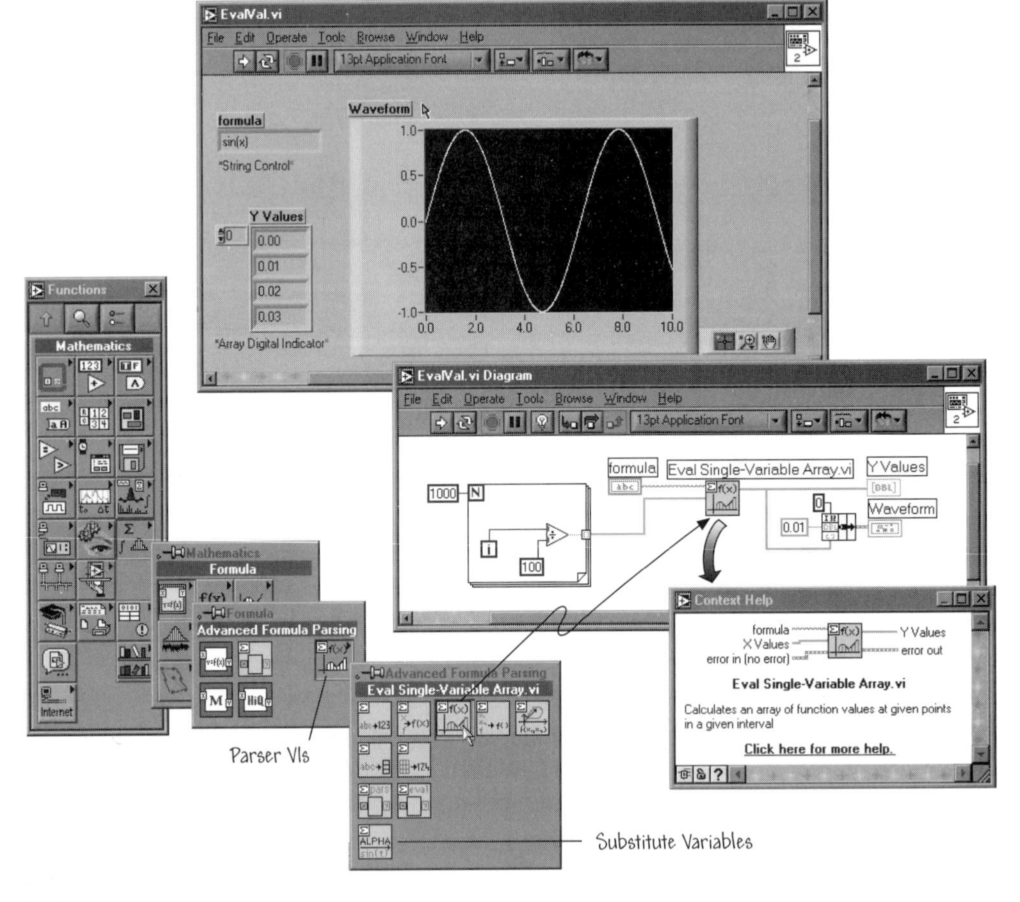

FIGURE 11.8
A VI that allows you to enter a formula on the front panel.

is the Eval Single-Variable Array.vi. It calculates an array of function values at given points in a given interval; that is,

$$y_i = f(x_i) ,$$

where $i = 1, \ldots, n$. Examining the block diagram in Figure 11.8, you can see that the For Loop is used to generate the x_i values, and $n = 1{,}000$. The points range from 0 to 9.99 in increments of 0.01.

After building the VI, return to the front panel and type in the expression $\sin(x)$ in the formula control and run the VI. The VI waveform will show the usual sine wave curve. The function has only one variable, which we called x—we could just as easily have input $\sin(z)$ or $\sin(y)$.

Change the formula to $3 * x\^{}2 + x * \log(x)$ and run the VI. Try another expression, such as $\text{step}(y) + \cos(y) * \sin(y)$. Then type in another expression of your choice in the formula control and run the VI. When you are finished, save the VI in User's Stuff and call it EvalVal.vi.

A working version of EvalVal.vi can be found in the Chapter 11 folder in the Learning directory. ◆

The Substitute Variables VI is another important VI in the parser library. This VI is used to substitute formulas for parameters in other formulas, resulting in more complex functions. This provides a great degree of control over the formulas themselves. Suppose that we define a "generic" formula as

$$\sin(A) + e^B .$$

We can then "substitute" for A and B to construct a more sophisticated function. We might, for example, use the Substitute Variables VI to obtain

$$\sin(\ln(x)) + e^{\cos x} .$$

The substitutions in this case are $A \rightarrow \ln(x)$ and $B \rightarrow \cos(x)$.

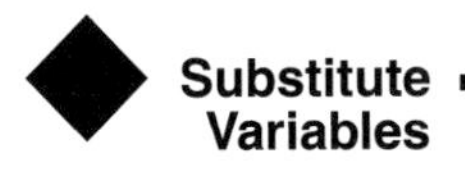

Substitute Variables

Consider the VI shown in Figure 11.9. This VI is similar to the VI that you developed in the previous example called EvalVal.vi, but includes the Substitute Variables VI. This allows you to enter a formula on the front panel, evaluate it, and graph the results. You can substitute formulas for the function parameters in the original formulas. Open a new VI and construct a VI using Figure 11.9 as a model. You might also consider opening EvalVal.vi and using it as a starting point for this new VI.

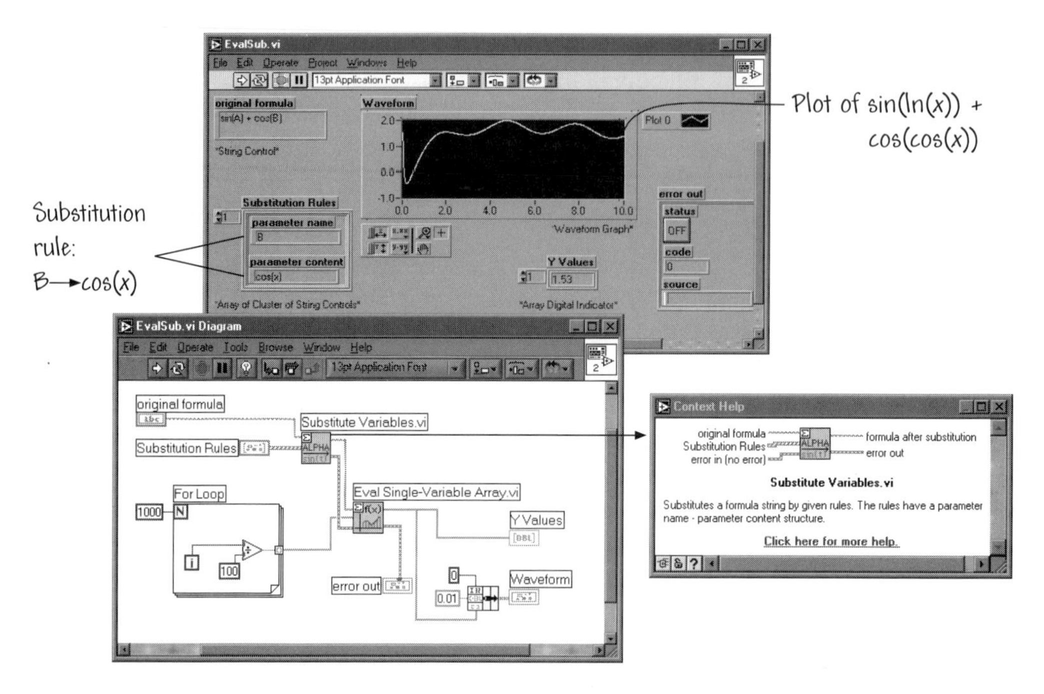

FIGURE 11.9
A VI that allows you to substitute formulas.

The parser VI at the center of this VI is Substitute Variables.vi; see Figure 11.8 to find it on the **Parser VIs** palette . This VI substitutes a formula string by given rules, as illustrated in Figure 11.9. The input **original formula** to the Substitute Variables.vi is the formula that you input on the front panel control. The input **Substitution Rules** specifies the substitutions to be made for the parameters in the original formula. Each element in the **Substitution Rules** array specifies a parameter and its associated substitution rule. The output **formula after substitution** is the resulting formula after the parameter substitutions specified in **Substitution Rules** have been made.

Include a For Loop to generate the x_i values, where $i = 1, \ldots, 1{,}000$. The points at which you evaluate the function should range from 0 to 9.99 in increments of 0.01. After building the VI, return to the front panel and type in the equation $\sin(A) + \cos(B)$ in **original formula**. Make the substitutions $A \rightarrow \ln(x)$ and $B \rightarrow \cos(x)$:

- In element 0 of the **Substitution Rules** control, type in
  - **parameter name**: A
  - **parameter content**: $\ln(x)$
- In element 1 of the **Substitution Rules** control, type in
  - **parameter name**: B
  - **parameter content**: $\cos(x)$

Run the VI and observe the waveform on the graph display.

- In element 0 of the **Substitution Rules** control, type in
  - **parameter name**: A
  - **parameter content**: step $(x - 1)$
- In element 1 of the **Substitution Rules** control, type in
  - **parameter name**: B
  - **parameter content**: square (x)

Run the VI and observe the waveform on the graph display. Try a few substitutions of your choosing.

When you are finished, save the VI in User's Stuff and call it EvalSub.vi.

A working version of EvalSub.vi *can be found in the* Chapter 11 *folder in the* Learning *directory.*

◆

11.4 DIFFERENTIAL EQUATIONS

In LabVIEW, you can solve linear and nonlinear ordinary differential equations (ODEs). Using one of seven VIs for solving first- and higher-order differential equations. Figure 11.10 shows the palette with the available ODE VIs.

The order of a differential equation is the order of the highest derivative in the differential equation.

Suppose that we have a set of first-order ordinary differential equations:

$$\dot{\mathbf{x}}(t) = \mathbf{f}(\mathbf{x}(t), \mathbf{u}(t)) \ ,$$

where $\mathbf{x}$ is the vector $(x_1 x_2 \cdots x_n)^T$, sometimes known as the *state vector*, and $\mathbf{u}$ is the vector $(u_1 u_2 \cdots u_m)^T$ of inputs to the system. When $\mathbf{u} = 0$, the differential equation is termed a *homogeneous* differential equation, and when

$$\mathbf{f}(\mathbf{x}(t), \mathbf{u}(t)) = \mathbf{A}\mathbf{x}(t) + \mathbf{B}\mathbf{u}(t) \ ,$$

the system is a system of *linear* ordinary differential equations.

An example of a *nonhomogeneous* system is

$$\begin{aligned} \dot{x}_1(t) &= x_1^2(t) + \sin x_2(t) + u_1(t) \\ \dot{x}_2(t) &= x_1(t) + x_2^3(t) + 3u_2(t) \ , \end{aligned}$$

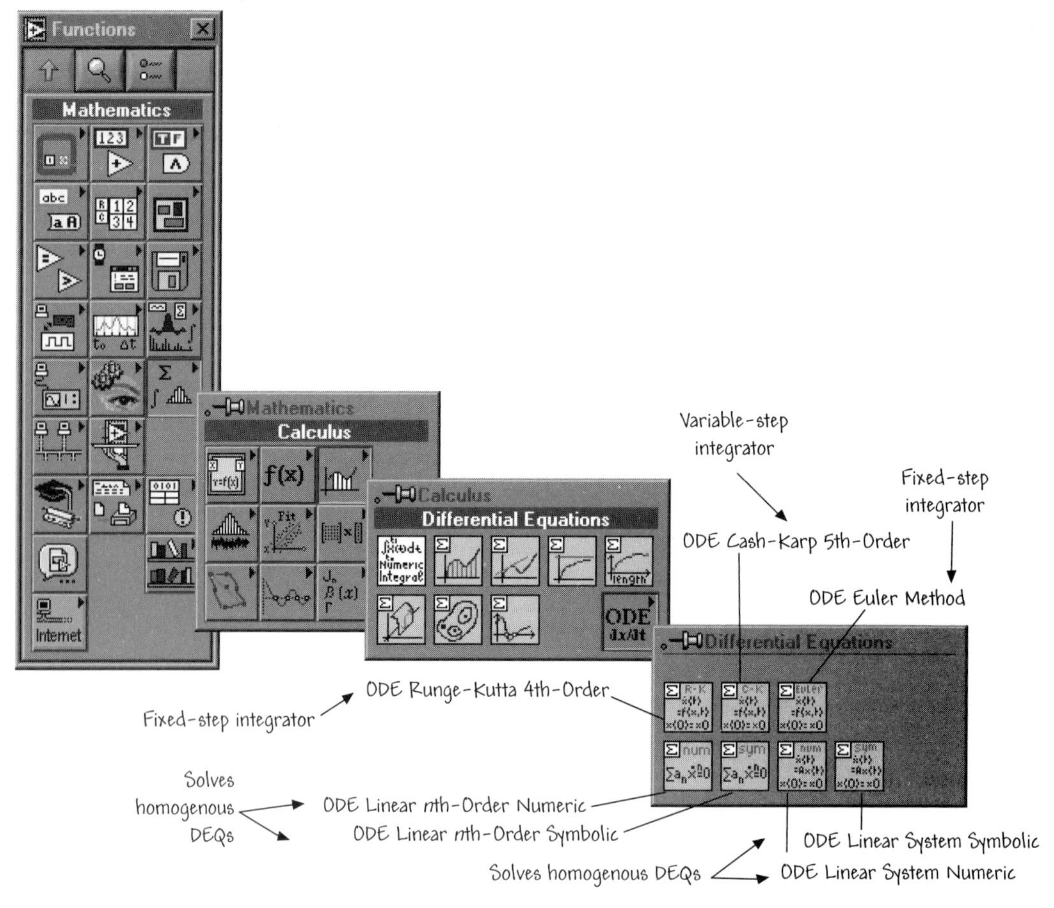

FIGURE 11.10
Solving differential equations using LabVIEW VIs.

where we write

$$\mathbf{x} = \begin{pmatrix} x_1 \\ x_2 \end{pmatrix} \quad \mathbf{u} = \begin{pmatrix} u_1 \\ u_2 \end{pmatrix} \quad \mathbf{f}(\mathbf{x}(t), \mathbf{u}(t)) = \begin{pmatrix} x_1^2(t) + \sin x_2(t) + u_1(t) \\ x_1(t) + x_2^3(t) + 3u_2(t) \end{pmatrix}$$

To compute a solution we need to specify the initial conditions. For a system with n first-order differential equations, we need to specify n initial conditions, $x_1(0)$, $x_2(0)$, ..., $x_n(0)$.

We can also represent physical systems by higher-order differential equations. An example of a second-order mass-spring-damper system is

$$m\frac{d^2y(t)}{dt} + b\frac{dy(t)}{dt} + ky(t) = g(t) \,,$$

where the system's parameters are m = mass, b = damping coefficient, k = spring constant, and the input function is $g(t)$. When $g(t) = 0$, the system is homogeneous. To compute the solution, we need two initial conditions: $y(0)$ and $\dot{y}(0)$.

You can describe an nth-order differential equation equivalently by n first-order differential equations. Consider the second-order DEQ presented above and define

$$x_1 = y \quad \text{and} \quad x_2 = \dot{y} \quad \text{and} \quad u = g(t) .$$

Taking time-derivatives of x_1 and x_2 yields

$$\begin{aligned} \dot{x}_1 &= \dot{y} = x_2 \\ \dot{x}_2 &= \ddot{y} = -\frac{b}{m}\dot{y} - \frac{k}{m}y + \frac{1}{m}g(t) = -\frac{b}{m}x_2 - \frac{k}{m}x_1 + \frac{1}{m}u(t) , \end{aligned}$$

or,

$$\dot{\mathbf{x}}(t) = \mathbf{A}\mathbf{x}(t) + \mathbf{B}u(t) = \begin{bmatrix} 0 & 1 \\ -\frac{k}{m} & -\frac{b}{m} \end{bmatrix} \mathbf{x}(t) + \begin{bmatrix} 0 \\ \frac{1}{m} \end{bmatrix} u(t)$$

LabVIEW has five VIs to solve sets of first-order differential equations, and two VIs to solve nth-order differential equations. Each VI uses a different numerical method for solving the differential equations. Each method has an associated *step size* that defines the time intervals between solution points. To solve DEQs of the form

$$\dot{\mathbf{x}}(t) = \mathbf{f}(\mathbf{x}(t), \mathbf{u}(t)) ,$$

use one of the following VIs:

- ODE Cash-Karp 5th-Order: Variable-step integrator that adjusts the step size internally to reduce numerical errors.
- ODE Euler Method: A very simple fixed-step integrator—that is, it executes fast, but the integration error associated with the Euler method is usually unacceptable in situations where solution precision is important.
- ODE Runge-Kutta 4th-Order: Fixed-step integrator that provides much more precise solutions than the Euler method.

To solve DEQs of the form

$$\dot{\mathbf{x}}(t) = \mathbf{A}\mathbf{x}(t)$$

where the coefficients of the matrix **A** are constant, use one of the following VIs:

- ODE Linear System Numeric: Generates a numerical solution of a homogeneous linear system of differential equations.

- ODE Linear System Symbolic: Generates a symbolic solution of a homogeneous linear system of differential equations.

By a symbolic solution, we mean that the output of the VI is actually a formula rather than a array of numbers. The solution is presented as a formula displayed on the VI front panel.

For solving homogeneous, higher-order differential equations of the form

$$a_0 \frac{d^n y}{dt^n} + a_1 \frac{d^{n-1} y}{dt^{n-1}} + \cdots + a_{n-1} y = 0 \,,$$

LabVIEW has two VIs:

- ODE Linear nth-Order Numeric: Generates a numeric solution of a linear system of nth-order differential equations.
- ODE Linear nth-Order Symbolic: Generates a symbolic solution of a linear system of nth-order differential equations.

Which VI should you use? A few general guidelines follow:

- Use the ODE Euler Method VI sparingly and only for very simple ODEs.
- For most situations, choose the ODE Runge-Kutta 4th-Order or the ODE Cash-Karp 5th-Order VI.
    - If you need the solution at equal intervals, choose the ODE Runge-Kutta 4th-Order VI.
    - If you are interested in a global solution and fast computation, choose the ODE Cash-Karp 5th-Order VI.

The objective of this exercise is to build a VI that solves a second-order differential equation that models the motion of a pendulum. You will see how incorporation of the Substitute Variables VI allows you to vary the model parameters. The pendulum model is given by

$$\frac{d^2\theta}{dt^2} + \frac{c}{ml}\frac{d\theta}{dt} + \frac{g}{l}\sin\theta = 0 \,,$$

where m is the mass of the pendulum, l is the length of the rod, $g = 9.8$ is the acceleration due to gravity, and θ is the angle between the rod and a vertical line passing through the point where the rod is fixed (that is, the equilibrium position).

The pendulum model is a homogeneous, nonlinear equation (notice the $\sin\theta$ term!)—you have three choices of VIs:

- ODE Cash-Karp 5th-Order
- ODE Euler Method
- ODE Runge-Kutta 4th-Order

The first step is to formulate the pendulum model as two first-order differential equations. All three integration VIs listed above are for first-order systems. This is achieved by making the following substitution:

$$x_1 = \theta \quad \text{and} \quad x_2 = \dot{\theta}\ .$$

Go ahead and convert the second-order DEQ to two first-order DEQs. You should end with the resulting system:

$$\dot{x}_1 = x_2$$
$$\dot{x}_2 = -\frac{c}{ml}x_2 - \frac{g}{l}\sin x_1$$

Construct a VI to simulate the motion of the pendulum. Use the front panel of the VI shown in Figure 11.11 as a guide.

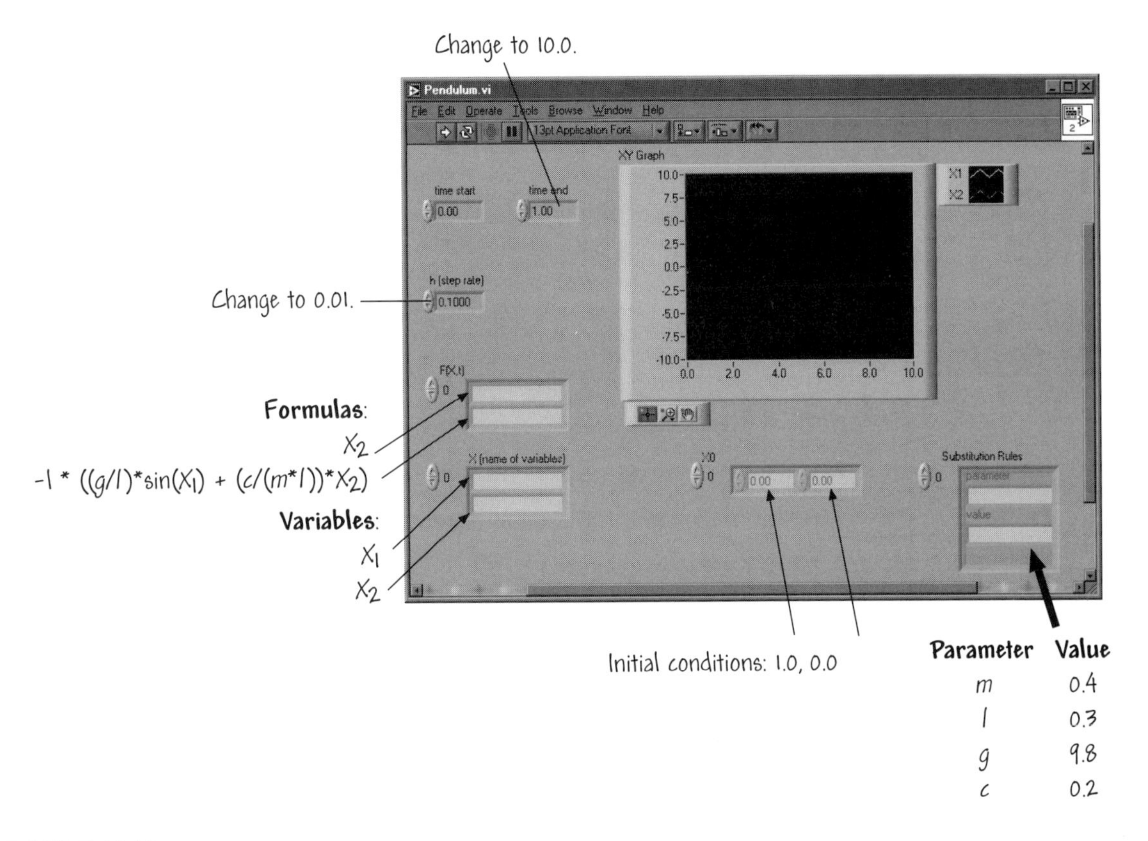

FIGURE 11.11
Simulating the motion of a pendulum—front panel.

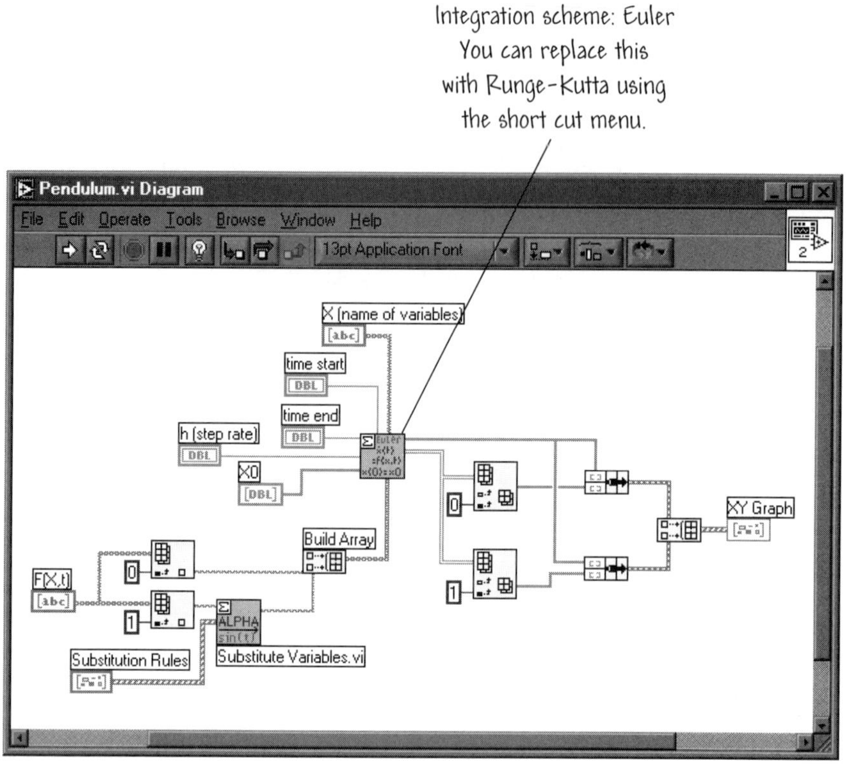

Integration scheme: Euler
You can replace this
with Runge-Kutta using
the short cut menu.

FIGURE 11.12
Simulating the motion of a pendulum—block diagram.

A block diagram is shown in Figure 11.12 that can be used as a guide in the VI development. To begin the development, use the ODE Euler Method VI. The inputs that you will need follow:

- **X**: An array of strings listing the dependent variables (x_1 and x_2).
- **time start**: The point in time at which to start the calculations ($t_0 = 0$).
- **time end**: The point in time at which to end the calculations ($t_f = 10$).
- **h**: The time increment at which to perform the calculations ($h = 0.01$).
- **X0**: The initial conditions ($x_1(0) = 1$, and $x_2(0) = 0$).

The pendulum differential equations are typed in the **F(x,t)** control. The VI uses the Substitute Variables VI as a vehicle for varying the pendulum parameters. This allows you to enter the pendulum model on the front panel and then to substitute numerical values for m, l, g, and c.

When the VI is ready to accept inputs, enter the right side of the pendulum model in the **F(X,t)** control. You will enter x_2 and $-(g/l)\sin x_1 - (c/ml)x_2$. For the substitution rules use

- In element 0 of the **Substitution Rules** control, type in
  - **parameter name**: m
  - **parameter content**: 0.4
- In element 1 of the **Substitution Rules** control, type in
  - **parameter name**: l
  - **parameter content**: 0.3
- In element 2 of the **Substitution Rules** control, type in
  - **parameter name**: g
  - **parameter content**: 9.8
- In element 3 of the **Substitution Rules** control, type in
  - **parameter name**: c
  - **parameter content**: 0.2

Run the VI and observe the waveform on the graph display. You should see a nice, stable response damping out around 6 seconds.

On the front panel, change the **h (step rate)** from 0.01 to 0.1. Run the VI. Did you detect a problem? The system response is no longer stable! Nothing has changed with the physical model, so this must be due to the Euler integration scheme. Now, switch to the block diagram and pop up on the ODE Euler Method VI and **Replace** it with the ODE Runge-Kutta 4th-Order VI. Both VIs have the same inputs and outputs. Run the VI. What happens? The expected smooth, stable response is obtained. This demonstrates the benefit of the Runge-Kutta method over the Euler method—you can run with larger time steps and obtain more accurate solutions.

Back on the front panel, switch the time step back to **h (step rate)** = 0.01. Run the VI and verify that the results remain essentially the same. Investigate the effect that varying the pendulum mass has on the response.

When you are finished, save the VI in User's Stuff and call it Pendulum.vi.

A working version of Pendulum.vi *can be found in the* Chapter 11 *folder in the* Learning *directory. If you run this VI, make sure to verify that all the input parameters are correct!*

◆

11.5 FINDING ZEROES OF FUNCTIONS

LabVIEW provides six VIs that can be used to compute zeroes of functions. The VIs can be used to determine the zeroes of general functions of the following form:

$$f(x, y) = 0$$
$$g(x, y) = 0$$

For example, you could use the VIs to find the zeroes of $\sin(x) + \cos(x)$ in the range $-10 \leq x \leq 10$.

Very often in the course of studying mathematics, engineering, business, and science it is necessary to compute the *zeroes* of a polynomial. A nth-order polynomial has the form

$$f(x) = x^n + a_{n-1}x^{n-1} + a_{n-2}x^{n-2} + \cdots + a_1 x + a_0 = 0 .$$

The zeroes of the polynomial are the values of x such that $f(x) = 0$. The zeroes are also known as the *roots* of the polynomial. For example, we discussed in previous sections that the characteristic equation associated with a $n \times n$ matrix $\mathbf{A}$ is an nth-order polynomial, and the zeroes of the characteristic equation are the eigenvalues of the matrix. Eigenvalues can be real and imaginary. But in this discussion, when we talk about zeroes of a function, we mean the real roots.

LabVIEW VIs are shown in Figure 11.13. There are seven VIs:

- Find All Zeroes of f(x): This VI determines all the zeroes of a 1D function in a specified interval.
- Newton-Raphson Zero Finder: Uses derivatives to assist in determining a zero of a 1D function in a specified interval.
- Nonlinear System Single Solution: Computes the zeroes of a nonlinear function, where an approximation is provided as input.
- Nonlinear System Solver: Computes the zeroes of a set of nonlinear functions.
- Polynomial Real Zero Counter: Determines the number of real zeroes of a polynomial in an interval without actually computing the zeroes.
- Ridders Zero Finder: This VI computes a zero of a function in a given interval, but the function must be continuous and when evaluated at the edges of the interval, the function must have different signs.
- Complex Polynomial Roots: Finds the complex roots of a complex polynomial.

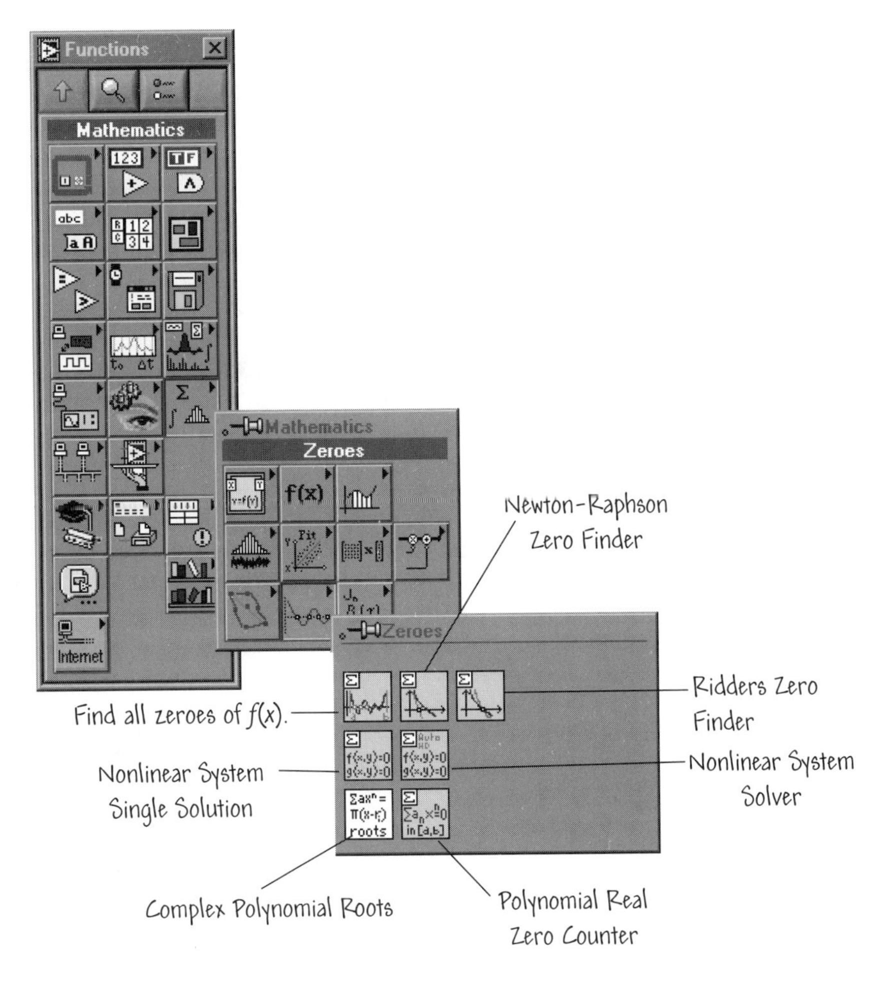

FIGURE 11.13
VIs for finding zeroes of functions.

Finding Zeroes of a Polynomial Function

The objective of this exercise is to build a VI to compute the zeroes of a polynomial function in a given interval. Construct a VI using the front panel and block diagram shown in Figure 11.14 as a guide.

Use the VI Find All Zeroes of f(x).vi to compute the zeroes. This VI is based on a numerical scheme that iterates on the interval to converge on the zeroes. Since it is an iterative scheme, it utilizes a search algorithm. You have two choices for the search—Ridders method and Newton-Raphson. As the search algorithm iterates, it can use either uniformly spaced function values or an optimal step size. These two inputs are wired as constants on the block diagram shown in Figure 11.14—you can change that to have them input from the front panel. You should experiment with the different search and step size possibilities to see

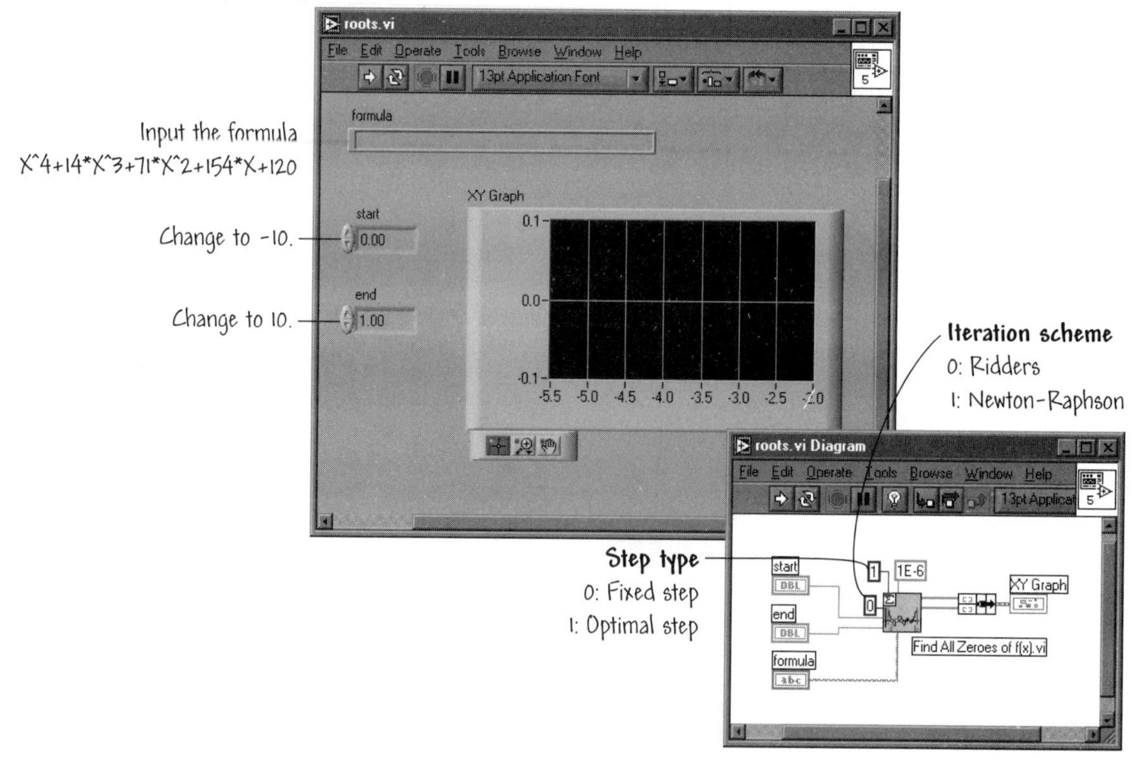

FIGURE 11.14
Computing the zeroes of a polynomial.

which you prefer. The inputs to the VI are:

- **formula**: Type in the formula.
- **start** and **end**: The endpoints of the search interval.

When your VI is ready to accept inputs, enter the following formula:

$$x^4 + 14x^3 + 71x^2 + 154x + 120 .$$

Set the endpoints to -10 and 10. Run the VI. Where are the zeroes? You should determine them to be $x = -5, -4, -3$, and -2. When you are finished, save the VI in User's Stuff and call it roots.vi.

A working version of roots.vi can be found in the Chapter 11 folder in the Learning directory.

◆

11.6 INTEGRATION AND DIFFERENTIATION

LabVIEW VIs for integration and differentiation are shown in Figure 11.15. The figure shows that there are many other VIs available for working with 1D

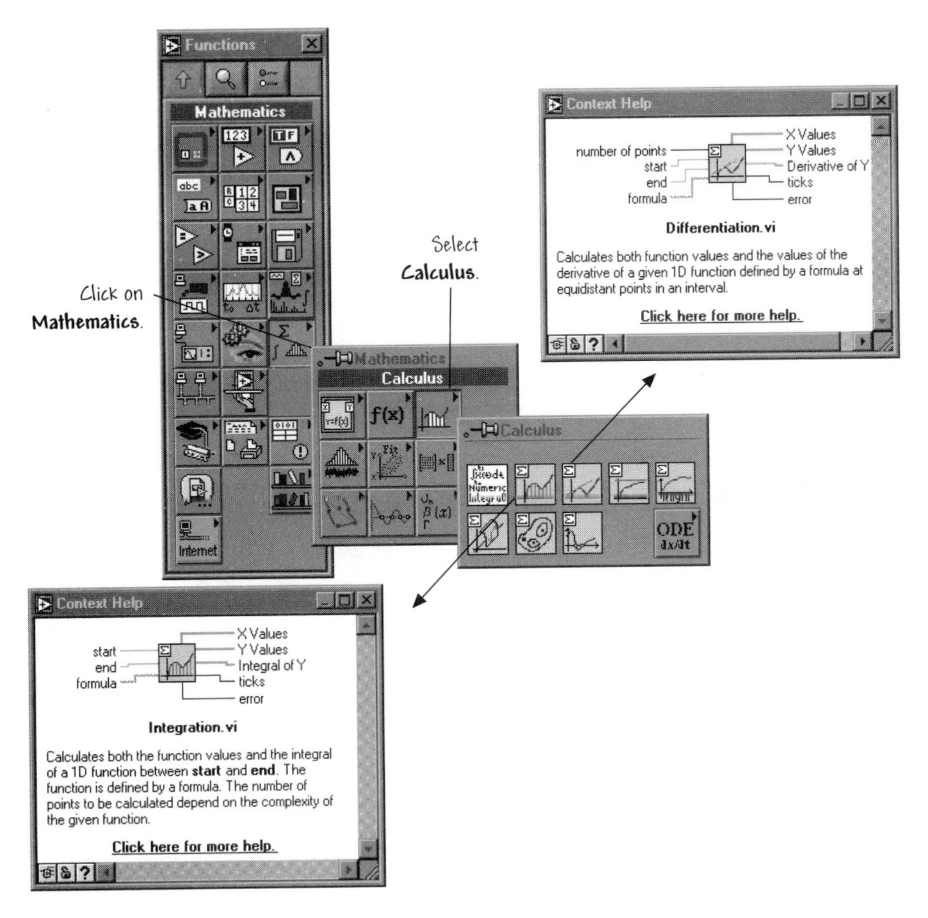

FIGURE 11.15
VIs for integration and differentiation.

functions—unfortunately we can not cover all the analysis capabilities of LabVIEW in this book.

A straightforward implementation of the integration and differentiation VIs is shown in Figure 11.16. You can find the VI shown in Figure 11.16 in the Chapter 11 folder within the Learning directory—it is called Derivative and Integration.vi. Open the VI and enter the formula $\sin(x)$ and run the VI. You will find that three plots appear on the graph, including the function, the derivative of the function at a number of points defined by the **number of points** input, and a plot of the integral of the function. Change the value of **end** from the default value of 1 to 6.28 and run the VI again. This time you should see one complete cycle of the sine wave, and the plot of the derivative of the sine function should appear as a cosine function. If you run the VI in **Continuous Run** mode, you can vary the parameter **number of points** and watch the sine function become smoother as the number increases and, conversely, become less smooth as the number of points decreases.

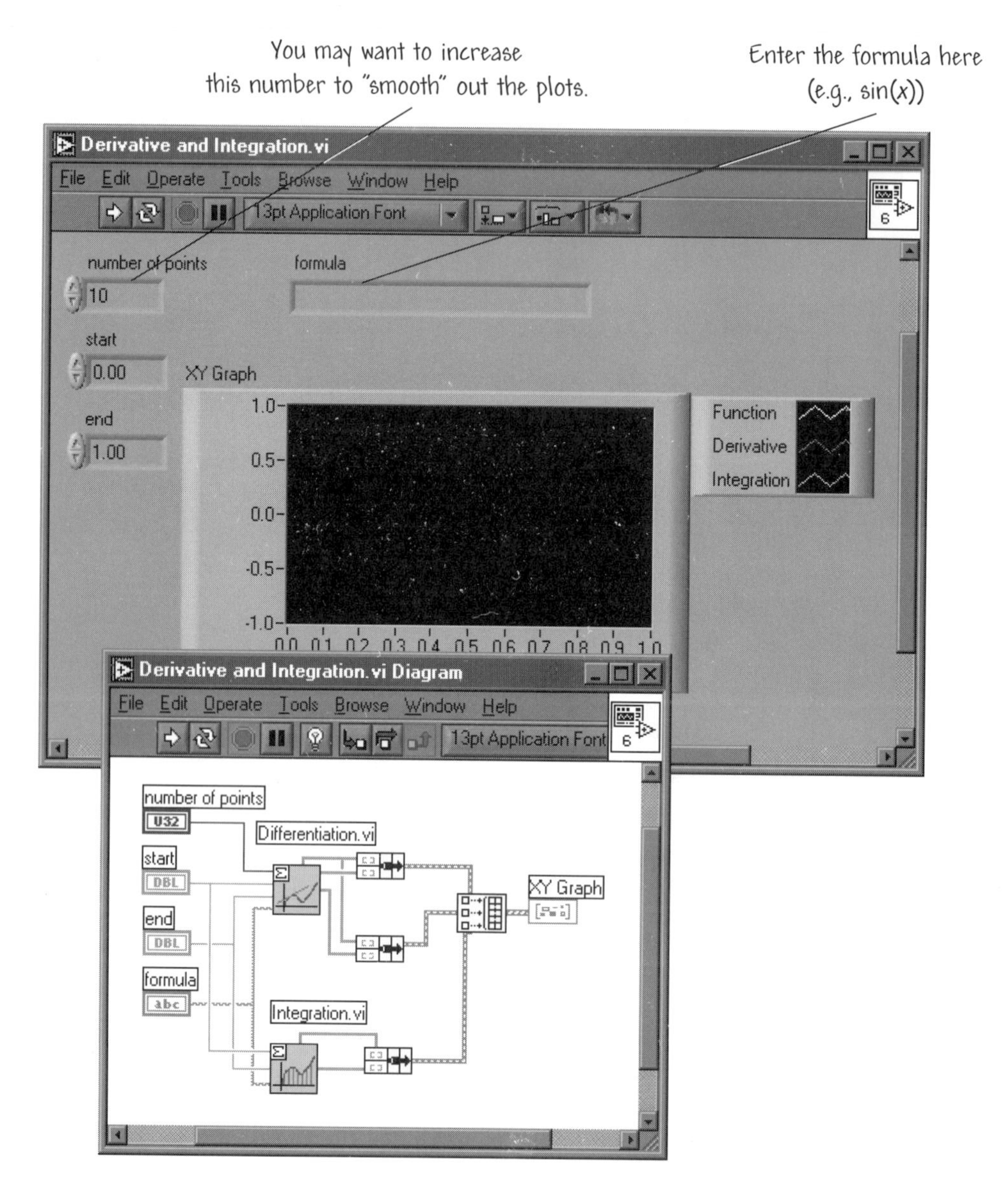

FIGURE 11.16
A VI for integration and differentiation of a 1D function.

11.7 SIGNAL GENERATION

You can use LabVIEW to generate signals for testing and other purposes when real-world signals are not available. You may also want to generate signals rather than rely on signals acquired from the real world whenever you need to accurately control the signal characteristics (such as magnitude, frequency and phase of periodic signals, and so on). In this section we discuss some of the possibilities for generating signals using VIs. The discussion begins by considering the notion of normalized frequency.

11.7.1 Normalized Frequency

In the digital signal world (and with many Signal Generation VIs) we often use the so-called **digital frequency** or **normalized frequency** (in units of cycles/sample) computed as

$$f = \text{digital frequency} = \frac{\text{analog frequency}}{\text{sampling frequency}}.$$

The analog frequency is generally measured in units of Hz (or cycles per second) and the sampling frequency in units of samples per second. The normalized frequency is assumed to range from 0.0 to 1.0 corresponding to a frequency range of 0 to the sampling frequency, denoted by f_s. The normalized frequency wraps around 1.0, so that a normalized frequency of 1.2 is equivalent to 0.2. As an example, a signal sampled at the **Nyquist frequency** (that is, at $f_s/2$) is sampled twice per cycle (that is, two samples/cycle). This corresponds to a normalized frequency of $1/2$ cycles/sample = 0.5 cycles/sample. Therefore, we see that the reciprocal of the normalized frequency yields the number of times that the signal is sampled in one cycle (more on sampling in Sec. 11.8.3).

The following VIs utilize frequencies given in normalized units:

1. Sine Wave
2. Square Wave
3. Sawtooth Wave
4. Triangle Wave
5. Arbitrary Wave
6. Chirp Pattern

When using these VIs, you will need to convert the frequency units given in the problem to the normalized frequency units of cycles/sample. The VI depicted in Figure 11.17 illustrates how to generate two cycles of a sine wave and then convert cycles to cycles/sample.

In the example shown in Figure 11.17, the number of cycles (2) is divided by the number of samples (50), resulting in a normalized frequency of $f = 2/50$ cycles/sample. This implies that it takes 50 samples to generate two cycles of the sine wave. What if the problem specifies the frequency in units of Hz (cycles/sec)? In this case, if you divide the frequency in Hz (cycles/sec) by the sampling rate given in Hz (samples/sec), you obtain units of cycles/sample:

$$\frac{\text{cycles/sec}}{\text{samples/sec}} = \frac{\text{cycles}}{\text{sample}}.$$

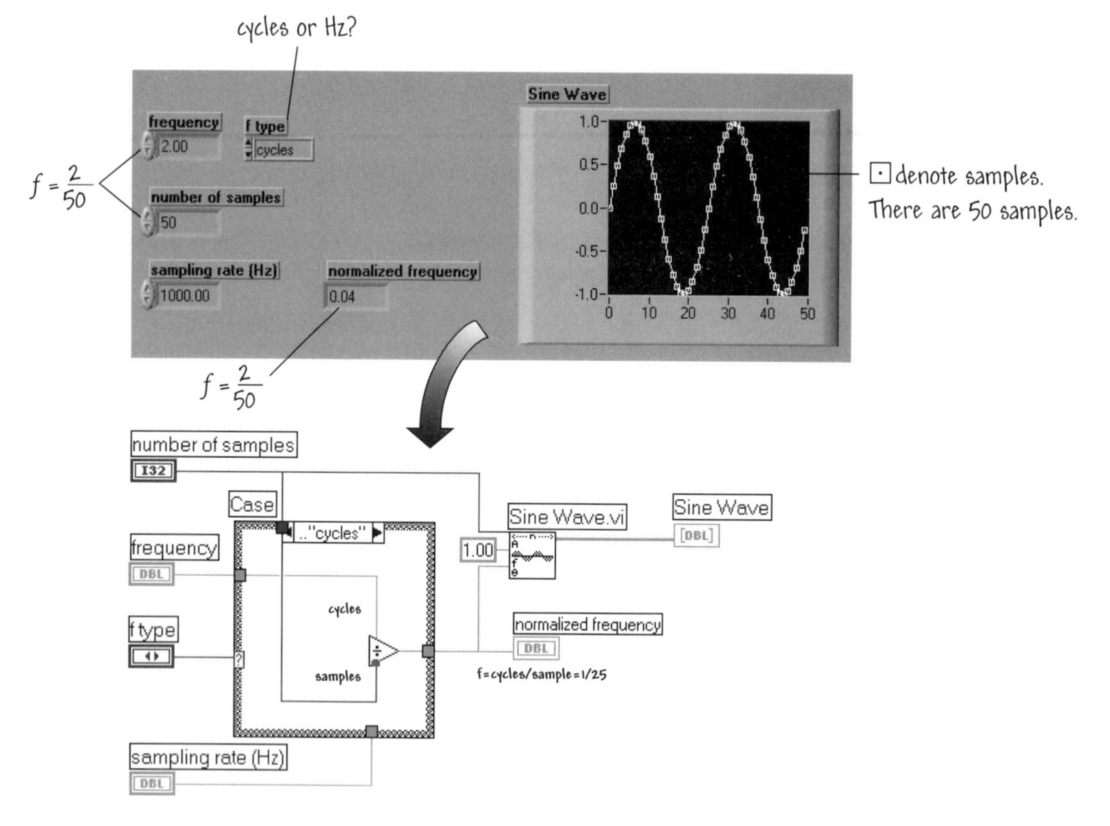

FIGURE 11.17
Generating two cycles of a sine wave and converting cycles to cycles/sample.

The illustration in Figure 11.18 shows a VI used to generate a 60 Hz sine signal and to compute the normalized frequency when the input is in Hz. The normalized frequency is found by dividing the frequency of 60 Hz by the sampling rate of 1,000 Hz to get the normalized frequency of $f = 0.06$ cycles/sample:

$$f = \frac{60}{1{,}000} = 0.06 \, \frac{\text{cycles}}{\text{sample}} \, .$$

Therefore, we see that it takes almost 17 samples to generate one cycle of the sine wave. The number 17 comes from computing the reciprocal of $f = 0.06$.

Normalized Frequencies

Open the VI called Normalized Frequency.vi located in the Chapter 11 folder in the Learning directory. The front panel and block diagram are shown in Figure 11.19. You can use this VI to experiment with calculating the normalized frequency when the input is in cycles and in Hz. Make sure that the **f type** is selected as Hz and run the VI. You should find that the normalized frequency is $f = 0.01$ with the default VI input values.

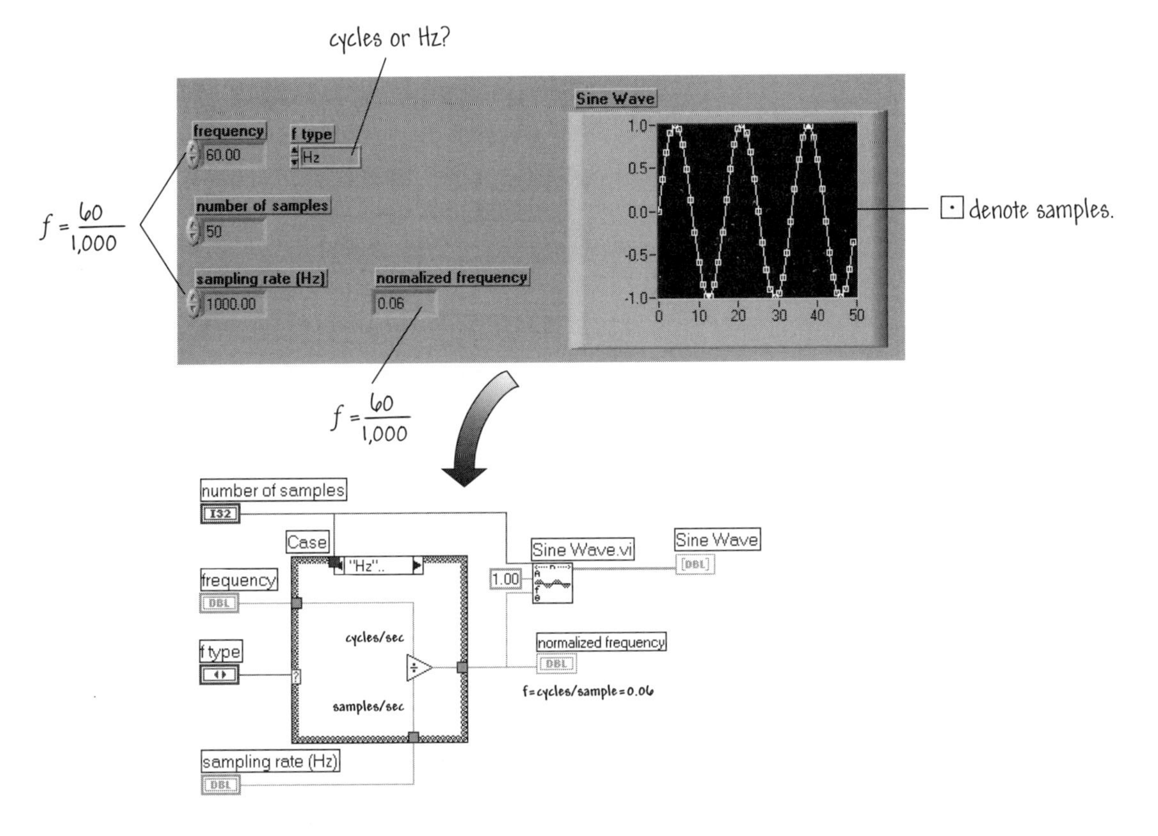

FIGURE 11.18
Generating a 60 Hz sine wave and computing normalized frequency.

Manually compute the normalized frequency for **f type** = Hz, **frequency** = 10, and **sampling rate (Hz)** = 1,000. Modify the VI input parameters accordingly and verify that you obtain the same answer as by hand. The answer should be $f = 0.01$, computed as

$$f = \frac{10}{1{,}000} = 0.01 \, \frac{\text{cycles}}{\text{sample}} \, .$$

You can also work in cycles (rather than Hz) by selecting **f type** to be cycles. Then, in this situation, the normalized frequency is computed as a ratio of **frequency** to **number of samples**.

◆

11.7.2 Wave and Pattern VIs

The basic difference in the operation of the Wave or Pattern VIs is whether or not the particular VI keeps track of the phase of its own generated signal internally. The Wave VIs keep track of phase internally; the Pattern VIs do not. You

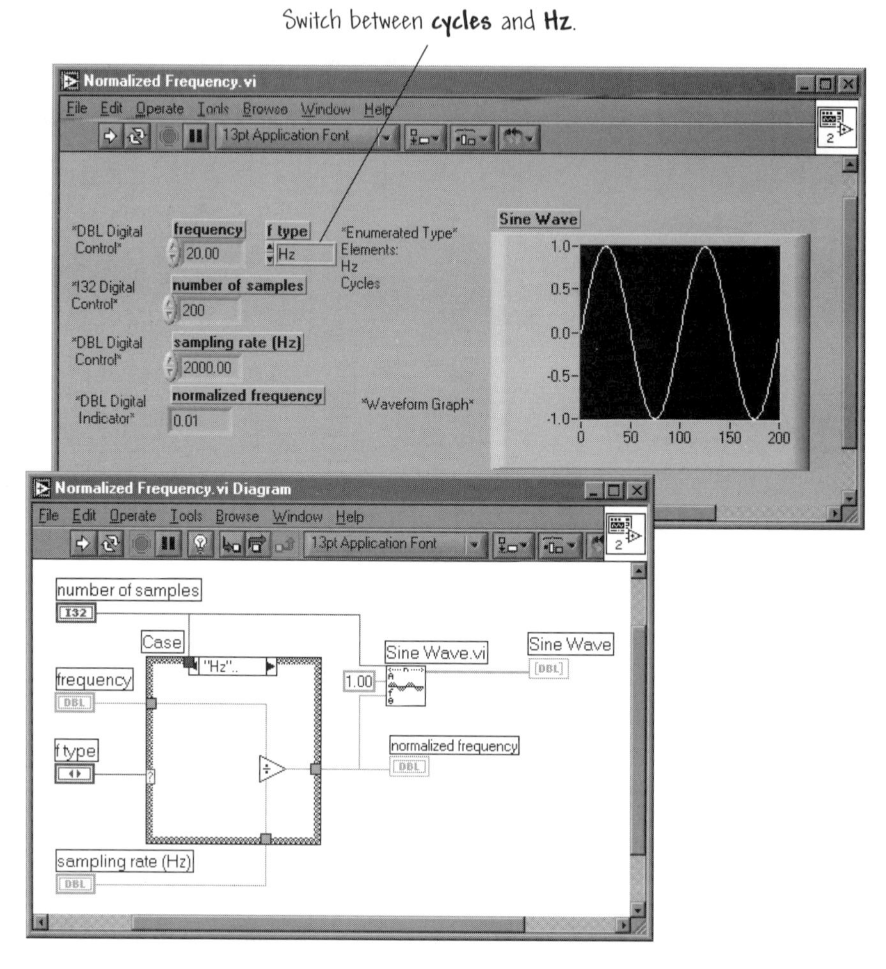

FIGURE 11.19
A VI to compute the normalized frequency.

can distinguish between the two types of VIs by recognizing that the VI names contain either the word *wave* or *pattern*, as illustrated in Figure 11.20.

The Wave VIs operate with normalized frequencies in units of cycles/sample. The only Pattern VI that uses normalized units is the Chirp Pattern VI.

Since the Wave VIs can keep track of the phase internally, they allow the user to control the value of the initial phase. The **phase in** control specifies the initial phase (in degrees) of the first sample of the generated waveform and the **phase out** indicator specifies the phase of the next sample of the generated waveform. In addition, a **reset phase** control dictates whether or not the phase of the first sample generated when the wave VI is called is the phase specified at the **phase in** control, or whether it is the phase available at the **phase out** control when the VI last executed. A TRUE value of reset phase sets the initial phase to **phase in**—a FALSE value sets it to the value of **phase out** when the VI last executed.

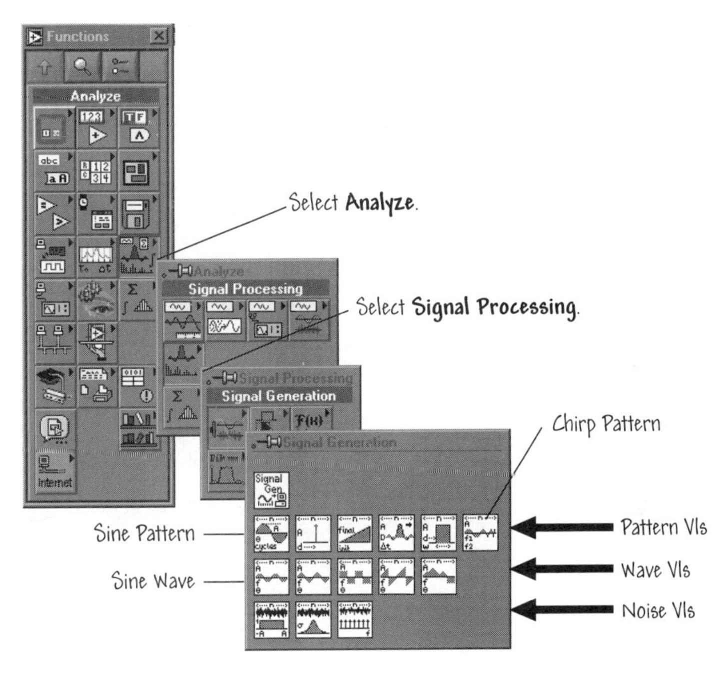

FIGURE 11.20
Signal generation VIs.

Practice with Signal Generation

In this exercise you will construct a VI that uses the two types of signal generation VIs. The front panel and block diagram for the VI is shown in Figure 11.21. The two main functions used on the block diagram are the Sine Pattern.vi and Sine Wave.vi. These functions are located on the **Signal Generation** palette, as shown in Figure 11.20. The VI also calculates and displays the normalized frequency.

Construct the VI using Figure 11.21 as a guide and run the VI when it is ready. Vary the VI inputs, paying particular attention to the effect that changing the control **phase in (degrees)** has on the Sine Pattern waveform graph. You should notice that the waveform begins to shift left and right as you vary the phase up and down. When you set the **reset phase** to TRUE (ON), the initial phase is reset to the value specified by **phase in** each time the VI is called in the loop; otherwise, the initial phase is set to the previous phase output. While th VI is running, set the **reset phase** to the ON position. The sine wave should be rendered stationary, whereas with the reset button set to the OFF position, the sine wave varies with the varying phase.

Stop the execution of the VI using the **Stop** button located at the bottom left of the VI front panel in Figure 11.21. When you are finished experimenting with the VI, save it as Signal Generation.vi in the Users Stuff folder in the Learning directory.

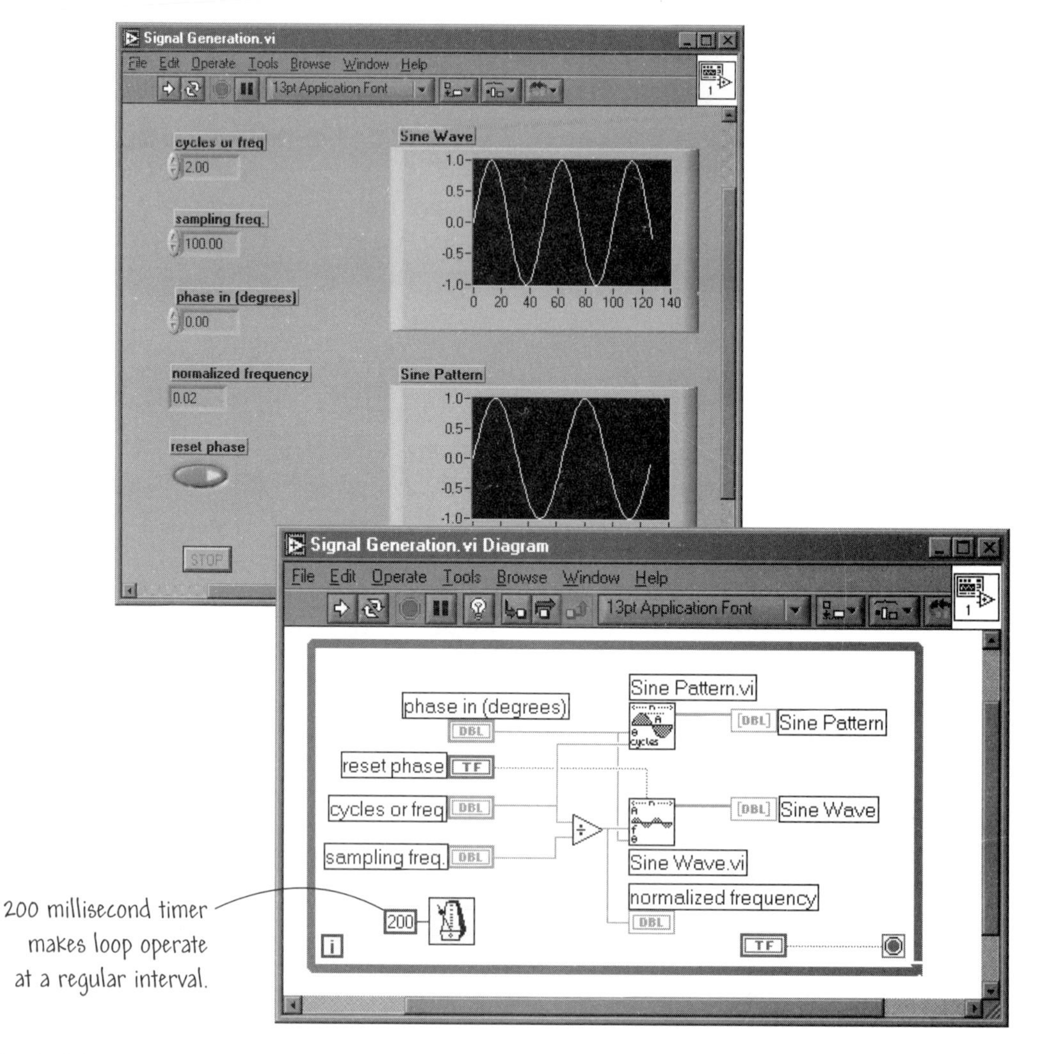

FIGURE 11.21
Using the Sine Wave and Sine Pattern VIs to construct a signal generation VI.

You can find a working version of Signal Generation.vi *in the* Chapter 11 *folder in the* Learning *directory.*

◆

11.8 SIGNAL PROCESSING

In this section we discuss three main topics: the Fourier transform (including the discrete Fourier transform and the fast Fourier transform), smoothing windows, and a brief overview of filtering.

11.8.1 The Fourier Transform

In Chapter 8 we covered the subject of data acquisition, where we discussed that fact that the samples of a measured signal obtained from the DAQ system are a time-domain representation of the signal, giving the amplitudes of the sampled signal at the sampling times. A significant amount of information is coded into the time-domain representation of a signal—maximum amplitude, maximum overshoot, time to settle to steady-state, and so on. The signal contains other useful information that becomes evident when the signal is transformed into the frequency domain. In other words, you may want to know the frequency content of a signal rather than the amplitudes of the individual samples. The representation of a signal in terms of its individual frequency components is known as the **frequency-domain representation** of the signal.

A common practical algorithm for transforming sampled signals from the time domain into the frequency domain is known as the **discrete Fourier transform**, or DFT. The relationship between the samples of a signal in the time domain and their representation in the frequency-domain is established by the DFT. This process is illustrated in Figure 11.22.

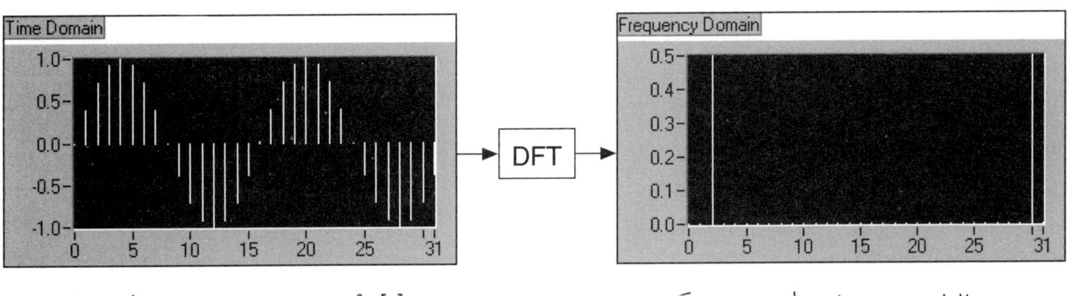

FIGURE 11.22
The DFT establishes the relationship between the samples of a signal in the time-domain and their representation in the frequency-domain.

If you apply the DFT to a time-domain signal represented by N samples of the signal, you will obtain a frequency-domain representation of the signal of length N. We denote the individual components of the DFT by $X(i)$. If the signal is sampled at the rate f_s Hz, and if you collect N samples, then you can compute the frequency resolution as $\Delta f = f_s/N$. This implies that the ith sample of the DFT occurs at a frequency of $i\Delta f$ Hz. We let the pth element $X(p)$ correspond to the Nyquist frequency. Regardless of whether the input signal is real or complex, the frequency-domain representation is always complex and contains two pieces of information—the amplitude and the phase.

For real-valued time-domain signals (denoted here by $x(i)$), the DFT is symmetric about the index $N/2$ with the following properties:

$$|X(i)| = |X(N-i)| \quad \text{and} \quad \text{phase}(X(i)) = -\text{phase}(X(N-i)) .$$

The magnitude of $X(i)$ is **even symmetric**, that is, symmetric about the vertical axis. The phase of $X(i)$ is **odd symmetric**, that is, symmetric about the origin. This symmetry is illustrated in Figure 11.23. Since there is repetition of information contained in the N samples of the DFT (due to the symmetry properties), only half of the samples of the DFT need to be computed, since the other half can be obtained from symmetry.

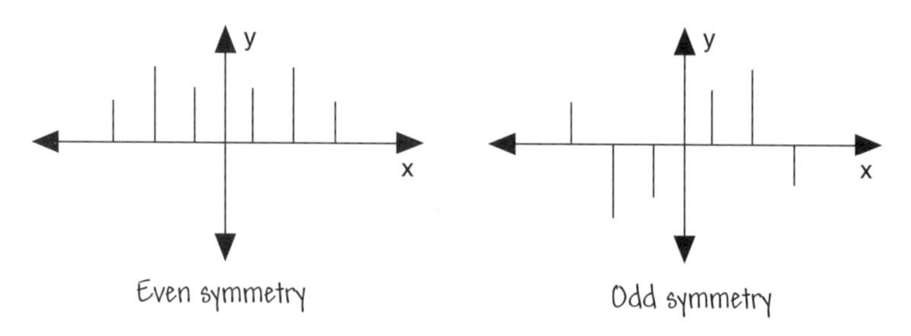

FIGURE 11.23
Even and odd symmetric signals.

Figure 11.24a depicts a *two-sided transform* for a complex sequence with $N=8$ and $p=N/2=4$. Since $N/2$ is an integer, the DFT contains the Nyquist frequency. When N is odd, $N/2$ is not an integer, and thus there is no component

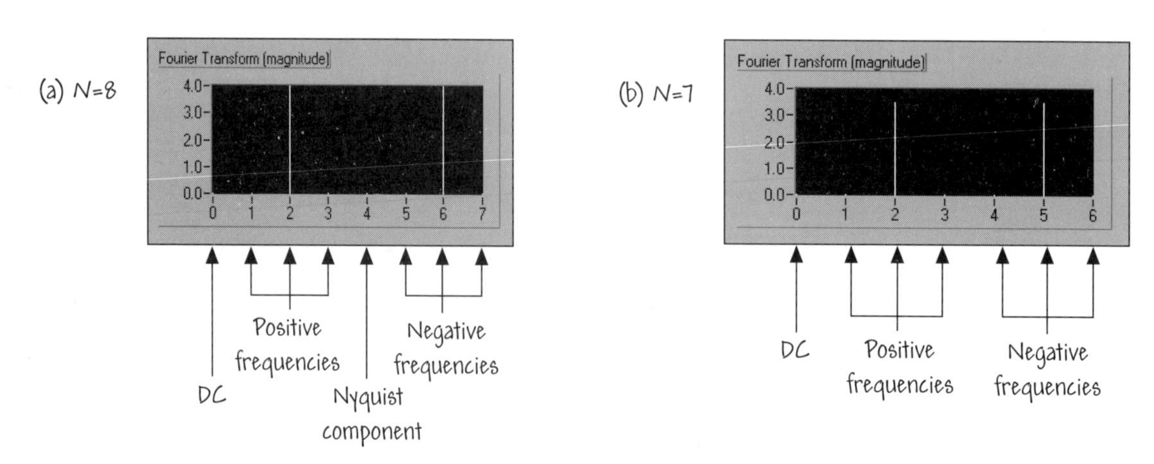

FIGURE 11.24
A two-sided transform representation of a complex sequence.

at the Nyquist frequency. Figure 11.24b depicts a two-sided transform when $N = 7$.

The computationally intensive process of computing the DFT of a signal with N samples requires approximately N^2 complex operations. However, when N is a power of 2, that is, when

$$N = 2^m \quad \text{for} \quad m = 1 \text{ or } 2 \text{ or } 3 \text{ or } \cdots ,$$

you can implement the so-called **fast Fourier transforms** (FFTs), which require only approximately $N\log_2(N)$ operations. In other words, the FFT is an efficient algorithm for calculating the DFT when the number of samples (N) is a power of 2.

Practice with FFT

In this example you will practice with FFTs by opening an existing VI and experimenting with the input parameters. Locate and open FFT_2sided.vi. You will find this VI in the Chapter 11 folder in the Learning directory. The front panel is shown in Figure 11.25.

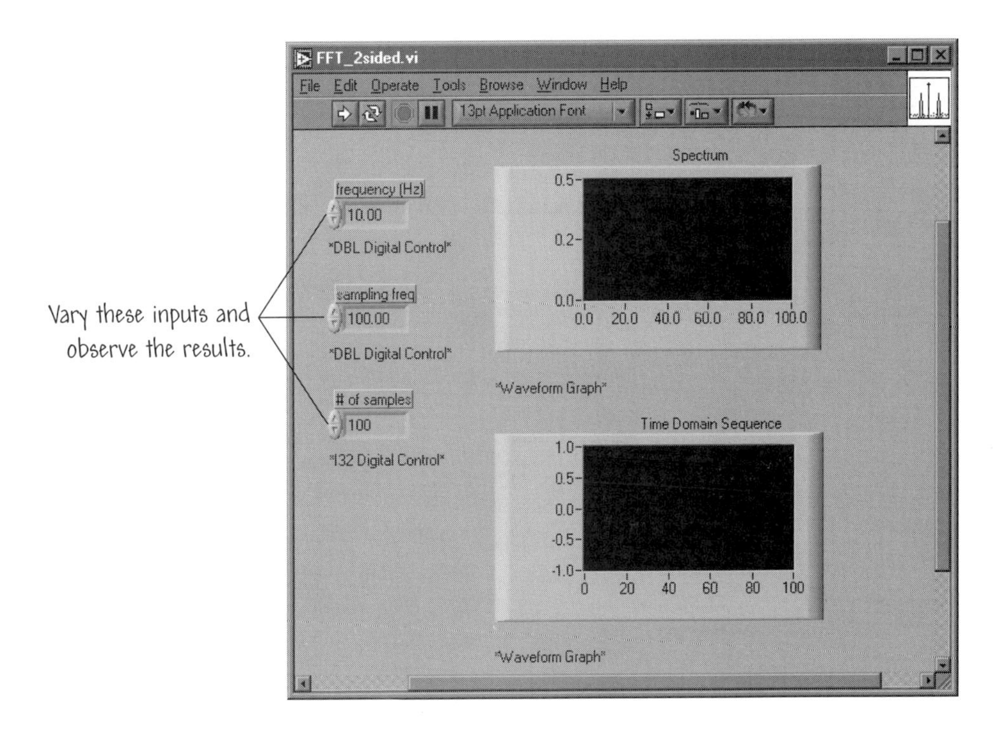

FIGURE 11.25
The front panel of FFT_2sided.vi.

The VI demonstrates how to use the FFT VI to analyze a sine wave of user-specified frequency. The VI block diagram, shown in Figure 11.26, contains three main VIs:

- Real FFT.vi: This VI computes the fast Fourier transform (FFT) or the discrete Fourier transform (DFT) of the input sequence. Real FFT.vi will execute FFT routines if the size of the input sequence is a power of 2. If the size of the input sequence is not a power of 2, then an efficient DFT routine is called.
- Sine Wave.vi: This VI generates an array containing a sine wave. The VI is located in Analyze≫Signal Processing≫Signal Generation subpalette of the **Functions** pallette.

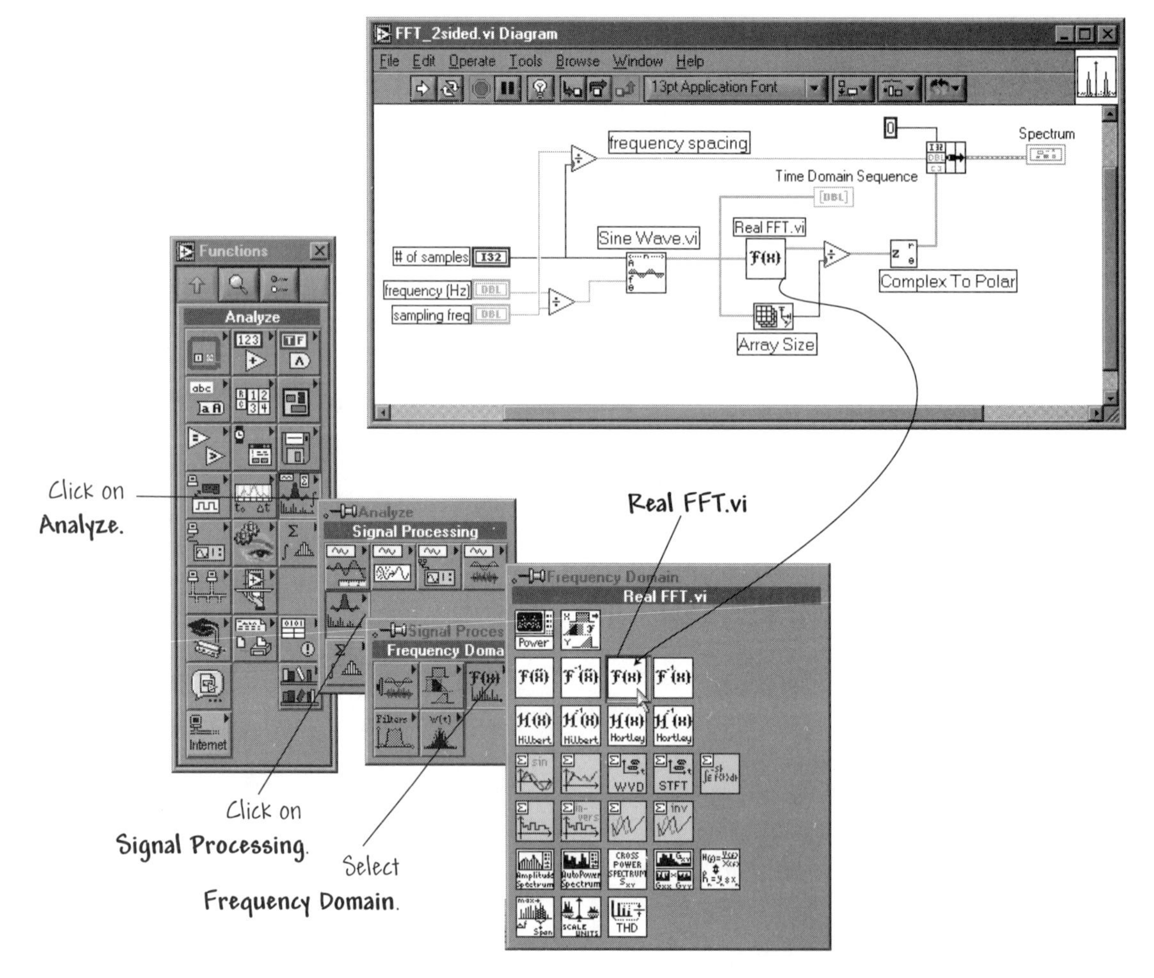

FIGURE 11.26
The block diagram of FFT_2sided.vi.

- Complex To Polar.vi: Separates a complex number into its polar components represented by magnitude and phase. The input can be a scalar number, a cluster of numbers, an array of numbers, or an array of clusters. In the case of Figure 11.26, the input is an array of numbers.

Run the VI and experiment with the input parameters. Run several numerical experiments using the **Continuous Run** mode. What happens to the spectrum when you vary the signal frequency? For example, set the input signal frequency to 10 Hz, the sampling frequency to 100 Hz, and the number of samples to 100. In this case, $\Delta f = 1$ Hz. The spectrum should have two corresponding peaks. Check this using the VI.

For another experiment, set the signal frequency to 50 Hz, the sampling frequency to 100 Hz, and the number of samples to 100, and run the VI. With the VI running in **Continuous Run** mode, set the sampling frequency to 101 and observe the effects on the time-domain sequence waveform and the corresponding spectrum. Now, slowly increase the sampling frequency and see what happens!

When you are finished, close the VI and do not save any changes.

11.8.2 Smoothing Windows

When using discrete Fourier transform methods to analyze a signal in the frequency domain, it is assumed that the available data of the time-domain signal represents at least a single period of a periodically repeating waveform. Unfortunately, in most realistic situations, the number of samples of a given time-domain signal available for DFT analysis is limited and this can sometimes lead to a phenomenon known as **spectral leakage**. To see this, consider a periodic waveform created from one period of a sampled waveform, as illustrated in Figure 11.27.

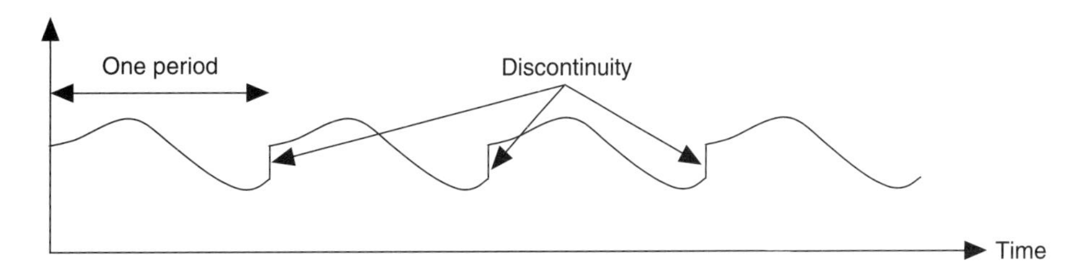

FIGURE 11.27
A periodic waveform created from one period of a sampled waveform.

The first period shown in Figure 11.27 is the sampled portion of the waveform. The sampled waveform is then repeated to produce the periodic waveform.

Sampling a noninteger number of cycles of the waveform results in discontinuities between successive periods! These discontinuities induced by the process of creating a periodic waveform lead to very high frequencies (higher than the Nyquist frequency) in the spectrum of the signal—frequencies that were not present in the original signal. Therefore, the spectrum obtained with the DFT will not be the true spectrum of the original signal. In the frequency domain, it will appear as if the energy at one frequency has "leaked out" into all the other frequencies, leading to what is known as spectral leakage.

Figure 11.28 shows a sine wave and its corresponding Fourier transform. The sampled time-domain waveform is shown in Graph 1 in the upper left corner.

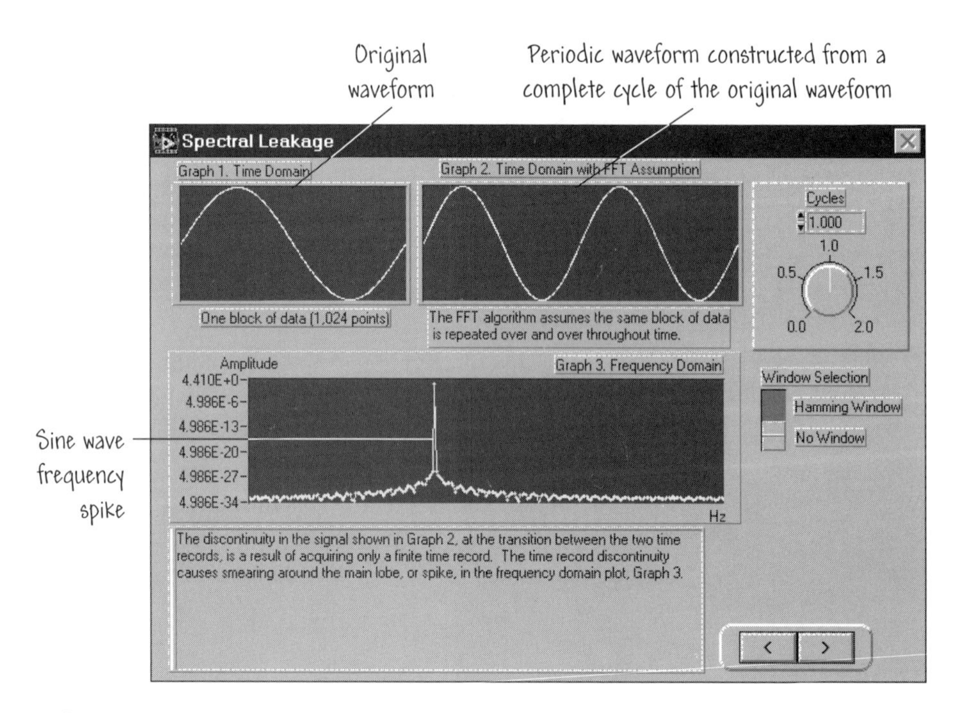

FIGURE 11.28
One complete period of a sine wave is repeated to obtain a periodic signal with no discontinuities. The corresponding Fourier transform shows no leakage.

In this case, the sampled waveform is an integer number of cycles of the original sine wave. The sampled waveform can be repeated in time, and a periodic version of the original waveform thereby constructed. The constructed periodic version of the waveform is depicted in Graph 2 in the upper middle section of Figure 11.28. The constructed periodic waveform does not have any discontinuities because the sampled waveform is an integer number of cycles of the original

Discontinuity due to using a noninteger number of cycles of the original waveform when constructing the periodic waveform

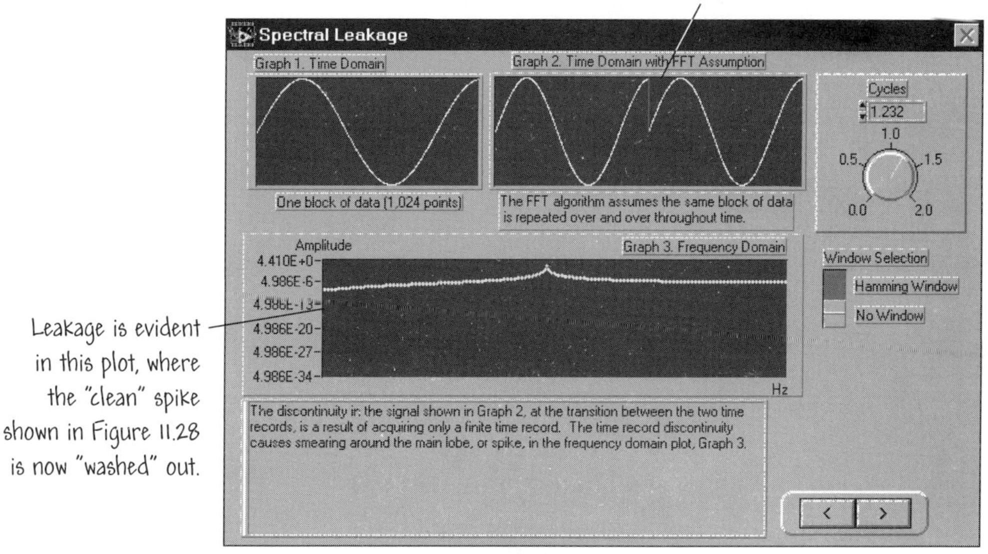

FIGURE 11.29
A portion of a sine wave period is repeated to obtain a periodic signal with discontinuities. The corresponding Fourier transform shows leakage.

waveform. The corresponding spectral representation of the periodic waveform is shown in Graph 3. Because the time record in Graph 2 is periodic, contains no discontinuities, and is an accurate representation of the true waveform, the computed spectrum is correct.

In Figure 11.29 a spectral representation of another periodic waveform is shown. However, in this case a noninteger number of cycles of the original waveform is used to construct the periodic waveform, resulting in the discontinuities in the waveform shown in Graph 2. The corresponding spectrum is shown in Graph 3. The energy is now "spread" over a wide range of frequencies—compare this result to Graph 3 in Figure 11.28. The smearing of the energy is called spectral leakage, as mentioned earlier.

Leakage results from using only a finite time sample of the input signal. One (unpractical) solution to the leakage problem is to obtain an infinite time record, from $-\infty$ to $+\infty$, yielding an ideal FFT solution. In practice, however, we are limited to working with a finite time record. A practical approach to the problem of spectral leakage is the so-called **windowing** technique. Since the amount of spectral leakage depends on the amplitude of the discontinuity, the larger the discontinuity, the more the leakage. Windowing reduces the amplitude of the

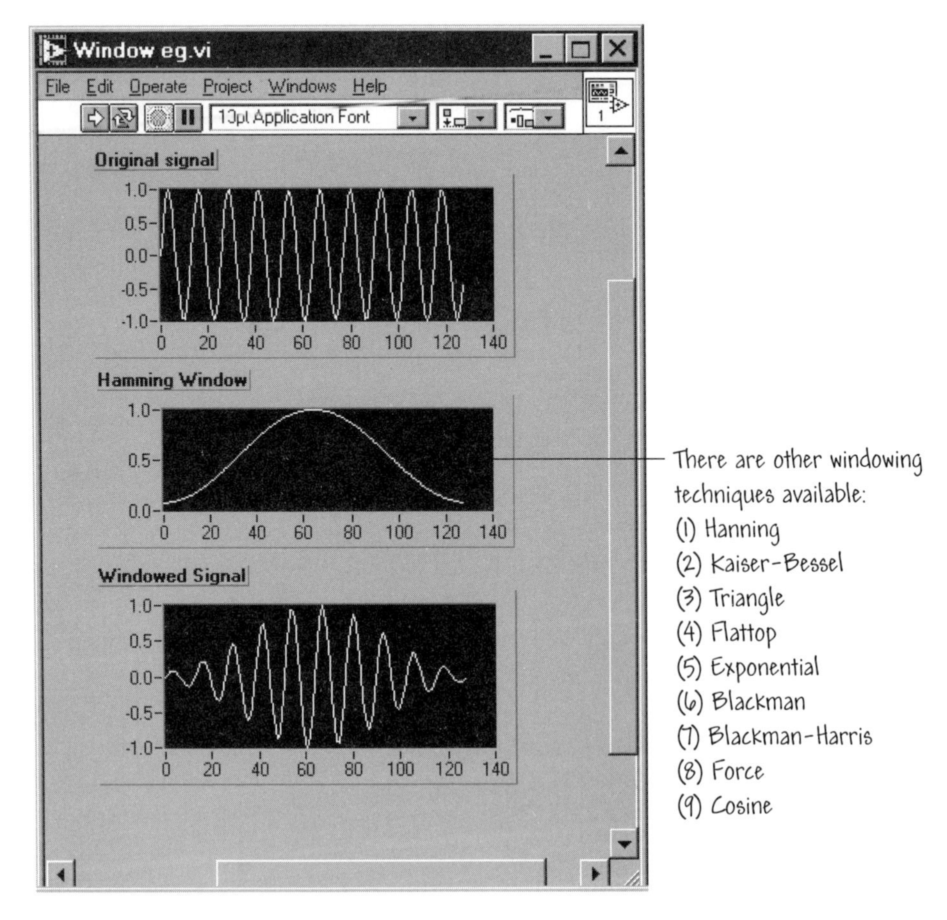

FIGURE 11.30
A sinusoidal signal windowed using a Hamming window.

discontinuities at the boundaries of each period by multiplying the sampled original waveform by a finite length window whose amplitude varies smoothly and gradually towards zero at the edges.

One such windowing technique uses the *Hamming window*, as illustrated in Figure 11.30. The sinusoidal waveform of the windowed signal gradually tapers to zero at the ends—see the bottom graph in Figure 11.30. When computing the discrete Fourier transform on data of finite length, you can use the windowing technique applied to the sampled waveform to minimize the discontinuities of the constructed periodic waveform. This approach will minimize the spectral leakage.

The Hamming Window

Open a new front panel and place four waveform graphs and one digital control, as shown in Figure 11.31.

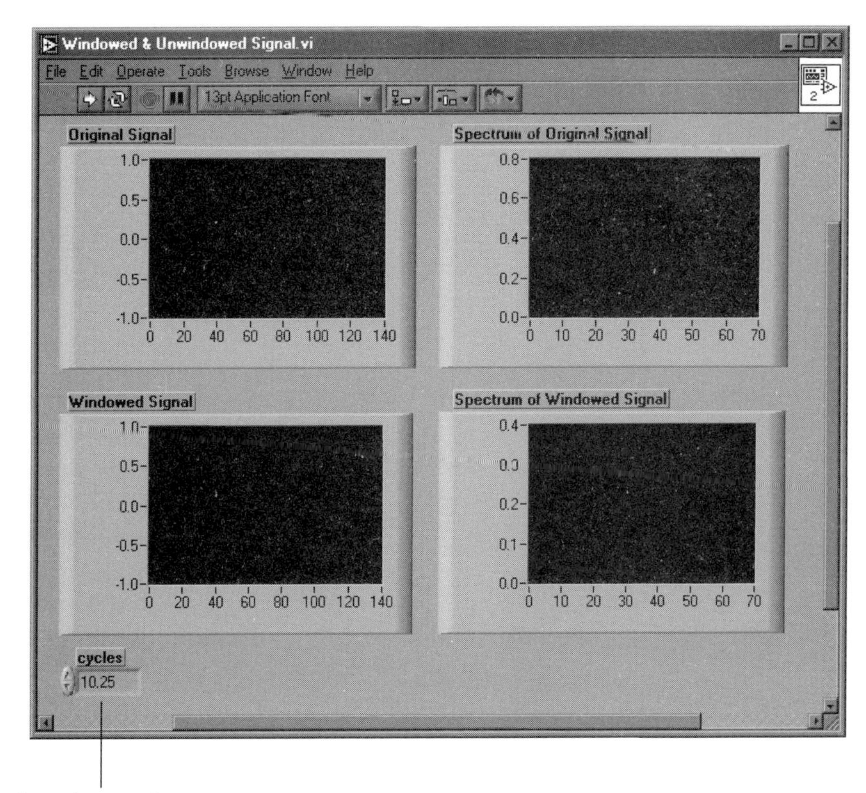

FIGURE 11.31
The front panel for a VI to investigate the use of windows.

Construct a block diagram using Figure 11.32 as a guide. In the block diagram, we use three main VIs:

- Hamming Window.vi: This VI applies the Hamming window to the input sequence. It is located in Functions≫Analyze≫Signal Processing≫ Windows, as shown in Figure 11.32. If we denote the input sequence as **X** (with n elements) and the output sequence of the Hamming window as **Y**, then

$$\mathbf{Y}(i) = \mathbf{X}(i)\,[0.54 - 0.46\cos\omega]$$

where

$$\omega = \frac{2\pi i}{n}.$$

- The Amplitude and Phase Spectrum.vi computes the amplitude spectrum of the windowed and nonwindowed input waveforms. You can find this

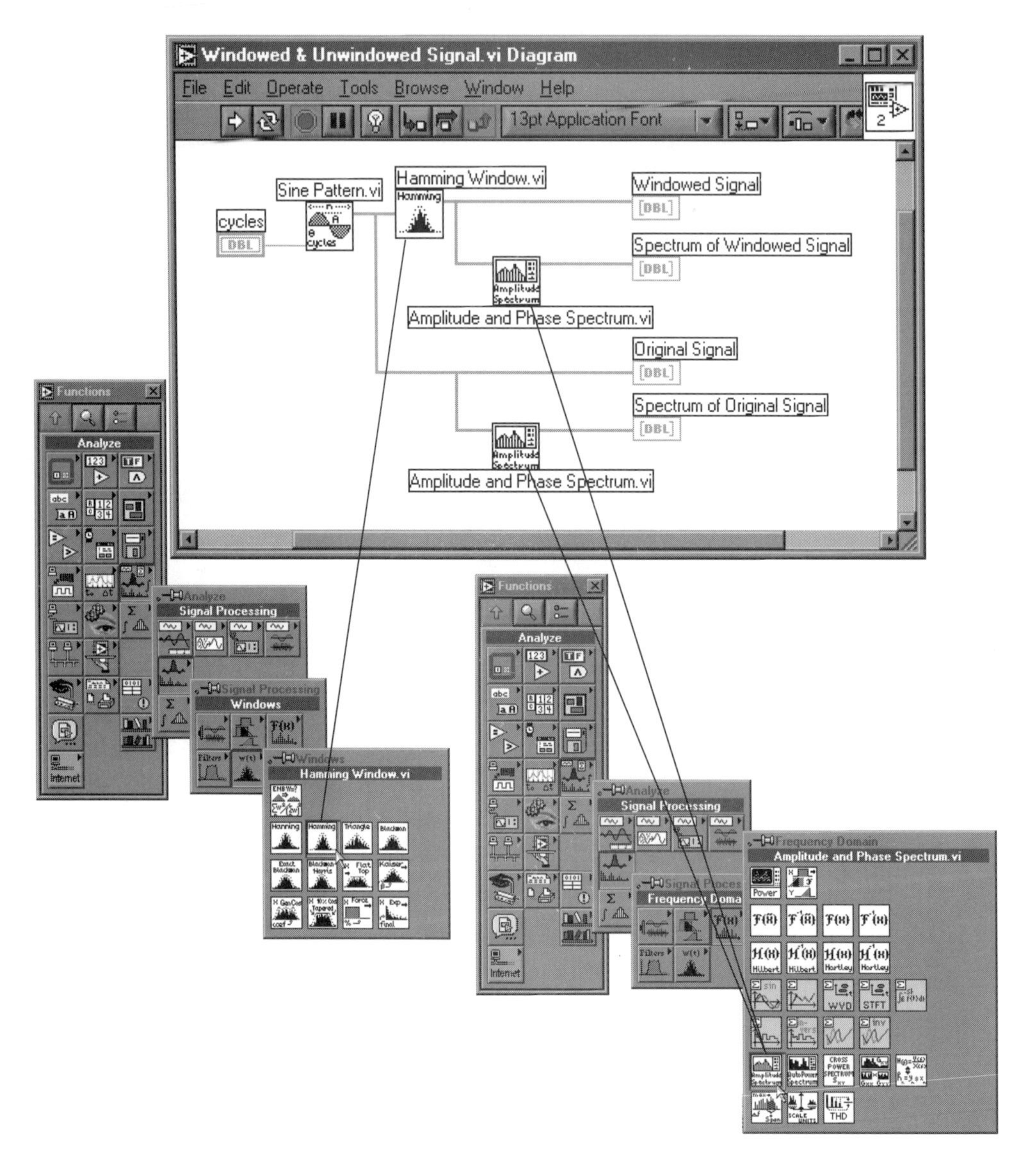

FIGURE 11.32
The block diagram for a VI to investigate the effect of windowing.

particular VI in the palette Functions≫Analyze≫Signal Processing≫Frequency Domain, as shown in Figure 11.32.

- The Sine Pattern.vi generates a sine wave with the number of cycles specified in the control labeled **cycles**. It is located in Functions≫Analyze≫Signal Processing≫Signal Generation.

As interesting numerical experiment, make the following two runs:

- Set the control **cycles** to 10. Since this is an integer number, when you repeat the waveform to construct a periodic waveform, you will not have any discontinuities. You should observe that the spectrum of the windowed and the nonwindowed waveforms are both centered at 10, and that the spectrum of the original signal displays no spectral leakage.

- Set the control **cycles** to 10.25. Since this is not an integer, you should observe that the spectrum of the windowed and the nonwindowed waveforms are different. The nonwindowed spectrum should show distinct signs of spectral leakage due to the discontinuities when constructing the periodic waveform. The windowed waveform, while not a perfect spike centered at 10, displays significantly less leakage.

Save your VI as Windowed & Unwindowed Signal.vi in the Users Stuff folder in the Learning directory.

You can find a working version of Windowed & Unwindowed Signal.vi *in Chapter 11 of the* Learning *directory.*

◆

11.8.3 Filtering

There are two main types of filters: analog and digital. In this section we will consider digital filters only. Why digital filters? Because digital filters

- are software programmable,
- are stable and predictable,
- do not drift with changes in external environmental conditions, and
- generally have superior performance-to-cost ratios compared to their analog counterparts.

LabVIEW can be used to control digital filter parameters (such as filter order, cutoff frequency, stopband and passbands, amount of ripple, and stopband attentuation). You can envision a LabVIEW-based DAQ system wherein data is acquired from external sources (such as an accelerometer sensor) and filtered in the VI software. The results can be easily analyzed and studied using the graphics provided by the G programming language and written to a spreadsheet (as we will discuss in the next chapter). The key is that by using LabVIEW you can utilize digital filters, allowing the VIs to handle the design issues, computations, memory management, and the actual data filtering.

The theory of filters is a rich and interesting subject, and one that cannot be dealt with here in any depth. Please refer to other reference materials for in-depth coverage of filtering theory.[1] But a brief discussion of terms is needed to give

1. A good source of information for LabVIEW users is *LabVIEW Signal Processing* by Mahesh L. Chugani, Abhay R. Samant, and Michael Cerna, Prentice Hall, Upper Saddle River, New Jersey, 1998.

you a better understanding of the filter parameters and how they relate to the VI inputs.

The **sampling theorem** states that a continuous-time signal can be reconstructed from discrete, equally spaced samples if the sampling frequency is at least twice that of the highest frequency in the time signal. The sampling interval is often denoted by δt. The **sampling frequency** is computed as the inverse of the sampling interval:

$$f_s = \frac{1}{\delta t}.$$

Thus, according to the sampling theorem, the highest frequency that the filter can process—the **Nyquist frequency**—is

$$f_{nyq} = \frac{f_s}{2}.$$

As an example, suppose that you have a system with sampling interval $\delta t = 0.01$ second. Then the sampling frequency is $f_s = \frac{1}{0.01} = 100$ Hz. From the sampling theorem we find that the highest frequency that the system can process is $f_{nyq} = \frac{f_s}{2} = 50$ Hz. If we expect that the signals that we need to process have components at frequencies higher than 50 Hz, then we must upgrade the system to allow for shorter sampling intervals, say for example, $\delta t = 0.001$ second. What is f_{nyq} in this case?

One main use of filters is to remove unwanted noise from a signal—usually if the noise is at high frequencies. Depending on the frequency range of operation, filters either pass or attenuate input signal components. Filters can be classified into the following types:

1. A **low-pass filter** passes low frequencies and attenuates high frequencies. The ideal low-pass filter passes all frequencies below the *cutoff frequency* f_c.
2. A **high-pass filter** passes high frequencies and attenuates low frequencies. The ideal high-pass filter passes all frequencies above f_c.
3. A **bandpass filter** passes a specified band of frequencies. The ideal bandpass filter only passes all frequencies between f_{c1} and f_{c2}.
4. A **bandstop filter** attenuates a specified band of frequencies. The ideal bandstop filter attenuates all frequencies between f_{c1} and f_{c2}.

The ideal frequency response of these filters is illustrated in Figure 11.33.

The frequency points f_c, f_{c1} and f_{c2} are known as the cutoff frequencies and can be viewed as filter design parameters. The range of frequencies that is passed through the filter is known as the **passband** of the filter. An ideal filter has a gain of one (0 dB) in the passband—that is, the amplitude of the output signal is the same as the amplitude of the input signal. Similarly, the ideal filter

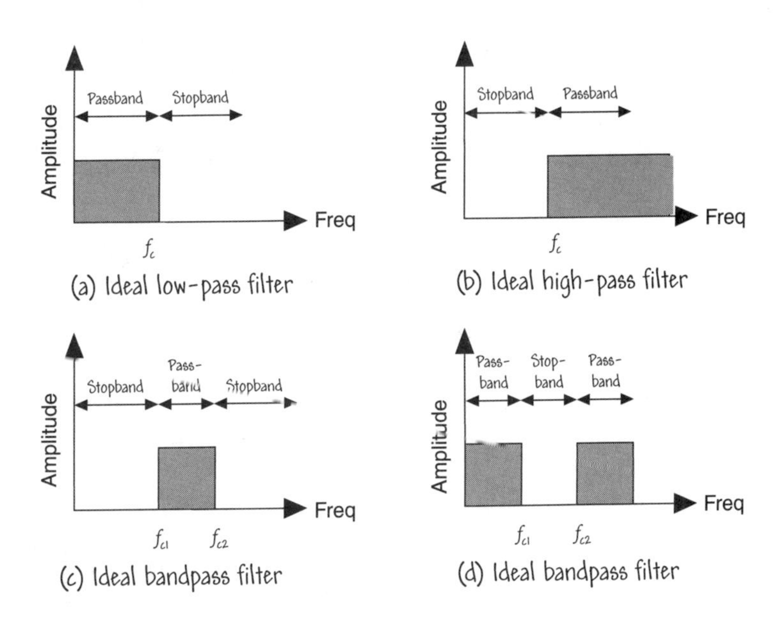

FIGURE 11.33
The ideal frequency response of common ideal filters.

completely attenuates the signals in the stopband—that is, the stopband attenuation is $-\infty$ dB. The low-pass and high-pass filters have one passband and one stopband. The range of frequencies that do not pass through the filter is known as the **stopband** of the filter. The stopband frequencies are rejected or attenuated by the filter. The passband and the stopband for the different types of filters are shown in Figure 11.33. The bandpass filter has one passband and two stopbands. Conversely, the bandstop filter has two passbands and one stopband.

Suppose you have a signal containing component frequencies of 1 Hz, 5 Hz, and 10 Hz. This input signal is passed separately through low-pass, high-pass, bandpass, and bandstop filters. The low-pass and high-pass filters have a cutoff frequency of 3 Hz, and the bandpass and bandstop filters have cutoff frequencies of 3 Hz and 8 Hz. What frequency content of the signal can be expected? The output of the filter in each case is shown in Figure 11.34. The low-pass filter passes only the signal at 1 Hz because this is the only component of the input signal lower than the 3 Hz cutoff. Conversely, the high-pass filter attentuates the 1 Hz component and passes a signal with components at 5 and 10 Hz. The bandpass filter passes only the signal component at 5 Hz, and the bandstop filter filters out the signal component at 5 Hz and passes a signal with frequency content at 1 and 10 Hz.

Ideal filters are not achievable in practice. It is not possible to have a unit gain (0 dB) in the passband and a gain of zero ($-\infty$ dB) in the stopband—there is always a *transition region* between the passband and stopband. A more realistic filter will have the passband, stopband, and transition bands as depicted in

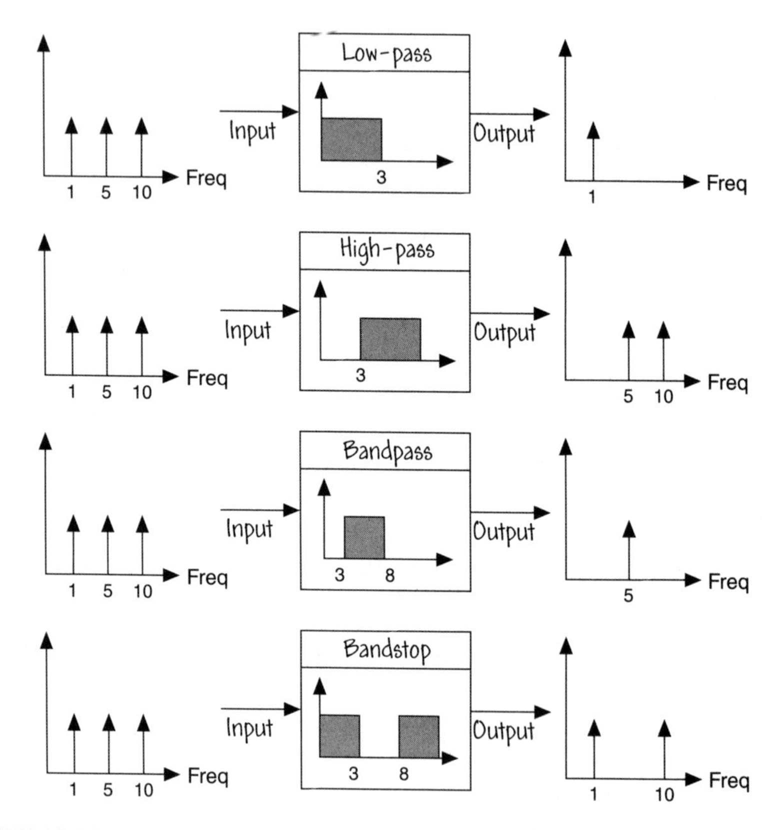

FIGURE 11.34
The output of the common filter in the case where the input signal contains component frequencies of 1 Hz, 5 Hz, and 10 Hz.

Figure 11.35. The variation in the passband is called the *passband ripple*; see the bandpass filter in Figure 11.35. The *stopband attenuation*, also depicted in Figure 11.35, cannot be infinite (as it would be for an ideal filter).

If we view the filter as a linear system, then we can consider the response of the system (that is, the filter) to various types of inputs.[2] One interesting input is the *impulse*. If the input to the digital filter is the sequence $x(0), x(1), x(2), \ldots,$ then the impulse is given by $x(0) = 1, x(1) = x(2) = \cdots = 0$. The **impulse response** of a filter (that is, the output of the filter when the input is an impulse) provides another classification system for filters. The Fourier transform of the impulse response is known as the **frequency response**. The frequency response of a system provides a wealth of information about the system, including how it will respond to periodic inputs at different frequencies. Therefore, the frequency

2. A good source of information on systems and system response to various inputs is *Modern Control Systems*, by Richard C. Dorf and Robert H. Bishop, Prentice Hall, Upper Saddle River, NJ, 2000.

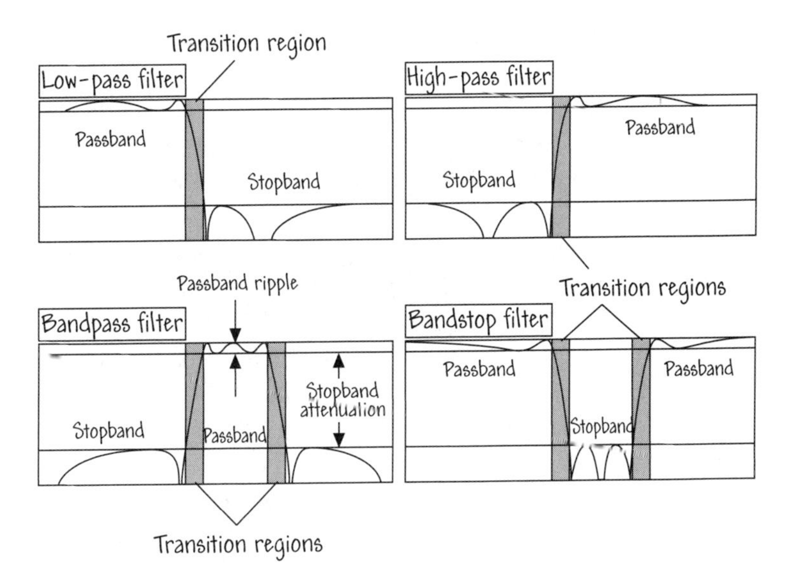

FIGURE 11.35
A realistic frequency response of a filter.

response will tell us about the filter characteristics: How does the filter respond in the passbands and stopbands? How accurately does the filter cut off and attenuate high-frequency components?

We can classify a filter based on its impulse response as either a **finite impulse response** (FIR) filter or an **infinite impulse response** (IIR) filter. For an IIR filter, the impulse response continues indefinitely (in theory), and the output depends on current and past values of the input signal and on past values of the output. In practical applications, the impulse response for stable IIR filters decays to near zero in a finite time. For an FIR filter the impulse response decays to zero in a finite time, and the output depends only on current and past values of the input signal. Take the example of processing noisy range measurements (that is, distance measurements) to a fixed target. Suppose that you want to determine the distance to the fixed object and you have available a ranging device (e.g., a laser ranging device) that is corrupted by random noise (as is the case for most realistic sensors!). One way to estimate the range is to take a series of range measurements $x(0), x(1), x(2), \ldots, x(k)$ and filter them by computing a running average:

$$x_{\text{ave}}(k) = \frac{1}{k} \sum_{i=1}^{k} x(i) \; .$$

The output of the filter is $x_{\text{ave}}(k)$. This is an FIR filter because the output depends only on previous values of the input $(x(0), x(1), x(2), \ldots, x(k-1)$ and on the current value of the input ($x(k)$). Now, we can rewrite the filter as

$$x_{\text{ave}}(k) = \frac{k-1}{k} x_{\text{ave}}(k-1) + \frac{1}{k} x(k) .$$

This is an IIR filter because the output depends on current and previous values of the input as well as on previous values of the output (that is, the $x_{\text{ave}}(k-1)$ term). Mathematically, the two filters provide the same output, but they are implemented differently. The FIR filter is sometimes referred to as a *nonrecursive* filter; the IIR filter is known as a *recursive* filter.

One disadvantage of IIR filters is that the phase response is nonlinear. You should use FIR filters for situations where a linear phase response is needed. A strong advantage of IIR filters are that they are recursive, thus reducing the memory storage requirements. The well-known Kalman filter, which can be viewed as an IIR filter (it is actually a bit more complicated to implement than an IIR filter), was used successfully to filter navigation data acquired by the Apollo spacecraft to find its way to the moon and to rendezvous around the moon for the long journey home.[3] LabVIEW provides many different types of filters, as shown in Figure 11.36.

The IIR filter VIs available in LabVIEW include the following:

- Butterworth: Provides a smooth response at all frequencies and a monotonic decrease from the specified cutoff frequencies. Butterworth filters are maximally flat—the ideal response of unity in the passband and zero in the stopband—but do not always provide a good approximation of the ideal filter response because of the slow rolloff between the passband and the stopband.
- Chebyshev: Minimizes the peak error in the passband by accounting for the maximum absolute value of the difference between the ideal filter and the filter response you want (the maximum tolerable error in the passband). The frequency response characteristics of Chebyshev filters have an equiripple magnitude response in the passband, monotonically decreasing magnitude response in the stopband, and a sharper rolloff than for Butterworth filters.
- Inverse Chebyshev: Also known as Chebyshev II filters. They are similar to Chebyshev filters, except that inverse Chebyshev filters distribute the error over the stopband (as opposed to the passband), and are maximally flat in the passband (as opposed to the stopband). Inverse Chebyshev filters minimize peak error in the stopband by accounting for the maximum absolute value of the difference between the ideal filter and the filter response you want. The frequency response characteristics are equiripple magnitude response in the stopband, monotonically decreasing magnitude response in the passband, and a rolloff sharper than for Butterworth filters.

---

3. A well-told story on the use of Kalman filters during Apollo can be found in *An Introduction to the Mathematics and Methods of Astrodynamics*, by R. H. Battin, AIAA Education Series, 1987.

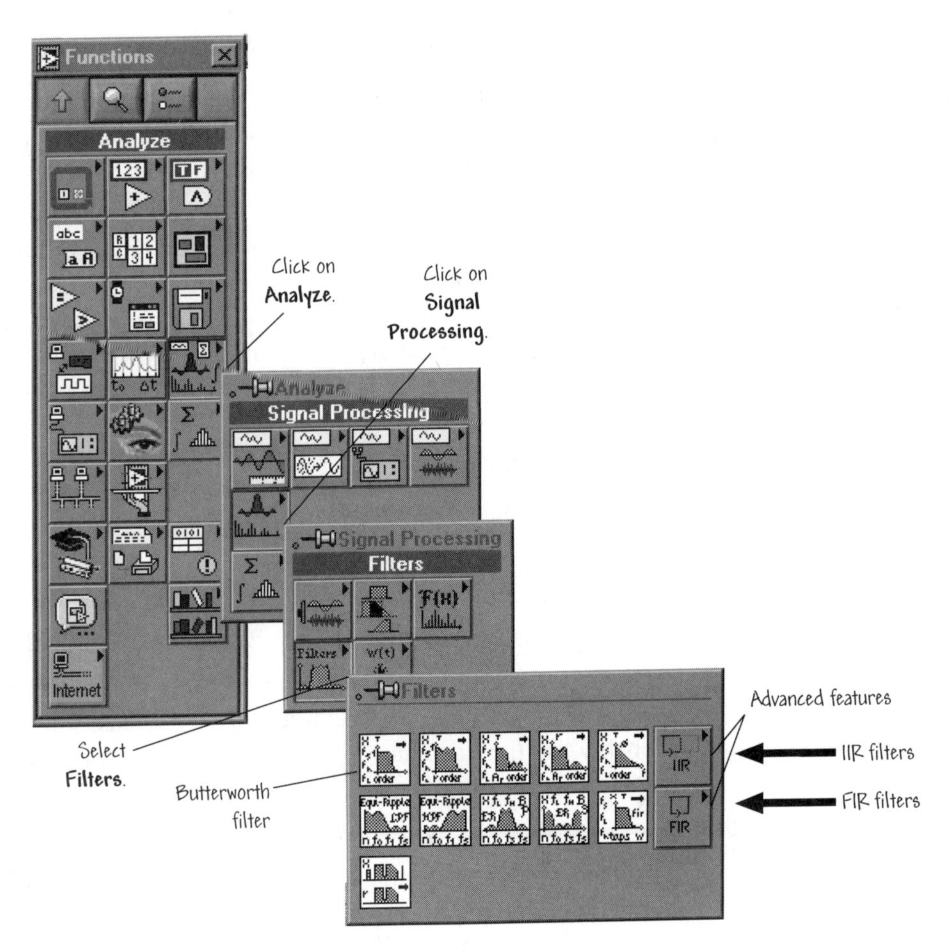

FIGURE 11.36
IIR and FIR filter choices are located in the **Functions** palette.

- Elliptic: Minimize the peak error by distributing it over the passband and the stopband. Equiripples in the passband and the stopband characterize the magnitude response of elliptic filters. Compared with the same-order Butterworth or Chebyshev filters, the elliptic design provides the sharpest transition between the passband and the stopband. For this reason, elliptic filters are widely used.
- Bessel: Can be used to reduce nonlinear phase distortion inherent in all IIR filters. Bessel filters have maximally flat response in both magnitude and phase. Furthermore, the phase response in the passband of Bessel filters, which is the region of interest, is nearly linear. Like Butterworth filters, Bessel filters require high-order filters to minimize the error.

The FIR filter VIs available in LabVIEW include the following:

- Windowed: The simplest method for designing linear-phase FIR filters is the window design method. You select the type of windowed FIR filter you want—low-pass, high-pass, bandpass, or bandstop—via input to the FIR Windowed Filter.vi.
- Optimum filters based on the Parks-McClellan algorithm: Offers an optimum FIR filter design technique that attempts to design the best filter possible for a given filter complexity. Such a design reduces the adverse effects at the cutoff frequencies. It also offers more control over the approximation errors in different frequency bands—control that is not possible with the window method. The VIs available include
  - Equiripple Low-pass
  - Equiripple High-pass
  - Equiripple Bandpass
  - Equiripple Bandstop

Which filter is best suited for your application? Obviously, the choice of filter depends on the problem at hand. Figure 11.37 shows a flowchart that can serve as a guide for selecting the best filter for your needs. Keep in mind that you will probably use the flowchart to determine several candidate filters, and you will have to experiment to make the final choice.

Open a new front panel and construct a front panel similar to the one shown in Figure 11.38. You will need to place one digital control, two vertical slides, and two waveform graphs. Label them according to the scheme shown in the figure.

Construct a block diagram similar to the one shown in Figure 11.38. In the block diagram, we use three main VIs:

- Butterworth.vi: This VI is used to filter the noise. It is located in Functions ≫Analyze≫Signal Processing≫Filters, as shown in Figure 11.36.
- The Uniform White Noise.vi generates a white noise that is added to the sinusoidal signal. You can find this particular VI in the palette Functions≫ Analyze≫Signal Processing≫Signal Generation.
- The Sine Pattern.vi generates a sine wave of the desired frequency. It is located in Functions≫Analyze≫Signal Processing≫Signal Generation.

With this VI you are generating 10 cycles of a sine wave (this value can be varied on the front panel), and there are 1,000 samples. Select a cutoff frequency of 25 Hz and a filter order of 5. Note that we did not previously discuss filter order—it is a measure of filter complexity and is related to the number of terms retained in the filter. Run the VI. Vary the cutoff frequency and observe the effects. What happens when the cutoff frequency is set to 50? Does the filtered

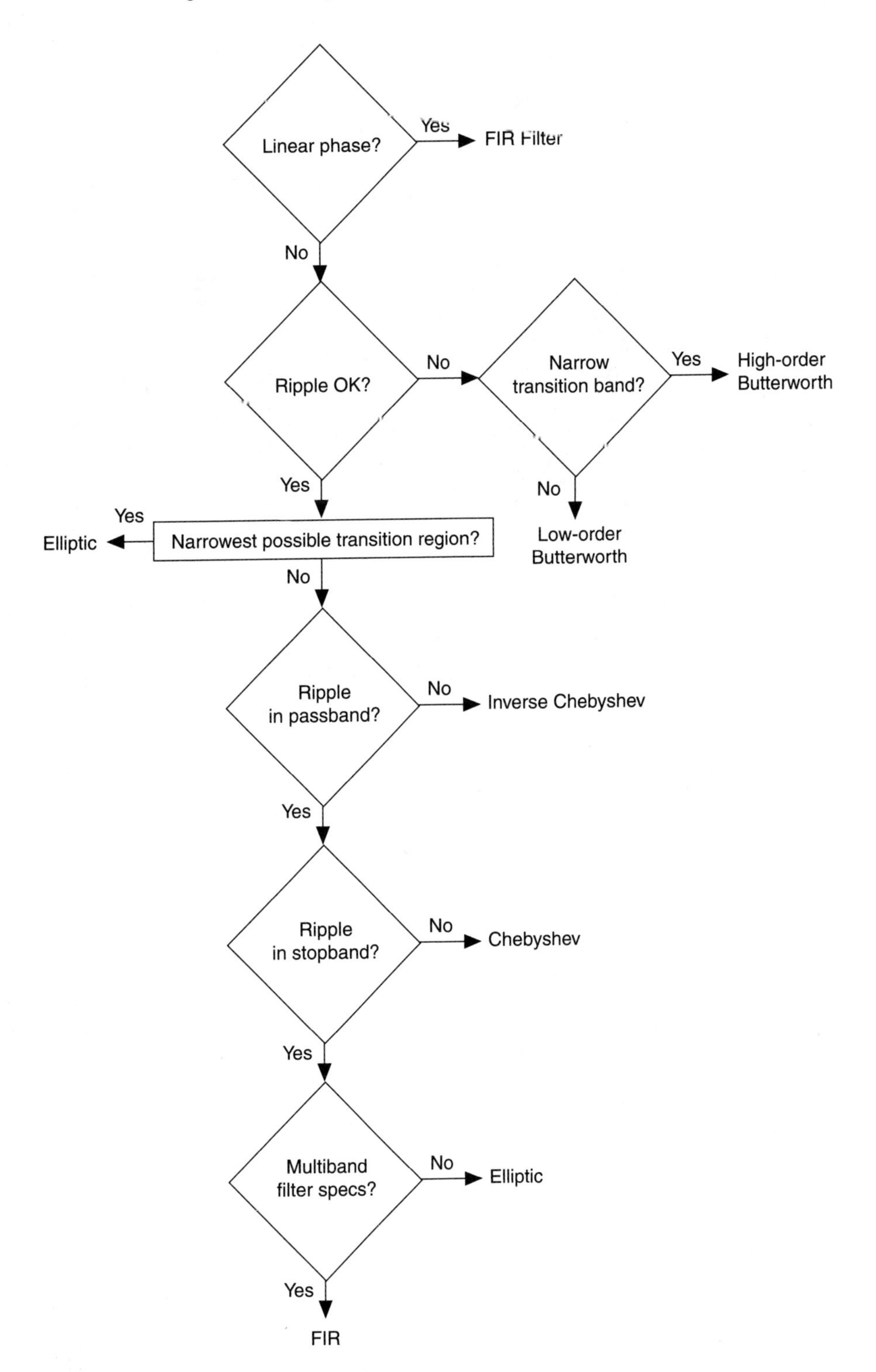

FIGURE 11.37
Flowchart that can serve as a guide for selecting the best filter.

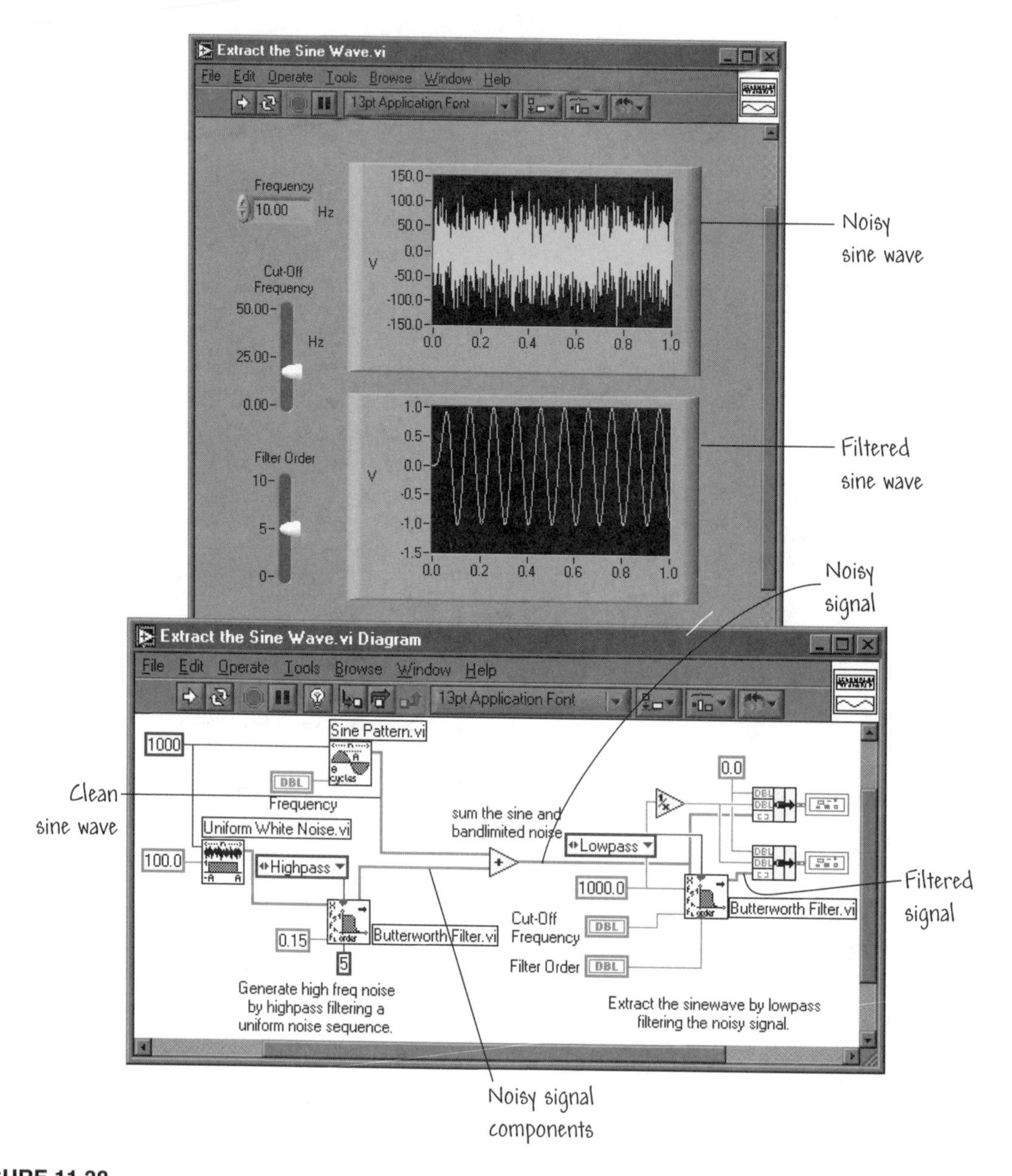

FIGURE 11.38
The front panel and block diagram for a VI to filter a noisy sine wave.

signal contain noise components? When you are finished exploring, save your VI as Extract the Sine Wave.vi in the Users Stuff folder.

You can find a working version of the VI above in folder Chapter 11 *in the* Learning *directory. It is called* Extract the Sine Wave.vi.

◆

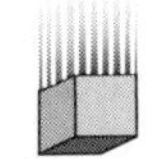

BUILDING BLOCK

11.9 BUILDING BLOCKS: MEASURING VOLUME

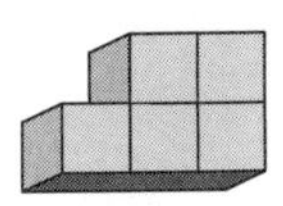

In the Chapter 9 "Building Block" exercise, you constructed a VI named Volume Data Saved.vi. In this exercise you will modify Volume Data Saved.vi by adding several analysis VIs. The VI that you developed in Chapter 9 should be have been saved in the Users Stuff folder.

The target front panel and block diagram are shown in Figure 11.39. The two additions are the functions Mean VI and Amplitude and Phase Spectrum VI, which compute the mean and amplitude spectrum of the measured volume data, respectively. The mean value is displayed in a numeric indicator and the amplitude spectrum is shown on a waveform graph.

Enter a value for the **Number of data points** on the front panel. A value of 100 is a reasonable number. Run the VI and observe the results—the curve fit of

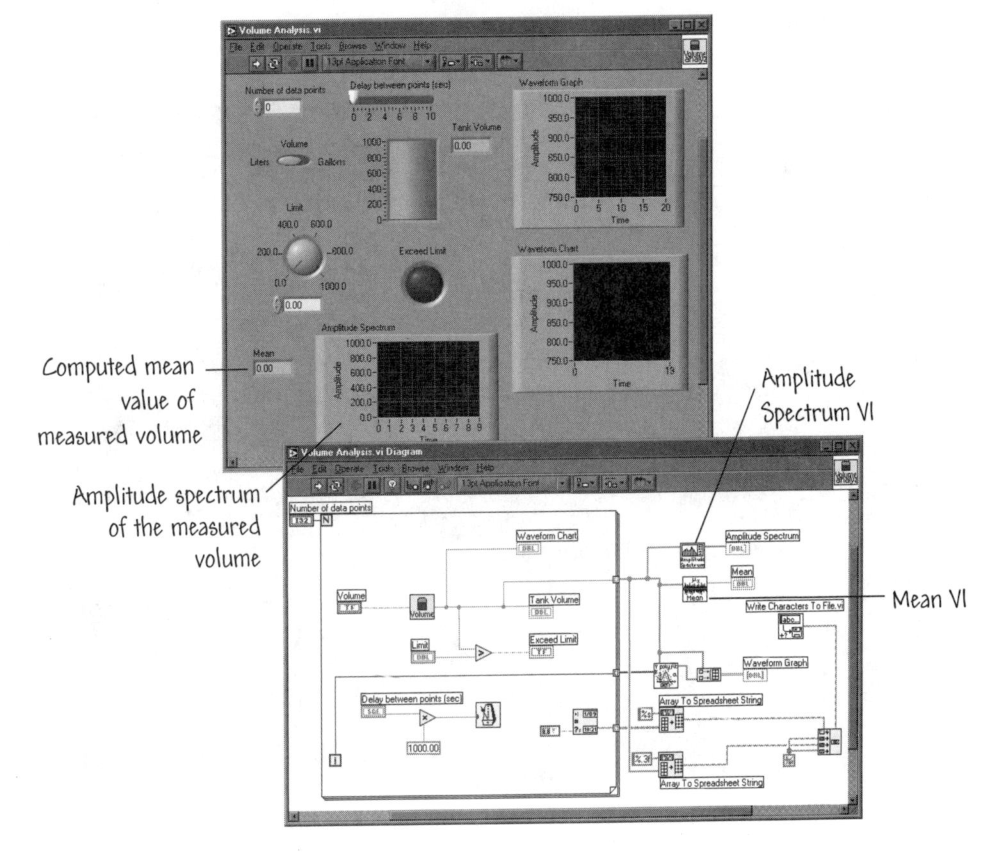

FIGURE 11.39
Analyzing the volume data.

the measured volume is superimposed on the raw data. Notice that the curve fit is much smoother than the raw volume data. Since the measured data is noisy, each time you run the VI, you will get different results. When you are finished with the VI, save it in the Users Stuff folder and name it Volume Analysis.vi.

11.10 RELAXING READING: EXPLORING THE SENSE OF SMELL

Fundamental questions exist about the sense of smell. Questions relating to signal transduction, the organization of neural systems, development, and behavior still remain unanswered. A great deal remains to be learned about the acute sensitivity of the olfactory system and its ability to distinguish among a very large number of smells.

The fruit fly Drosophila is highly sensitive to a wide variety of odorants and offers several advantages as an organism in which to study the sense of smell. Its olfactory system is relatively simple and can be easily analyzed in vivo. Researchers have developed an automated data acquisition and analysis system to study the olfactory receptor neurons of fruit flies for Yale University's Depart-

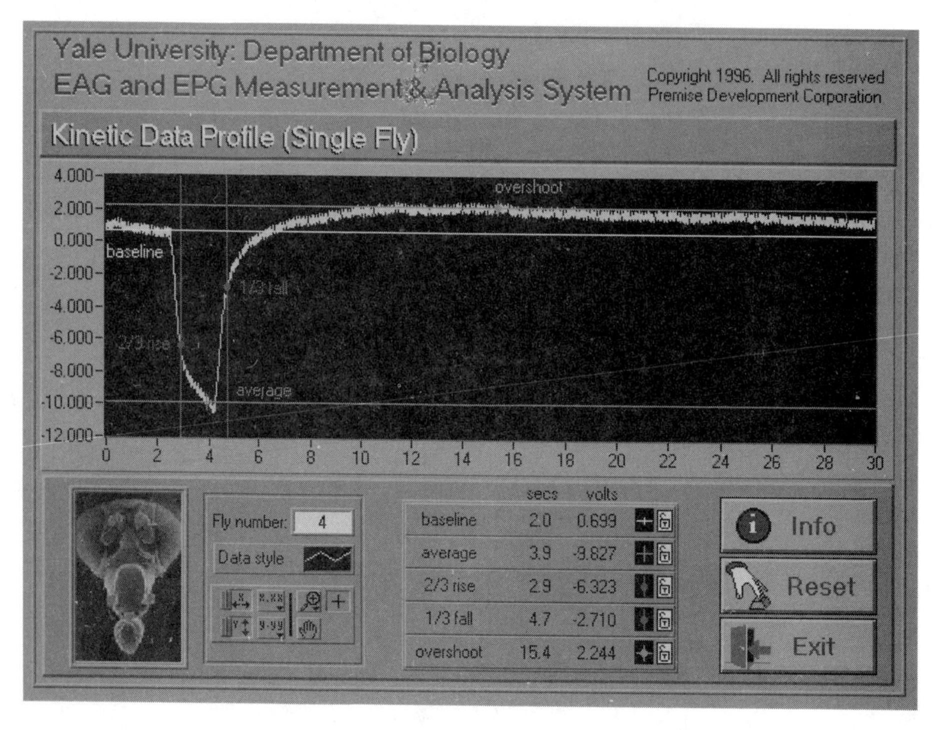

FIGURE 11.40
The graphical user interface (GUI) makes it easy for researchers to select and analyze data, as shown in this example of the kinetic data profile of a fly.

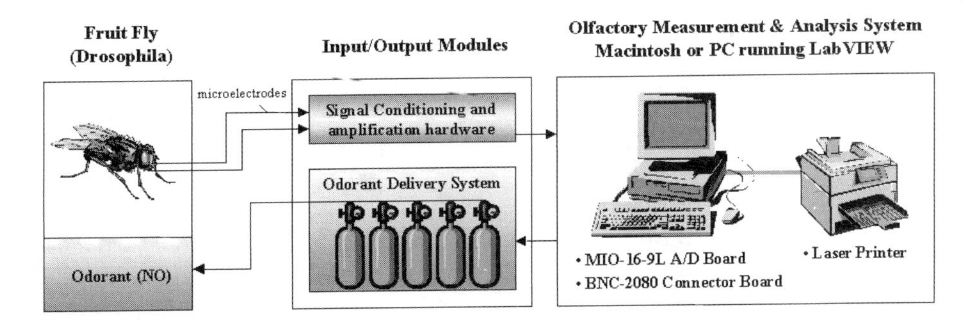

FIGURE 11.41
Diagram of the fruit fly olfactory measurement and analysis system.

ment of Biology. With a LabVIEW-based system, researchers save time and money by using a virtual instrumentation solution for their test and measurement application. The Olfactory Measurement and Analysis System improved the accuracy, efficiency, and analytical capabilities of studying the function and behavior of olfactory neuron receptors.

Microelectrodes are connected to the fruit fly's antennae, and sequential "snapshots" are collected with the National Instruments NB-MIO-16 DAQ board to ensure that all the sensors and signal-conditioning equipment were functioning properly and to establish baseline measurements. After waiting a preset period of time, the computer initiated a digital output signal to begin odorant stimulus to the fly. The LabVIEW development environment made it easy to control the respective valves of the odorant delivery system. Although the application performs a wide array of analyses, all data was stored in standard ASCII format, allowing easy transport to spreadsheet applications.

Researchers have the flexibility to combine various sets of flies in an organized and methodical manner. One of the primary analysis routines was "kinetic analysis," in which the operator may select an individual fly and view parameters (in both numeric and graphic formats) such as the baseline, 2/3 rise time, 1/3 fall time, overshoot, peak amplitude, and duration.

For more information on this application of LabVIEW, read the *User Solutions* entitled "An Automated LabVIEW and DAQ System Helps Scientist Explore Smell" by Joseph Adam and Eroc Rosow, which can be found on the National Instruments website, or contact:

Joseph Adam
Premise Development Corporation
36 Cambridge Street
West Hartford, CT 06110
web: http://www.connix.com/ premise

11.11 SUMMARY

LabVIEW provides a great computational environment for analysis of signals and systems. In this chapter, we presented some applications of the many VIs available for analysis of signals, systems, functions, and systems of equations. The material was intended to motivate you to look further into the capabilities of the G programming language in developing your own VIs for solving linear algebraic systems of equations, curve fitting, integrating ordinary differential equations, computing function zeroes, computing derivatives of functions, integrating functions, generating and analyzing signals, computing discrete Fourier transforms, and filtering signals.

KEY TERMS

Bandpass filter: A system that passes a specified band of frequencies.

Bandstop filter: A system that attenuates a specified band of frequencies.

Condition number: A quantity used in assessing the accuracy of solutions to systems of linear algebraic equations. The condition-number can vary from 0 to ∞—a matrix with a condition number near 1 is closer to being nonsingular than a matrix with a very large condition number.

Determinant: A characteristic number associated with an $n \times n$ square matrix that is computed as a function of the minors and cofactors of the matrix. When the determinant is identically equal to zero, then we say that the matrix is singular.

Digital frequency: Computed as the analog frequency divided by the sampling frequency. Also known as the **normalized frequency**.

Discrete Fourier transform (DFT): A common practical algorithm for transforming sampled signals from the time domain into the frequency domain.

Even symmetric signal: A signal symmetric about the y-axis.

Fast Fourier transforms (FFT): The FFT is a fast algorithm for calculating the DFT when the number of samples (N) is a power of 2.

Frequency-domain representation: The representation of a signal in terms of its individual frequency components.

Impulse response: The output of a system (e.g., a filter) when the input is an impulse.

FIR filter: A finite impulse response filter in which the impulse response decays to zero in a finite time, and the output depends only on current and past values of the input signal.

Frequency response: The Fourier transform of the impulse response.

G Math Toolkit: A toolkit that enables you to interface real-world measurements to mathematical analysis algorithms.

High-pass filter: A system that passes high frequencies and attenuates low frequencies.

Homogeneous DEQ: A differential equation that has no input driving function (that is, $u(t) = 0$).

IIR filter: An infinite impulse response filter in which the impulse response continues indefinitely (in theory), and the output depends on current and past values of the input signal and on past values of the output.

Low-pass filter: A system that passes low frequencies and attenuates high frequencies.

Nyquist frequency: The highest frequency that a filter can process, according to the sampling theorem.

Odd symmetric signal: A signal symmetric about the origin.

Passband: The range of frequencies that is passed through a filter.

Sampling theorem: The statement that a continuous-time signal can be be reconstructed from discrete, equally spaced samples if the sampling frequency is at least twice that of the highest frequency in the time signal.

Spectral leakage: A phenomenon that occurs when you sample a noninteger number of cycles, leading to "artificial" discontinuities in the signal that manifest themselves as very high frequencies in the DFT/FFT spectrum, appearing as if the energy at one frequency has "leaked out" into all the other frequencies.

Stopband: The range of frequencies that do not pass through a filter.

Trace: The sum of the diagonal elements of an $n \times n$ square matrix.

Windowing: A method used to reduce the amplitude of sampled signal discontinuities by multiplying the sampled original waveform by a finite-length window whose amplitude varies smoothly and graduates towards zero at the edges.

EXERCISES

E11.1 Several window algorithms have been implemented in LabVIEW, including Hanning, Hamming, Triangle, Blackman, Exact Blackman, Blackman-Harris, Kaiser, and Flat Top. To observe the effects of each algorithm, open the Window Comparison.vi from Examples\Analysis\windxmpl.llb and experiment with the VI.

E11.2 Several different types of IIR filters have been implemented in LabVIEW, including Elliptic, Bessel, Butterworth, Chebyshev, and inverse Chebyshev. To experiment with each filter type , open the IIR Filter Design.vi, which is located in Examples\Analysis\fltrxmpl.llb. Run the VI and examine each filter design. You can choose the filter type: bandpass, bandstop, low-pass, or high-pass.

E11.3 In Section 11.8 of this chapter, you learned about signal processing. You can dynamically determine the frequency, power spectrum, and amplitude spectrum of a signal that can change as the VI executes. Open Dynamic Signal Analyzer (sim).vi from Examples\Analysis\measxmpl.llb and experiment with the VI.

PROBLEMS

P11.1 Complete the crossword puzzle.

Across

1. A quantity used is assessing the accuracy of solutions to systems of linear algebraic equations.
3. The Fourier transform of the impulse response.
4. A phenomenon that occurs when you sample a noninteger number of cycles, leading to artificial discontinuities.
5. A filter in which the impulse response decays to zero in a finite time and the output depends only on current and past values of the input signal.
9. The output of a system when the input is an impulse.
10. A method used to reduce the amplitude of sampled signal discontinuities by multiplying the sampled original waveform by a finite-length window.
12. The range of frequencies that is passed through a filter.

Down

2. The highest frequency that a filter can process, according to the sampling theorem.
6. The range of frequencies that do not pass through a filter.
7. A filter in which the impulse response continues indefinitely.
8. A system that passes a specified band of frequencies.
11. A toolkit that enables you to interface real-world measurements to mathematical analysis algorithms.

CHAPTER 12

Other LabVIEW Applications

In this chapter we discuss several LabVIEW applications that may be of interest to you as you begin to master LabVIEW and want to further utilize the power of virtual instruments. The material presented here is intended to stimulate your curiosity about VI Servers, DataSocket connections, and HiQ. Did you know that you can automatically generate reports with HiQ and pass that report to Microsoft Word, Excel spreadsheets, and PowerPoint presentations? Or that you can control remote VIs and send data over a network using VI Server and DataSocket connections. With LabVIEW you can interact with e-mail, web pages, FTP and Telnet. These topics are introduced in this chapter, and you are pointed to other information sources from which you can continue to discover the many uses of LabVIEW. Did you ever want to sample your computer sound card and "watch" your favorite music CD?

GOALS

1. Read about displaying and controlling VIs on the Internet and over networks.
2. Understand the basic capabilities of HiQ to visualize data and generate reports.
3. Discuss how to acquire and generate waveforms with a sound card and how to control sound files.

12.1 VI SERVERS

You can control VIs and send data over a network using VI Server and DataSocket. But under what situations would you use DataSocket versus VI Server? The brief discussions that follow are intended to serve as a guide to point you to the best approach for your situation—the details of using each method are beyond the scope of this book, but good documentation exists both on-line and in hardcopy from National Instruments. If you are a **Macintosh** user, only VI Server is available to you. If you are a **Windows** user, both methods are available, but DataSocket exists only on 32-bit Windows platforms—Windows 2000/9X/NT.

The **VI Server** controls any VIs remotely (or locally), including DAQ, GPIB, or any other VI. Basically, what happens is that a VI running on the local machine runs a VI on a remote machine that controls the remote device. The VI Server VIs exist on the first row of the **Application Control** palette, as shown in Figure 12.1.

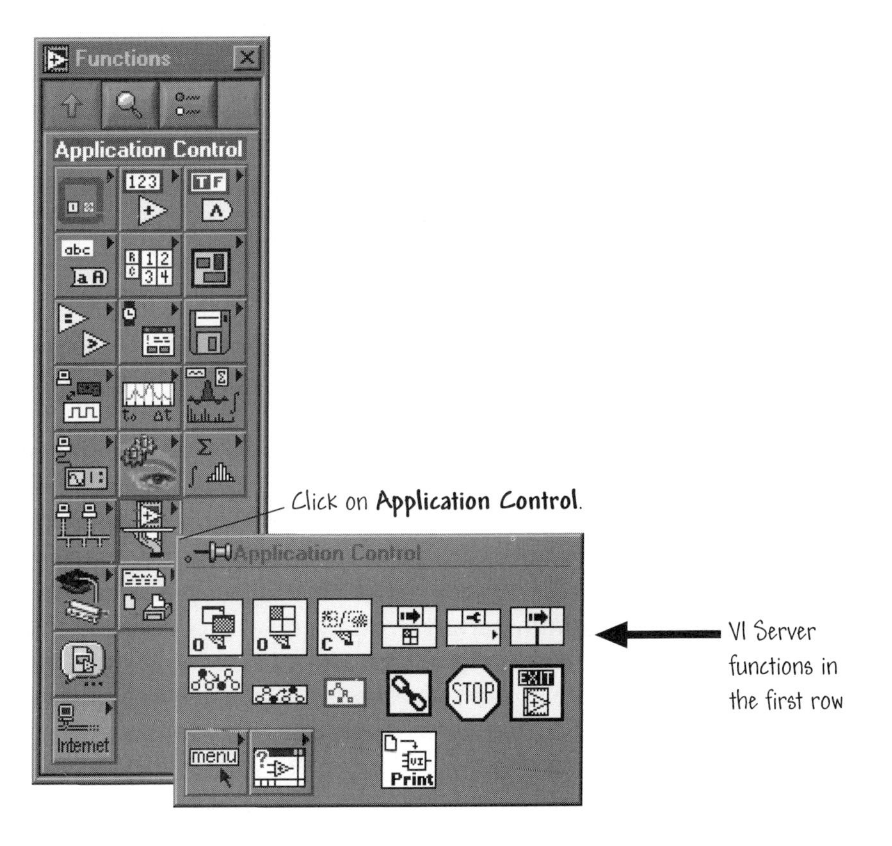

FIGURE 12.1
The VI Server functions.

Besides being able to control more than DAQ remote devices, one advantage of using VI Server is that it can be run on any computer platform that LabVIEW supports—so you can use it on the **Macintosh** and in **Windows**. Another advantage is that VI Server has security—you can set up the IP addresses that have access to the remote machines. On the other hand, with VI Server you must have LabVIEW on the local and remote computers.

One characteristic of using VI Server is that you cannot see the remote VI front panel. What difference does this make? Well, one use of VI Server is to change the appearance of the remote VIs at runtime without worrying about the low-level (network) protocol. For example, you could remove the toolbars and scrollbars of the remote VI at runtime. However, the changes would not be visible to you at the local computer site. You can send and receive data, but the remote front panel is not accessible visually.

12.2 DATASOCKET CONNECTIONS

DataSocket technology is an open technology that allows you to interact with a variety of servers and to exchange information between an application, like LabVIEW, and a number of different data sources or targets. Sources and targets include other applications, files, web servers, and FTP servers. Often, sources and targets are located on a different computer. DataSocket pulls together established communication protocols for measurement and automation in much the same way a Web browser pulls together different Internet technologies. You can specify DataSocket sources and targets (connections) using URLs (Uniform Resource Locators). The structure used to pass data is enhanced for measurement and automation applications. For instance, you can pass information such as acquisition rate, login name, and timestamp as well as the measured data.

DataSocket technology only exists on 32-bit ***Windows 2000/9x/NT*** *platforms.*

12.2.1 Communication Protocols

URLs use communication protocols to transfer data. The protocol you use in a URL depends on the type of data you want to publish and how you configure your network. With DataSocket, you can publish and subscribe to the following data types:

- **Raw text**–To deliver a string to a string indicator.
- **Tabbed text**–To publish data in arrays, as in a spreadsheet.
- **.wav data**–To publish a sound to a VI or function.
- **Variant data**–To subscribe to data from another application.

There are five communication protocols that you can use with DataSocket connections:

- **DataSocket Transport Protocol (dstp)**–This is the DataSocket native protocol. When you use this protocol, the VI communicates with the DataSocket server. You must provide a named tag for the data, which is appended to the URL. An example is
  - dstp://servername.com/numericdata,

  where numericdata is the named tag. The DataSocket connection uses the named tag to address a particular data item on a DataSocket server. To use this protocol, you must run a DataSocket server.
- **(Windows) OLE for Process Control (opc)**–This protocol is designed specifically for sharing real-time production data, such as data generated by industrial automation operations. An example is
  - opc:\National Instruments.OPCTest\item1

  To use this protocol, you must run an OPC server.
- **(Windows) logos**–This is an internal National Instruments technology for transmitting data between the network and your local computer. An example protocol URL is
  - logos://computer_ name/process/data_item_name
- **File Transfer Protocol (ftp)**–You can use this protocol to specify a file from which to read data. An example protocol URL is
  - ftp://ftp.natinst.com/datasocket/ping.wav
- **file**–You can use this protocol to provide a link to a local or network file that contains data. As an example, you could use
  - c:\mydata\ping.wav

DataSocket does not support Telnet protocols.

12.2.2 Using DataSocket on the Front Panel

DataSocket technology provides access to several input and output mechanisms from the front panel through the **DataSocket Connection** dialog box or from the block diagram with the DataSocket Read and Write functions. In this section we concentrate on using DataSocket on the front panel; the next section will consider using DataSockets on the block diagram. You publish (write) or subscribe to (read) data by specifying a URL (as discussed in the previous section) in much the same way you specify URLs in a Web browser. We say that you have

published data when you share the data of a front panel object with other users. Conversely, we say that users **subscribe** to the data when they retrieve the published data and view it on their front panel. For example, if you want to share the data in a thermometer indicator on the front panel with other computers on the Web, publish the thermometer data by specifying a URL in the **DataSocket Connection** dialog box, as illustrated in Figure 12.2. Users on other computers subscribe to the data by placing a thermometer on their front panel and selecting the URL in the **DataSocket Connection** dialog box.

Front panel controls or indicators can publish or subscribe to data through their own individual DataSocket connections. Front panel DataSocket connections publish only the data, not a graphic of the front panel control, so the VIs that subscribe through a DataSocket connection can perform their own operations on the data. You can set the value of a front panel control directly on the front panel and then publish the data, or you can build a block diagram, wire the output of a VI or function to an indicator, and publish the data from that indicator.

A typical scenario using DataSocket connections with controls and indicators is publishing a value from a front panel control for other users to subscribe to through a control or indicator. For example, if you place a knob on your front panel, a user on another computer can subscribe to the data and use it in a control wired to one of their own subVIs, or they can view the numerical value of the knob in an indicator. Similarly, you can publish a value that appears in a front panel indicator on your VI so another user can subscribe to the data and view the data in a control or indicator on their own front panel, or use the results as data

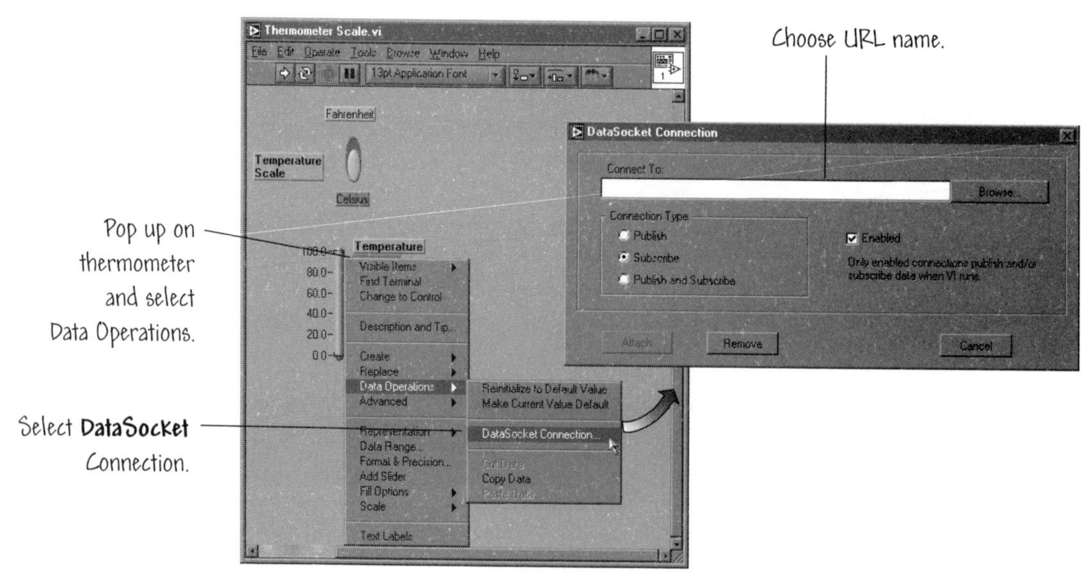

FIGURE 12.2
Publishing temperature data using DataSocket connections on the front panel.

in a control wired to an input in a subVI or function. For example, if you have a VI that calculates the mean temperature and displays the temperature in a thermometer on the front panel, you can publish the data so that another user can subscribe to and view the same temperature data.

You also be on the receiving end of the data, by subscribing to a value from a front panel indicator on someone elses VI to view data in a front panel indicator on your VI that appears in a control or indicator on the front panel of another VI. If you subscribe to a value from a front panel control on someone elses VI, you can view their data in a front panel control on your own. If you subscribe to the data with a control, you can use the data in your VI by wiring the control to an input of a subVI or function.

Another typical scenario for using DataSocket connections is to publish from and subscribe to a front panel control so users can manipulate a control on the front panel of your VI from the front panels of their VIs. When you run the VI, the front panel control on your VI retrieves the current value that another VI or application published through the DataSocket connection. When a user changes the control value on his or her front panel, the DataSocket connection publishes the new value to the front panel control of your VI. If you then manipulate the value of your front panel control, your VI publishes the value to the front panels of other users.

Front panel objects that subscribe to data do not have to be the same kind of objects that publish the data. However, the front panel objects must be the same data type or, if they are numeric data, they must coerce. For example, if you use a digital indicator to subscribe to temperature data from a thermometer on another VI, your digital indicator can be an integer if the thermometer is a floating-point number, since the data can be coerced.

12.2.3 Reading and Writing Live Data through the Block Diagram

Front panel DataSocket connections are primarily intended for sharing live data. To read data in local files, FTP servers, or Web servers, on the block diagram, you can programmatically read or write data using the DataSocket functions located on the Functions≫Communication≫DataSocket palette, as shown in Figure 12.3.

An example of using DataSocket Write function to write live data through a DataSocket connection programmatically is shown in Figure 12.4a. In the example, we illustrate how to write a numeric value. The DataSocket Write function is polymorphic, so you can wire most data types to the data input. An example of using DataSocket Read function to read live data through a DataSocket connection programmatically is shown in Figure 12.4b. In the example, we illustrate how to shows how to read data and convert it to a double-precision floating-point number. If you do not specify a type, the data output of DataSocket Read returns variant data, which you must manipulate with the Variant to Data function

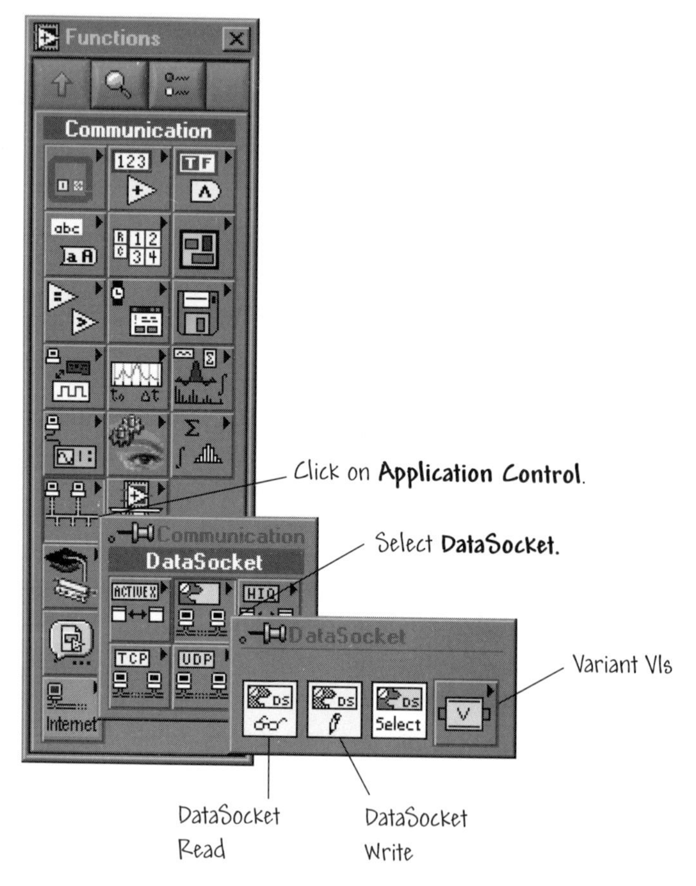

FIGURE 12.3
The DataSocket palette.

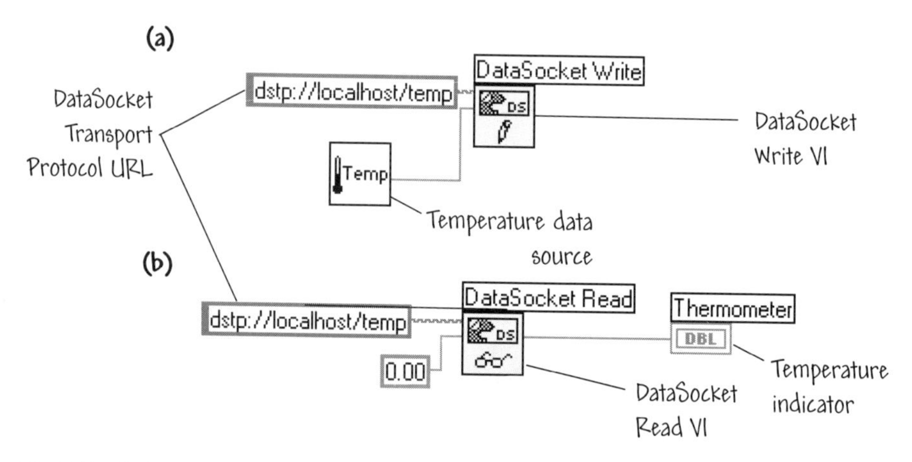

FIGURE 12.4
Writing and Reading live data using DataSocket functions on the block diagram.

located on the **Functions≫DataSocket≫Variant** palette (see Figure 12.3). In some cases, you must convert variant data to LabVIEW data, as discussed in the next section.

12.2.4 Variant Data

In some cases, the VI or other application that programmatically reads the data cannot convert the data back into its original data type, such as when you subscribe to data from another application. Also, you might want to add an attribute to the data, such as a timestamp or warning, which the data types do not permit. In these cases, use the To Variant function located on the **DataSocket≫Variant** palette to programmatically convert the data you write to a DataSocket connection to variant data (see Figure 12.3). Figure 12.5 shows a block diagram that continually acquires a temperature reading, converts the data to variant data, and adds a timestamp as an attribute to the data.

When another VI reads the live data, the VI must convert the variant data to a data type it can manipulate. Figure 12.6 shows a block diagram that continually reads temperature data from a DataSocket connection, converts the variant data into a temperature reading, retrieves the timestamp attribute associated with each reading, and displays the temperature and the timestamp on the front panel.

These two examples on the use of variant data are representative of the types of situations that you may face as you begin to venture into using the DataSocket technology. You can many interesting examples on the use of DataSocket connections in the library **examples/comm/datasktx.llb**.

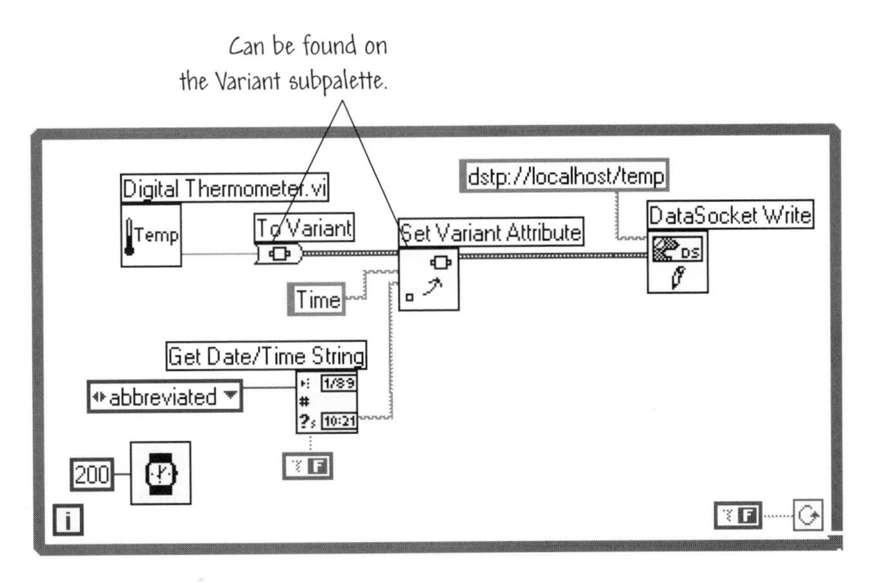

FIGURE 12.5
Converting Live Temperature Data to Variant Data.

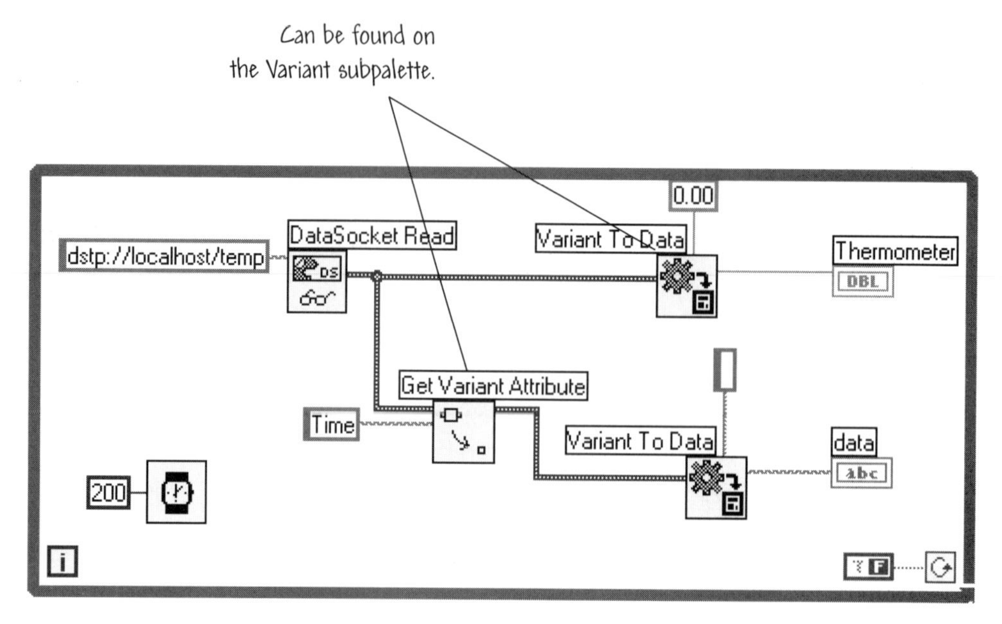

FIGURE 12.6
Converting Live Variant Data to Temperature Data.

12.3 VISUALIZING DATA AND GENERATING REPORTS WITH HIQ

HiQ is a data visualization, report generation, and analysis software program included on the *LabVIEW Student Edition* CD. The HiQ environment is illustrated in Figure 12.7. Refer to Chapter 1 for HiQ installation instructions. Once HiQ has been properly installed on your computer, you can access and work in the HiQ environment as a stand-alone application, or you can launch HiQ from LabVIEW, as shown in Figure 12.8. Two HiQ manuals are available in HiQ\Manuals: *Getting Started with HiQ* and the *HiQ Reference Manual*. Of course, comprehensive on-line help is available from within the HiQ environment.

HiQ uses a virtual **notebook** as the basic interface. The notebook has many of the components of a typical engineering or scientific notebook—pages, sections, and tabs. On a single notebook page, you can arrange information in a free-form manner much as with a typical drawing program. Each piece of information that you place on a notebook page is known as an object, and you can put many types of objects on a page, including text, numerics, and graphs.

When might you use HiQ? If you want to display LabVIEW-generated data on 2D or 3D graph, you can pass data from LabVIEW to HiQ using the HiQ VIs shown in Figure 12.8, or save the data to a file that HiQ can access later. If you have a three-dimensional graph on a HiQ notebook page, you can rotate it, zoom in or out, or make changes to the graph.

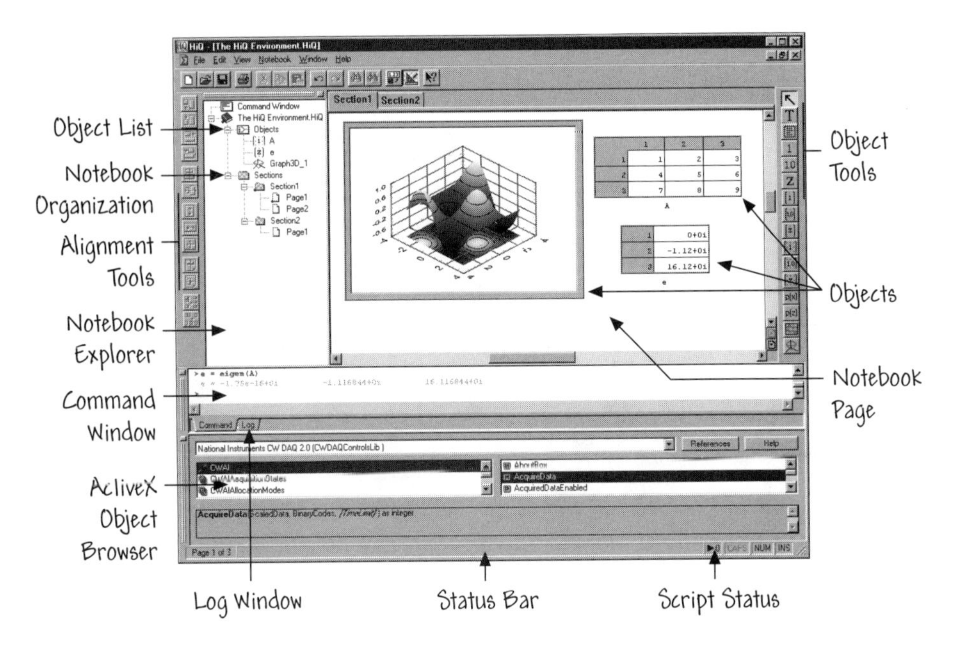

FIGURE 12.7
The HiQ environment.

HiQ provides an extensive library of analysis and visualization functions that you access through the command window or a **HiQ-Script** object on the notebook page. These functions allow you to visualize data in 3D objects, solve systems of ordinary differential questions, and solve many other analytical and visualization problems. The **HiQ Command Window** provides an interactive interface for experimenting with the analysis and graphics capabilities using HiQ-Script. For example, you can study a mathematical formula graphically by varying input values and immediately viewing the results on a graph. The Command Window exists at about the midsection of the notebook page, as shown in Figure 12.7. HiQ also includes a comprehensive set of functions from the mathematical disciplines, including the following:

- Linear Algebra
  - Matrix algebra using a natural and intuitive syntax in HiQ-Script.
  - Matrix norms, decompositions and transformations.
  - Eigenvalue and eigenvector analysis and linear system solvers.
- Data Fitting and Interpolation
  - Line, exponential, polynomial, and Gauss function data fitting.
  - Single- and multiple-dimension nonlinear data fitting and linear data fitting using a set of basis functions.
  - Polynomial and spline interpolation.

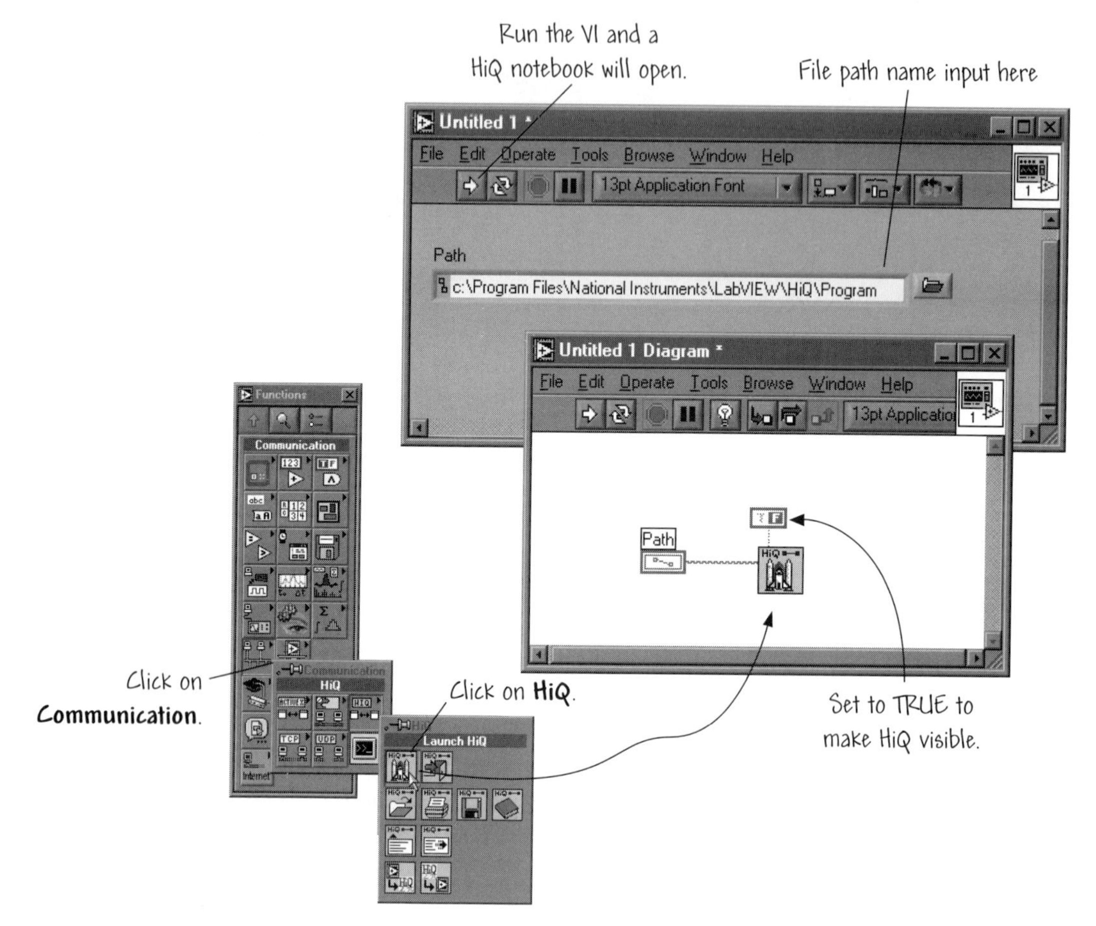

FIGURE 12.8
Launching HiQ from LabVIEW.

- Optimization
  - Nonlinear single- and multiple-dimension optimization
  - Linear programming
- Statistics
  - Computation of the mean, standard deviation, variance, covariance, and correlation.
  - Cumulative distribution and probability density functions of several types.
- Differential Equations
  - Nonlinear initial-value problems—including stiff systems—and linear and nonlinear boundary-value problems.

- Nonlinear Analysis
  - Roots of polynomials and nonlinear functions, and a solution for a system of nonlinear equations.
- Integration
  - Data integration, nonlinear function integration, and polynomial integration
- Differentiation
  - Numeric multiple derivatives, partial derivatives, and polynomial derivatives.
- Polynomials
  - Polynomial algebra using a natural and intuitive syntax in HiQ-Script.
  - Polynomial math—inverse, greatest common divisor, and least common multiple.
  - Integration and differentiation.

One important use of HiQ notebooks is for analyzing and visualizing data. This data may have been acquired from external sources using DAQ VIs, or perhaps it was generated with LabVIEW. In any case, the data is usually stored in a file on your hard drive. When you need to import text or binary data from a file, you can take advantage of the HiQ Import Wizard. You can use the default import settings in most cases, but Import Wizard does allow you to define the data format in your file.

You can transfer your data from LabVIEW to HiQ in one of two ways:

- Use the HiQ **ActiveX** VIs (**Windows**) or **AppleEvents** VIs (**Macintosh**).
- Save the data to a file and then read that file into HiQ.

ActiveX allows you to customize your notebook to include any ActiveX object. For example, you can access the Internet and download data from the web (or from FTP sites) and then use HiQ functions to analyze and graph the data. You can embed ActiveX objects directly into your HiQ notebook. You can, for example, embed Microsoft Word documents, Excel spreadsheets, and PowerPoint presentations in your notebook. These documents (spreadsheets and presentations) can then be edited and stored within HiQ. Conversely, you can embed a HiQ notebook into a Word document, Excel spreadsheet, or PowerPoint presentation and that HiQ notebook can be edited within the other application.

12.4 PLAYING WITH SOUND

To see how to acquire information with a sound card, open and run SoundCard Auto Power Spectrum.vi from the examples\sound\sndExample.adv.llb.

Place your favorite music CD into the player and "watch" and listen to the sounds. To see how to generate waveforms with a sound card, run Sim Phone.vi and Snd Morse Code Generator.vi from the examples\sound\sndExample.adv.llb.

You can also control sound files with LabVIEW. If you want to play a sound (.wav) file, you open a .wav file and pass the waveform and sound format information to the same VIs used to generate waveforms. To see how to play a sound file, open the Wave File Player.vi found in examples\sound\sndExample.llb.

To record a sound file, you use the same VIs as for acquiring a waveform. After you acquire the waveform, you save it along with the sound format information to a .wav file. To see how to record a sound file, open the Record Wave File.vi located in examples\sound\sndExample.llb.

12.5 RELAXING READING: INFANT VISION ASSESSMENT

The Southern College of Optometry created a new, state-of-the-art clinic for testing the visual function of infants with eye or neurological diseases, or those at risk for developing visual disorders. A clinic of this type needs to be capable of performing some fairly complex tasks.

- **Versatility**—the DAQ system had to be able to measure a very wide range of recordings from infants. The types of recordings included visual, electrodiagnostic tests (such as evoked potential measurements of the brain's neural activity in response to visual input) and electroretinograms (a measure of the electrical activity of the neurons within the retina). Many different types of evoked potentials had to be recorded to determine the infants' visual acuity, the presence of eye or neurological disease, or maldevelopment of the visual system that could lead to strabismus (eye turn) and amblyopia (developmental loss of vision). Other tests included eye-movement recordings to assess binocular visual functioning, and stereoscopic psychophysical tasks to measure the infants' developing ability to perceive depth.
- **Ease of use**—all the tests needed to be operated by a single clinician who not only had to run the software, but also had to monitor the infants' direction of gaze and attentiveness, as well as the quality of the recorded data.
- **Speed**—the tests needed to run fast, because infants have short attention spans!
- **Expandability**—new developments in infant testing had to be added to the clinic quickly and easily. Some of the conducted tests were so new that scientists had only conducted them in infant research laboratories.

The infant vision tests involved displaying an animated visual display, up to 60 frames/s, while simultaneously recording the infant's physiological responses

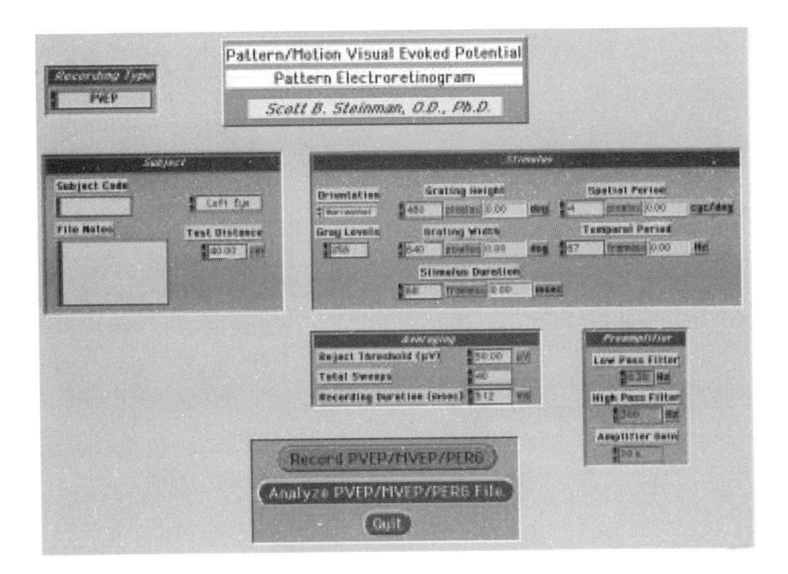

FIGURE 12.9
The LabVIEW graphical instrument panel makes it easy for clinicians to conduct vision tests while observing the infant.

to the visual stimulus at 1,000 Hz sampling rates using the NB-MIO-16X board, and then displaying the recorded data. The clinic was constructed using two interconnected Macintosh Quadra computers. One computer is dedicated solely to data acquisition and analysis using a National Instruments NB-MIO-16X analog input board and LabVIEW software. The front panel is shown in Figure 12.9. The second computer is dedicated solely to displaying rapidly animated visual stimuli by means of small C++ programs that perform either palette animation or frame animation. With two Macintosh computers, the clinician can perform data acquisition and animation simultaneously at the highest possible rate.

AppleEvents is the glue that ties everything together. A schematic diagram of the system is depicted in Figure 12.10. Once the infant is ready, the test begins with a mouse click. An AppleEvent is sent by the LabVIEW program across an Ethernet cable to the stimulus display program on the second Macintosh. The stimulus display computer acts as a "slave" to the recording computer. The AppleEvent it receives tells the display program what to display and when. The precise time locking of the visual display and data acquisition is ensured by sync signals sent across digital I/O lines between NB-DIO-96 boards within each computer.

LabVIEW and the Macintosh computer were both critical to the development of the infant vision clinic and to its continued operation. LabVIEW reduced the development time, opened up a wider choice of recording and analysis techniques, and provided AppleEvent tools for the intercomputer communication that made the simultaneous high-speed visual stimulus display and data acquisition

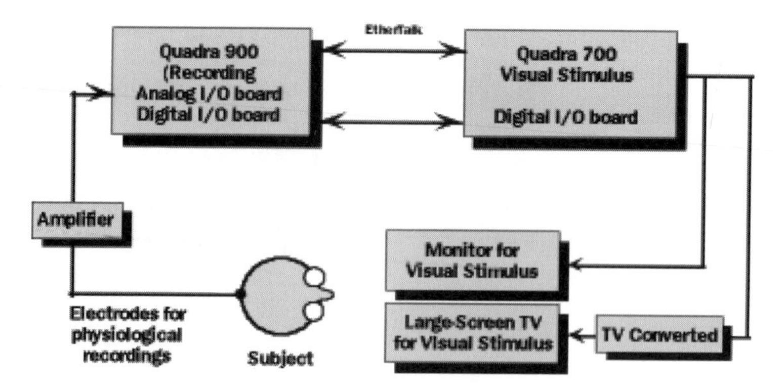

FIGURE 12.10
The infant clinic was built with two Macintosh computers linked by NB-DIO-96 digital I/O boards and an EtherTalk network cable.

possible. The infant vision software demonstrates that LabVIEW is not just a laboratory research or engineering tool—it can also solve real-world problems in such unlikely places as infant vision clinics.

For more information on this application of LabVIEW, read the *User Solutions* entitled "Infant Vision Assessment with LabVIEW on Linked Macintosh Computers," which can be found on the National Instruments website, or contact:

Dr. Scott B. Steinman
Chair, Department of Biomedical Sciences
Director of Research Programs
Southern College of Optometry
1245 Madison Avenue
Memphis, TN 38104
e-mail: steinman@sco.edu

12.6 SUMMARY

In this chapter we discussed some LabVIEW applications to help you utilize the power of virtual instruments. The material presented here was intended to stimulate your curiosity about the topics of RDA, VI Server, DataSocket, and the Internet Toolkit. It is beyond the scope of this introduction to LabVIEW to go into much detail on these advanced topics, but it is appropriate to point out to you the many interesting LabVIEW topics that remain for your study. Yes, you can use LabVIEW and HiQ in conjunction with MATLAB, Microsoft Word, Excel spreadsheets, and PowerPoint presentations, and yes, you can interact with the web, e-mail, ftp, and much more!

KEY TERMS

ActiveX: Windows technology that allows you to share data and objects between applications.

AppleEvent: Macintosh technology that allows you to share data between applications.

DataSocket: Technology that allows you to share live data with other VIs on the Web.

HiQ command window: A window where you type HiQ Script to get immediate results.

HiQ Script: An intuitive HiQ programming language that includes an extensive library of analysis and visualization functions.

Publish: When you share the data of a front panel object with other users.

Subscribe: When users retrieve published data and view it on their front panel.

Notebook: The workspace in HiQ that stores, organizes, and displays all the components of an analysis and visualization problem.

URL(Uniform Resource Locators): A web address.

VI Server: Mechanism for controlling VIs and applications programmatically. It can also be used to control VIs or applications remotely as well as on a local machine.

EXERCISES

E12.1 From the LabVIEW startup screen, click on **Search Examples** and then click on the **Communication** link and then the TCP/IP link. When this is done, click on **VI Server** and **VI Client** links. These two VIs illustrate how you can open and run VIs on the same machine or across a network using the VI Server technology. Before running these VIs, select Tools≫Options≫VI Server: Configuration. Check the TCP/IP Protocol. This enables your computer to use the VI Server technology through a TCP/IP network. If you do not have another machine with LabVIEW that is connected to a network, then you do not need to change the **Machine Access List** on the Server-VIServer Example.vi and **Server Names** on the Client-VIServer example.vi. If you do want these VIs to communicate across a network, open Server-VIServer Example.vi on one machine and Client-VIServer example.vi on another machine. Update **Machine Access List** and **Server Names** with the complete TCP/IP address of each machine. Run Server-VIServer Example.vi first then Client-VIServer example.vi. Select one or more VIs listed in **server: station** of Client-VIServer example.vi. If selected, these VIs will run on the server machine, and data is returned that is displayed on the client machine.

E12.2 On the LabVIEW startup screen, select **Search Examples** and follow the path to Communication\DataSocket\Data Socket Reader. This VI uses the DataSocket technology to acquire information from web and FTP sites. Copy one of the addresses at the bottom of the window and paste it into the string control at the top of the window. Run the VI and observe the results. When you are done, press the **Stop** button.

E12.3 In order to complete this exercise, you must have installed HiQ from the LabVIEW Student Edition CD. On the LabVIEW startup screen, click **Open VI** and select Examples\Comm\hiq\HIQAutomationExample.vi. This VI shows how you can pass data to HiQ and display that information on a notebook. The data is created from constants in the block diagram. This VI also reads that same information back from the notebook and displays the information on the VI front panel. You can print the HiQ notebook from LabVIEW if you desire.

PROBLEMS

P12.1 On the LabVIEW startup screen, click **Search Examples** and follow the path to and select Advanced≫Sound Control. Open the **Wave File Player**. This example shows how to play a wav file. When the VI opens up, click on **Run**. Then click on the folder icon near the top of the VI. This opens a dialog box. Navigate to the examples/comm/dsdata folder. Choose the tone.wav and select **Save**. Then hit the play button on the player and you will hear the tone.

P12.2 Complete the crossword puzzle.

Across

2. An intuitive HiQ programming language that includes an extensivelibrary of analysis and visualization functions.
4. Technology that allows you to share live data with other VIs on the web.
7. The workspace in HiQ that stores, organizes, and displays all the components of an analysis and visualization problem.
9. Mechanism for controlling VIs and applications programmatically.

Down

1. When users retrieve published data and view it on their front panels.
2. A window where you type HiQ Script to get immediate results.
3. When you share the data of a front panel object with other users.
5. Macintosh technology that allows you to share data between applications.
6. Windows technology that allows you to share data and objects between applications.
8. (Uniform Resource Locators) A web address.

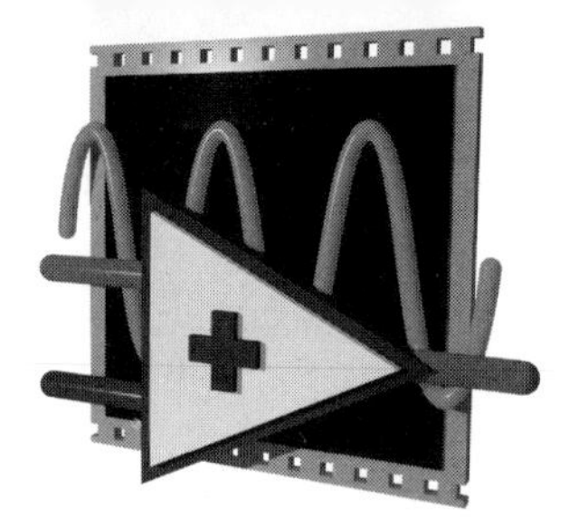

INDEX